KU-095-597

# THERMOHYDRAULICS OF TWO-PHASE SYSTEMS FOR INDUSTRIAL DESIGN AND NUCLEAR ENGINEERING

## SERIES IN THERMAL AND FLUIDS ENGINEERING

JAMES P. HARTNETT and THOMAS F. IRVINE, JR., Editors
JACK P. HOLMAN, Senior Consulting Editor

Cebeci and Bradshaw **Momentum Transfer in Boundary Layers**
Chang **Control of Flow Separation: Energy Conservation, Operational Efficiency, and Safety**
Chi **Heat Pipe Theory and Practice: A Sourcebook**
Delhaye, Giot, and Riethmuller **Thermohydraulics of Two-Phase Systems for Industrial Design and Nuclear Engineering**
Eckert and Goldstein **Measurements in Heat Transfer, 2d edition**
Edwards, Denny, and Mills **Transfer Processes: An Introduction to Diffusion, Convection, and Radiation, 2d edition**
Fitch and Surjaatmadja **Introduction to Fluid Logic**
Ginoux **Two-Phase Flows and Heat Transfer with Application to Nuclear Reactor Design Problems**
Hsu and Graham **Transport Processes in Boiling and Two-Phase Systems, Including Near-Critical Fluids**
Hughes **An Introduction to Viscous Flow**
Kestin **A Course in Thermodynamics, revised printing**
Kreith and Kreider **Principles of Solar Engineering**
Lu **Introduction to the Mechanics of Viscous Fluids**
Moore and Sieverding **Two-Phase Steam Flow in Turbines and Separators: Theory, Instrumentation, Engineering**
Nogotov **Applications of Numerical Heat Transfer**
Richards **Measurement of Unsteady Fluid Dynamic Phenomena**
Siegel and Howell **Thermal Radiation Heat Transfer, 2d edition**
Sparrow and Cess **Radiation Heat Transfer, augmented edition**
Tien and Lienhard **Statistical Thermodynamics, revised printing**
van Stralen and Cole **Boiling Phenomena**
Wirz and Smolderen **Numerical Methods in Fluid Dynamics**

### PROCEEDINGS

Asanuma **Flow Visualization**
Durst, Tsiklauri, and Afgan **Two-Phase Momentum, Heat and Mass Transfer in Chemical, Process, and Energy Engineering Systems**
Hoogendoorn and Afgan **Energy Conservation in Heating, Cooling, and Ventilating Buildings: Heat and Mass Transfer Techniques and Alternatives**
Keairns **Fluidization Technology**
Spalding and Afgan **Heat Transfer and Turbulent Buoyant Convection: Studies and Applications for Natural Environment, Buildings, Engineering Systems**
Zarić **Thermal Effluent Disposal from Power Generation**

# THERMOHYDRAULICS OF TWO-PHASE SYSTEMS FOR INDUSTRIAL DESIGN AND NUCLEAR ENGINEERING

*Edited by*

**J. M. Delhaye**

*Service des Transferts Thermiques*
*Centre d'Etudes Nucléaires de Grenoble, France*

**M. Giot**

*Département Thermodynamique et Turbomachines*
*Université Catholique de Louvain, Belgium*

**M. L. Riethmuller**

*von Karman Institute for Fluid Dynamics*
*Rhode-Saint-Genèse, Belgium*

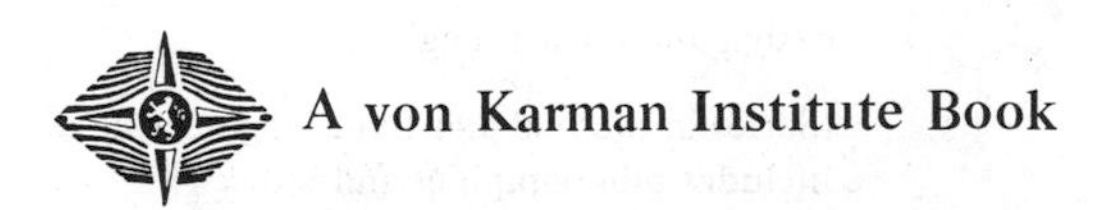

**A von Karman Institute Book**

**Hemisphere Publishing Corporation**

Washington New York London

**McGraw-Hill Book Company**

New York St. Louis San Francisco Auckland Bogotá
Hamburg Johannesburg London Madrid Mexico
Montreal New Delhi Panama Paris São Paulo
Singapore Sydney Tokyo Toronto

05654568

THERMOHYDRAULICS OF TWO-PHASE SYSTEMS FOR INDUSTRIAL DESIGN AND NUCLEAR ENGINEERING

Copyright © 1981 by Hemisphere Publishing Corporation. All rights reserved. Printed in the United States of America. No part of this publication may be reproduced, stored in a retrieval system, or transmitted, in any form or by any means, electronic, mechanical, photocopying, recording, or otherwise, without the prior written permission of the publisher.

1 2 3 4 5 6 7 8 9 0 BRBR 8 9 8 7 6 5 4 3 2 1 0

**Library of Congress Cataloging in Publication Data**
Main entry under title:

Thermohydraulics of two-phase systems for industrial design and nuclear engineering.

(Series in thermal and fluids engineering)
Includes bibliographies and index
1. Nuclear reactors–Fluid dynamics. 2. Nuclear reactors–Cooling. 3. Two-phase flow. 4. Heat–Transmission. I. Delhaye, J. M., date. II. Giot, M., date. III. Riethmuller, M. L., date.
TK9202.T47 1980 621.48′31 80-14312
ISBN 0-07-016268-9

This book was set in Press Roman by Hemisphere Publishing Corporation. The editors were Diane Heiberg, Honey Sauberman, and Judith B. Gandy; the production supervisor was Rebekah McKinney; and the typesetter was Wayne Hutchins. Braun-Brumfield, Inc. was printer and binder.

D
621.4022
THE

# CONTENTS

# CONTRIBUTORS

J. COSTA
Service des Transferts Thermiques
Centre d'Etudes Nucléaires de Grenoble, France

J. M. DELHAYE
Service des Transferts Thermiques
Centre d'Etudes Nucléaires de Grenoble, France

M. GIOT
Département Thermodynamique et Turbomachines
Université Catholique de Louvain, Belgium

D. GRAND
Service des Transferts Thermiques
Centre d'Etudes Nucléaires de Grenoble, France

Y. Y. HSU
Division of Reactor Safety Research
Nuclear Regulatory Commission, Washington, D.C., U.S.A.

G. YADIGAROGLU
Department of Nuclear Engineering
University of California, Berkeley, U.S.A.

# INTRODUCTORY REMARKS

A thorough understanding of two-phase flow phenomena is now available for scientists and engineers working in nuclear and chemical engineering. However, very few books have been published in this area so far; thus the lecture series held at the von Karman Institute for Fluid Dynamics in 1978 offered an opportunity to present a comprehensive and up-to-date review of this field. This course had been prepared with the aim of giving the participants a sound and consistent approach to two-phase flows and heat transfer. Particular attention was paid to the tutorial aspect of the lectures and to the practical aspects of complex nuclear engineering problems. These efforts were continued during the preparation of this volume, which was intended to be more than a simple collection of lecture notes.

This book shows how difficult engineering issues can be solved by using the experimental and theoretical knowledge accumulated during the past 20 years. Examples of problems encountered in nuclear engineering are given in the first two chapters. At the end of the book, it is shown how these problems can be handled by means of calculations based on developments presented in the other chapters.

On behalf of the authors, the editors would like to express their special indebtedness to the following people for supplying useful material on thermohydraulics and computer codes and for their assistance in preparing the lecture notes: P. Bell, M. Charles, C. Coing, E. de Gorriti, P. Denys, J. Lambrechts, N. Maran, M. Mast, P. Pasteels, A. Smeers, L. Thompson, and L. S. Tong.

*J. M. Delhaye*
*M. Giot*
*M. L. Riethmuller*

# FOREWORD

Just as it has been said in mathematics that most of the real problems are nonlinear, so in the combined realm of fluid mechanics and heat transfer a large fraction of the real problems involve more than one fluid phase. The vital area of energy production, using both conventional and alternative sources, relies heavily on phase change processes such as convective boiling, combustion of fuel droplets, and film condensation. The safety of nuclear reactors, which is currently an active research area in laboratories throughout the world, can be verified only by an improved understanding of combined heat and mass transfer under transient conditions in two-phase (and even multiphase) flows. Many areas of industrial production, such as petroleum refining and production, chemical and food processing, and treatment of wastes, depend heavily on multiphase flow. The subject is too broad to be covered in detail in a single volume, or indeed in several volumes. What has been attempted here is a fundamental treatment of the thermohydraulics of two-phase flow. Clear and authoritative descriptions are given of the basic principles of two-phase flow and heat transfer, which will be useful to both the graduate student and the practicing engineer. This treatment has been provided by well-recognized specialists, all of whom also have close connections to the nuclear reactor field. The practical applications are therefore tied closely to the safety and operation of both light-water reactors and liquid-metal fast-breeder reactors.

*S. G. Bankoff*

# PREFACE

This publication is based on the lecture notes for a course on two-phase flows in nuclear reactors presented at the von Karman Institute for Fluid Dynamics in 1978. It was attended by more than 80 students, scientists, and practicing engineers from several countries who are involved in the design, development, research, and operation of nuclear reactors.

The material, presented as a comprehensive and unified view made available to general industry by specialists working in the field, is in the form of a high-level textbook requiring a basic knowledge of fluid mechanics and heat transfer. It is derived from the latest nuclear developments and recent advances in thermohydraulics of two-phase systems.

A large fraction of this material is useful for the assessment of the safety margin of nuclear reactors, a topic that is now more than ever important for design engineers. It aims at mathematical modeling and the development of computer codes, which is the direction of progress if the new generation of measuring instruments can be successfully developed.

In Chap. 1, Hsu briefly discusses two-phase flow heat transfer problems related to pressurized-water reactor and boiling-water reactor loss of coolant accidents (LOCAs) and emergency core cooling. In Chap. 2, Costa describes some two-phase flow problems encountered in the design and safety analysis of liquid-metal fast-breeder reactors.

Delhaye's purpose in Chap. 3 is to describe the different configurations of gas-liquid pipe flows, to examine the transition phenomena between the flow patterns, and to propose a set of flow maps allowing the determination of the flow pattern, given the system parameters.

Apart from the problems encountered in single-phase flow, further difficulties appear when attempts are made to measure quantities in two-phase flows. They result from the presence of the two phases and from flow confinement. Despite these difficulties, several techniques have a fair degree of accuracy. However, new sophisticated techniques need further substantiation. In Chap. 4, Delhaye describes the principles of a few methods that are considered as well established.

The next five chapters are devoted to the derivation of two-phase flow equations. Local instantaneous equations are presented in detail by Delhaye in Chap. 5. The balance laws for each phase are expressed in terms of partial differential equations; for the interface, they are formulated in terms of jump conditions. Two illustrative examples are considered: the dynamics of a spherical gas bubble and of a liquid film on an inclined plane. In Chap. 6, Giot develops simple and fundamental ideas for the basic phenomena related to nucleation in simple situations excluding cavitation and vapor explosions. Instantaneous space-averaged equations derived by Delhaye in 1968 are given in Chap. 7. An example is given for the case of an isothermal stratified flow of two inviscid incompressible fluids in a horizontal channel with emphasis on the interface behavior in the absence of phase change. In Chap. 8 Delhaye proposes a simpler method than that used by Ishii in 1975 to establish time-averaged equations. Finally, Delhaye derives the composite-averaged equations in Chap. 9, after demonstrating the identity of the space-time- and time-space-averaging operators. All practical problems of two-phase flow in channels are dealt with using these equations.

Although the two-phase flow model is the most satisfactory one in theory, its high degree of complexity has to be reduced by considering particular flow evolutions. Delhaye presents different solutions to this problem in Chap. 10.

Chapter 11, written by Giot, is primarily a review of the most popular overall correlations for predicting friction pressure drops in single channels. Flows in rod bundles involve important heterogeneities in cross sections. Consequently, transverse momentum equations are needed. Furthermore, it is difficult to isolate the specific role of friction. As explained by Grand in Chap. 12, the estimation of pressure drop is thus often based on crude assumptions such as in the mixed flow model. In Chap. 13, Giot gives a brief presentation of pressure drops in sudden enlargements and contractions, orifices, bends, and tees.

The purpose of Chap. 14, presented by Hsu, is to recommend, based on the current state of knowledge, a set of heat transfer equations and their related switching logics for the "best estimate model" code. Blowdown and reflood stages are considered. In each case, the range of condition, sensitivities, correlations to be recommended and their data bases, and special features are discussed.

In reactor safety research, condensation heat transfer has been by far less studied than boiling heat transfer. In Chap. 15, Hsu tentatively recommends equations for condensation of bubbles and jets, subject to future verification.

Successful models of the regime transitions in boiling heat transfer are presented by Yadigaroglu in Chap. 16, as well as state-of-the-art correlations for predicting these models. Physical mechanisms are described with emphasis on applications to the analysis of steady-state and transient normal–as well as accident–conditions in nuclear reactors.

The potential for instabilities exists in all two-phase flow systems including nuclear power systems. These instabilities and their associated propagation phenomena and kinetic waves are studied by Yadigaroglu in Chap. 17 with emphasis on pressure drop excursions and oscillations and density wave oscillations. Some current instability concerns in nuclear reactors are discussed.

The calculation of critical flow rates is important for the design of equipment involving two-phase flows, especially for LOCA studies. Chapter 18 by Giot discusses several theoretical models and reviews experimental results.

Finally, Chaps. 19 and 20, written by Hsu and Grand, respectively, show how to obtain the solutions of complex problems occurring in nuclear engineering.

All of these chapters will definitely contribute to a thorough understanding of two-phase flows and heat transfer in the power and process industries. As such, this book will constitute a reference tool for engineers and scientists working in these fields.

*Jean J. Ginoux*

# THERMOHYDRAULICS OF TWO-PHASE SYSTEMS FOR INDUSTRIAL DESIGN AND NUCLEAR ENGINEERING

CHAPTER
# ONE

# TWO-PHASE FLOW PROBLEMS IN PRESSURIZED-WATER REACTORS

**Y. Y. Hsu**

## 1 INTRODUCTION

Among the reactor safety issues for light-water reactors (LWR), the most important are those concerning the hypothetical loss of coolant accident (LOCA). The LOCA issue is important because it is considered the limiting factor in reactor safety and thus is classified as the design-base accident. Consequently, the major efforts of LWR safety research are directed at providing codes with a best-estimate capability to predict the consequences of LOCAs.

Because a LOCA is basically a thermohydraulic phenomenon, it is natural for us to find many topics of two-phase flow heat transfer in reactor safety research. In the following section, we will discuss these subjects as related to pressurized-water reactor (PWR) and boiling-water reactor (BWR) LOCAs and emergency core cooling (ECC).

## 2 PWR LOCA–ECC

There are three major stages in a PWR LOCA, that is, blowdown, ECC refill, and reflood, as shown in Fig. 1. We will discuss the two-phase flow in each stage.

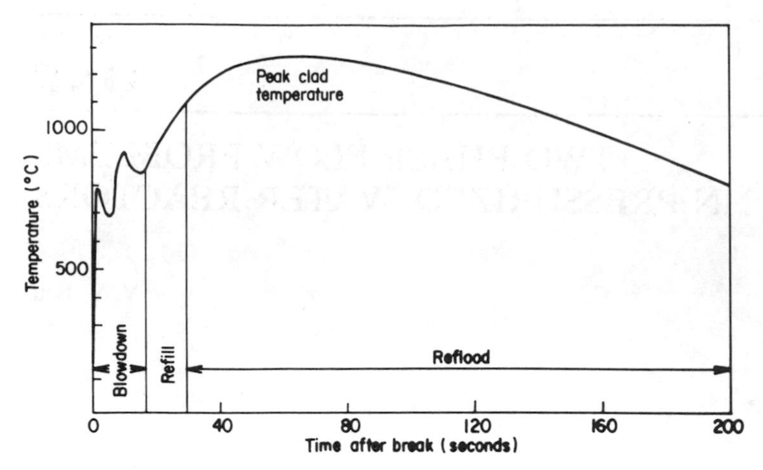


**Fig. 1** Conservative estimate of cladding temperature for a double-ended cold-leg break (guillotine) LOCA.

## 2.1 Blowdown Stage

The questions to be answered during the blowdown stage are: How fast are we losing coolant and how fast does the clad temperature rise? Thus the two areas of interest are the critical discharge rate from the break and in-core heat transfer.

**2.1.1 Critical discharge rate** The critical discharge rate of a two-phase flow differs from that of single-phase flow in that the flow rate in two-phase flow is highly dependent on the flow pattern. Thus void fraction and upstream-flow geometry are important. Although there is no major heat transfer between the solid and the fluid, there is interfacial exchange of heat and mass between liquid and vapor, which determines the compressibility of the mixture. For more detailed information see Henry (1968), Hsu and Graham (1976), and Chap. 18.

**2.1.2 In-core heat transfer** In-core heat transfer is discussed in detail in Chap. 14. For purposes of this chapter it is sufficient to say that during blowdown, the coolant flow rate goes through flow reversal, rereversal, and source stagnation, while quality changes drastically during the whole period; therefore the clad may experience critical heat flux (CHF) and periods of post-CHF heat transfer. Determination of clad temperature through the proper choices of heat transfer coefficient is one of the major aims of reactor safety research.

## 2.2 Refill

During refill, the emergency core cooling system (ECCS) injects coolant water into the vessel through the cold leg and downcomer (Fig. 2). This water comes into direct contact with the escaping steam. The encounter first takes place in the pipeline and then in the downcomer.

**2.2.1 Steam-water mixing** The steam-water mixing in the pipeline is intermittent. The transient condensation rate at the water-steam interface decreases with time. At first contact, fast condensation creates a vacuum, causing the slugs in the pipeline to move away from the downcomer. However, as the condensation rate decreases, pressure builds up to push the slug downstream again. The interaction of transient condensation and the inertia of the flow results in a strong flow oscillation with a large amplitude that could cause severe knocking in valves and pipelines. The phenomenon is shown in Fig. 3. Discussion of these effects can be found in Block (1976). The model shown in Fig. 3 is highly simplified. We still need to know more precisely the transient condensation in a slug flow and the effect of noncondensable gases on heat transfer.

**2.2.2 ECC bypass in the downcomer** When water enters the downcomer, it interacts violently with the countercurrent flow of steam, causing the two-phase mixture to have a chaotic flow pattern. Because water is injected horizontally from the intact cold leg and a steam-water mixture escapes from the broken leg, the water flows in both circumferential and vertical directions. At the interface, steam bubbles collapse on condensation with a violent implosion. In general, if there were no condensation, the flooding limit for countercurrent air-water flow can be determined by Wallis' flooding criteria relating $j_f^*$ and $j_g^*$ or $J_f^*$ and $J_g^*$, as shown in Fig. 4. Even so, there is still a question of which length should be used for scaling (Crowley et al., 1976). With condensation, there is a strong indication of hysteresis effects; that is, the flooding limit appears to depend on whether the flow is steam into water or water into steam (Fig. 5) (Crowley et al., 1977). The determination of condensation rate and the effect of the condensation rate on flow pattern are still problems.

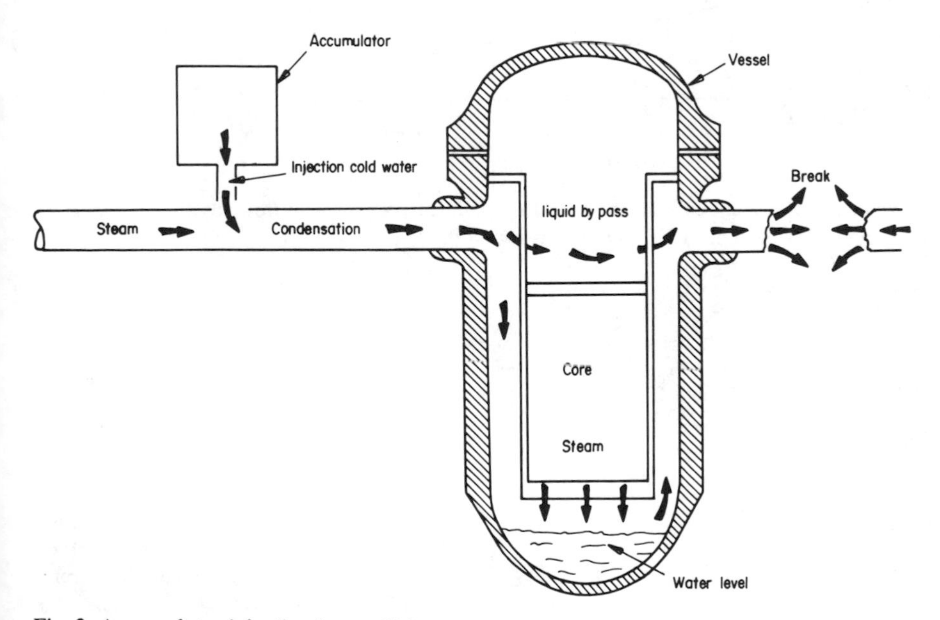


**Fig. 2** Accumulator injection into cold leg.

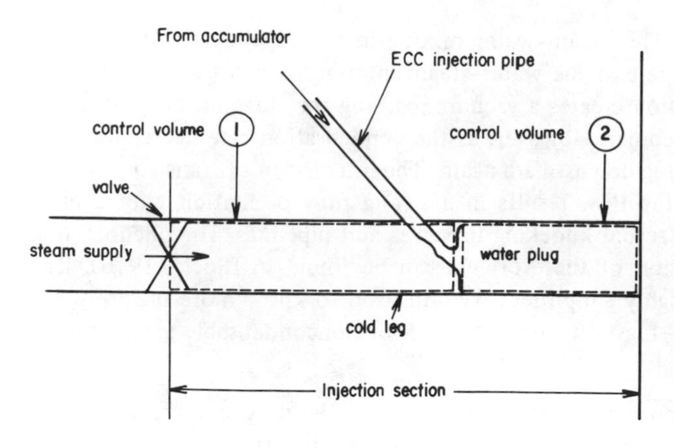


**Fig. 3** Model for analysis of cold-leg oscillations (Block, 1976).

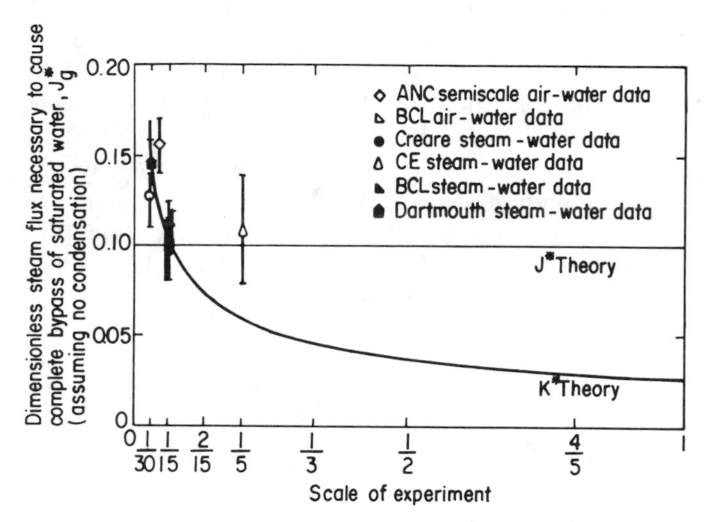


**Fig. 4** Summary of present experimental results for dimensionless steam flow to cause bypass at several scales (Crowley et al., 1976). ($J^*$ theory: $J_g^* = 0.10$; $K^*$ theory: $\mathrm{Ku}_g^* = 0.10$.)

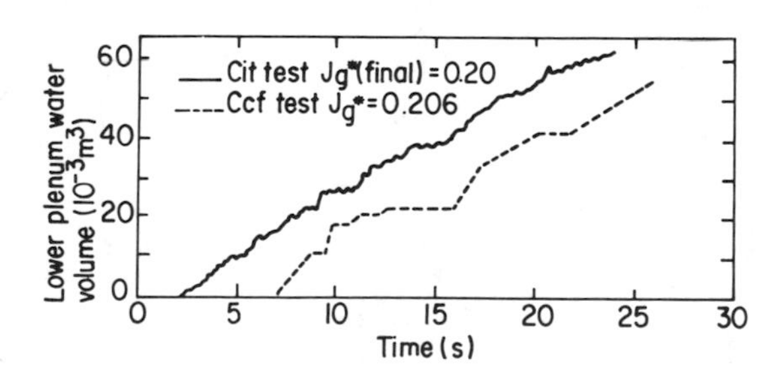


**Fig. 5** Comparison of lower-plenum filling behavior in condensation-induced transient and countercurrent flow test (Crowley et al., 1977).

## 2.3 Reflood

During the reflood stage, water enters the core from the lower plenum. As it progresses, steam is formed, carrying with it many drops. At a given station, the clad will successively see steam cooling, dispersed droplet flow, a slowly increasing presence of water (that is, a decreasing void fraction), the quench front, and finally, submersion in water without boiling. The various flow regimes and heat transfer modes in a reflood channel are shown in Fig. 6.

Reflood thermohydraulic equations will be discussed in detail in Chap. 14, and we will now discuss other problems not usually considered by reflood thermohydraulic codes.

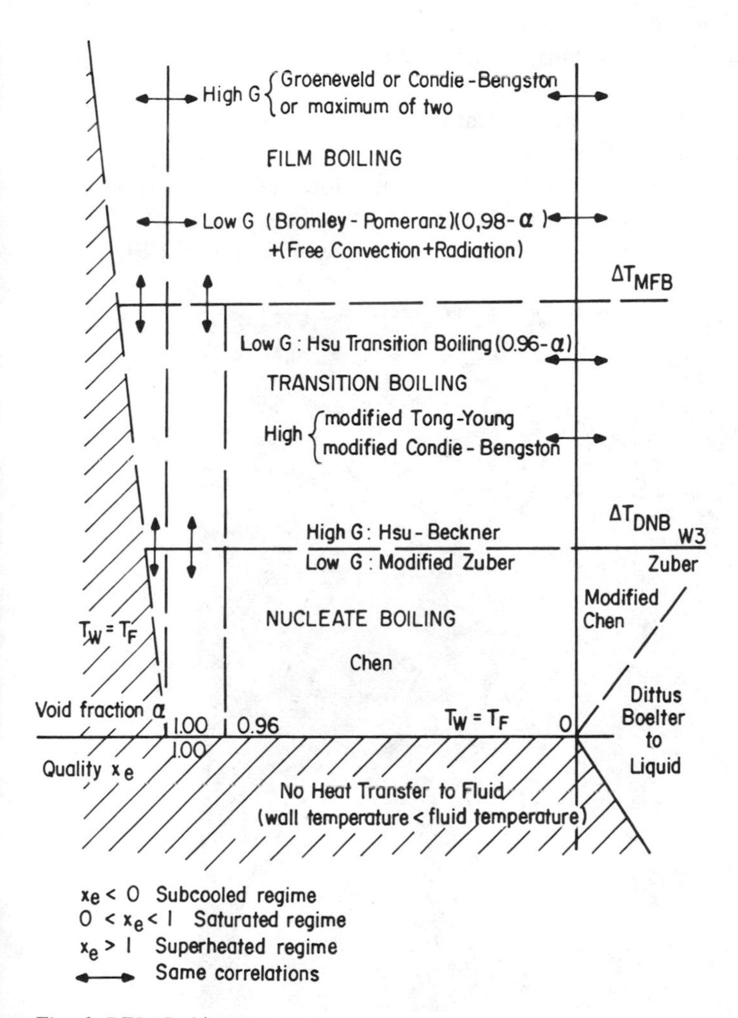


Fig. 6 RELAP 4/MOD 6 reflood heat transfer correlations and their regions of application (see Chap. 14).

**2.3.1 Entrainment in core** In any reflood code, it is assumed that entrainment takes place at the top of the froth front; that is, below the froth front water is the continuous phase, and above the froth front steam is the continuous phase. In reality, there seem to be two types of drops. The majority of drops do seem to be carried by steam, but there are some drops that seem to shoot up like cannonballs. This second kind may be a segment of slug flow propelled upward by a rapidly expanding steam bubble. Furthermore, grid spacers seem to change the entrainment distribution pattern (Fig. 7). More microscopic studies are needed.

**2.3.2 Quenching from the bottom flood** It is assumed in the code that the quench front moves upward uniformly around a rod. In reality, the front moves nonuniformly and intermittently. The nonuniformity results from the interaction of the nonuniform coolant flow distribution and the nonuniform gap distribution in a rod. The intermittency is caused by slug flow. Quench-front movement is yet to be properly modeled.

**2.3.3 Chimney effect and internal flow circulation** Because of the nonuniform power distribution in a bundle, the axial enthalpy increase is higher in the high-power zone, thus resulting in a flow of higher quality. This higher flow gives an updraft, which sucks in flow from the lower-power zone. This is the so-called chimney effect. If the upward flow is such that $j_g$ breaks the countercurrent flow limitation (CCFL) in the

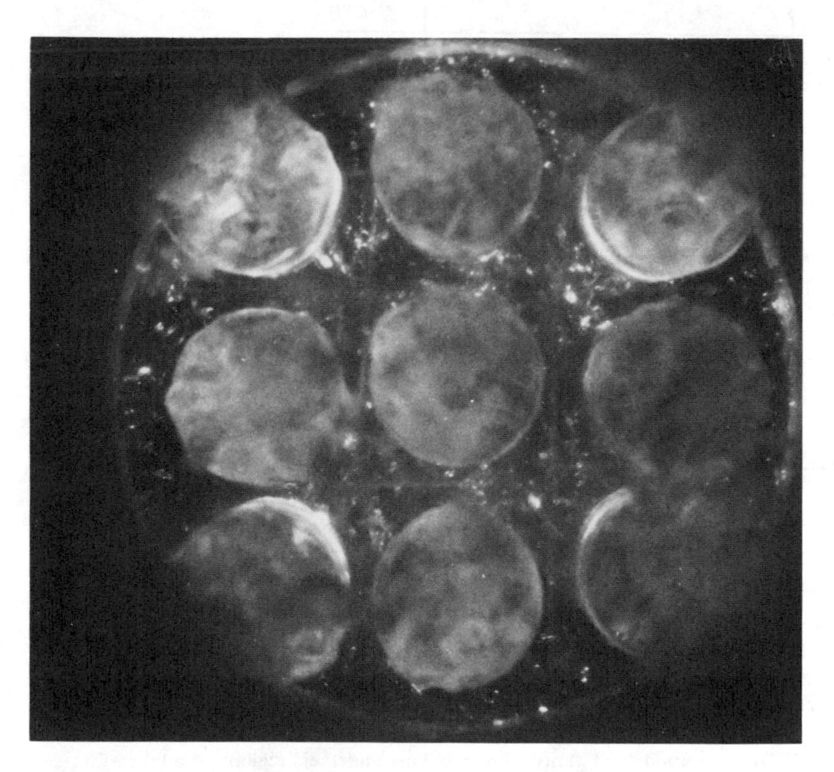

**Fig. 7** View at exit of grid (Smith, 1978).

| Typical reactor conditions | |
|---|---|
| regions | quench level |
| I | 0.60 m |
| II | 1.20m |
| III | 1.80m |

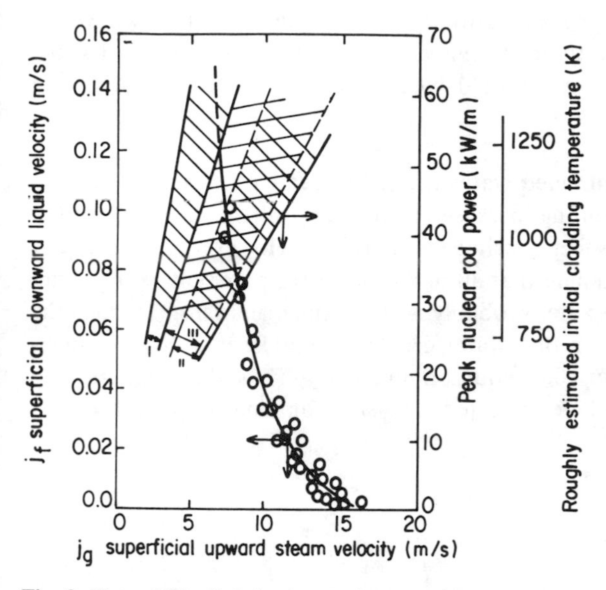


**Fig. 8** Water fallback behavior at the top of core, with approximated relationship between steam velocities and rod and reactor conditions. Shaded areas represent typical ranges of superficial upward steam velocities as a function of peak rod power. The line passing through data points represents the CCFL. Thus for the same liquid velocity, steam velocity in the high-power zone will penetrate upward (Tong, 1978).

high-power zone and $j_f$ breaks the CCFL in the lower-power zone (Fig. 8), an internal circulation results with steam droplet flow going up in the high-power zone and water falling back from the upper plenum in the low-power zone. Such phenomena cannot be observed in a small core but can be expected in a large test core. Such tests will be conducted in Germany and Japan as part of the Germany–United States–Japan three-dimensional-test cooperative program.

**2.3.4 Fallback and top quench** As mentioned in Sec. 2.3.3, the water carryout from the core can fall back to the core. This fallback will cause quenching of the rod from the top. Top quenching is also expected if upper-head injection is incorporated as part of the ECC system. Quenching of solids by falling films has been described in numerous studies; however, most of these studies assume a falling film without interference by steam generated at the quench front. In the subchannels of a bundle, the flow path is limited; thus the steam flow will have to interfere with falling-water flow. The amount of steam flow that goes to a neighboring subchannel and the amount that rises directly are a function of the fallback area. More studies are needed.

**2.3.5 Entrainment–deentrainment in the upper plenum** In the upper plenum of a reactor, the space is crowded with internal structures. In terms of entrainment, these structures serve two contradictory functions: the large surface area of the structures serves to collect water deentrained from the flow; and the reduced-flow cross-sectional area (compared with an empty upper plenum) serves to preclude much slowdown of steam flow, which is unfavorable to deentrainment. The net effect of internal structures on entrainment will have to be studied by both bench models and large-scale tests.

**2.3.6 Steambinding** When the entrained water enters the steam generator, it is evaporated by the stored heat. The resulting increase of pressure upstream slows down the entrance of reflood water. This effect is called steambinding. However, the backup of pressure is usually calculated assuming that all of the carryover water enters the steam generator. In reality, most of the water probably is deentrained and falls back into the core. Furthermore, it is well known that water droplets can exist in a flow of superheated steam well past the point of 100% equilibrium quality. Thus, the water entering the steam generator may not all be evaporated. The steambinding pressure drop still needs to be properly evaluated.

## 3 CONCLUDING REMARKS

Although much has been learned in the past few years about LOCA phenomena, many research problems still need to be solved. One major handicap for any thermohydraulic research is the lack of adequate two-phase flow instrumentation, and consequently, the uncertainty of two-phase flow measurements (flow rates, density, and quality). The lack of adequate two-phase flow instrumentation is especially serious for transient in-core measurements. Until a new generation of in-core instrumentation becomes available, all the flow parameters are inferred from code calculations and interpolated between inlet and exit measurements.

The qualitative uncertainty not only affects the accuracy of prediction but also reduces confidence in the ability to postulate and select models, which in turn reduces confidence in the reliability of codes used to infer the flow parameters in the first place. To break such a vicious circle, one has to rely on the development of new and better instrumentation and on better insight of phenomena through well-designed bench tests. Thus, the procedure is to develop better models through bench tests, model analysis, and better instrumentation; verify each model through separate effect and medium-size tests; and finally, in large-scale, integrated system tests verify the code encompassing all the models.

## NOMENCLATURE

| | |
|---|---|
| $D_h$ | hydraulic diameter |
| $G$ | mass velocity |
| $g$ | gravitational acceleration |
| $j$ | superficial velocity |
| $j_f^*$ | liquid dimensionless velocity defined as $j_f^* \triangleq j_f \rho_f^{1/2} (gD_h \, \Delta\rho)^{-1/2}$ |
| $J_f^*$ | liquid dimensionless velocity defined as $J_f^* \triangleq j_f \rho_f^{1/2} (gL \, \Delta\rho)^{-1/2}$ |
| $j_g^*$ | gas dimensionless velocity defined as $j_g^* \triangleq j_g \rho_g^{1/2} (gD_h \, \Delta\rho)^{-1/2}$ |
| $J_g^*$ | gas dimensionless velocity defined as $J_g^* \triangleq j_g \rho_g^{1/2} (gL \, \Delta\rho)^{-1/2}$ |
| $\mathrm{Ku}_g^*$ | Kutateladze number: $\mathrm{Ku}_g^* \triangleq j_g \rho_g^{1/2} [g(\sigma/g \, \Delta\rho)^{1/2} \, \Delta\rho]^{-1/2} \equiv j_g \rho_g^{1/2} (g\sigma \, \Delta\rho)^{1/4}$ |
| $L$ | annulus circumference |
| Re | Reynolds number |
| $T$ | temperature |
| $x$ | quality |
| $z$ | elevation |
| $\alpha$ | void fraction |
| $\rho$ | density |
| $\Delta\rho$ | density difference between liquid and steam |
| $\sigma$ | surface tension |

**Subscripts**

| | |
|---|---|
| DNB | departure of nucleate boiling |
| $e$ | equilibrium |
| $F$ | fluid |
| $f$ | liquid |
| $g$ | gas |
| MFB | minimum film boiling |
| sat | saturation |
| $v$ | vapor |
| $W$ | wall |

## REFERENCES


Block, J. A., Emergency Cooling Water Delivery to the Core Inlet of PWRs during LOCA, Creare TM-529, 1976.

Crowley, C. J., Block, J. A., and Rothe, P. H., An Evaluation of ECC Penetration Data Using Two Scaling Parameters, Creare TN-233, 1976.

Crowley, C. J., Wallis, G. B., and Rothe, P. H., Preliminary Analysis of Condensation-Induced Transients, Creare TN-271, 1977.

Henry, R. E., A Study of One and Two Component, Two-Phase Critical Flow at Low Qualities, ANL-7430, 1968.

Hsu, Y. Y. and Graham, R. W., *Transport Processes in Boiling and Two-Phase Systems*, chap. 11, Hemisphere, Washington, D.C., 1976.

RELAP 4/MOD 6, A Computer Program for Transient Thermal-hydraulic Analysis of Nuclear Reactor and Related Systems, User's Manual, EG&G CDAP TR 003, 1978.

Smith, R. V., University of Wichita, Wichita, KS, private communication, 1978.

Tong, L. S., Heat Transfer in Reactor Safety, in *Heat Transfer 1978*, vol. 6, Keynote Papers, Hemisphere, Washington, D.C., 285–309, 1978.

CHAPTER

# TWO

# TWO–PHASE FLOW PROBLEMS IN LIQUID–METAL FAST–BREEDER REACTORS

J. Costa

## 1 INTRODUCTION

### 1.1 Brief Historical Survey of Liquid-Metal Fast-Breeder Reactors

Liquid-metal fast-breeder reactors (LMFBR) have a long history that can be illustrated by two important events:

1. Clementine, the first fast-neutron reactor, went critical in 1946 in the United States.
2. In 1951 the first nuclear-generated electricity was delivered by Experimental Breeder Reactor I (EBR I) in the United States; the power was 200 kWe.

In Table 1 characteristics of these two and other early fast-neutron reactors are presented. All were experimental reactors and either are no longer operating or have been profoundly modified. Their characteristics were very different with respect to both the fuel (metal, oxide, uranium, or plutonium) and the coolant (mercury, sodium potassium, or sodium).

In Table 2 characteristics of the nine fast-neutron reactors presently in operation are presented. The reactors cover a thermal power range from 7.5 to 1000 MWth. In

**Table 1 Early liquid-metal-cooled fast reactors[a]**

| Country | Reactor name | Power MWth | Power MWe | Fuel | Coolant | Years of operation |
|---|---|---|---|---|---|---|
| United States | Clementine | 0.025 | | Pu | Hg | 1946-1953 |
| | EBR I | 1.2 | 0.2 | $UO_2$ | NaK | 1951-1964 |
| | Fermi | 300 | 61 | U(10% Mo) | Na | 1965-1972 |
| | SEFOR | 20 | | $(U, Pu)O_2$ | Na | 1969-1972 |
| Soviet Union | BR 5 | 5 | | $PuO_2$, PuC | Na | 1958-1971 |
| United Kingdom | DFR | 60 | 15 | U(7% Mo) | NaK | 1959-1977 |
| France | Rapsodie | 20 | | $(U, Pu)O_2$ | Na | 1967-1970 |

[a]Data mainly from *Nucl. Eng. Int.* (1977).

addition to the experimental or irradiation reactors the first power-station prototypes (BN 350, PFR, and Phenix) make their appearance. Table 3 describes fast-breeder reactors under construction or planned. In addition to the irradiation and prototype reactors the first commercial reactors (Superphenix) appear.

## 1.2 LMFBR Characteristics

With fast-breeder reactors it is possible to extract from natural uranium about 50 times more energy than from $^{235}U$ only. This capability explains the interest in this type of reactor. However, there is no unanimous opinion about fast-breeder reactors on the part of different countries because of differences in energy, industrial, economic, and

**Table 2 Liquid-metal-cooled fast reactors now operating[a]**

| Country | Reactor name | Power MWth | Power MWe | Fuel | Coolant P-pool type, L-loop type | First year of operation |
|---|---|---|---|---|---|---|
| United States | EBR II | 60 | 20 | U | Na(P) | 1963 |
| Soviet Union | BR 10 | 7.5 | | $PuO_2$ | Na | 1973 |
| | BOR 60 | 60 | 10 | $UO_2$ | Na(L) | 1971 |
| | BN 350 | 1000 | 350-200[b] | $UO_2$ | Na(L) | 1973 |
| United Kingdom | PFR | 600 | 250 | $(U, Pu)O_2$ | Na(P) | 1974 |
| France | Rapsodie Fortissimo | 40 | | $(U, Pu)O_2$ | Na(L) | 1970 |
| | Phenix | 560 | 250 | $(U, Pu)O_2$ | Na(P) | 1973 |
| Japan | Joyo | 50 | | $(U, Pu)O_2$ | Na(L) | 1977 |
| West Germany | KNK II | 60 | 20 | $(U, Pu)O_2$ | Na(L) | 1977 |

[a]Data mainly from *Nucl. Eng. Int.* (1977).
[b]Desalting.

**Table 3 Liquid-metal-cooled fast-breeder reactors planned or under construction[a]**

| Country | Reactor name | Power | | Fuel | Coolant P-pool type, L-loop type | Probable first year of operation |
|---|---|---|---|---|---|---|
| | | MWth | MWe | | | |
| United States | FFTF | 400 | | $(U, Pu)O_2$ | Na(L) | 1979 |
| | CRBR? | 1000 | 350 | $(U, Pu)O_2$ | Na(L) | 1983? |
| Soviet Union | BN 600 | 1500 | 600 | $(U, Pu)O_2$ | Na(P) | 1978 |
| United Kingdom | CDFR | 3000 | 1300 | $(U, Pu)O_2$ | Na(P) | ? |
| France | Superphenix | 3000 | 1200 | $(U, Pu)O_2$ | Na(P) | 1982 |
| Japan | MONJU | 700 | 300 | $(U, Pu)O_2$ | Na(L) | 1984 |
| West Germany | SNR 300 | 750 | 300 | $(U, Pu)O_2$ | Na(L) | 1981 |
| | SNR 2 | 3000 | 1300 | $(U, Pu)O_2$ | Na(L) | ? |
| Italy | PEC | 130 | | $(U, Pu)O_2$ | Na(L) | 1978 |
| India | FBTR | 40 | 15 | $(U, Pu)O_2$ | Na(L) | ? |

[a]Data mainly from *Nucl. Eng. Int.* (1977).

political situations. Furthermore, environmentalists are opposed to fast-breeder reactors because of some of the reactors' physical characteristics (plutonium, core not in its more reactive configuration, sodium, etc.).

Sodium was chosen as the heat transfer fluid, first because of its small neutron-capture cross section and second because of its heat transport capabilities. Steam was abandoned because of its large neutron-capture cross section that reduces the breeding ratio too much. Helium slows the neutrons less than sodium, which is good for the breeding ratio, but brings with it many technological unknowns.

Present fast-breeder reactor designs all have approximately the same coolant and fuel nature and have intermediate or secondary sodium circuits between the sodium primary circuits (radioactive) and the water-steam circuits. They are of two types, however:

Loop type: SNR 300, MONJU, CRBR, etc.
Pool type: PFR, Phenix, Superphenix, BN 600, etc.

In the pool type, all the radioactive primary sodium is contained in the main vessel (Fig. 1); in the loop type the secondary sodium circuit is outside the reactor vessel (Fig. 2).

In the core, where the power density is very high (about 1 MW/l), the fuel pins [$(U, Pu)O_2$] have a diameter of about 7 mm, their height is about 1 m, and they are packed in bundles of 217 or 271 per subassembly. The active zone is surrounded by lower, peripheral, and upper blankets ($UO_2$). The inlet temperature of the sodium is about 400°C, and the temperature difference through the core is about 160°C.

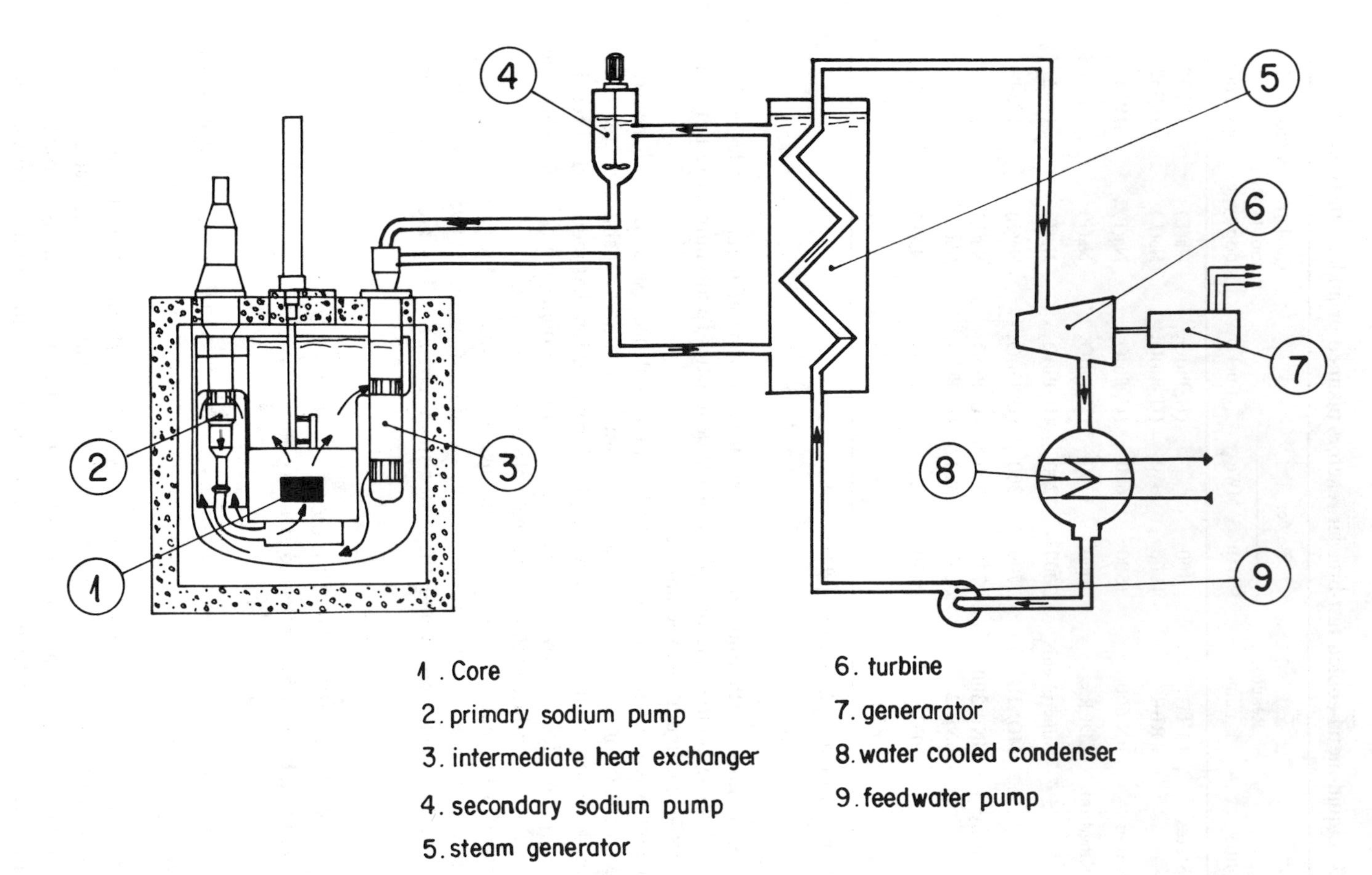


**Fig. 1** Simplified drawing of a pool-type LMFBR (modified from Phenix, 1971).

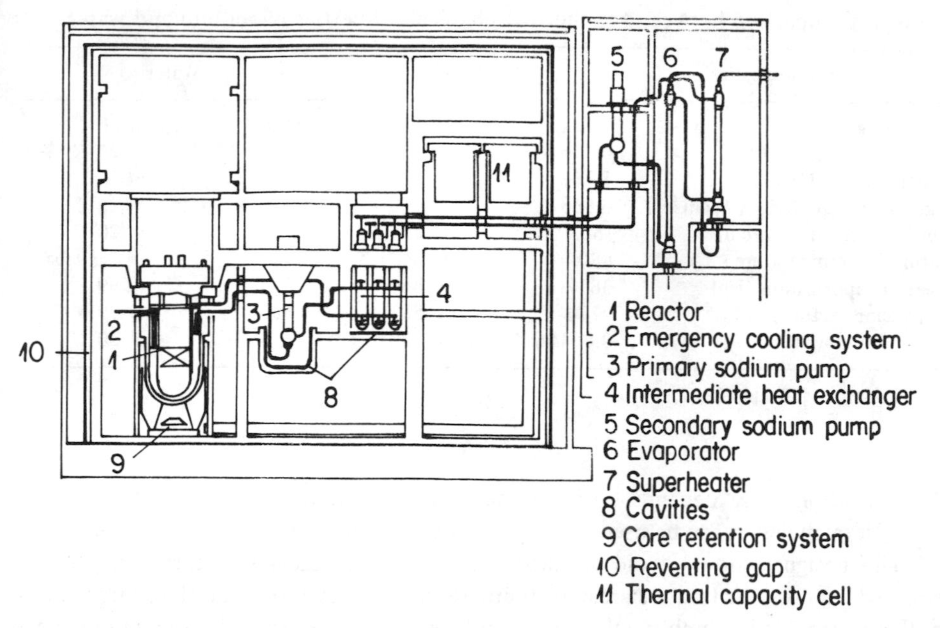


**Fig. 2** Simplified drawing of SNR 300 (loop-type LMFBR) (Traub, 1976).

## 1.3 Physical Properties of Sodium Compared with Water

Liquid metals are renowned for their high thermal diffusivity. In Table 4 some physical properties of sodium, of interest for heat transfer problems, are compared with those of water.

# 2 TWO-PHASE FLOW PROBLEMS IN LMFBRS UNDER NORMAL OPERATING CONDITIONS

Under normal conditions, in the temperature range 400–600°C under 1–2 bar, sodium is highly subcooled. Consequently, there are no design problems relevant to sodium boiling. However, cavitation, which should be avoided because of its damaging effects, is a two-phase flow problem that occurs and will be discussed below. Unlike the sodium circuits, the steam-water circuits, especially the steam generators, present classical examples of two-phase flow problems.

## 2.1 Cavitation

Cavitation is present in many industrial installations (Knapp et al., 1970). It can be described as the formation and collapse of bubbles in a flowing fluid. By using proper means, it can be observed in a transparent pipe downstream of a diaphragm or a venturi. There is a wide range of cavitation levels between onset of cavitation (a few tiny bubbles per second) to supercavitation (with separated liquid–vapor flows).

**Table 4 Comparison between the thermal physical properties of sodium and water**

| Physical properties | Sodium[a] | | Water | |
|---|---|---|---|---|
| Density ($g/cm^3$) | 0.832 | at 500°C | 0.998 | at 20°C |
| Dynamic viscosity (P) | $0.236 \times 10^{-2}$ | at 500°C | $1.01 \times 10^{-2}$ | at 20°C |
| Specific heat (J/g °C) | 1.262 | at 500°C | 4.171 | at 20°C |
| Thermal conductivity (W/cm °C) | 0.668 | at 500°C | $0.598 \times 10^{-2}$ | at 20°C |
| Surface tension (dyn/cm) | 156.7 | at 500°C | 72.8 | at 20°C |
| Saturation temperature (°C) | 880 | under 1 bar | 99.6 | under 1 bar |
| Heat of vaporization (J/g) | 3877 | at 880°C | 2258 | at 99.6°C |
| Expansion factor ($\rho_l/\rho_v$) | 2776 | at 880°C | 1630 | at 99.6°C |
| Prandtl number (Pr) | 0.00446 | at 500°C | 7 | at 20°C |

[a]Golden and Tokar (1967).

Cavitation noise depends on both the nature of the fluid and the nature of the pipes. Nevertheless it can be used to detect and characterize the different regimes.

The designer should avoid cavitation because mechanical structures are rapidly damaged by erosion. In a fast reactor there are risks of cavitation, on the sodium side, in the pumps, subassembly inlet, etc., and on the water side, in the pumps and throttling devices used for flow control.

Therefore for LMFBRs, as for other industrial installations, the two main questions are

1. What are the conditions for cavitation inception?
2. What are the damaging effects of cavitation?

Cavitation in water is of growing concern and is studied in laboratories all over the world. Here the survey will be limited to problems specific to sodium.

Because cavitation is the formation of a vapor bubble in a flowing fluid, the first cavitation inception criterion used is the Thoma number:

$$\sigma \triangleq \frac{P - P_v}{\frac{1}{2}\rho V^2}$$

where $P$ = pressure
$V$ = velocity in a given section
$P_v$ = saturation pressure
$\rho$ = density

The Thoma number allows a comparison between the saturation pressure $P_v$ and the actual pressure in the fluid.

This Thoma number is independent of the nature of the fluid and has been used to make predictions for sodium using measurements on water test sections. In fact, comparisons of the Thoma numbers measured on two identical test sections, one in water, the other in sodium (Bonnin et al., 1971) showed good agreement in a first approximation; if they are examined in more detail, however, differences of as much as 20% (Figuet, 1977) are evident because of the different nucleation situations in the two fluids. Because water tests are much easier to perform and less costly than sodium tests, it is particularly important to explain these differences and so be able to predict cavitation inception in sodium from water tests.

A study of this type was initiated by Figuet (1977). The evolution of a small, spherical gas bubble in a transient pressure field was calculated taking into account the surface tension, viscosity, and heat and mass transfer through the interface. These calculations show in this simple case the influence of the above parameters on the Thoma number. An attempt has been made to improve these predictions by taking into account not only the pressure of the mean flow but also the pressure fluctuations.

## 2.2 Two-Phase Flow Problems in LMFBR Steam Generators

The steam generator is a major component in a nuclear power station. It transfers heat from the secondary sodium to the feedwater, which is preheated, evaporated, and superheated to produce high-temperature steam used in the turbines (Fig. 3) (Budney and Marvosh, 1976).

Several designs of sodium-heated steam generators have been considered: double-walled tubes to avoid sodium-water reactions from a leak caused by corrosion; natural circulation or forced convection systems; once-through designs with straight (SNR, Fig. 4) or coiled (Superphenix, Fig. 5) tubes; superheater and resuperheater separated from the evaporator (Phenix).

The main differences between sodium-heated steam generators and PWR steam generators result from

1. The thermodynamic conditions, that is, steam pressure and temperature are higher for the sodium-heated steam generators ($T \cong 500°C$, $P \cong 175$ bar), conditions close to the classical fossil fuel power station.
2. The fact that in LMFBRs, steam is generated inside tubes heated from outside by sodium, whereas in PWRs, steam is generated outside tubes heated by the primary coolant.
3. The nature of the heating fluid: sodium has a much higher heat transfer capability,

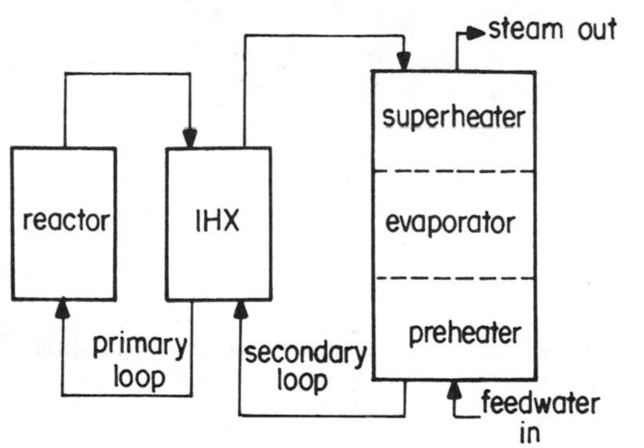


**Fig. 3** Flow diagram for a once-through steam generator (IHX, intermediate heat exchanger).

**Fig. 4** Straight-tubes steam generator (modified from Ünal et al., 1977).

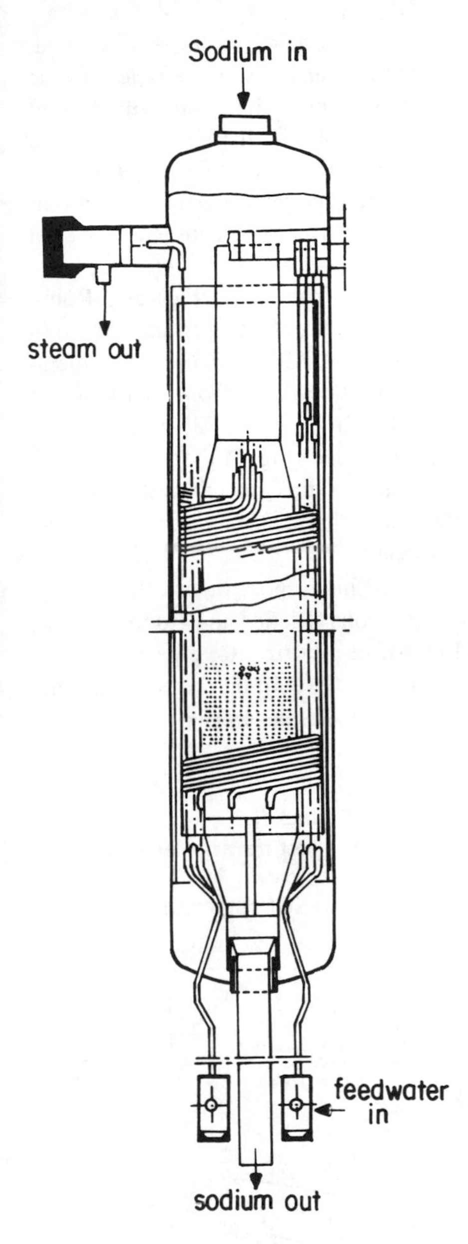


**Fig. 5** Helically coiled tubes steam generator (modified from Lavalerie and Bouju, 1972).

with the associated possibility of severe thermal gradients under transient conditions.

The designer is faced with many problems, such as the choice of the concept, differential expansion, vibrations, corrosion, purity of the fluids, leak detection, repair or replacement, and stable operation. Only the thermohydraulic problems in the water circuits, which involve a lot of two-phase flow phenomena, will be examined.

To determine the effective tube area one needs the heat transfer coefficient in the different zones of the flow: liquid, nucleate boiling, film boiling, and superheated steam. General two-phase flow knowledge, based on the LWR thermal core analysis, can be used as a first approximation, but it is generally agreed that few experimental results in the parameter range are related to sodium-heated steam generators and that specific tests are necessary to take into account the particular characteristics of each steam-generator design.

Examples of calculations are given by Budney and Marvosh (1976) and Robin (1972) (Fig. 6). One concern is the transition between nucleate boiling and film boiling [the conditions of departure from nucleate boiling (DNB)] because of temperature fluctuations, associated with the fluctuations of the dryout front of the liquid film, producing thermal fatigue of the duct. Capolunghi et al. (1973) measured the heat transfer in long vertical straight tubes. They concluded (Fig. 7) that in long once-through steam generators with very high length–diameter ratios and relatively low heat flux the usual DNB correlations obtained for PWR core thermal analysis, show noticeable errors (up to 50%) and may not be directly applied.

Furthermore, when there are horizontal parts in the evaporator or coiled tubes, the transition to film boiling is affected by stratifications (Ricque and Roumy, 1970) (Fig. 8) or secondary flows (Kozeki, 1973) (Fig. 9). In a horizontal pipe the dryout of the liquid film can be obtained for a small vapor quality ($\cong 0$) on the upper generatrix of the tube and for a very high quality ($\cong 1$) on the lower generatrix.

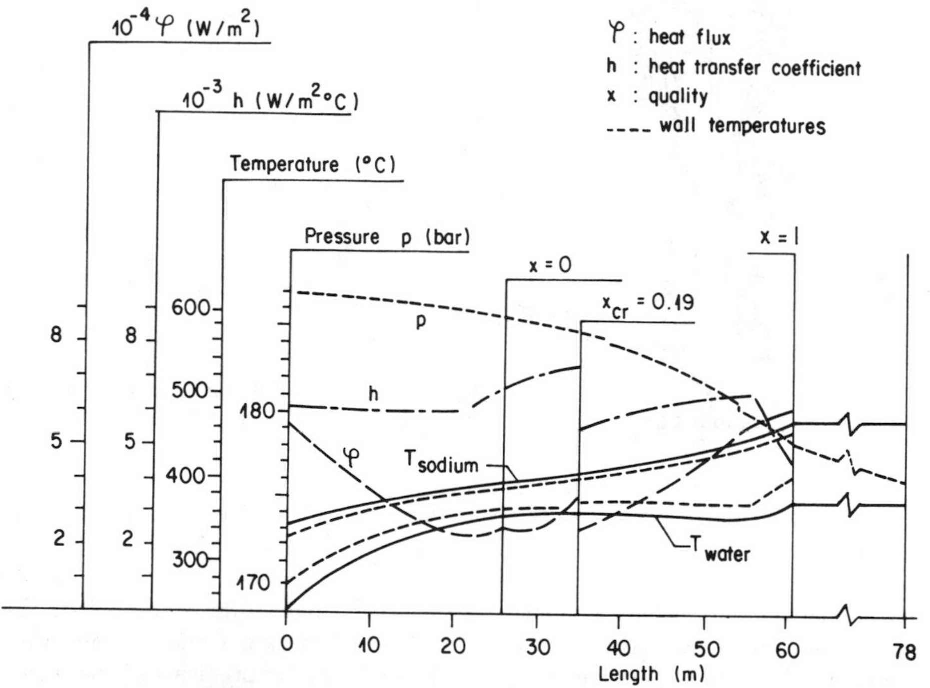


**Fig. 6** Phenix steam-generator thermal parameters at nominal power (modified from Robin, 1972).

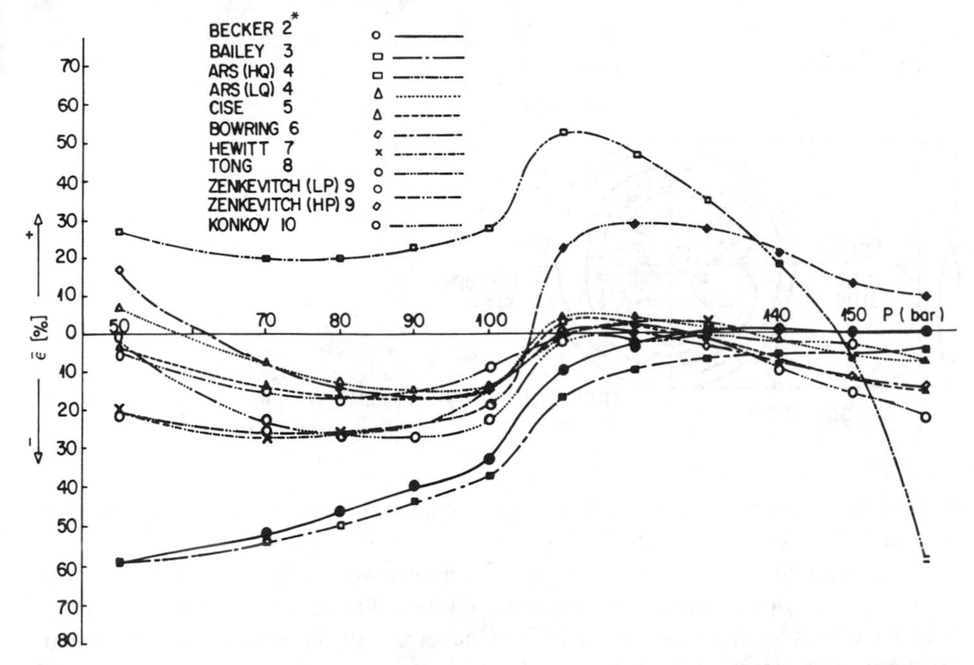


**Fig. 7** Comparison of burnout correlations as functions of Comitato Nazionale per l'Energia Nucleare experimental data (percentage differences compared with $P$) for $P > 120$ bar; Becker correlation gives the best previsions; *references as given in Capolunghi et al. paper (Capolunghi et al., 1973).

An accurate prediction of the pressure drop along the two-phase flow is required because the quality of the vapor depends on the pressure. The Martinelli-Nelson correlation or the modified Darcy equation is generally used. The problem is more complicated when the steam generator is not a straight tube and when pressure drops in bends or pipe expansions or contractions have to be evaluated.

Another important problem with sodium-heated steam generators is that under certain circumstances, for instance, low power, they can experience dynamic instabilities, which might disturb the distribution of the flow among the tubes and create

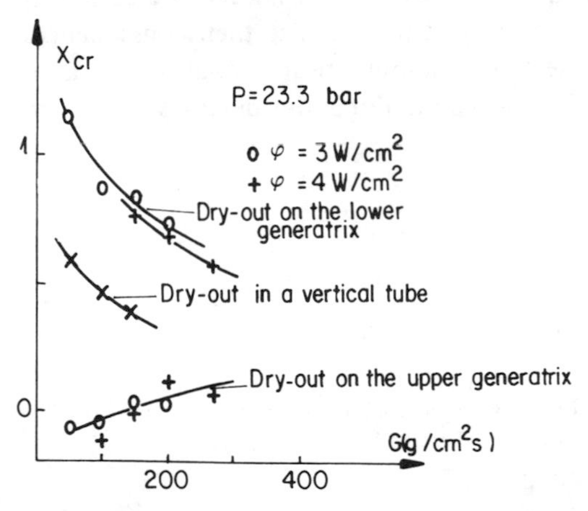


**Fig. 8** Dryout quality in a horizontal tube as related to mass velocity (Ricque and Roumy, 1970).

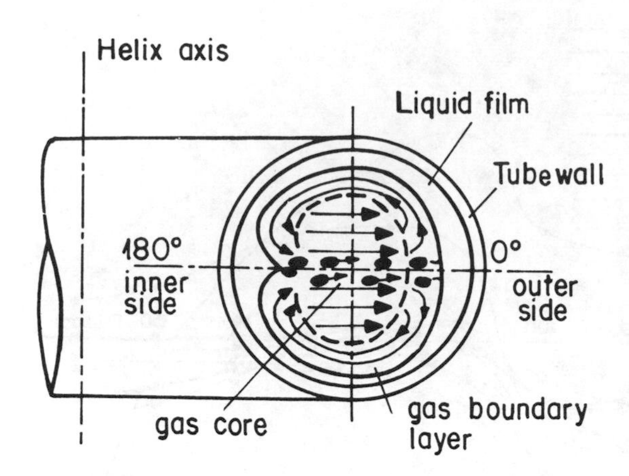


Fig. 9 Exaggerated model of an annular two-phase flow in a helically coiled tube (Kozeki, 1973).

thermal cycles in the mechanical structures. Figure 10 illustrates this type of temperature oscillation: large amplitude (60°C) and low frequency (0.1 Hz).

The different types of instabilities have been reviewed by Bouré et al. (1973). The large-amplitude, low-frequency oscillations are referred to as density wave instabilities. They have been studied experimentally by Ünal et al. (1977) on a full-scale module of the SNR 300 steam generator (Fig. 4), and more theoretically by Suzuoki and Yamakawa (1976), who developed a computer code model of them.

To conclude, another two-phase flow problem must be mentioned, that is, the large hydrogen bubble dynamics produced by a sodium–water reaction after the rupture of a tube (Lacroix et al., 1967).

## 3 TWO-PHASE FLOW PROBLEMS IN LMFBR SAFETY ANALYSIS

### 3.1 Introduction to Two-Phase Safety Problems

Detailed safety analyses are in progress to estimate whether fast reactors can be built and operated with an acceptable risk for the public. At the design level all precautions are taken to eliminate the risk of malfunctions and to minimize their consequences should they occur. Furthermore, consequences of hypothetical core-disruptive accidents (HCDA) are discussed and taken into account. For more details see Graham

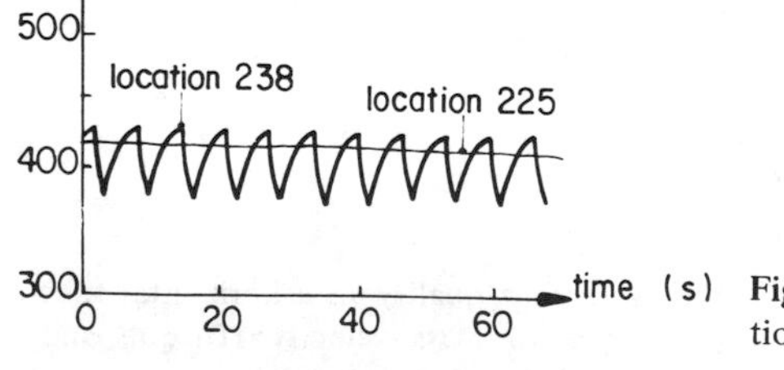


Fig. 10 Typical example of density wave oscillations (Ünal et al., 1977).

(1971) and Fauske (1977). Pump failure without scram is one of the possible origins of these HCDAs.

Because the core response to this perturbation and the perturbation itself depend on the size, design, and characteristics of the reactor, there are as many accident trees as reactors. Therefore, no HCDA will be presented in detail; only the phenomena encountered during this accident will be discussed. More details concerning a particular reactor, Superphenix, will be given later in the section on sodium boiling.

The principal LMFBR characteristics affecting safety are

1. High power density of the core: any mismatch between the power generated by the fuel and the heat removed can lead to rapid change in the temperature.
2. The core is not in its more reactive configuration; as a result, power excursions may occur, depending on the relative motion of the core materials.

To introduce the two-phase flow problems, we shall divide the hypothetical accident into several steps:

1. Initial phase
   a. Initiation: fuel temperature rises and the coolant is in single-phase flow.
   b. Saturation temperature of sodium is reached, and there is sodium boiling; the geometry is still intact; because of the void effect, the power changes and could increase.
2. Transition phase
   a. Clad reaches melting temperature, causing clad relocation and consequences on coolant flow and power.
   b. Fuel reaches melting temperature causing fuel relocation, with consequences on power.
3. Disassembly phase
   a. Fuel vaporization and/or fuel coolant interaction.
4. Postaccident phase
   a. Core debris cooling.

Each phase defines the initial conditions for the following one, and in a mechanistic description of an HCDA they should all be linked. In practice, pessimistic initial conditions are taken for each phase for which independent calculations are then made. The following sections will illustrate two-phase flow problems encountered in the above sequences.

In this safety evaluation the big question is the phenomenology; therefore the problems are treated differently from the problems related to LWR thermal core performances.

## 3.2 Sodium Boiling

Sodium boiling studies (Costa, 1977), which have been conducted for more than 10 yr, cover the various abnormal cooling situations usually considered in the safety analysis for LMFBRs: blockage in the inlet nozzle of the subassembly, local blockage in the bundle, loss of flow (pump coastdown), and power excursion.

To illustrate the sodium boiling question, this presentation is limited to the problem of boiling phenomena occurring during slow transients at moderate power, such as those occurring during a pump coastdown without scram, considered as a reference hypothetical accident for Superphenix (Robert et al., 1976). The recent reviews of Peppler (1977) and Han and Fontana (1977) illustrate other problems.

The two main questions involving boiling phenomena in the above accident concern predictions of the voiding rate of the core and of the dryout conditions. The voiding rate is important because the void coefficient is positive in the central zone of the core; the dryout conditions determine the onset of clad melting. Problems of this sort are common to all the LMFBRs under development in different countries; nevertheless specific studies are necessary because of differences in the concept (loop or pool type, inertia of the pumps), in the design of the subassembly (grid or helical wire spacers), and so on.

The particular characteristics of the thermohydraulic studies presented below are that they deal with a long pump coastdown time (one-tenth of the nominal flow after about 10 min), accompanied by a reduction of power (about one-third of the nominal power after 10 min), and they involve a subassembly with wire-wrap spacers. Saturation conditions, based on bulk calculations, are reached about 10 min after the onset of the pump rundown (see Fig. 11).

The method of approach that has been used can be characterized by the two following points:

1. To proceed from simple to more complex technological, experimental, and analytic problems, the subassembly was first simulated by a single-channel geometry; the bundle effect was considered of second order and was studied later.
2. As far as feasible, use was made of the knowledge available on general boiling and two-phase flow obtained from water studies for LWRs (Mondin and Séméria, 1969).

**3.2.1 Single-channel approximation: Description of the boiling phenomena during a pump coastdown without scram** As can be seen in Fig. 11, the saturation condition is

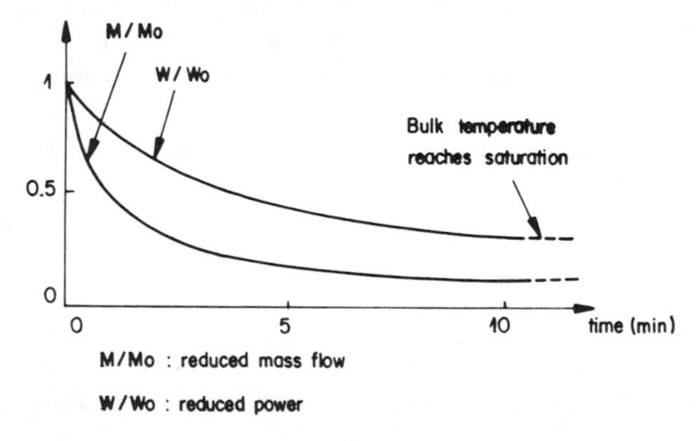


**Fig. 11** The pump coastdown accident for a LMFBR of the Superphenix type.

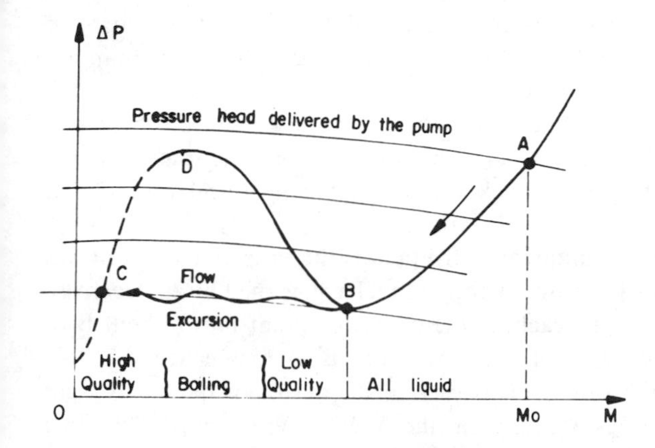


**Fig. 12** The flow excursion during a pump coastdown.

reached during a very slow transient so that consideration of the stability of two-phase flow becomes relevant.

The shape of the channel characteristic pressure-drop curve (so-called *S* curve) and the evolution with time of the pressure head available (see Fig. 12) suggest the likelihood of a particular flow instability of the Ledinegg type called the *flow excursion* (Bouré et al., 1973).

The *S* curve represents for a series of steady-state regimes the pressure drop ($\Delta P$) across the channel as a function of mass flow rate $M$ for given inlet temperature, heat input, and pressure in the outlet plenum. Curves of that sort were obtained with single-channel test sections representing as closely as possible the hydraulic, thermal, and boundary conditions of a real subassembly (the heated length, as well as the downstream breeder blanket and shielding zones, was represented).

In Fig. 12, the AB part of the curve corresponds to the all-liquid regimes and the BDC region to the boiling regimes. The two main qualitative results shown by these curves are

1. Stable boiling of sodium exists in forced convection; no significant superheat was observed. When sodium boils in such a test section, it behaves like an ordinary boiling fluid (many bubbles were observed).
2. During the boiling runs a small amount of vapor is at the top of the heated zone and a large amount is in the shielding zone where flashing can occur; sometimes the critical flow conditions are reached at the outlet.

Concerning the problem of superheat, it should be noted that in the past high superheat values have been measured in stagnant sodium. Because the resulting voiding of the channel was high, superheat has become a safety problem and has been the subject of considerable research. It has been very difficult to draw many conclusions because nucleation involves many parameters that are not always under the control of the different experimenters. A recent review of the superheat problem in sodium has been presented by Kottowski and Savatteri (1977).

For some steady-state boiling experiments a method utilizing three electromagnetic flowmeters was used to measure the quality and the void fraction. The results, shown in Fig. 13, are in good agreement with the Lockhart-Martinelli void correlation (Costa, 1977).

As shown in Fig. 12, during the rundown of the pumps the operating point moves from A to B, and the decrease of the flow in the channel is governed by the decrease of the pressure head. When point B is reached, the operating point moves from B to the new stable point C. The reduction of the flow from B to C is called the flow excursion. In general, because the flow at C is very small, the vapor quality is high; between B and C, dryout as well as a change in the flow pattern occurs. This flow excursion is an irreversible process. For instance let us assume a small decrease in the pressure head $\Delta P$ from $P_B$; this induces the flow excursion to C. If the pressure is then increased, the working point moves on the CD part of the *S* curve up to D and then from D to the AB part of the curve. Thus it is during this flow excursion that the two phenomena critical for safety, namely, the voiding of the channel and dryout, occur.

It is crucial to know how fast the flow decreases from B to C. Because the quantitative description of this flow excursion was not available, a special program was undertaken to understand it and to obtain experimental data to validate the computer codes. From the series of experiments performed (Schmitt, 1974), covering a wide range of parameters, different heated lengths, downstream geometries, and power levels, it can be concluded that the flow excursion is a rather slow process, lasting for a period on the order of several seconds (Fig. 14). Two distinct phases can be observed during flow excursion. In the first phase, there is a progressive voiding of the channel accompanied by a quiet boiling regime (bubbling). The second phase starts with the local dryout of the pin; during this phase, the flow pattern is of the chugging type with expulsions. The main reason for the rather slow decrease of the flow during the flow excursion is that the thermal inertia of the structural material and the pins is large compared with the thermal inertia of the sodium.

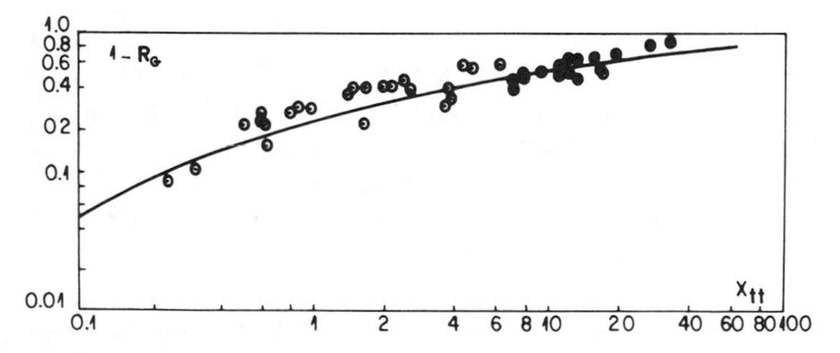


**Fig. 13** Comparison between measured liquid fraction and the Lockhart-Martinelli correlation. Filled circles, author's data; circled dot, Fauske's data; curve, Lockhart-Martinelli correlation (Costa, 1977).

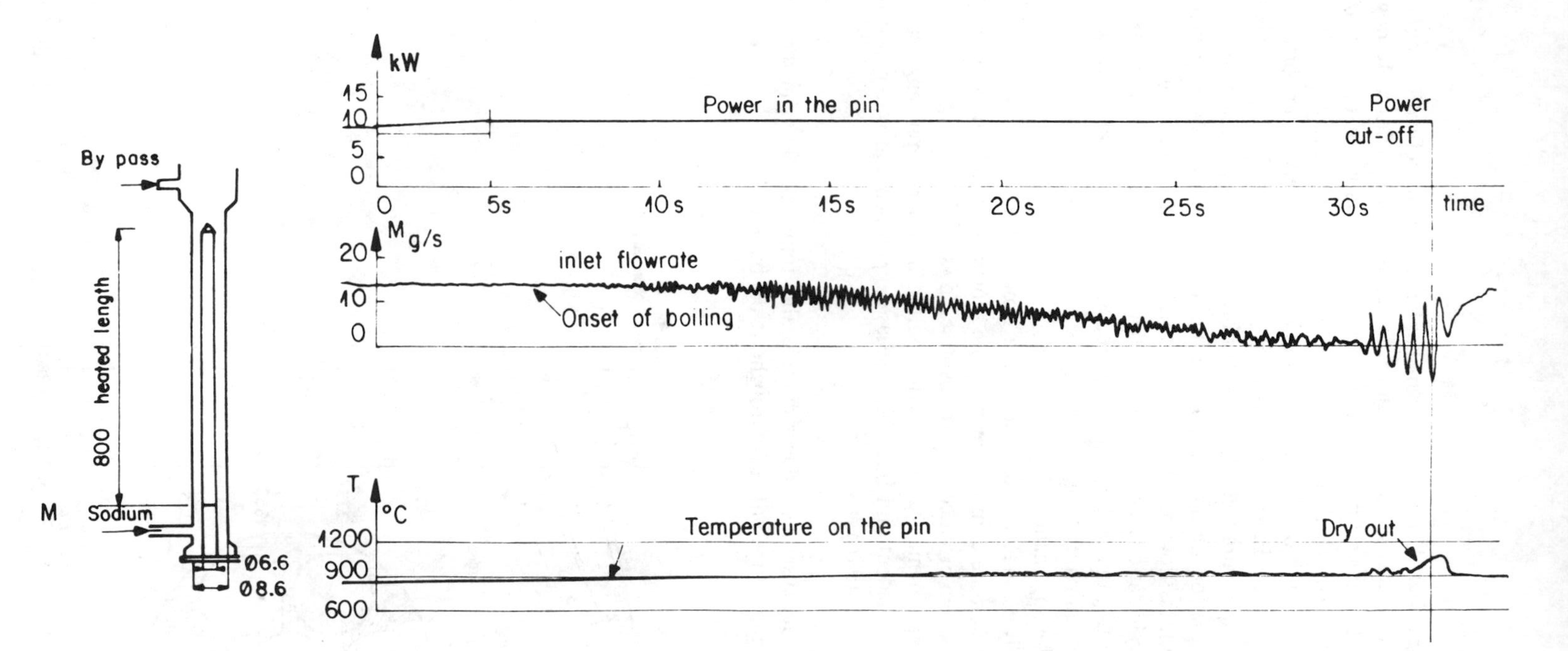


Fig. 14 Typical result of flow excursion (Schmitt, 1974).

**3.2.2 The bundle effect** Although great care has been taken to define the characteristics of the single-channel test section, a major uncertainty still remains as to how far the results obtained in such a simplified geometry are representative of a subassembly.

Some preliminary calculations using the subchannel mixing code Flica showed that the flow diversion around a local boiling zone could be very important and lead to a rapid dryout. The Flica code (Plas, 1975), developed for light-water reactors, described the thermohydraulics of a subassembly using a subchannel analysis and thus calculated the variation of the velocity profile. Figure 15 shows some results of the evolution of the mass flow profile along a seven-pin bundle with boiling in the central zone. For the calculations the bundle has been simulated by three interconnected channels representing the central, intermediate, and outer subchannels. Downstream of the location of boiling inception, a significant reduction of the mass flow rate can be observed in the boiling channel.

This flow diversion is comparable to the flow excursion described in Sec. 3.2.1, but the problem is more complex because the subchannels are connected and because the helical-wire system produces large transverse flows.

From the experiments carried out in sodium in a parameter range valid for the pump coastdown accident and from single-phase experiments in water, it can be concluded that (Menant, 1976)

1. The helical-wire system produces considerable transverse flows in the bundle not only near the wrapper tube but also throughout the bundle.

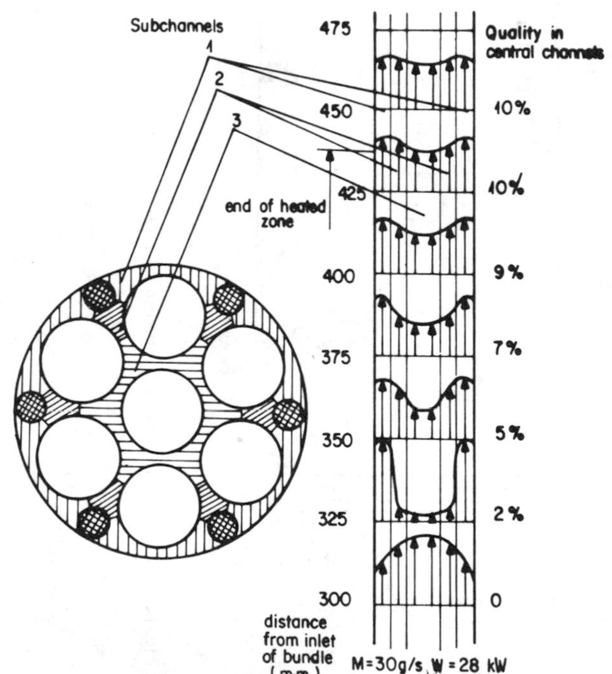


**Fig. 15** Mass flow rate profiles in a seven-pin bundle at different levels (Flica calculations).

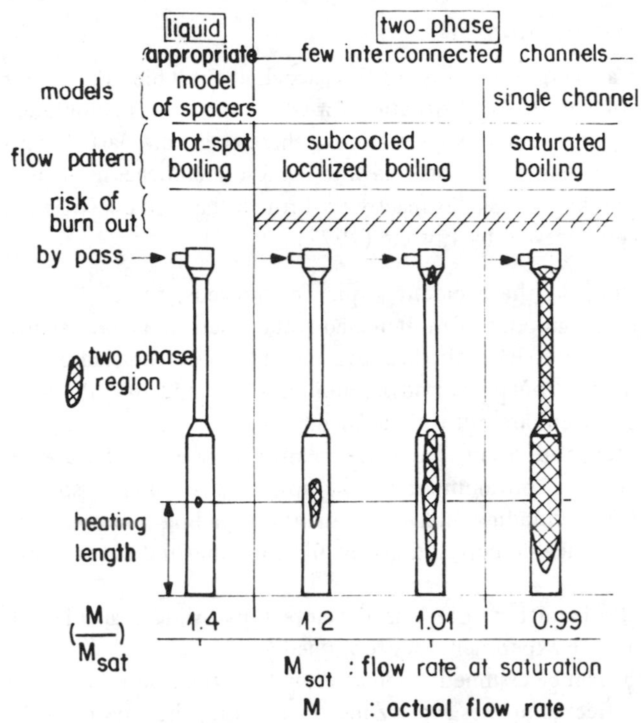


**Fig. 16** First stages of boiling in a 19-pin bundle during slow flow transients.

2. Although this mixing is high, significant hot spots exist so that (Fig. 16) localized boiling can be initiated in the 19-pin bundle for a mass flow 40% higher than that necessary to obtain bulk boiling with bundle-average conditions. In this first type of boiling regime, the void fraction is small, and the vapor does not migrate through the bundle. Although this boiling regime has no great influence on the void fraction (or the reactivity), it is important because, along with other factors, it reduces any risk of high superheat in the bundle.
3. The second boiling regime is still subcooled and is located in the hottest subchannels of the subassembly.
4. The third regime is saturated boiling and corresponds to the boiling regime observed on the single-channel geometry (with vapor not only around the heated pins but also in the downstream region).
5. The subcooled boiling regime between hot-spot boiling and saturation boiling covers a wide range of flows. The void fraction is not very large so that it has not a great influence on reactivity, but local dryout can be achieved very easily. The subcooled boiling regime is important because the failure and melting of the pin cladding can occur while the subassembly is still under bundle-average subcooled conditions.

**3.2.3 Boiling detection** Because sodium boiling occurs early in an accident, many attempts have been made to use this boiling noise to detect an accident while it is still in an initial phase.

## 3.3 Clad and Fuel Melting and Motion

Between the initial phase of a hypothetical loss of flow accident for which the core is still in its normal configuration and the final situation for which the core is completely melted down there are many intermediate stages. Among these stages are partial melting of the clad and the fuel; these melting stages have to be described because of their effects on reactivity (just as boiling influences reactivity through the void coefficient). This problem has recently been reviewed by Epstein (1977).

**3.3.1 Clad melting and motion** As the accident sequence proceeds, the clad of the fuel pins melts progressively. The amount of molten clad is the result of a heat balance between the heat generated by the fuel and the heat evacuated through the clad by the coolant. The problem is that the coolant is under a boiling phase and that the mass flow and the heat transfer coefficient are not well defined.

The motion of the molten clad results from the balance of two effects: gravity acting on the molten material and entrainment by the upward streaming of sodium vapor. This is the problem of the flooding and entrainment of a falling film by an upward stream of gas. There are available correlations on film flooding and entrainment that can be used for calculations.

A one-dimensional single-channel model was developed by Ishii et al. (1976). Fairly good agreement with in-pile experiments was obtained.

The following events can then be assumed: soon after melting the clad is entrained by the vapor flow, the clad freezes in the upper zone of the core, the vapor flow is reduced by the blockage, and the clad drains and plugs the lower part of the subassembly.

**3.3.2 Volumetric boiling in an open pool** Another two-phase flow problem appears: fuel melting will follow the clad melting and there will be boiling of the fuel-steel mixture subjected to volumetric heat generation (the boiling point of steel is close to the melting point of the fuel, 2800°C). The recriticality problem depends on the behavior of this boiling pool; therefore the flow regime is important.

A significant contribution to this problem was made by Fauske (1975), who used the Kutateladze criterion for stability of two-phase flows. He proposed a map with the transitions from bubbly to churn-turbulent to dispersed droplets (with no slug flow because the fuel must be in a dispersed phase to avoid superheating).

There are many other two-phase flow problems in this transition phase, but only two have been mentioned for illustration.

## 3.4 Molten Fuel–Coolant Interaction

Violent vaporizations have been observed, for instance, in the paper or foundry industries and in some water nuclear reactor experiments (SL1, SPERT), when a hot molten material, "the fuel," comes in contact with a coolant whose boiling point is well below the temperature of the hot material. This process is physical rather than chemical, and the rapid energy transfer needed to produce the violent vaporization is achieved through a fine fragmentation and dispersion of the hot material into the coolant. In such a process a significant fraction of the thermal energy stored in the hot material can be transformed into mechanical energy (from thermodynamic calculations the maximum efficiency could be about 30% with molten $UO_2$ and sodium).

One of the key safety issues for LMFBRs is thus to know whether the contact between molten fuel and sodium during a hypothetical core-meltdown accident could lead to an energetic fuel–coolant interaction and to evaluate what the consequences on the reactor containment would then be.

Of all the experiments performed in the world up to now only two, described in Amblard et al. (1970) and Armstrong et al. (1976), gave an energetic vaporization. The experiments were on a small scale and in no way prototypic. Nevertheless, a good understanding of this physical process is necessary for the safety evaluation of LMFBRs. Significant progress has been made recently and has been reviewed by Board and Caldarola (1977) and Henry and Cho (1977). But there are still several theories in competition to explain this complex phenomenon, which involves not only two-phase flow problems but also hydrodynamic, three-phase flows. To illustrate this point a brief presentation of the detonation model (Board et al., 1975) will be given.

In the detonation model for fuel–coolant interactions, in which an analogy is made with the detonation of chemical explosives, three stages are postulated (Fig. 17):

1. Fuel and coolant become coarsely intermixed.
2. A trigger mechanism is assumed to result in a shock wave.
3. The shock wave traveling through the coarse mixture causes fine fragmentation mixing, the resulting rapid heat transfer and vaporization being able to sustain the propagation of the shock wave.

In fact there is still the possibility that the propagation front is a small pressure wave that triggers the interaction, followed by fragmentation and heat transfer at a relatively low rate. At the present time several laboratories are trying to extrapolate the mathematical treatment of the theory of chemical detonation to a thermal two-phase detonation.

## 4 CONCLUSION

This chapter has described some two-phase flow problems encountered in the design and safety analysis of LMFBRs, but there are many others, such as

Gas entrainment from the free surface
Fission gas release
Fuel vaporization
Fuel freezing
Debris bed cooling
Sodium-water chemical reaction

Some of these problems are classical liquid–gas two-phase flow problems (sodium boiling) but most of them are more complex because the flow boundaries are not well

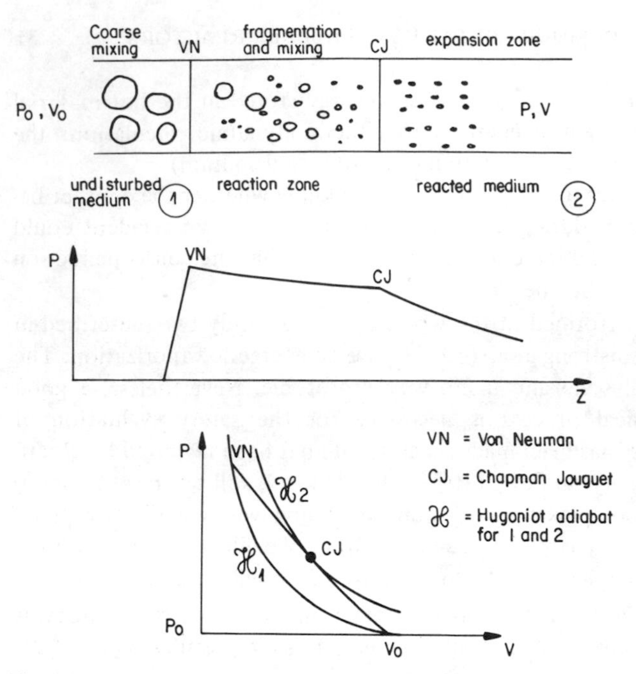


**Fig. 17** Detonation model for fuel–coolant interactions.

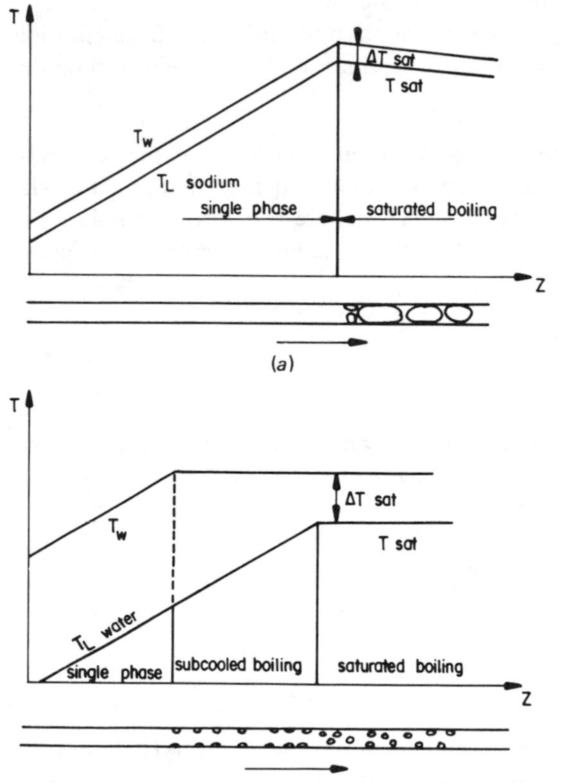


**Fig. 18** Boiling in a heated channel, comparison between sodium and water. (*a*) Sodium boiling in a channel. (*b*) Water boiling in a channel.

defined and change with time (for instance, clad melting, fuel relocation, and sodium-fuel interaction) and also because often more than one liquid phase and one gas phase are present (for instance, clad motion, sodium-fuel interaction, etc.)

That the coolant is sodium does not radically change the nature of the two-phase flow phenomena compared with that in water reactors, but nevertheless there are differences owing to the high thermal conductivity of sodium. These differences are illustrated by the absence of subcooled boiling in sodium (Fig. 18).

Although two-phase flow phenomena are now generally well understood, progress lies in the direction of creating more precise mathematical models and developing computer codes for the more complex problems mentioned above, such as sodium fuel interaction.

## NOMENCLATURE

| | |
|---|---|
| $G$ | mass velocity |
| $h$ | heat transfer coefficient |
| $M$ | mass flow rate |
| $P$ | pressure |
| $P_v$ | saturation pressure |
| $\Delta P$ | pressure drop across a channel |
| Pr | Prandtl number |
| $R_G$ | gas fraction |
| $T$ | temperature |
| $V$ | velocity; specific volume |
| $W$ | heat input |
| $x$ | quality |
| $\bar{\epsilon}$ | percentage difference |
| $\rho$ | density |
| $\sigma$ | Thoma number |
| $\varphi$ | heat flux |

**Subscripts**

| | |
|---|---|
| cr | critical |
| $l$ | liquid |
| $L$ | liquid |
| sat | saturation |
| $v$ | vapor |
| $W$ | wall |

## REFERENCES

Amblard, M., Séméria, R., Vernier, P., Gouzy, A., and Najuc, M., Contact entre du Bioxyde d'Uranium Fondu et un Réfrigérant (Sodium et Eau), CEA rapport TT 96, Centre d'Etudes Nucléaires de Grenoble, 1970.

Armstrong, D. R., Goldfuss, G. T., and Gebner, R. H., Explosive Interaction of Molten $UO_2$ and Liquid Sodium, ANL-76-24, 1976.

Board, S. J. and Caldarola, L., Fuel Coolant Interaction in Fast Reactors, in *Thermal and Hydraulic Aspects of Nuclear Reactor Safety*, vol. 2, *LMFBR*, eds. O.C. Jones and S. G. Bankoff, pp. 195–222, ASME, New York, 1977.

Board, S. J., Hall, R. W., and Hall, R. S., Detonation of Fuel Coolant Explosions, *Nature*, vol. 254, no. 5498, pp. 319–321, 1975.

Bonnin, J., Bonnafoux, R., and Gicquel, J., Comparaison des Seuils d'Apparition de la Cavitation dans un Tube de Venturi dans l'Eau et le Sodium Liquide, *Electr. Fr. Bull. Dir Etud. et Rech. Sér. A*, no. 1, pp. 5–12, 1971.

Bouré, J. A., Bergles, A. E., and Tong, L. S., Review of Two-Phase Flow Instability, *Nucl. Eng. Des.*, vol. 25, pp. 165–192, 1973.

Budney, G. S. and Marvosh, J. V., Steam generators, in *Sodium NaK Engineering Handbook*, vol. 2, *Sodium Flow, Heat Transfer, Intermediate Heat Exchangers and Steam Generators*, ed. O. J. Foust, pp. 273–381, Gordon and Breach, New York, 1976.

Capolunghi, F., Cumo, M., Ferrari, G., Leo, R., and Vaccaro, G., An Experimental Study on Heat Transfer in Long, Sub-critical Once-through Steam Generators, in *Reactor Heat Transfer*, ed. M. Dalle Donne, *Proc. Int. Meet., Karlsruhe, Oct. 9–11*, pp. 373–403, Gesellschaft für Kernforschung mbH, Karlsruhe, 1973.

Costa, J., Contribution of the Grenoble Heat Transfer Laboratory to the Study of Sodium Boiling during Long Pump Coastdown Times in Subassemblies with Wire-Wrap Spacers, in *Thermal and Hydraulic Aspects of Nuclear Reactor Safety*, vol. 2, *LMFBR*, eds. O. C. Jones and S. G. Bankoff, pp. 155–169, ASME, New York, 1977.

Epstein, M., Melting, Boiling and Freezing–The "Transition Phase" in Fast Reactor Safety Analysis, in *Thermal and Hydraulic Aspects of Nuclear Reactor Safety*, vol. 2, *LMFBR*, eds. O. C. Jones and S. G. Bankoff, pp. 171–193, ASME, New York, 1977.

Fauske, H. K., Boiling Flow Regime Maps in LMFBR HCDA Analysis, *Trans. Am. Nucl. Soc.*, vol. 22, pp. 385–386, 1975.

Fauske, H. K., An Overview Including Implications of Alternate Fuel Cycles, in *Thermal and Hydraulic Aspects of Nuclear Reactor Safety*, vol. 2, *LMFBR*, eds. O. C. Jones and S. G. Bankoff, pp. 1–17, ASME, New York, 1977.

Figuet, J., Etude de la Cavitation Initiée par des Tourbillons–Cavitation dans un Venturi: Analogie entre l'Eau et le Sodium Liquide, Ecole Nationale Supérieure des Techniques Avancées, rapport de recherche 084, 1977.

Golden, G. H. and Tokar, J. V., Thermophysical Properties of Sodium, ANL 7323, 1967.

Graham, J., *Fast Reactor Safety*, Academic, New York, 1971.

Han, J. T. and Fontana, M. H., Blockages in LMFBR Fuel Assemblies, in *Thermal and Hydraulic Aspects of Nuclear Reactor Safety*, vol. 2, *LMFBR*, eds. O. C. Jones and S. G. Bankoff, pp. 51–121, ASME, New York, 1977.

Henry, R. E. and Cho, D. H., An Evaluation of Potential for Energetic Fuel Coolant Interactions in Hypothetical LMFBR Accidents, in *Thermal and Hydraulic Aspects of Nuclear Reactor Safety*, vol. 2, *LMFBR*, eds. O. C. Jones and S. G. Bankoff, pp. 223–237, ASME, New York, 1977.

Ishii, M., Chen, W. L., and Grolmes, M. A., Molten Clad Motion Model for Fast Reactor Loss of Flow Accidents, *Nucl. Sci. Eng.*, vol. 60, pp. 435–451, 1976.

Knapp, R. T., Hammitt, F. G., and Daily, J. W., *Cavitation*, McGraw-Hill, New York, 1970.

Kottowski, H. M. and Savatteri, C., Evaluation of Sodium Incipient Superheat Measurements with Regard to the Importance of Various Experimental and Physical Parameters, *Int. J. Heat Mass Transfer*, vol. 20, pp. 1281–1300, 1977.

Kozeki, M., Film Thickness and Flow Boiling for Two-Phase Annular Flow in Helically Coiled Tube, in *Reactor Heat Transfer*, ed. M. Dalle Donne, *Proc. Int. Meet., Karlsruhe, Oct. 9–11*, pp. 351–372, Gesellschaft für Kernforschung mbH, Karlsruhe, 1973.

Lacroix, A., Lions, N., and Robin, M., Etudes de Sûreté des Générateurs de Vapeur Sodium–Eau–Méthode de Calcul et Résultats Expérimentaux, *Conférence Internationale sur la Sûreté des*

*Réacteurs à Neutrons Rapides*, Aix-en-Provence, Sept. 19–22, 1967, ed. G. Deniélou, paper 6-16, Commissariat à l'Energie Atomique, Paris, 1967.

Lavalerie, C. and Bouju, J. L., Etudes Hydrauliques et Thermiques sur une Maquette de Générateur de Vapeur, S.H.F., XII° Journées de l'Hydraulique, Paris, question 2, 1ère partie, rapport 7, 1972.

Menant, B., Quelques Particularités des Ecoulements de Sodium Liquide et Bouillant dans des Grappes d'Aiguilles Chauffantes, Thèse de docteur ingénieur, Université Scientifique et Médicale et Institut National Polytechnique, Grenoble, 1976.

Mondin, H., and Séméria, R., Boiling and Two-Phase Flow of Sodium, Paper presented at European Two-Phase Flow Group Meet., Karlsruhe, June 2–5, 1969.

*Nucl. Eng. Int.*, suppl. vol. 22, no. 258, p. 30, 1977.

Peppler, W., Sodium Boiling in Fast Reactors: "A state of the art review," in *Thermal and Hydraulic Aspects of Nuclear Reactor Safety*, vol. 2, *LMFBR*, eds. O. C. Jones and S. G. Bankoff, pp. 123–153, ASME, New York, 1977.

Phenix, Centrale Nucléaire Prototype à Neutrons Rapides, Commissariat à l'Energie Atomique, Paris, 1971.

Plas, R., Flica III, Programme pour l'Etude Thermohydraulique de Réacteurs et de Boucles d'Essais, note CEA-N-1802, 1975.

Ricque, R. and Roumy, R., Echange Thermique en Double Phase dans des Tubes Verticaux ou Horizontaux, in *Heat Transfer 1970*, vol. 5, eds. U. Grigull and E. Hahne, paper B5-12, Elsevier, Amsterdam, 1970.

Robert, E., Lucenet, G., Leduc, J., and Chalot, E., Main Safety Features of the Superphenix Project, in *Proc. ANS/ENS Int. Meet. Fast Reactor Safety Related Physics, Chicago, Oct. 5–8*, CONF-761001, 1976.

Robin, M. G., Transmission de Chaleur et Pertes de Pression dans les Générateurs de Vapeur Chauffés par Circulation de Sodium, *Rev. Gén. Therm.*, no. 132, pp. 1189–1208, 1972.

Schmitt, F., Contribution Expérimentale et Théorique à l'Etude d'un Type Particulier d'Ecoulement Transitoire de Sodium en Ebullition. La Redistribution de Débit, Thèse de docteur ingénieur, Université Scientifique et Médicale et Institut National Polytechnique, Grenoble, 1974.

Suzuoki, A. and Yamakawa, M., Studies of Thermal-hydrodynamic Flow Instability, *Bull. JSME*, vol. 19, no. 132, pp. 619–625, 1976.

Traub, K., Safety Design of SNR 300, in *Proc. ANS/ENS Int. Meet. Fast Reactor Safety Related Physics, Chicago, Oct. 5–8*, CONF-761001, 1976.

Ünal, H. C., Van Gasselt, M. L. G., and Ludwig, P. W. P. H., Dynamic Instabilities in Tubes of a Large Capacity, Straight-Tube, Once-through Sodium Heated Steam Generator, *Int. J. Heat Mass Transfer*, vol. 20, pp. 1389–1399, 1977.

CHAPTER

# THREE

# TWO-PHASE FLOW PATTERNS

J. M. Delhaye

A two-phase mixture flowing in a pipe can exhibit several interfacial geometries, such as bubbles, slugs, and films. However, this geometry is not always clearly defined, which prevents the flow patterns from being precisely and objectively described.

In single-phase flows, laminar and turbulent flows are modeled differently. Laminar flows can be described by instantaneous quantities, the solutions of the Navier-Stokes equations, whereas turbulent flows are described by time- or statistical-averaged quantities that are the solutions of a system involving the Reynolds equations and some closure equations.

Likewise, in two-phase flows, the flow patterns must be known to model the physical phenomenon as closely as possible. It is obviously impossible to describe bubbly flows and annular flows with good accuracy by the same model. It is far better to adopt two different models, each one fitting the individual flow description. Nevertheless, this approach is still difficult because of the transition zone between two flow patterns and the lack of physical knowledge to describe these buffer zones.

In addition to the random character of each flow configuration, two-phase flows are never fully developed. In fact, the gas phase expands because of the pressure drop along the pipe. This expansion may lead to a modification of the flow structure, such as an evolution from bubbly flow to slug flow. The flow pattern depends also on the singularities occurring along the pipes, such as bends, junctions, air-injection devices, and so on.

The parameters that govern the occurrence of a given flow configuration are numerous, and it seems hopeless to try to represent all the transitions on a two-dimensional flow chart. Among these various parameters, one can select

1. Volumetric flow rates of each phase
2. Pressure
3. Heat flux at the wall
4. Densities and viscosities of each phase
5. Surface tension
6. Pipe geometry
7. Pipe characteristic dimension
8. Angle of the pipe with respect to the horizontal plane
9. Flow direction (upward, downward, cocurrent, countercurrent)
10. Inlet length
11. Phase-injection devices

The last two areas of concern are of the utmost importance, and one always has to keep in mind that a flow map is proposed for given conditions and that it must be used for the same conditions.

The purpose of this chapter is to describe the different configurations of gas–liquid pipe flows, to examine the transition phenomena between the flow patterns, and to propose a set of flow maps allowing the determination of the flow pattern, given the system parameters. The chapter starts with the definitions of the basic quantities describing two-phase pipe flows and a survey of the experimental techniques available for recognizing the flow structure.

# 1 DESCRIBING PARAMETERS OF TWO-PHASE PIPE FLOWS

Given the fluctuating character of two-phase flows, averaging operators have to be introduced. These operators, which act on space or time domains, are studied in detail in Chaps. 7, 8, and 9.

## 1.1 Phase Density Function

The presence or absence of phase $k (k = 1, 2)$ at a given point $\mathbf{x}$ and a given time $t$ is characterized by the unit value (or zero value) of a phase density function $X_k(\mathbf{x}, t)$ defined as follows:

$$X_k(\mathbf{x}, t) \triangleq \begin{cases} 1 & \text{if point } \mathbf{x} \text{ pertains to phase } k \\ 0 & \text{if point } \mathbf{x} \text{ does not pertain to phase } k \end{cases} \tag{1}$$

The phase density function is a binary function analogous to the intermittency function used in single-phase flow.

## 1.2 Instantaneous Space-averaging Operators

Instantaneous field variables may be averaged over a line, an area, or a volume, that is, over an $n$-dimension domain ($n = 1, 2$, or 3 for a segment, an area, or a volume, respectively). For instance, in a pipe flow, the field variable can be averaged over a diameter, a chord, a plane cross section, or a finite control volume. At a given time, this $n$-dimension domain $\mathfrak{D}_n$ can be divided into two subdomains $\mathfrak{D}_{kn}$ pertaining to each phase ($k = 1, 2$),

$$\mathbf{x} \in \mathfrak{D}_{kn} \qquad \text{if } X_k(\mathbf{x}, t) = 1 \qquad \forall\, \mathbf{x} \in \mathfrak{D}_{kn}$$

Consequently two different instantaneous space-averaging operators are introduced:

$$\sphericalangle\ \ \triangleright_n \triangleq \frac{1}{\mathfrak{D}_n} \int_{\mathfrak{D}_n} d\mathfrak{D}_n \tag{2}$$

and

$$<\ >_n \triangleq \frac{1}{\mathfrak{D}_{kn}} \int_{\mathfrak{D}_{kn}} d\mathfrak{D}_n \tag{3}$$

The instantaneous space fraction $R_{kn}$ is defined as the average over $\mathfrak{D}_n$ of the phase density function $x_k(\mathbf{x}, t)$,

$$R_{kn} \triangleq \sphericalangle X_k \triangleright_n = \frac{\mathfrak{D}_{kn}}{\mathfrak{D}_n} \tag{4}$$

This definition leads directly to the usual instantaneous space fraction

1. Over a segment:

$$R_{k1} = \frac{L_k}{\Sigma_{k=1,2} L_k} \tag{5}$$

   where $L_k$ is the cumulated length of the segments occupied by phase $k$
2. Over a surface:

$$R_{k2} = \frac{\mathfrak{a}_k}{\Sigma_{k=1,2}\, \mathfrak{a}_k} \tag{6}$$

   where $\mathfrak{a}_k$ is the cumulated area occupied by phase $k$
3. Over a volume:

$$R_{k3} = \frac{\mathcal{V}_k}{\Sigma_{k=1,2}\, \mathcal{V}_k} \tag{7}$$

   where $\mathcal{V}_k$ is the volume occupied by phase $k$

The instantaneous volumetric flow rate $Q_k$ through a pipe cross section of area $\mathfrak{a}$ is defined by

$$Q_k \triangleq \int_{\mathcal{A}_k} w_k \, d\mathcal{A} = \mathcal{A} R_{k2} \langle w_k \rangle_2 \tag{8}$$

where $w_k$ is the axial component of the velocity of phase $k$.

The instantaneous mass flow rate $M_k$ is given by

$$M_k \triangleq \int_{\mathcal{A}_k} \rho_k w_k \, d\mathcal{A} = \mathcal{A} R_{k2} \langle \rho_k w_k \rangle_2 \tag{9}$$

where $\rho_k$ is the density of phase $k$.

## 1.3 Local Time-averaging Operators

Local field variables can be averaged over a time interval $[t - T/2; t + T/2]$. As for single-phase turbulent flow, this time interval of magnitude $T$ must be chosen large enough compared with the time scale of turbulence fluctuations and small enough compared with the time scale of the overall flow fluctuations. This is not always possible, and a thorough discussion of this delicate question can be found in papers by Delhaye and Achard (1977, 1978).

If we consider a given point $\mathbf{x}$ in a two-phase flow, phase $k$ passes this point intermittently, and a field variable $f_k(\mathbf{x}, t)$ associated with phase $k$ is a piecewise continuous function. Denoting by $T_k(\mathbf{x}, t)$ the cumulated residence time of phase $k$ within the interval $T$, we can define two different local time-averaging operators:

$$\overline{\quad} \triangleq \frac{1}{T} \int_{[T]} \quad dt \tag{10}$$

and

$$\overline{\quad}^{X} \triangleq \frac{1}{T_k} \int_{[T_k]} \quad dt \tag{11}$$

The local time fraction $\alpha_k$ is defined as the average over $T$ of the phase density function $X_k$,

$$\alpha_k(\mathbf{x}, t) = \overline{X_k(\mathbf{x}, t)} = \frac{T_k(\mathbf{x}, t)}{T} \tag{12}$$

## 1.4 Commutativity of Averaging Operators

Considering all the definitions given previously, one can easily derive the following identity:

$$\overline{R_{kn} \langle f_k \rangle_n} \equiv \langle\!\langle \alpha_k \overline{f_k}^{X} \rangle\!\rangle_n \tag{13}$$

A particular case for Eq. (13) is obtained by taking $f_k \equiv 1$, which leads to

$$\overline{R_{kn}} \equiv \langle\!\langle \alpha_k \rangle\!\rangle_n \tag{14}$$

Note that identities (13) and (14) are valid for segments ($n = 1$), areas ($n = 2$), or volumes ($n = 3$).

As a consequence, the time-averaged volumetric and mass flow rates can be expressed as

$$\overline{Q_k} = \mathfrak{a}\overline{R_{k2}\langle w_k \rangle_2} \equiv \mathfrak{a} \langle\!\langle \alpha_k \overline{w_k}^X \rangle\!\rangle_2 \tag{15}$$

and

$$\overline{M_k} = \mathfrak{a}\overline{R_{k2}\langle \rho_k w_k \rangle_2} \equiv \mathfrak{a} \langle\!\langle \alpha_k \overline{\rho_k w_k}^X \rangle\!\rangle_2 \tag{16}$$

## 1.5 Qualities

The mass velocity $\bar{G}$ is defined by

$$\bar{G} \triangleq \frac{\bar{M}}{\mathfrak{a}} \tag{17}$$

where $\bar{M}$ is the time-averaged total mass flow rate.

The (true) quality $x$ is defined as the ratio of the gas mass flow rate to the total mass flow rate:

$$x \triangleq \frac{\overline{M_G}}{\overline{M_G} + \overline{M_L}} = \frac{\overline{M_G}}{\bar{M}} \tag{18}$$

It is currently impossible to measure or calculate with high precision the quality of a liquid-vapor mixture flowing in a heated channel and withstanding a phase change. Nevertheless, a fictitious quality, the so-called equilibrium or thermodynamic quality, can be calculated by assuming that both phases are flowing under saturation conditions, that is, that their temperatures are equal to the saturation temperature corresponding to their common pressure (see Chap. 18).

## 1.6 Volumetric Quantities

The volumetric quality $\beta$ is defined as the ratio of the gas volumetric flow rate to the total volumetric flow rate,

$$\beta \triangleq \frac{\overline{Q_G}}{\overline{Q_G} + \overline{Q_L}} \tag{19}$$

The local volumetric flux $j_k$ is a local time-averaged quantity defined by

$$j_k \triangleq \overline{X_k w_k} = \alpha_k \overline{w_k}^X \tag{20}$$

Its area average $J_k$ over the total cross-sectional area $\mathfrak{a}$ is a space-time-averaged quantity called the superficial velocity. This quantity is defined by

$$J_k \triangleq \langle\!\langle \overline{X_k w_k} \rangle\!\rangle_2 = \langle\!\langle j_k \rangle\!\rangle_2 \tag{21}$$

This quantity is directly related to the volumetric flow rate. In fact we have, taking Eq. (13) into account,

$$J_k = \measuredangle \alpha_k \overline{w_k}^X \measuredangle_2 = \overline{R_{k2} < w_k >_2} = \frac{\overline{Q_k}}{a} \tag{22}$$

If the density of phase $k$ is constant, the superficial velocity $J_k$ can be expressed in terms of the quality $x_k$ and the mass velocity $\bar{G}$ as follows:

$$J_k = \frac{\overline{M_k}}{\rho_k} = \frac{x_k \bar{G}}{\rho_k} \tag{23}$$

The mixture superficial velocity $J$ is defined as the sum of the superficial velocities of each phase:

$$J \triangleq J_1 + J_2 \tag{24}$$

One can show that $J$ is the velocity of a cross-sectional plane through which the total volumetric flow rate is equal to zero.

## 2 EXPERIMENTAL DETERMINATION OF FLOW PATTERNS

Flow patterns can be recognized on an integral basis by observing the flow itself or a photograph or by looking at movies. These techniques are necessary but not sufficient because they rarely provide quantitative information. Very often they are supplemented by an examination of the statistical properties of a local quantity, such as the phase density function, the instantaneous line fraction, or the instantaneous wall pressure.

### 2.1 Integral Techniques

If the pipe walls are transparent, the flow pattern can be determined by taking photographs or normal or high-speed movies. When the walls are opaque, one can use X-ray photography or cinematography. These optical techniques are described in detail by Cooper et al. (1964) and Arnold and Hewitt (1967).

When motion pictures are taken, one can focus on a given zone of the test section or one can track a specific item of the flow such as a bubble, a gas plug, or a roll wave. Specific items can be tracked by using a mirror that can rotate at an adjustable speed (Fig. 1). The internal structure of the flow appears clearly, and it has even been possible to measure the speed of waves in an annular flow (Hewitt and Lovegrove, 1970).

Flow patterns in forced convection boiling can be observed directly by using specific geometries (Fig. 2) or transparent heated walls (Fig. 3). If the walls are opaque, X-rays or neutron techniques can be employed (Bennett et al., 1965). When it is based on a visual examination, the determination of the transition between two flow configurations is a qualitative concept that depends on the observer. This is why other methods are used, which are based on the statistical properties of a local fluctuating quantity.

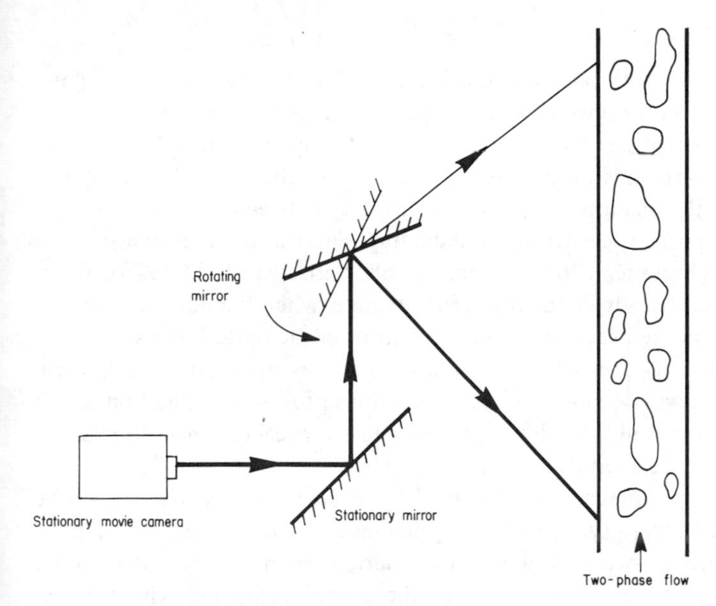


**Figure 1** Bubble tracker used by Roumy (1969).

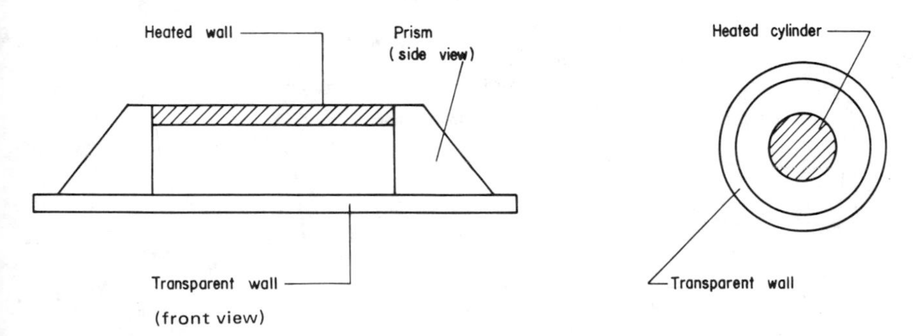


**Figure 2** Visualization of forced convection boiling in channels of particular geometries.

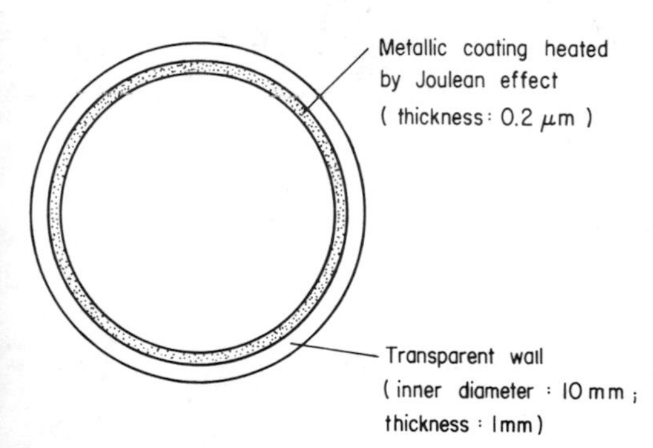


**Figure 3** Visualization of forced convection boiling in heated wall channels.

## 2.2 Local Techniques

More specifically, flow patterns can be classified according to the properties of the probability density functions or power spectrum densities of some variables.

At a given point, the phase density function $X_k$ can be determined by a small probe located at that point. This probe must be sensitive to the phase that surrounds its extremity, hence to the changes of a physical property between one phase and the other. When the liquid phase is electrically conducting, one can use the change in the electrical conductivity (Lackmé, 1967; Raisson, 1968; Serizawa et al., 1975). When the liquid phase is not electrically conducting, for example, when the liquid is a Freon, one can use the changes in the heat transfer coefficient or in the optical index.

The line gas fraction $R_{G1}$ can be measured by an X-ray attenuation technique. Jones and Zuber (1975) gave probability density functions of $R_{G1}$ as a function of the flow pattern for an air-water mixture flowing upward in a vertical channel of rectangular cross section. Some of their results appear in Fig. 4.

The fluctuations of the pressure at the wall also enable the flow patterns to be classified. Hubbard and Dukler (1966) looked at the power density spectra of the pressure signals for horizontal air-water flow at atmospheric pressure. They classified the flow structures according to the distribution of the energy compared with the frequency. Figure 5 shows some typical results.

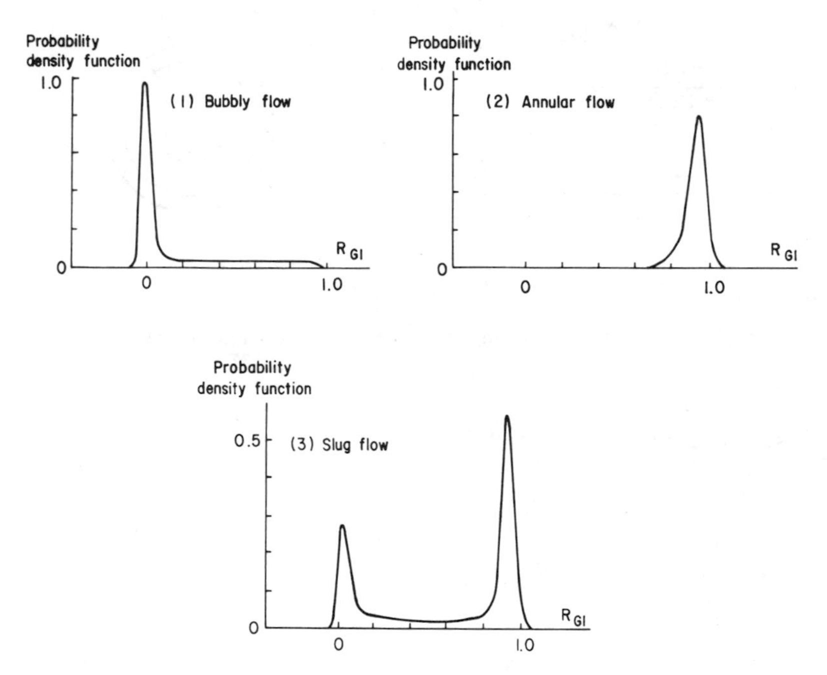


**Figure 4** Probability density functions for $R_{G_1}$ (Jones and Zuber, 1975).

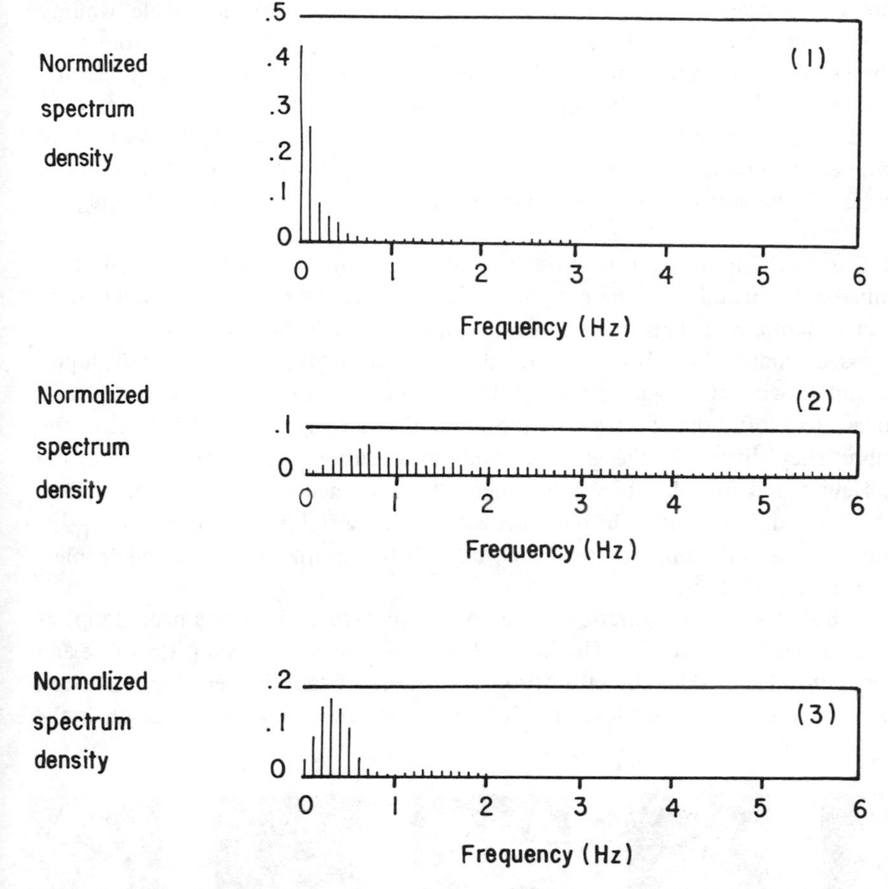


**Figure 5** Power density spectra of pressure signals (Hubbard and Dukler, 1966). (1) Separated flow. (2) Dispersed flow. (3) Intermittent flow.

# 3 UPWARD COCURRENT FLOWS IN A VERTICAL PIPE

If the pipe axis is positively oriented in the upward direction, an upward cocurrent flow will be such that

$$\overline{Q_G} > 0 \qquad \overline{Q_L} > 0$$

## 3.1 Description

The main flow patterns encountered in a vertical pipe are shown in Fig. 6. Bubbly flow is certainly the most widely known configuration, although at high velocity its milky appearance prevents it from being easily recognized. Bubbles are spherical only if their diameters do not exceed 1 mm, whereas beyond 1 mm their shape is variable. Roumy (1969) distinguished two bubbly flow patterns. In the independent bubble configuration bubbles are spaced and noninteracting with each other. In contrast, in the packed configuration bubbles are crowded together and interact strongly with each other.

Slug flow is composed of a series of gas plugs. The head of a gas plug is generally blunt, whereas its end is flat with a bubbly wake. A simple, visual observation reveals that the liquid film surrounding a gas plug moves downward with respect to the pipe wall.

Given a constant liquid flow rate, an increase of the gas flow rate leads to a lengthening and a breaking of the gas plugs. The flow pattern evolves toward an annular flow in a chaotic way. This transition configuration is called a churn flow.

Dispersed annular flow is characterized by a central gas core loaded with liquid droplets and flowing at a higher velocity than the liquid film that clings to the wall. Droplets are torn off from the crest of the waves that propagate on the surface of the liquid film. They diffuse in the gas core and can eventually impinge onto the film surface. Hewitt and Roberts (1969) evidenced the existence of a wispy annular flow where the liquid droplets gather into clouds within the central gas core (Fig. 7).

Finally, if the wall temperature is high enough to vaporize the film, the droplets will constitute a mist flow.

Figure 8 shows the configurations taken by a liquid–vapor flow in a heated pipe as a function of the wall heat flux. The liquid enters the pipe at a constant flow rate and at a temperature lower than the saturation temperature. When the heat flux increases, the vapor appears closer and closer to the pipe inlet. The local boiling length is the

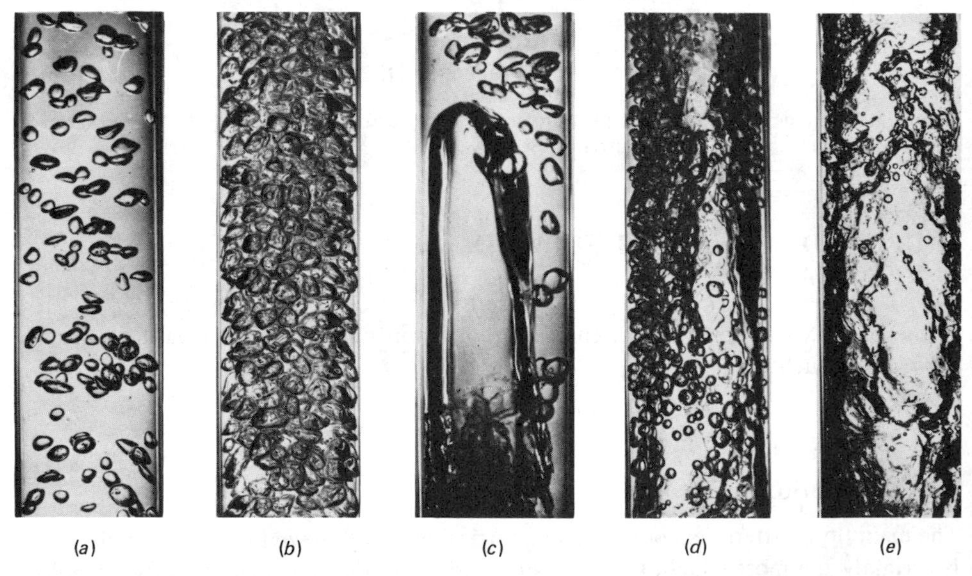


**Figure 6** Upward cocurrent flow in a vertical, 32-mm-diameter pipe; air–water flow pattern (Roumy, 1969). (*a*) Independent bubbles. (*b*) Packed bubbles. (*c*) Slug flow. (*d*) Churn flow. (*e*) Annular flow.

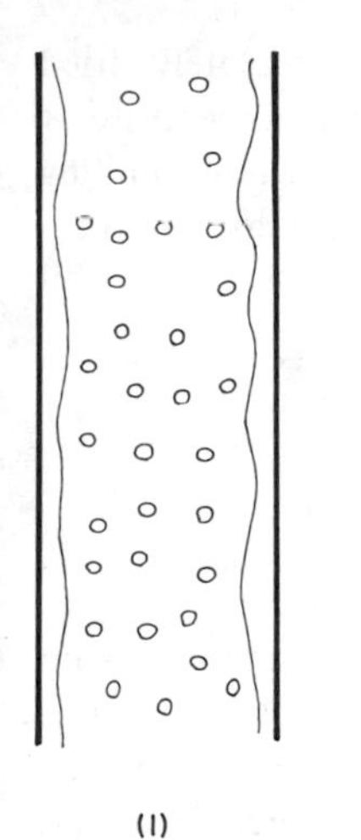


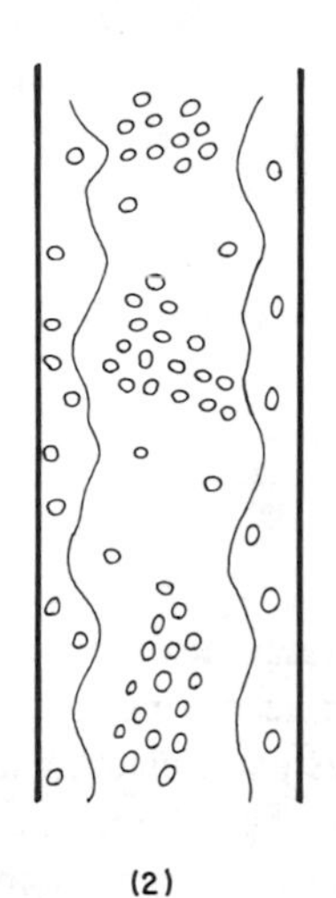


**Figure 7** Annular flows. (1) Dispersed. (2) Wispy.

extent of pipe at which the bubbles are generated at the wall and condense in the liquid core where the liquid temperature is still lower than the saturation temperature. The vapor is produced by two mechanisms: (1) wall nucleation and (2) direct vaporization on the interfaces located in the flow itself. The latter increases in importance as one proceeds along the pipe. There is progressively less liquid between the wall and the interfaces. Consequently the thermal resistance decreases, as does the wall temperature, which brings the wall nucleation to a stop. In annular flow, the liquid-film flow rate decreases through evaporation and entrainment of droplets, although some droplets

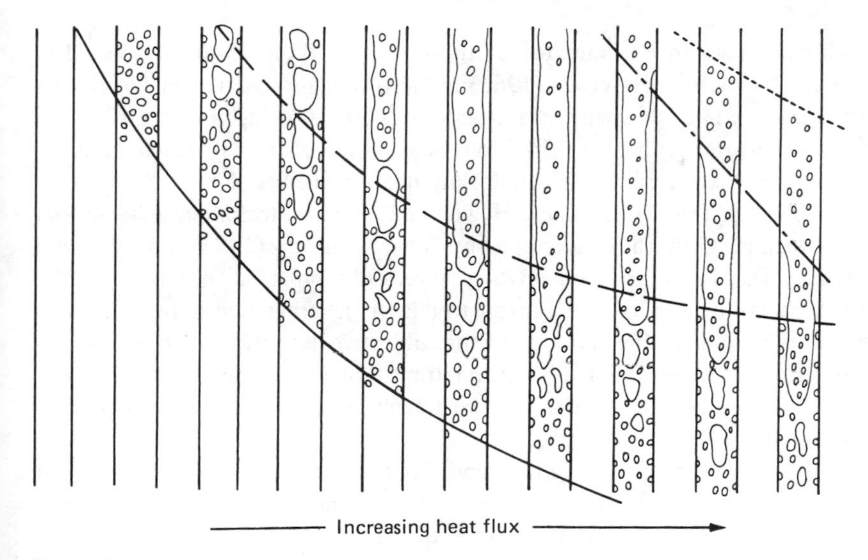


**Figure 8** Convective boiling in a heated channel at a constant liquid flow rate. (——) Onset of nucleate boiling. (— — —) End of nucleate boiling. (— - - - -) Dryout. (- - - - -) Limit of the superheated vapor region (Hewitt and Hall-Taylor, 1970).

are redeposited. In heat-flux-controlled systems, when the film is completely dried out, the wall temperature rises very quickly and can exceed the melting temperature of the wall. This phenomenon has received several names such as dryout, burnout, boiling crisis, or critical heat flux. Note that the film dryout in an annular flow does not provide the only explanation for a boiling crisis, that is, a sudden increase of the wall temperature.

## 3.2 Flow Maps

A flow map is a two-dimensional representation of the flow pattern existence domains. The coordinate systems are different according to the authors, and so far there is no agreement on the best coordinate system. Selecting a series of flow maps is not easy, and no recommendation can be made because no method has yet proved entirely adequate. Nevertheless the following charts can give some expectations, which, however, may be contradictory from one method to the other.

Hewitt and Roberts' map (1969) shown in Fig. 9, is the most widely used chart for air-water and steam-water flows. It was originally established for an air-water mixture flowing in a pipe 31.2 mm in diameter at a pressure varying from 1.4 to 5.4 bar. The coordinate system is as follows:

Abscissa: $$\rho_L J_L^2 = \frac{\bar{G}^2(1-x)^2}{\rho_L} \tag{25}$$

Ordinate: $$\rho_G J_G^2 = \frac{\bar{G}^2 x^2}{\rho_G} \tag{26}$$

These coordinates have to be evaluated at the pressure of the observed zone. The steam-water results of Bennett et al. (1965) are also well represented in the Hewitt and Roberts diagram. They deal with steam-water mixtures flowing in a pipe 12.7 mm in diameter at a pressure varying from 34.5 to 69 bar. In the coordinate calculation the mass quality $x$ is approximated by the equilibrium quality (see Sec. 1.5).

Taitel and Dukler (1977) compared Hewitt and Roberts' technique with several other methods, among them those developed by Govier and Aziz (1972) and Oshinowo and Charles (1974). The comparison revealed several discrepancies not only on the quantitative aspects but also on the general trends of the transition curves. The discrepancies can be easily explained because of the subjective definitions of the flow patterns and the single representation for several transitions. Consequently, Taitel and Dukler went back to a physical analysis of the transitions and came up with the following conclusions.

1. For transition between bubbly flow and slug flow or between bubbly flow and churn flow (Fig. 10), the ordinate is the ratio $J_L/J_G$ of the liquid superficial velocity to the gas superficial velocity. The abscissa is the quantity

$$\frac{J_G \rho_L^{1/2}}{[g(\rho_L - \rho_g)\sigma]^{1/4}}$$

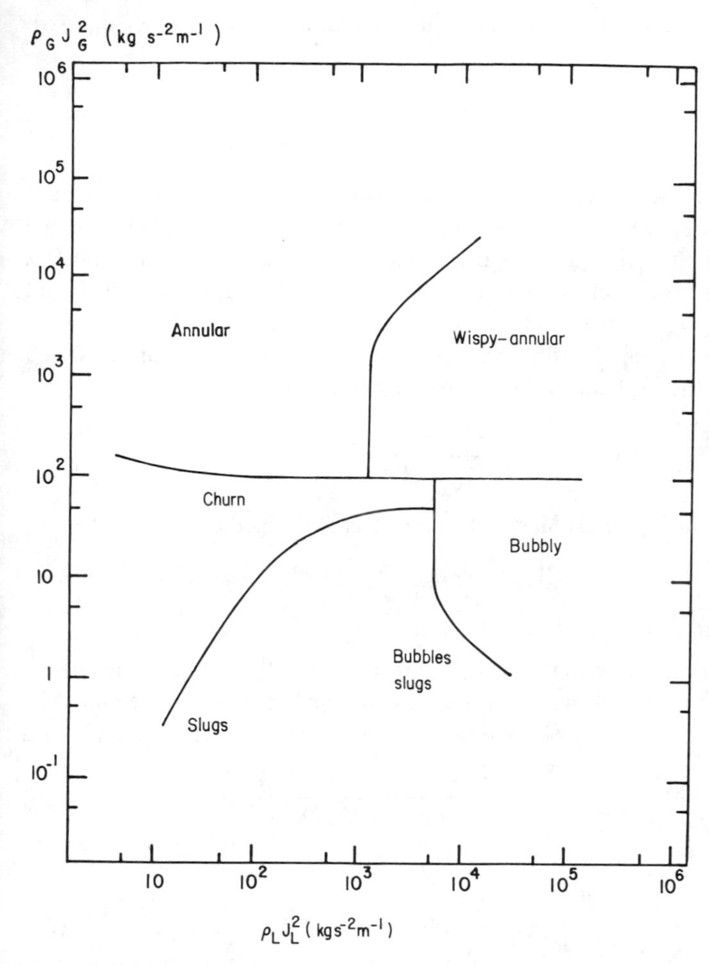


**Figure 9** Hewitt and Roberts' map (air–water flow up to 5.9 bar; steam–water flow up to 69 bar).

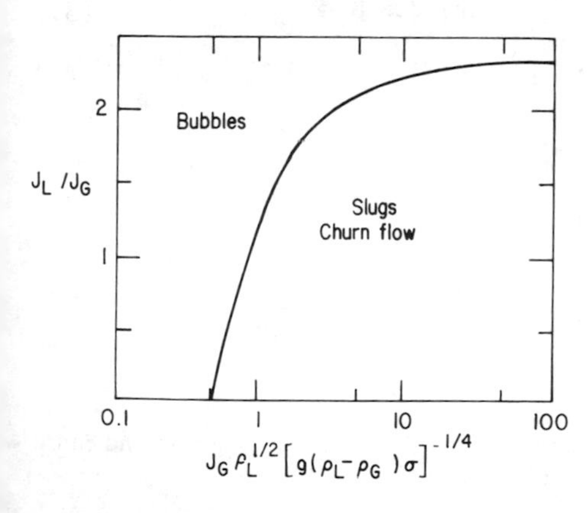


**Figure 10** Transition between bubbly flow and slug flow or between bubbly flow and churn flow (Taitel and Dukler, 1977).

where $\sigma$ is the surface tension. The equation of the transition curve based on a theoretical analysis is written

$$\frac{J_L}{J_G} = 2.34 - 1.07 \frac{[g(\rho_L - \rho_G)\sigma]^{1/4}}{J_G \rho_L^{1/2}} \tag{27}$$

2. For transition between slug flow and churn flow (Fig. 11), the volumetric quality $\beta$ is the ordinate, whereas the abscissa is the quantity $J/\sqrt{gD}$, where $J$ is the mixture superficial velocity [Eq. (24)], $g$ the acceleration caused by gravity, and $D$ is the pipe diameter. The transition line has a complex equation that involves the liquid Reynolds number $J_L D/\nu_L$, where $\nu_L$ is the liquid kinematic viscosity.

3. For transition between slug flow and annular flow or between churn flow and annular flow (Fig. 12), the ordinate is the Kutateladze number

$$\text{Ku} = \frac{J_G \rho_G^{1/2}}{[g(\rho_L - \rho_G)\sigma]^{1/4}}$$

whereas the abscissa is the Lockhart-Martinelli parameter $X$ defined by

$$X \triangleq \left[\frac{(dp/dz)_L}{(dp/dz)_G}\right]^{1/2} \tag{28}$$

where $(dp/dz)_L$ and $(dp/dz)_G$ are the frictional pressure drops of the liquid and the gas that would have been observed if these fluids were flowing alone in the same pipe. The equation of the transition line was obtained from theoretical considerations and reads

$$\frac{J_G \rho_G^{1/2}}{[g(\rho_L - \rho_G)\sigma]^{1/4}} = 3.09 \frac{(1 + 20X + X^2)^{1/2} - X}{(1 + 20X + X^2)^{1/2}} \tag{29}$$

which simplifies to

$$\frac{J_G \rho_G^{1/2}}{[g(\rho_L - \rho_G)\sigma]^{1/4}} = 3.09 \qquad \text{for } X \ll 1 \tag{30}$$

and to

$$\frac{J_G \rho_G^{1/2}}{[g(\rho_L - \rho_G)\sigma]^{1/4}} = \frac{30.9}{X} \qquad \text{for } X \gg 1 \tag{31}$$

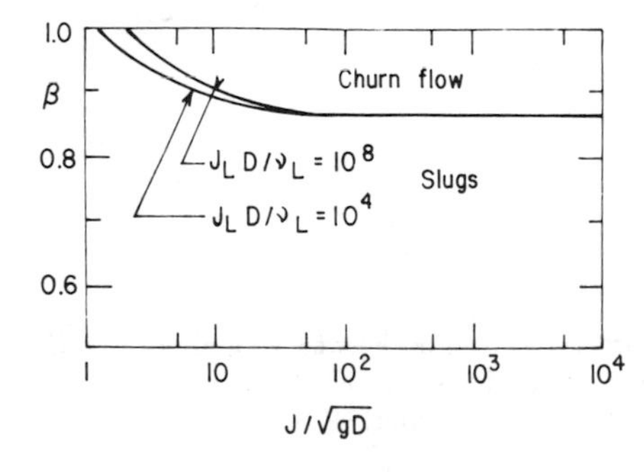


**Figure 11** Transition between slug flow and churn flow (Taitel and Dukler, 1977).

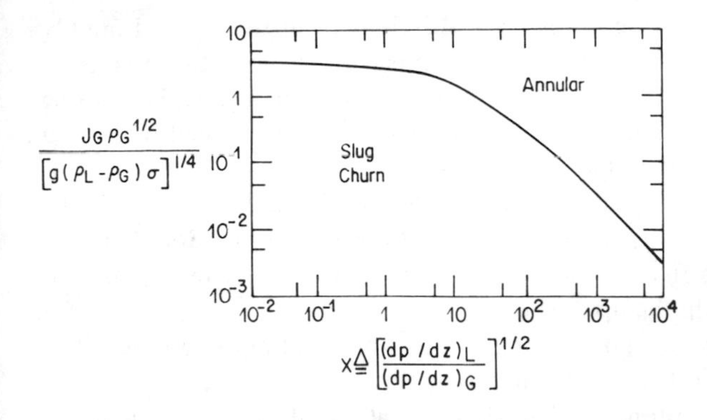


**Figure 12** Transition between slug flow and annular flow or between churn flow and annular flow (Taitel and Dukler, 1977).

To use Taitel and Dukler's method, one needs first to use Fig. 10, then Fig. 12 if the flow is not a bubbly flow, and finally Fig. 11 if the flow is neither bubbly nor annular.

# 4 DOWNWARD COCURRENT FLOWS IN A VERTICAL PIPE

If the pipe axis is positively oriented in the upward direction, a downward cocurrent flow will be such that

$$\overline{Q_G} < 0 \qquad \overline{Q_L} < 0$$

## 4.1 Description

So far the most comprehensive studies of downward cocurrent flow patterns are from Oshinowo and Charles (1974). These authors distinguished six different flow configurations (Fig. 13).

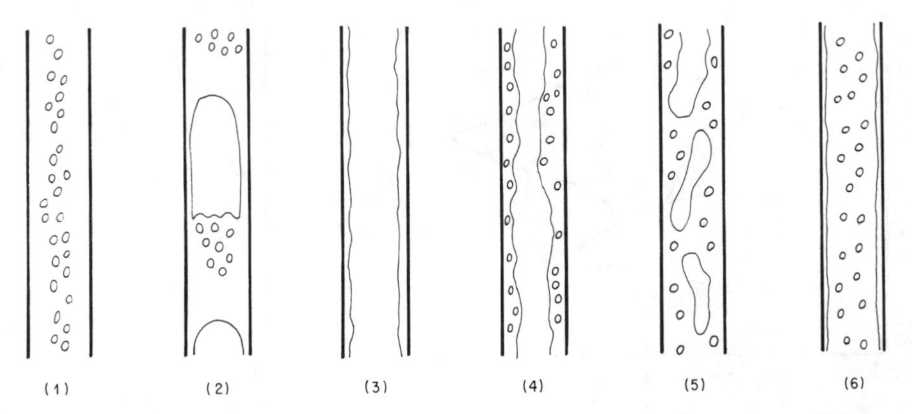


**Figure 13** Downward cocurrent flow in a vertical pipe; air–water flow pattern (Oshinowo and Charles, 1974). (1) Bubbles. (2) Slugs. (3) Falling film. (4) Bubbly falling film. (5) Churn. (6) Dispersed annular.

The downward bubbly flow structure is quite different from the upward bubbly flow configuration. In the latter case, bubbles are spread over the entire pipe cross section, whereas in the downward flow bubbles gather near the pipe axis. This coring effect is similar to the phenomenon observed in a flow of liquid loaded with solid particles having a density smaller than the liquid density.

When the gas flow rate is increased, the liquid flow rate being held constant, the bubbles agglomerate into large gas pockets. The top of these gas plugs is dome shaped, whereas the lower extremity is flat with a bubbly zone underneath. This slug flow is generally more stable than it is in the upward case.

The annular configuration can take several forms. For small liquid and gas flow rates, a liquid film flows down the wall (falling film flow). If the liquid flow rate is higher, bubbles are entrained within the film (bubbly falling film). When liquid and gas flow rates are increased, a churn flow appears and can evolve into a dispersed annular flow for a very high gas flow rate.

## 4.2 Flow Maps

Oshinowo and Charles (1974) proposed a chart (Fig. 14) that was obtained by using their own experimental data. They studied two-component mixtures of air and different liquids flowing in a pipe 25.4 mm in diameter at a pressure of around 1.7 bar. They chose as abscissa and ordinate the quantities $\mathrm{Fr}/\sqrt{\Lambda}$ and $\sqrt{\beta(1-\beta)}$, respectively, which are calculated at the test section pressure and temperature. The Froude number Fr is defined by the following relation:

$$\mathrm{Fr} \triangleq \frac{(J_G + J_L)^2}{gD} \tag{32}$$

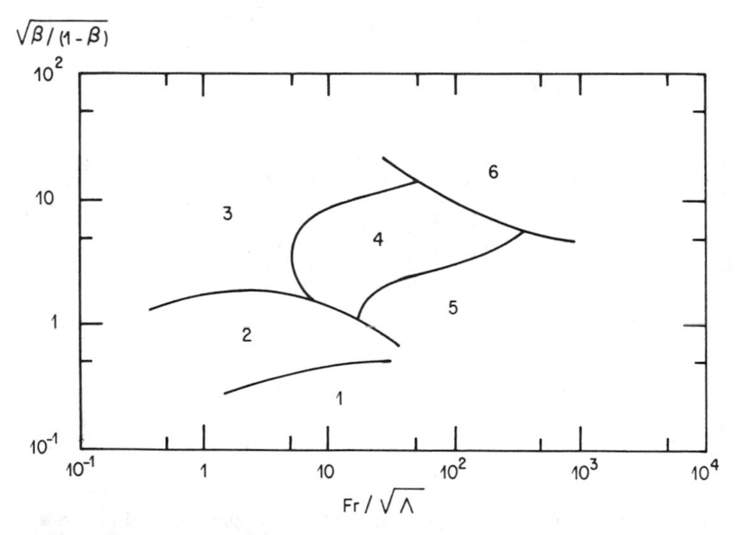


**Figure 14** Oshinowo and Charles (1974) flow map. 1, bubbles; 2, slugs; 3, falling film; 4, bubbly falling film; 5, churn; 6, dispersed annular.

where $g$ is the acceleration caused by gravity and $D$ is the pipe diameter. $\Lambda$ is a coefficient that takes account of the liquid physical properties. It is defined as follows:

$$\Lambda \triangleq \frac{\mu_L}{\mu_W}\left[\frac{\rho_L}{\rho_W}\left(\frac{\sigma}{\sigma_W}\right)^3\right]^{-1/4} \tag{33}$$

where $\mu$ = liquid viscosity
$\rho$ = density
$\sigma$ = surface tension
$W$ = water at 20°C and 1 bar

# 5 COCURRENT FLOWS IN A HORIZONTAL PIPE

The number of possible flow patterns in a horizontal pipe is higher than that in a vertical pipe. The greater number of possible flow patterns results from the effect of gravity, which tends to separate the phases and to create a horizontal stratification.

## 5.1 Description

Alves (1954) proposed the classification shown in Fig. 15. In the bubbly flow configuration bubbles are moving in the upper part of the pipe. When the gas flow rate is increased, bubbles coalesce and a plug flow takes place. For low liquid and gas flow rates a stratified flow appears with a smooth interface. At higher gas rates, waves propagate along the interface (wavy flow) and can reach the top wall of the pipe giving rise to a slug flow. Finally at high gas flow rates and low liquid flow rates an annular flow can exist with a thicker film in the lower part of the pipe. These flow pattern names differ depending on the author. As an example, Taitel and Dukler (1976) classed plug flows and slug flows under the same category (intermittent flow).

Figure 16 shows the evolution of a vaporizing flow in a horizontal pipe. The liquid enters the heated pipe with a low flow rate and at a temperature slightly lower than the saturation temperature. Note that the upper part of the tube can dry out periodically and then suffer a sudden increase in wall temperature. If the wall temperature is high enough, the wall dries out completely and the liquid droplets form a mist flow.

## 5.2 Flow Maps

Baker's diagram is still commonly used, especially in petroleum industries and for condenser design. The coordinates used by Baker were simplified by Bell et al. (1970), and the resulting map is shown in Fig. 17. Quantities $\overline{G_G}$ and $\overline{G_L}$ are the superficial gas and liquid mass velocities defined by the following relations:

$$\overline{G_G} \triangleq \frac{\overline{M_G}}{\mathfrak{a}} \qquad \overline{G_L} \triangleq \frac{\overline{M_L}}{\mathfrak{a}} \tag{34}$$

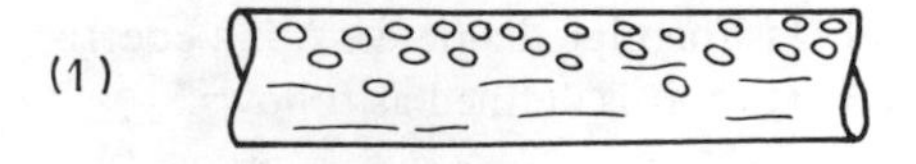


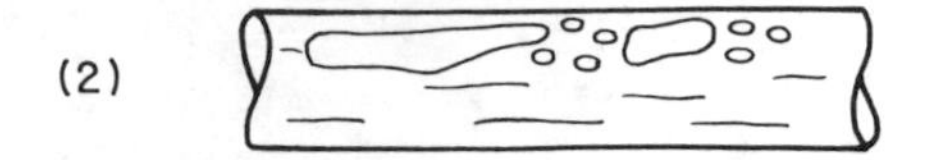


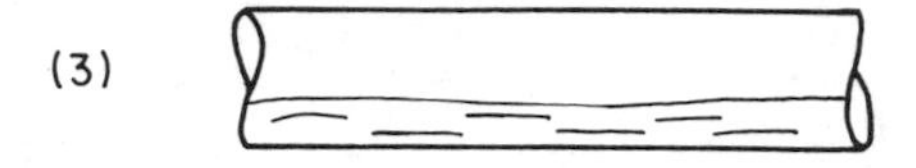

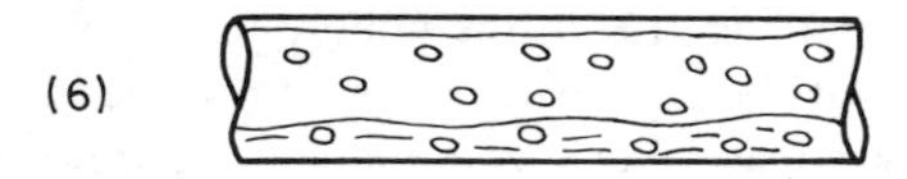


**Figure 15** Flow patterns in a horizontal pipe. (1) Bubbly flow. (2) Plug flow. (3) Stratified flow. (4) Wavy flow. (5) Slug flow. (6) Annular flow.

where $\overline{M_G}$ and $\overline{M_L}$ are the gas and liquid mass flow rates and $\boldsymbol{a}$ is the pipe cross-sectional area. Coefficients $\lambda$ and $\psi$ depend on the physical properties of the fluids and are defined as follows:

$$\lambda \triangleq \left(\frac{\rho_G}{\rho_A}\frac{\rho_L}{\rho_W}\right)^{1/2} \tag{35}$$

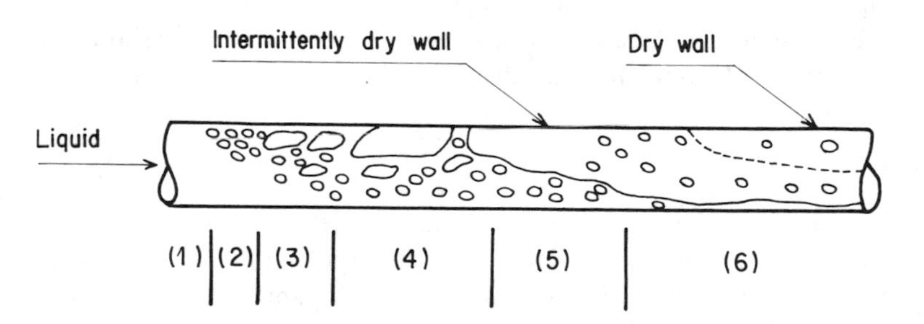


**Figure 16** Evolution of flow pattern in a horizontal evaporator tube, where segment 1 designates the liquid single-phase flow; segment 2, bubbly flow; segment 3, plug flow; segment 4, slug flow; segment 5, wavy flow; and segment 6, annular flow.

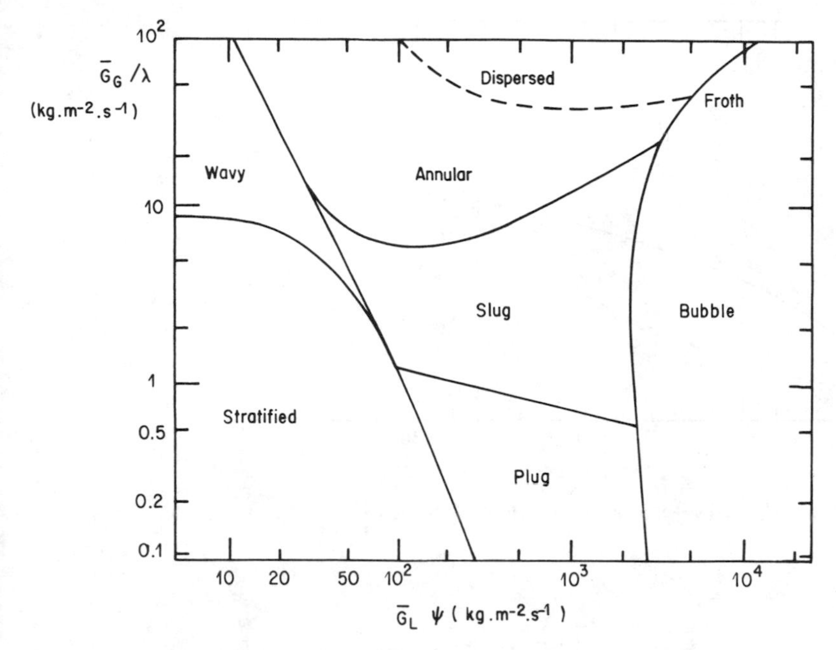


**Figure 17** Modified Baker diagram (Bell et al., 1970).

$$\psi \triangleq \frac{\sigma_{WA}}{\sigma}\left[\frac{\mu_L}{\mu_W}\left(\frac{\rho_W}{\rho_L}\right)^2\right]^{1/3} \tag{36}$$

Subscripts $A$ and $W$ refer, respectively, to the physical properties of air and water at 1 bar and 20°C. Consequently, for air-water flows at 1 bar and 20°C we have $\lambda \equiv 1$ and $\psi \equiv 1$. For steam-water flows, $\lambda$ and $\psi$ are given in Fig. 18 as functions of the saturation pressure (Collier, 1972).

Mandhane et al. (1974) came up with a map based on 5935 data points, 1178 of which concern air-water flows. Its coordinates are the superficial velocities $J_L$ and $J_G$ calculated at the test section pressure and temperature. This map (Fig. 19) is valid for the parameter ranges given in Table 1.

The general trends of Mandhane's map were found by Taitel and Dukler (1976) by a theoretical analysis of the flow pattern transitions. The theoretical chart proposed by Taitel and Dukler (Fig. 20) uses different coordinate systems according to the transitions considered.

Curve 1. $F$ related to $X$ between stratified or wavy flow and intermittent flow with

$$F \triangleq \left(\frac{\rho_G}{\rho_L - \rho_G}\right)^{1/2} \frac{J_G}{(Dg\cos\alpha)^{1/2}} \tag{37}$$

$$X \triangleq \left[\frac{(dp/dz)_L}{(dp/dz)_G}\right]^{1/2} \tag{38}$$

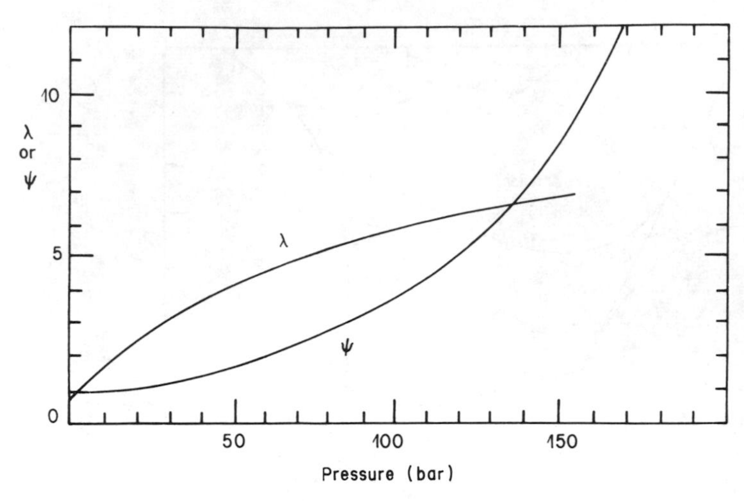


**Figure 18** Values of λ and ψ for steam–water flows (Collier, 1972).

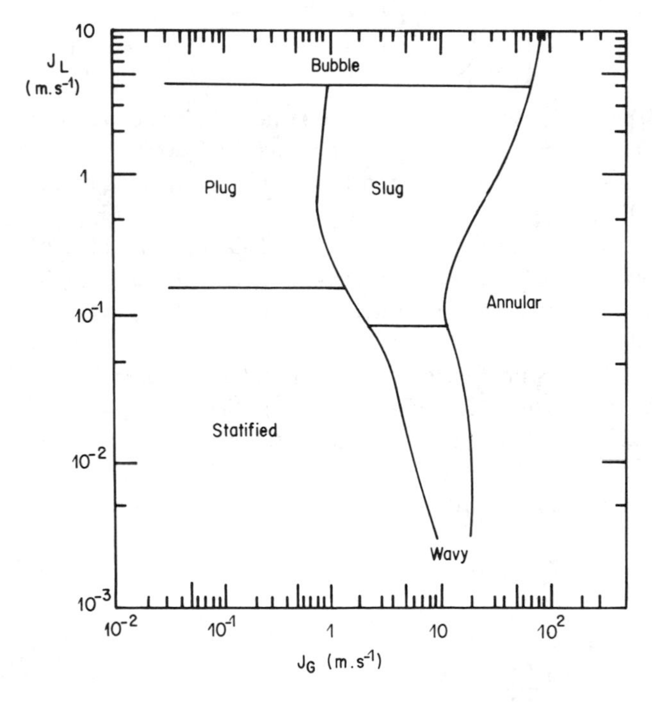


**Figure 19** Flow map proposed by Mandhane et al. (1974).

**Table 1 Parameter ranges for the map proposed by Mandhane et al. (1974)**

| Property | Parameter range |
|---|---|
| Pipe inner diameter | 12.7–165.1 mm |
| Liquid density | 705–1009 kg/m$^3$ |
| Gas density | 0.80–50.5 kg/m$^3$ |
| Liquid viscosity | $3 \times 10^{-4}$–$9 \times 10^{-2}$ Pl |
| Gas viscosity | $10^{-5}$–$2.2 \times 10^{-5}$ Pl |
| Surface tension | 24–103 mN/m |
| Liquid superficial velocity | 0.09–731 cm/s |
| Gas superficial velocity | 0.04–171 m/s |

where $D$ is the pipe diameter and $\alpha$ is the pipe inclination angle with respect to the horizontal plane (positive for downward flow). The quantities $(dp/dz)_L$ and $(dp/dz)_G$ are the frictional pressure drops of the liquid and gas that would have been observed if these fluids were flowing alone in the pipe.

Curve 2. $X$ constant between bubbly or intermittent flows and annular flows.

Curve 3. $K$ related to $X$ between stratified flow and wavy flow, with $K$ defined by

$$K \triangleq \left[\frac{\rho_G J_G^2 J_L}{(\rho_L - \rho_G) g \nu_L \cos\alpha}\right]^{1/2} \tag{39}$$

where $\nu_L$ is the liquid kinematic viscosity.

Curve 4. $T$ related to $X$ between bubbly flow and intermittent flow, with $T$ defined by

$$T \triangleq \left[\frac{|dp/dz|_L}{(\rho_L - \rho_G) g \cos\alpha}\right]^{1/2} \tag{40}$$

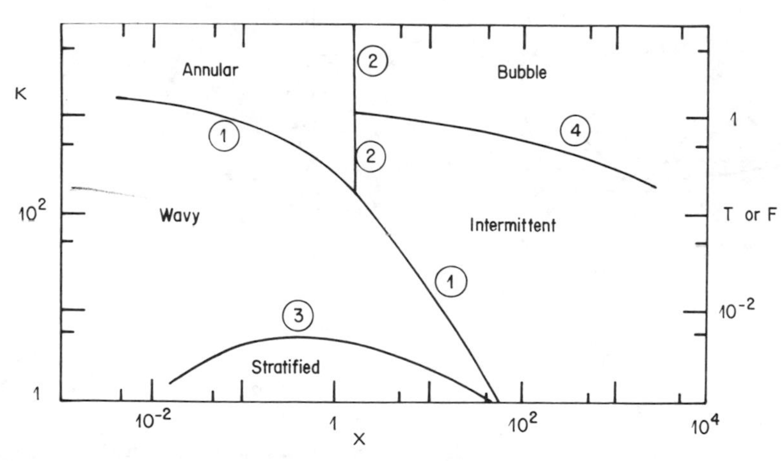


**Figure 20** Theoretical chart proposed by Taitel and Dukler (1976), where $F$ is related to $X$ for curve 1, $X$ is constant for curve 2, $K$ is related to $X$ for curve 3, and $T$ is related to $X$ for curve 4.

The theoretical transitions obtained by Taitel and Dukler are compared in Fig. 21 with Mandhane's map for an air–water mixture flowing in a 25-mm pipe at 1 bar and 25°C. Finally, we can note that Taitel and Dukler's method takes the pipe diameter into account.

## 6 FLOW PATTERNS IN A ROD BUNDLE

Fuel elements of boiling-water reactors and pressurized-water reactors are generally assembled into rod bundles. Heat is extracted from the rods by water flowing parallel to their axes. Rod-bundle geometries are described in detail by Lahey and Moody (1977) for boiling-water reactors and by Tong and Weisman (1970) for pressurized-water reactors. To our knowledge, only one paper deals with flow patterns in rod bundles (Bergles, 1969). This is not surprising, given the difficulty in introducing measurement devices into the limited space between the rods.

Bergles (1969) used resistive probes (see Sec. 2.2) to determine the flow patterns at different points in a given cross section of a four-rod bundle (Fig. 22). Each rod was composed of a 46-cm-long nonheated zone and a 61-cm-long heated zone. Spacers located at 76 mm upstream and 25 mm downstream of the heated portion enabled the rods to be positioned with respect to the housing. Three resistive probes were placed in the interior subchannel and in the corner subchannel, as indicated in Fig. 22 at 12.5

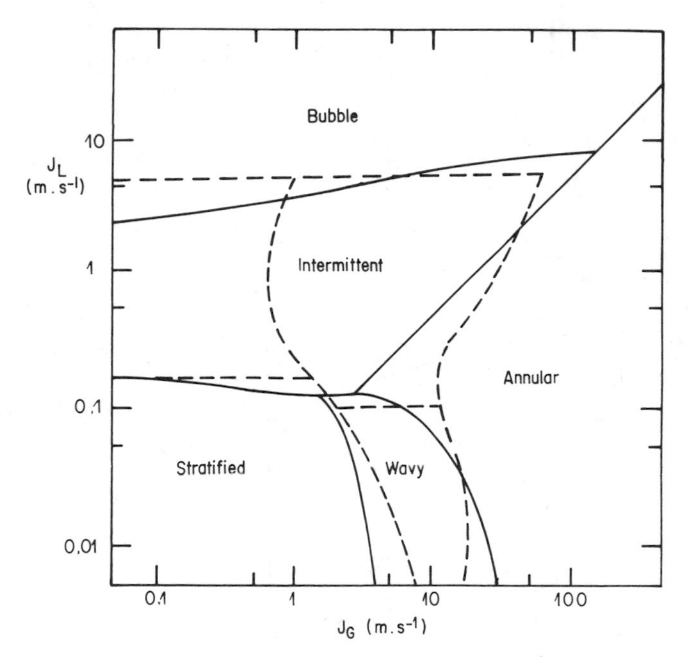


**Figure 21** Horizontal air–water flow at 1 bar and 25°C, where the pipe inner diameter is 25 mm. The solid lines indicate theoretical transitions according to Taitel and Dukler (1976); the dashed lines indicate data from the experimental diagram of Mandhane et al. (1974).

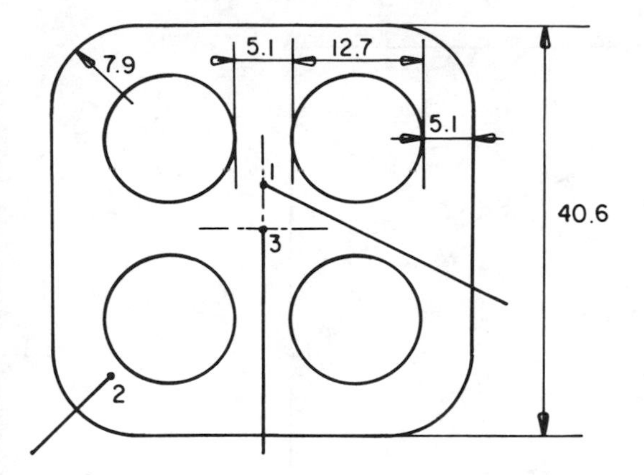


**Figure 22** Dimensions (mm) of the four-rod bundle tested by Bergles (1969).

mm upstream of the end of the heated length. The flow patterns were determined for a steam-water mixture at 69 bar.

Two cases were considered, the first without heating (Figs. 23-25) and the second with heating of the rods. Figure 23 shows a comparison between the flow pattern limits in the interior subchannel (probes 1 and 3). Results are given in $\bar{G}$ related to $x_{eq}$ diagram where $\bar{G}$ is the mass velocity [Eq. (17)] and $x_{eq}$ is the thermodynamic quality calculated for the rod bundle as a whole. Actually, this representation is doubtful because $\bar{G}$ and $x_{eq}$ are integral quantities that do not refer to the particular subchannel under consideration. Nevertheless, this representation was adopted because of the impossibility of obtaining subchannel quantities directly. On the axis of the rod bundle (probe 3) changes in flow patterns occur at lower qualities than within the gap between

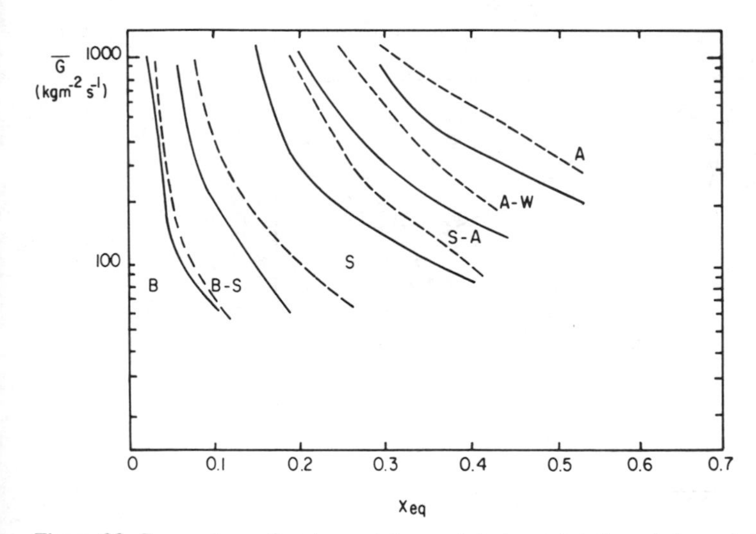


**Figure 23** Comparison of regimes at two points in an interior subchannel. The solid lines indicate data from probe 3; the dashed lines indicate data from probe 1. B, bubbles; B-S, bubble slug; S, slug; S-A, slug annular; A-W, annular waves; A, annular.

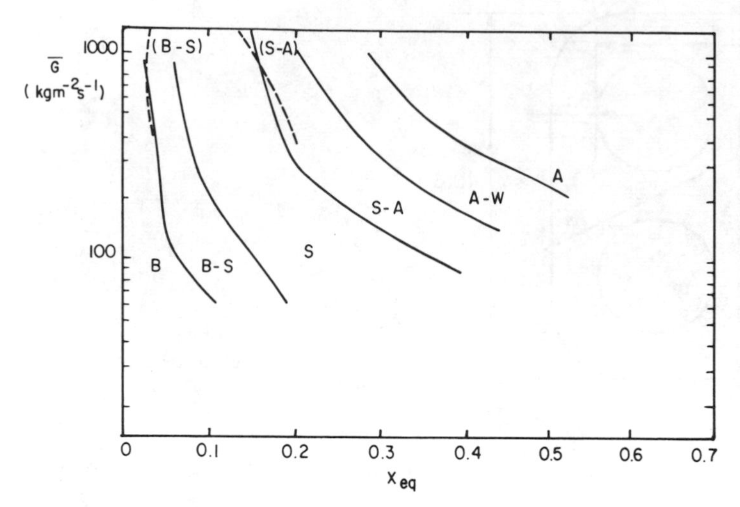


**Figure 24** Comparison of regimes for interior subchannel and circular tube. Solid lines indicate data from probe 3; dashed lines indicate data from a 10.2 mm tube at 69 bar. B, bubbles; B-S, bubble slug; S, slug; S-A, slug annular; A-W, annular waves; A, annular.

two rods (probe 1), indicating that the liquid tends to accumulate in the restricted area.

The interior-subchannel flow regimes were compared with the flow patterns obtained in a circular tube of the same hydraulic diameter, at the same pressure. If the interior subchannel is assumed to be limited by the rod surface and by the planes connecting their axes, the hydraulic diameter is found to be 12.6 mm. Figure 24 shows a

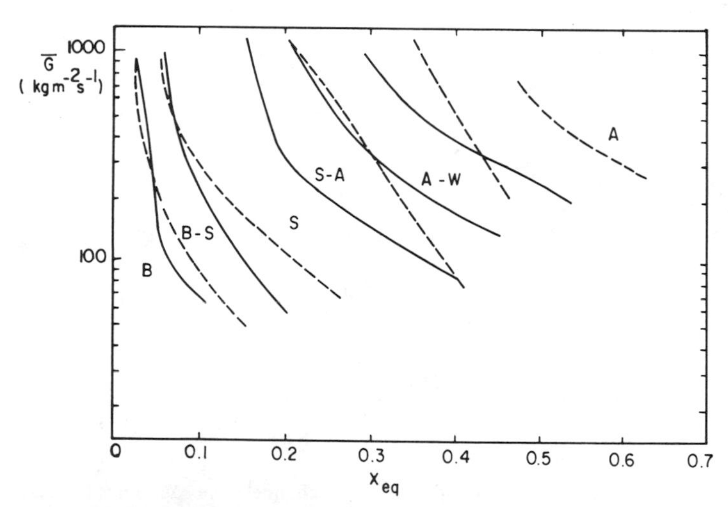


**Figure 25** Comparison of regimes for interior and corner subchannels. Solid lines indicate data from probe 3; dashed lines indicate data from probe 2. B, bubbles; B-S, bubble slug; S, slug; S-A, slug annular; A-W, annular waves; A, annular.

comparison between the flow pattern transitions in the interior subchannel and those obtained in a tube, 10.2 mm in diameter, for the same length-to-hydraulic-diameter ratio. The agreement is good but should be considered a coincidence because the rod bundle results were obtained from integral values of the mass velocity and quality instead of subchannel values.

Finally, flow patterns in the interior subchannel (point 3) and in the corner subchannel (point 1) are compared in Fig. 25. All the transitions in the corner subchannel appear at higher qualities, which indicates a liquid accumulation in this area. Hence, Figs. 23 and 25 show clearly that different flow patterns can coexist in a given rod bundle cross section. Limited data obtained with heated rods tend to support this conclusion.

## 7 FLOODING AND FLOW REVERSAL

Flooding and flow reversal are basic mechanisms encountered in several phenomena occurring in nuclear reactor thermohydraulics, for example, (1) rewetting from the top or countercurrent flow limiting conditions (Chan and Grolmes, 1975; Tien, 1977), (2) motion of molten cladding material (Grolmes et al., 1974), (3) downcomer flow pattern during an emergency core cooling (Block and Schrock, 1977).

Although more and more experiments are carried out on flooding and flow reversal, no adequate theories have been provided so far. However, some empirical correlations have been developed, but they can be used only in their range of validity. This seems to be a trivial statement, but it appears that these correlations are occasionally applied to operating conditions far beyond their range of validity. Generally the ensuing results are not good, which is not surprising given the lack of physical bases for such correlations.

### 7.1 Description

Figure 26 shows a vertical tube with a liquid injection device made up of a porous wall. At the beginning of the experiment there is no gas flow, and a liquid film flows down

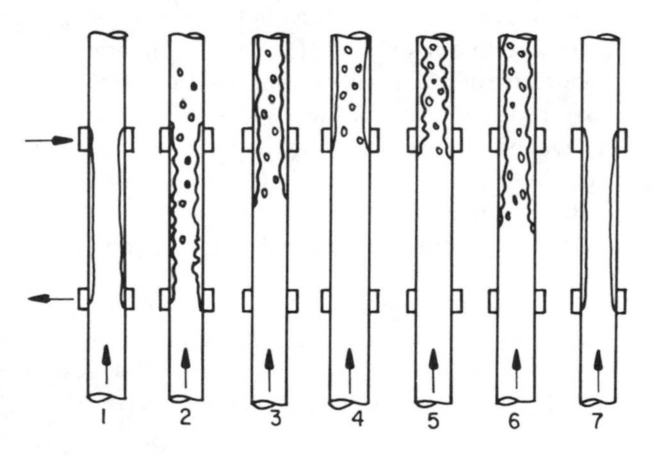


**Figure 26** Flooding and flow reversal experiment. See text for explanation of figure parts.

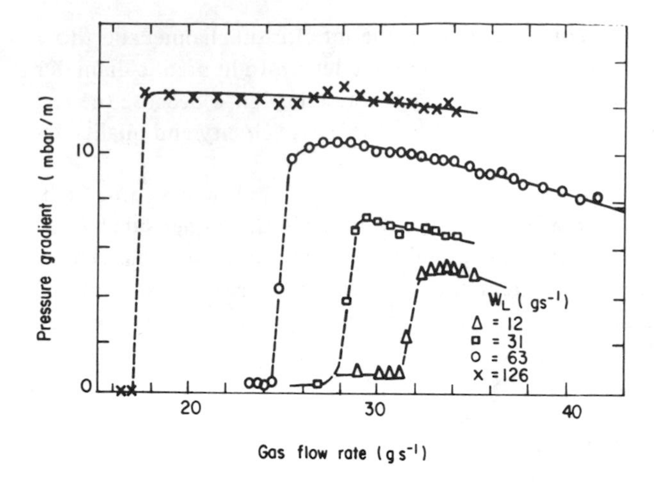


**Figure 27** Pressure drop as related to gas flow rate (Dukler and Smith, 1976).

the tube at a constant flow rate. If gas flows upward at a low rate, a countercurrent flow takes place in the tube (Fig. 26, pt. 1). When the gas flow rate is increased, the film thickness remains constant and equal to the Nusselt thickness (Hewitt and Wallis, 1963). This thickness $e$, calculated by assuming that the flow is laminar, reads

$$e = \left( \frac{3\nu_L Q_L}{\pi D g} \right)^{1/3} \tag{41}$$

where $\nu_L$ = liquid kinematic viscosity
$Q_L$ = liquid volumetric flow rate
$D$ = pipe diameter
$g$ = acceleration from gravity

This relation is valid for liquid Reynolds numbers $\mathrm{Re}_L$ up to 4000 with

$$\mathrm{Re}_L \triangleq \frac{4Q_L}{\pi D \nu_L} \tag{42}$$

For a higher gas flow rate the liquid film becomes unstable, and waves of large amplitude appear. Droplets are torn from the crests of the waves and are entrained with the gas flow above the liquid injection level. These droplets diffuse in the gas core; some of them impinge onto the wall and give rise to a liquid film (Fig. 26, pt. 2). Meanwhile the pressure drop in the tube above the liquid injection level increases sharply as shown in Fig. 27 (Dukler and Smith, 1976). This constitutes the best experimental evidence of the flooding phenomenon. When the gas flow rate is increased further, the tube section below the liquid injection dries out progressively (Fig. 26, pt. 3). No net liquid downflow appears for a certain gas flow rate. However, this transition value is not well defined. At higher gas flow rates, churn flow or annular flow appears in the upper tube (Fig. 26, pt. 4).

Suppose we decrease the gas flow rate. For a given value the liquid film becomes unstable and large-amplitude waves appear on the interface. The pressure drop increases and the liquid film tends to fall below the liquid injection device. This is called the flow reversal phenomenon (Fig. 26, pt. 5). When the gas flow rate decreases, the liquid film flows downward (Fig. 26, pt. 6) and the upper tube dries out for a not-well-defined value of the gas flow rate. The whole experiment shown in Fig. 26 can be summarized in the diagram of Fig. 28.

## 7.2 Experimental Correlations

**7.2.1 The flooding phenomenon** Wallis (1969) defined the following nondimensional superficial velocities:

$$J_G^* \triangleq \frac{J_G \rho_G^{1/2}}{[gD(\rho_L - \rho_G)]^{1/2}} \tag{43}$$

$$J_L^* \triangleq \frac{J_L \rho_L^{1/2}}{[gD(\rho_L - \rho_G)]^{1/2}} \tag{44}$$

where $J_G$ and $J_L$ are the superficial velocities given by Eq. (21). Note that $J_G^*$ and $J_L^*$ are analogous to Froude numbers and represent the ratio of the inertial forces to the gravity forces. The liquid viscosity $\mu_L$ is taken into account by Wallis through a Grashof number, defined as

$$\mathrm{N}_L \triangleq \left[\frac{\rho_L g D^3 (\rho_L - \rho_G)}{\mu_L^2}\right]^{1/2} \tag{45}$$

which represents the ratio of the gravity forces to the viscous forces. According to Wallis, the flooding points can be correlated by the following formula:

$$J_G^{*1/2} + m J_L^{*1/2} = C \tag{46}$$

where $m$ and $C$ are given in Figs. 29 and 30.

1. When gravity forces are far more important than viscous forces, $\mathrm{N}_L$ is high, and we have

$$m = 1$$

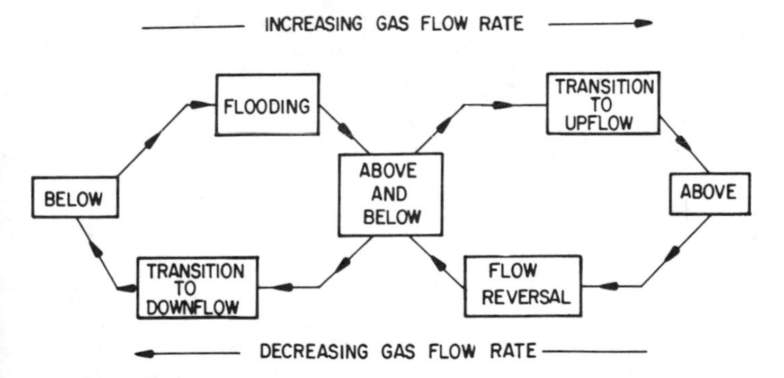


**Figure 28** Flooding and flow reversal. Location of the liquid phase above and/or below the liquid injection zone.

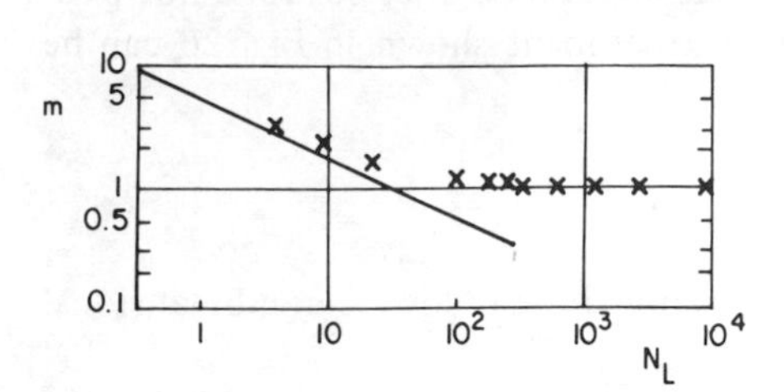


Figure 29 Value of $m$ as a function of $N_L$ (Wallis, 1969).

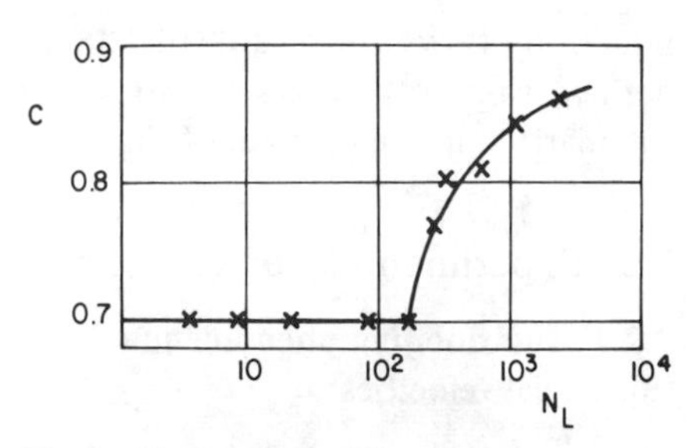


Figure 30 Value of $C$ as a function of $N_L$ (Wallis, 1969).

$$0.88 < C < 1 \qquad \text{for round-edged tubes (Fig. 31, pt. 1)}$$

$$C = 0.725 \qquad \text{for sharp-edged tubes (Fig. 31, pt. 2)}$$

Parameter $C$ depends on the inlet and outlet conditions (Shires and Pickering, 1965), a fact that has been recognized by all the authors.

2. When gravity forces can be neglected with respect to viscous forces, $N_L$ is small and we have

$$m = 5.6 N_L^{-1/2}$$

$$C = 0.725 \qquad \text{for round-edged tubes}$$

Despite its fairly good agreement with a relatively extended set of data, the Wallis flooding correlation has a major drawback. It does not take the length effect into account even though this effect can be very strong, as shown in Fig. 32. Suzuki and Ueda (1977) tried to overcome this snag but they came up with only four different correlations that apply to four different lengths!

**7.2.2 The flow reversal phenomenon** Wallis (1969) proposed the following correlation:

$$J_G^* = 0.5 \tag{47}$$

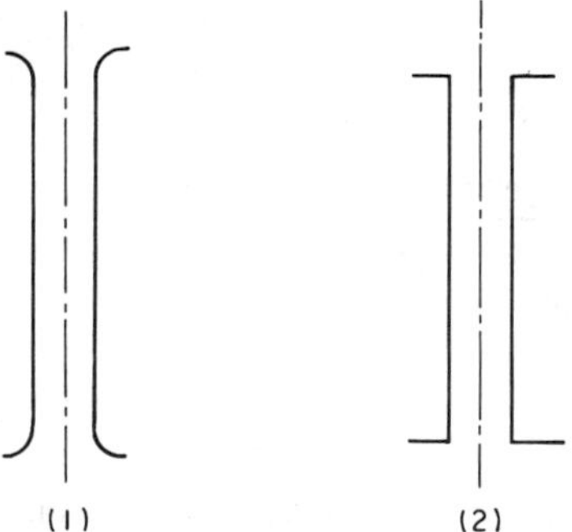


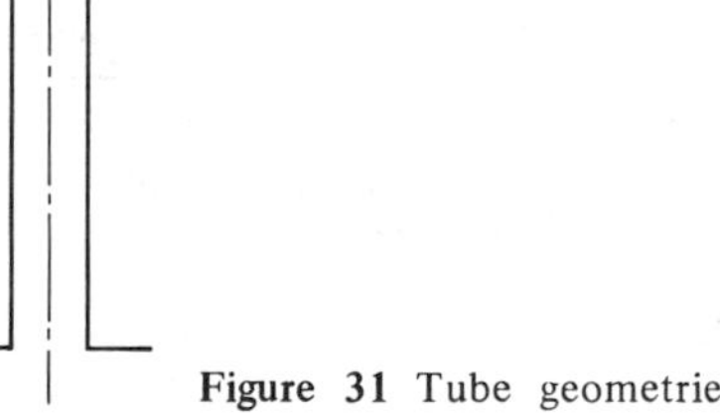


Figure 31 Tube geometries. (1) Round-edged tube. (2) Sharp-edged tube.

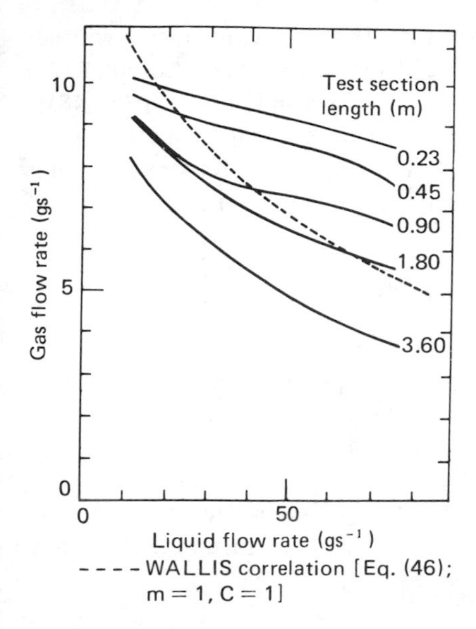


**Figure 32** Length effect on flooding curve.

The major disadvantage of this correlation lies in the diameter effect that it involves. Actually the experimental results of Pushkina and Sorokin (1969) did not show any diameter influence, as shown in Fig. 33.

Pushkina and Sorokin (1969) correlated their results fairly well by the following expression:

$$\mathrm{Ku} \triangleq \frac{J_G \rho_G^{1/2}}{[g\sigma(\rho_L - \rho_G)]^{1/4}} = 3.2 \tag{48}$$

Ku is the Kutateladze number, which represents the ratio of the gas inertial forces acting on capillary waves whose characteristic length is $[\sigma/(\rho_L - \rho_G)g]^{1/2}$. Pushkina and Sorokin ran their experiments with air and water flowing in tubes 6–309 mm in diameter. In addition they checked their correlation against results obtained for other fluids by other authors. Note that for air and water Eq. (48) yields a limiting air velocity of

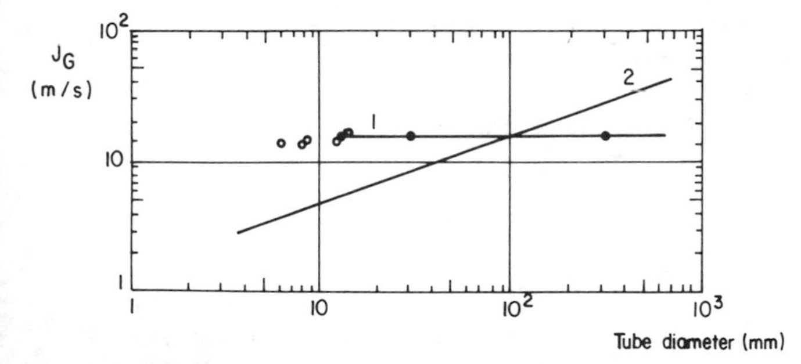


**Figure 33** Diameter effect in flow reversal. Line 1 indicates the Pushkina and Sorokin (1969) results and line 2 indicates the Wallis correlation [Eq. (47)].

about 15.8 m/s at ambient pressure and temperature. In recent experiments with air and water flowing in large-diameter tubes up to 254 mm, Richter and Lovell (1977) confirmed the validity of the criterion proposed by Pushkina and Sorokin [Eq. (48)].

Facing the issue of diameter influence, Wallis tried to reconcile the Pushkina and Sorokin criterion [Eq. (48)] with his own correlation [Eq. (47)]. As a result, Wallis and Kuo (1976) proposed to correlate the Kutateladze number Ku as a function of a nondimensional diameter $D^*$ defined by

$$D^* \triangleq D\left[\frac{(\rho_L - \rho_G)g}{\sigma}\right]^{1/2} \tag{49}$$

Note that $D^*$ is analogous to a Bond number, which represents the ratio of gravity forces to surface-tension forces. In Fig. 34 the curves $J_G^* = \text{const}$ are parabolas

$$\text{Ku} = J_G^* \sqrt{D^*} \tag{50}$$

However, Wallis and Kuo (1976) and Richter and Lovell (1977) pointed out that more experiments are needed to clarify the effects of length, diameter, surface tension, contact angle, and viscosity.

## 7.3 Theoretical Approaches

The flooding phenomenon has been handled analytically by several authors. The onset of flooding has been explained by four different mechanisms:

1. Occurrence of a standing wave on the liquid film (Shearer and Davidson, 1965)
2. Liquid film instabilities (Jameson and Cetinbudaklar, 1969; Cetinbudaklar and Jameson, 1969; Kusuda and Imura, 1974; Imura et al., 1977)
3. No net flow in the liquid film (Grolmes et al., 1974)
4. Droplet entrainment from the liquid film (Dukler and Smith, 1976)

The flow reversal phenomenon, studied by Wallis and Makkenchery (1974) and by Wallis and Kuo (1976), has been shown to be strongly connected to the hydrodynamics of hanging films. Although a great effort has been made to understand these phenomena, no definite theoretical work has appeared so far, and the issue is still open.

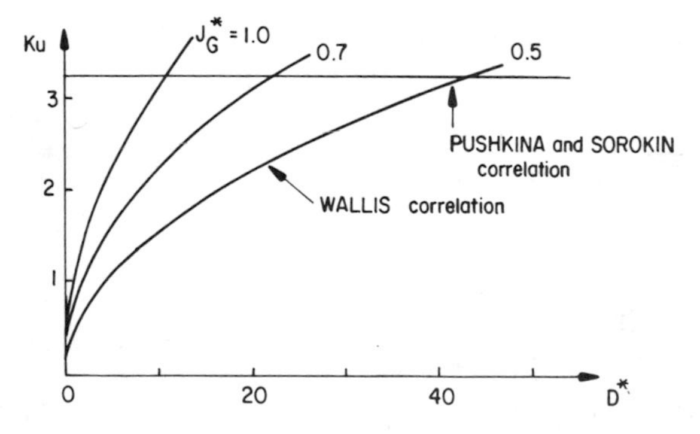


**Figure 34** Wallis and Kuo (1976) graph for the onset of flow reversal.

## NOMENCLATURE

| | |
|---|---|
| $\mathfrak{a}$ | area |
| $C$ | Wallis parameter [Eq. (46)] |
| $\mathfrak{D}$ | domain |
| $D$ | pipe diameter |
| $D^*$ | nondimensional diameter [Eq. (49)] |
| $e$ | film thickness |
| $F$ | Taitel and Dukler parameter [Eq. (37)] |
| Fr | Froude number [Eq. (32)] |
| $f$ | arbitrary function |
| $\bar{G}$ | mass velocity [Eq. (17)] |
| $g$ | acceleration resulting from gravity |
| $J$ | superficial velocity [Eq. (21)] |
| $J^*$ | nondimensional superficial velocity [Eqs. (43) and (44)] |
| $j$ | local volumetric flux [Eq. (20)] |
| $K$ | Taitel and Dukler parameter [Eq. (39)] |
| Ku | Kutateladze number [Eq. (48)] |
| $L$ | length |
| $M$ | mass flow rate [Eq. (9)] |
| $m$ | Wallis parameter [Eq. (46)] |
| N | Grashof number [Eq. (45)] |
| $p$ | pressure |
| $Q$ | volumetric flow rate [Eq. (8)] |
| $R$ | space fraction [Eqs. (4)–(7)] |
| Re | Reynolds number [Eq. (42)] |
| $T$ | time interval; Taitel and Dukler parameter [Eq. (40)] |
| $t$ | time |
| $\mathcal{V}$ | volume |
| $w$ | axial velocity component |
| $X$ | phase density function [Eq. (1)]; Lockhart-Martinelli parameter [Eqs. (28) and (38)] |
| $x$ | quality |
| $\mathbf{x}$ | position vector |
| $z$ | axial coordinate |
| $\alpha$ | local time fraction [Eq. (12)]; pipe inclination angle |
| $\beta$ | volumetric quality [Eq. (19)] |
| $\Lambda$ | Oshinowo and Charles parameter [Eq. (33)] |
| $\lambda$ | Baker parameter [Eq. (35)] |
| $\mu$ | viscosity |
| $\nu$ | kinematic viscosity |
| $\rho$ | density |
| $\sigma$ | surface tension |
| $\psi$ | Baker parameter [Eq. (36)] |

**Subscripts**

| | |
|---|---|
| $A$ | air at 20°C and 1 bar |
| eq | equilibrium |
| $G$ | gas |
| $k$ | phase index ($k = 1, 2$ or $G, L$) |
| $L$ | liquid |
| $n$ | dimension index |
| $W$ | water at 20°C and 1 bar |

**Averaging operators**

| | |
|---|---|
| $\langle\!\langle\ \rangle\!\rangle_n$ | space-averaging operator over $\mathfrak{D}_n$ [Eq. (2)] |
| $\langle\ \rangle$ | space-averaging operator over $\mathfrak{D}_{kn}$ [Eq. (3)] |
| $\overline{\phantom{x}}$ | time-averaging operator over $[T]$ [Eq. (10)] |
| $\overline{\phantom{x}}^{X}$ | time-averaging operator over $[T_k]$ [Eq. (11)] |

## REFERENCES


Alves, G. E., Cocurrent Liquid-Gas Flow in a Pipeline Contactor, *Chem. Eng. Prog.*, vol. 50, no. 9, pp. 449–456, 1954.

Arnold, C. R. and Hewitt, G. F., Further Developments in the Photography of Two-Phase Gas-Liquid Flow, *J. Photo. Sci.*, vol. 15, pp. 97–114, 1967.

Bell, K. J., Taborek, J., and Fenoglio, F., Interpretation of Horizontal In-Tube Condensation Heat Transfer Correlations with a Two-Phase Flow Regime Map, *Chem. Eng. Prog. Symp. Ser.*, vol. 66, no. 102, pp. 150–163, 1970.

Bennett, A. W., Hewitt, G. F., Kearsey, H. A., Keeys, R. K. F., and Lacey, P. M. C., Flow Visualization Studies of Boiling at High Pressure, *Inst. Mech. Eng. Proc. 1965–1966*, vol. 180, pt. 3C, pp. 260–270, 1965.

Bergles, A. E., Two-Phase Flow Structure Observations for High Pressure Water in a Rod Bundle, in *Two-Phase Flow and Heat Transfer in Rod Bundles*, ed. V. E. Schrock, pp. 47–55, ASME, New York, 1969.

Block, J. A. and Schrock, V. E., Emergency Cooling Water Delivery to the Core Inlet of PWR's during LOCA, in *Thermal and Hydraulic Aspects of Nuclear Reactor Safety*, vol. 1, *Light Water Reactors*, eds. O. C. Jones and S. G. Bankoff, pp. 109–132, ASME, New York, 1977.

Cetinbudaklar, A. G. and Jameson, G. J., The Mechanism of Flooding in Vertical Countercurrent Flow, *Chem. Eng. Sci.*, vol. 24, pp. 1669–1680, 1969.

Chan, S. H. and Grolmes, M. A., Hydrodynamically-Controlled Rewetting, *Nucl. Eng. Des.*, vol. 34, pp. 307–316, 1975.

Collier, J. G., *Convective Boiling and Condensation*, McGraw-Hill, New York, 1972.

Cooper, K. D., Hewitt, G. F., and Pinchin, B., Photography of Two-Phase Gas/Liquid Flow, *J. Photo. Sci.*, vol. 12, pp. 269–278, 1964.

Delhaye, J. M. and Achard, J. L., On the Use of Averaging Operators in Two-Phase Flow Modeling, in *Thermal and Hydraulic Aspects of Nuclear Reactor Safety*, vol. 1, *Light Water Reactors*, eds. O. C. Jones and S. G. Bankoff, pp. 289–332, ASME, New York, 1977.

Delhaye, J. M. and Achard, J. L., On the Averaging Operators Introduced in Two-Phase Flow Modeling, in *Transient Two-Phase Flow, Proc. CSNI Specialists Meet., Aug. 3 and 4, 1976, Toronto*, eds. S. Banerjee and K. R. Weaver, vol. 1, pp. 5–84, AECL, 1978.

Dukler, A. E. and Smith, L., Two-Phase Interactions in Countercurrent Flow Studies of the Flooding Mechanism, U.S. NRC contract AT (49-24) 0194, summary rept. 1, 1976.

Govier, G. W. and Aziz, K., *The Flow of Complex Mixtures in Pipes*, Van Nostrand-Reinhold, New York, 1972.

Grolmes, M. A., Lambert, G. A., and Fauske, H. K., Flooding in Vertical Tubes, *Symp. on Multiphase Flow Systems, Glasgow, Inst. Chem. Eng. Symp. Ser.* 38, paper A4, 1974.

Hewitt, G. F. and Hall-Taylor, N. S., *Annular Two-Phase Flow*, Pergamon, New York, 1970.

Hewitt, G. F. and Lovegrove, P. C., A Scanning Device for Velocity Measurement and Its Application to Wave Studies in Annular Two-Phase Flow, *J. Phys. E*, vol. 3, pp. 6–8, 1970.

Hewitt, G. F. and Roberts, D. N., Studies of Two-Phase Flow Patterns by Simultaneous X-ray and Flash Photography, AERE-M 2159, 1969.

Hewitt, G. F. and Wallis, G. B., Flooding and Associated Phenomena in Falling Film Flow in a Tube, AERE-R 4022, 1963.

Hubbard, M. G. and Dukler, A. E., The Characterization of Flow Regimes for Horizontal Two-Phase Flow: 1. Statistical Analysis of Wall Pressure Fluctuations, *Proc. 1966 Heat Transfer and Fluid Mechanics Inst.*, eds. M. A. Saad and J. A. Miller, pp. 100–121, Stanford University Press, Stanford, Calif., 1966.

Imura, H., Kusuda, H., and Funatsu, S., Flooding Velocity in a Counter-current Annular Two-Phase Flow, *Chem. Eng. Sci.*, vol. 32, pp. 79–87, 1977.

Jameson, G. J. and Cetinbudaklar, A., Wave Inception by Air Flow Over a Liquid Film, *Cocurrent Gas-Liquid Flow*, eds. E. Rhodes and D. S. Scott, pp. 271–282, Plenum, New York, 1969.

Jones, O. C. and Zuber, N., The Interrelation between Void Fraction Fluctuations and Flow Patterns in Two-Phase Flow, *Int. J. Multiphase Flow*, vol. 2, pp. 273–306, 1975.

Kusuda, H. and Imura, H., Stability of a Liquid Film in a Countercurrent Annular Two-Phase Flow (Mainly on a Critical Heat Flux in a Two-Phase Thermosyphon), *Bull. JSME*, vol. 17, no. 114, pp. 1613–1618, 1974.

Lackmé, C., Structure et Cinématique des Ecoulements Diphasiques à Bulles, CEA-R-3203, 1967.

Lahey, R. T. and Moody, F. J., *The Thermal-Hydraulics of a Boiling Water Nuclear Reactor*, American Nuclear Society, 1977.

Mandhane, J. M., Gregory, G. A., and Aziz, K., A Flow Pattern Map for Gas-Liquid Flow in Horizontal Pipes, *Int. J. Multiphase Flow*, vol. 1, pp. 537–553, 1974.

Oshinowo, T. and Charles, M. E., Vertical Two-Phase Flow. Part 1. Flow Pattern Correlations, *Can. J. Chem. Eng.*, vol. 52, pp. 25–35, 1974.

Pushkina, O. L. and Sorokin, Y. L., Breakdown of Liquid Film Motion in Vertical Tubes, *Heat Transfer Sov. Res.*, vol. 1, no. 5, pp. 56–64, 1969.

Raisson, C., Hydrodynamique de la Double-Phase à Haute Pression: Evolution des Configurations d'Ecoulement jusqu'à l'Echauffement Critique, Thèse de docteur ingénieur, Faculté des Sciences, Université de Grenoble, 1968.

Richter, H. J. and Lovell, T. W., The Effect of Scale on Two-Phase Countercurrent Flow Flooding in Vertical Tubes, U.S. NRC contract AT (49-24)-0329, final rept., Thayer School of Engineering, Dartmouth College, Hanover, N.H., 1977.

Roumy, R., Structure des Ecoulements Diphasiques Eau-air. Etude de la Fraction de Vide Moyenne et des Configurations d'Ecoulement, CEA-R-3892, 1969.

Serizawa, A., Kataoka, I., and Michiyoshi, I., Turbulence Structure of Air-Water Bubbly Flow–I. Measuring Techniques, *Int. J. Multiphase Flow*, vol. 2, pp. 221–233, 1975.

Shearer, C. J. and Davidson, J. F., The Investigation of a Standing Wave Due to Gas Blowing Upwards over a Liquid Film, Its Relation to Flooding in Wetted-Wall Columns, *J. Fluid Mech.*, vol. 22, pt. 2, pp. 321–335, 1965.

Shires, G. L. and Pickering, A. R., The Flooding Phenomenon in Countercurrent Two-Phase Flow, *Proc. Symp. in Two-Phase Flow*, vol. 2, pp. B501–B538, University of Exeter, United Kingdom, 1965.

Suzuki, S. and Ueda, T., Behaviour of Liquid Films and Flooding in Countercurrent Two-Phase Flow. Part 1. Flow in Circular Tubes, *Int. J. Multiphase Flow*, vol. 3, pp. 517–532, 1977.

Taitel, Y. and Dukler, A. E., A Model for Predicting Flow Regime Transitions in Horizontal and Near Horizontal Gas-Liquid Flow, *AIChE J.*, vol. 22, no. 1, pp. 47–55, 1976.

Taitel, Y. and Dukler, A. E., Flow Regime Transitions for Vertical Upward Gas-Liquid Flow: A

Preliminary Approach Through Physical Modeling, AIChE 70th Annual Meet., New York, Session on Fundamental Research in Fluid Mechanics, 1977.

Tien, C. L., A Simple Analytical Model for Countercurrent Flow Limiting Phenomena with Vapor Condensation, *Lett. Heat Mass Transfer*, vol. 4, pp. 231–238, 1977.

Tong, L. S. and Weisman, J., *Thermal Analysis of Pressurized Water Reactors*, American Nuclear Society City, 1970.

Wallis, G. B., *One-dimensional Two-Phase Flow*, McGraw-Hill, New York, 1969.

Wallis, G. B. and Kuo, J. T., The Behaviour of Gas-Liquid Interfaces in Vertical Tubes, *Int. J. Multiphase Flow*, vol. 2, pp. 521–536, 1976.

Wallis, G. B. and Makkenchery, S., The Hanging Film Phenomenon in Vertical Annular Two-Phase Flow, *J. Fluids Eng.*, pp. 297–298, 1974.

CHAPTER

# FOUR

# TWO-PHASE FLOW INSTRUMENTATION

**J. M. Delhaye**

Because of the need for measurements of practical and fundamental parameters, there has been a great deal of study of measuring techniques used in two-phase gas-liquid flows. A decade ago, an in-depth review of two-phase flow instrumentation was edited by LeTourneau and Bergles (1969) that included 368 references. More recently, several detailed summaries of measurement techniques were given for particular two-phase flow applications (Table 1).

In addition to the problems encountered in any measuring method used in single-phase flow [amply described in the works of Katys (1964), Doebelin (1966), Benedict (1969), Bradshaw (1971), Considine (1971), Dowdell (1974), and Richards (1977)], further difficulties appear when attempts are made to measure quantities in two-phase flow. Most of these difficulties are related to the presence of the two phases, but others are related to the conditions of flow confinement. These difficulties are exemplified in the introduction of the paper by Jones and Delhaye (1976). Difficulties from the presence of two phases may include the hydrodynamic response time owing to the interaction between probes and interfaces, the entrapment of gas bubbles in the manometric lines, the occurrence of significant fluctuations in the quantities to be measured, cavitation effects, and so on. In addition to these are difficulties related to flow geometries, such as opaque metallic walls imposed by temperature and pressure conditions or small hydraulic diameters such as in rod bundles.

Despite these difficulties, several techniques can be used to measure practical or

**Table 1 Recent reviews on two-phase flow instrumentation**

| Authors | Major area |
|---|---|
| Brockett and Johnson (1976) | Water reactor safety studies |
| Hewitt and Lovegrove (1976) | Two-phase flow experimental studies |
| Jones and Delhaye (1976) | Transient and statistical aspects of two-phase flows |
| Leonchik et al. (1976) | Dispersed flows |
| Moore (1976) | Wet steam flows |

fundamental parameters with a fair degree of accuracy. The purpose of this chapter is to present the principles of a few methods that can be considered well established. For reviews covering a broader spectrum the reader is referred to Table 1.

# 1 PHOTON ATTENUATION TECHNIQUES

Radiation attenuation methods furnish the line and area void fractions $R_{G1}$ and $R_{G2}$, which were defined in Chap. 3. They are the most frequently used void-fraction measurements and can provide detailed information on the flow pattern (Jones, 1973; Jones and Zuber, 1975).

## 1.1 Absorption Law

A beam of monochromatic collimated photons of incident intensity $I_0$, traversing a substance of thickness $e$ and density $\rho$, has an emerging intensity $I$ given by the following exponential absorption law:

$$I = I_0 \exp\left(-\frac{\mu}{\rho}\rho e\right) \tag{1}$$

The quantity $\mu/\rho$ is the specific absorption coefficient of the material. The coefficient generally decreases piecewise with the energy of the photons and is independent of the physical state of the substance (solid, liquid, or gas). Tables giving the values of this coefficient for different materials can be found in McMaster et al. (1969), Storm and Israel (1970), and Stukenbroeker et al. (1970). Figure 1 shows the mass absorption coefficient $\mu/\rho$ as a function of the incident beam energy for water.

## 1.2 Line Void-Fraction Measurements

When a collimated beam is used, the radiation is absorbed by the wall and the two phases in a series mode. At a given time the attenuation of the beam is given by

$$I = I_0 \exp(-\mu_P e_P) \exp[-\mu_L(1-R_{G1})\,d] \exp(-\mu_G R_{G1} d) \tag{2}$$

where $e_P$ = total wall thickness

$d$ = distance between the walls

$\mu_P, \mu_G, \mu_L$ = absorption coefficients of the wall, gas, and liquid, respectively

1. For low-pressure gas-liquid flows at ambient temperature (Galaup, 1975) the absorption by the gas can be considered negligible compared with that of the liquid. As a result, it is necessary to measure only intensities $I_G$ and $I_L$ corresponding, respectively, to the channel filled with gas and filled with liquid. The instantaneous line void fraction is then easily obtained, provided $I_0$ is a constant:

$$R_{G1} = \frac{\log I/I_L}{\log I_G/I_L} \tag{3}$$

2. For high-pressure steam-water flows we have

$$\frac{\mu_L}{\rho_L} = \frac{\mu_G}{\mu_L} \tag{4}$$

but the steam density is no longer negligible compared with that of the water. Consequently a calibration in conditions as close as possible to experimental conditions will be necessary (Martin, 1969, 1972; Réocreux, 1974).

## 1.3 Area Void-Fraction Measurements

The instantaneous area void fraction $R_{G2}$ can be measured by two methods:

1. In the one-shot technique (Nyer, 1969; Gardner et al., 1970; Charlety,

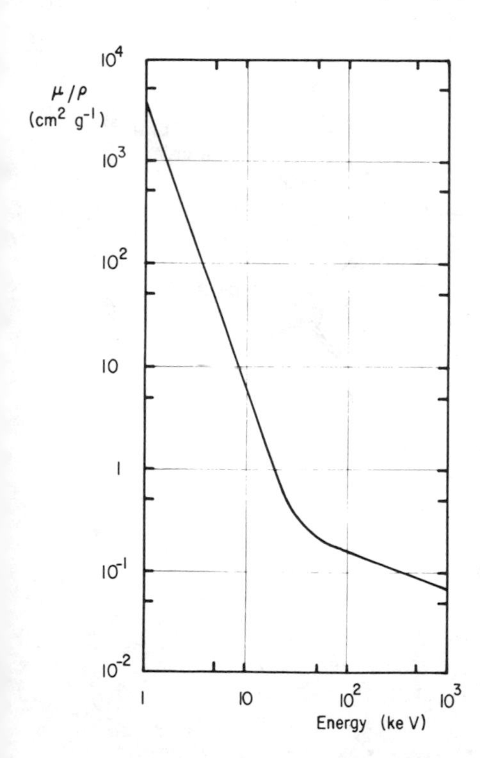


**Figure 1** Mass absorption coefficient as a function of the incident beam energy for water.

1971) the radiation crosses the whole of the tube cross section containing the two-phase mixture. The beam height must be small compared with the tube diameter to restrict the measurement to a given cross section. Calibration with lucite mockups is generally necessary because of the lack of a specific absorption law.

2. The multibeam gamma densitometer has been used for a few years by laboratories involved in water reactor safety studies. Figure 2 shows a schematic of the three-beam densitometer used by Banerjee et al. (1978). A 25-Ci cesium-137 source provides three $\gamma$-ray beams directed through the pipe in the same cross-sectional plane. The beams are attenuated according to the line void fractions, and the measurement of the attenuation of each beam can be used to determine the area void fraction $R_{G2}$. This technique requires a model connecting the area void fraction, the flow pattern, and the three measured line void fractions.

The time-averaged area void fraction $\overline{R_{G2}}$ can be determined from a profile of time-averaged chordal void fraction $\overline{R_{G1}}$. In a pipe of circular cross section we have

$$\overline{R_{G2}} = \frac{1}{\pi R^2} \int_{y=-R}^{y=R} 2\sqrt{R^2 - y^2}\ \overline{R_{G1}}(y)\, dy \tag{5}$$

where $R$ is the pipe radius and $y$ is the distance from the axis of the pipe to the photon beam. Note that if the flow is axisymmetric, the local void fraction $\alpha_G$ is a solution of the following Abel integral equation:

$$\int_y^R \frac{r\alpha_G(r)}{\sqrt{r^2 - y^2}}\, dr = \overline{R_{G1}}(y)\sqrt{R^2 - y^2} \tag{6}$$

where $r$ is the radial coordinate.

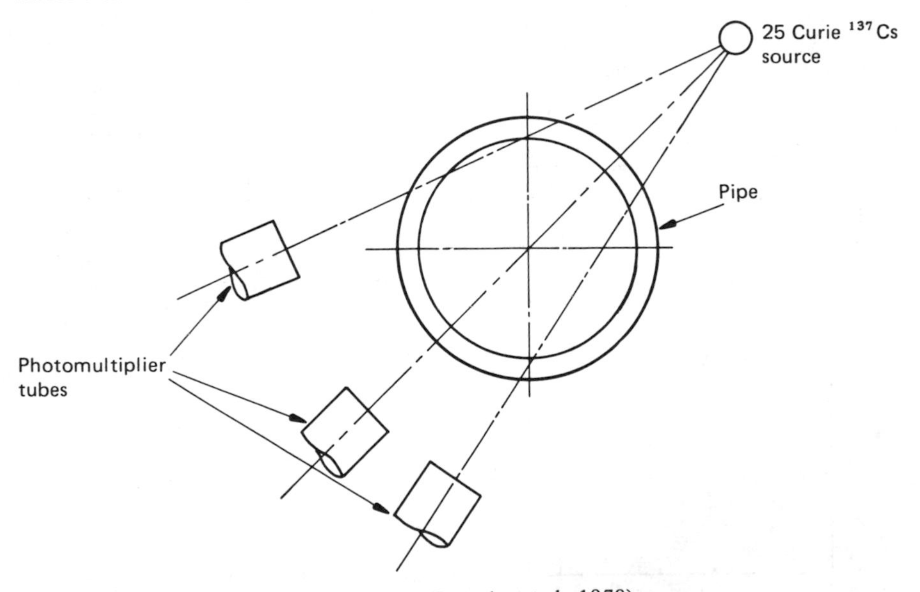


**Figure 2** Three-beam gamma densitometer (Banerjee et al., 1978).

## 1.4 Errors Resulting from Fluctuating Voids

In steady-state two-phase flow the emerging intensity $I$ (in photons per second) is generally measured over a certain period $\theta$ (10 s to 1 min). Consequently the measurement gives the number $N$ of emerging photons counted over the time interval $\theta$,

$$N \triangleq \int_{t-\theta/2}^{t+\theta/2} I\,dt \qquad \text{in photons} \tag{7}$$

Because the average of an exponential is not equal to the exponential of the average, the measurement does not give in fact the void fraction $\overline{R_G}$ averaged over the time interval $\theta$. This error can be considerable for highly fluctuating flows. The deviation can range from 0.05 for churn flow up to 0.20 for slug flow. However, the measurement of time-averaged void fractions in fluctuating flows can be handled by one of the following methods:

1. The averaged void fraction $\overline{R_G}$ can be determined from the probability density function of the instantaneous void fraction $R_G$ obtained for a short counting time, for example, 0.05 s. This technique has been reported in a series of papers by Hancox et al. (1972), Harms and Laratta (1973), and Laratta and Harms (1974).
2. Levert and Helminski (1973) used a dual energy method based on the differential absorption of radiation energy.
3. Log amplifiers were used by Jones (1973) to linearize the instantaneous signal. The only requirement is that the response time of the amplifier be smaller than the lower period of the physical fluctuations.

## 1.5 Choice of Radiation

A radiation is characterized by its energy spectrum and its intensity. The contrast is defined by the relationship

$$c \triangleq \frac{I_G}{I_L} \tag{8}$$

where $I_G$ and $I_L$ correspond, respectively, to the channel filled with gas and filled with liquid. For gas-liquid flow at low pressure,

$$c = \exp \mu_L d \tag{9}$$

The greater the contrast, the higher the sensitivity of the measurement. To increase the contrast, the mass absorption coefficient $\mu_L/\rho_L$ must be increased and therefore the energy decreased (Fig. 1).

On the other hand, the photon emission fluctuation leads to the following statistical relationship:

$$\frac{\Delta N}{N} = \left(\frac{1}{N}\right)^{1/2} \tag{10}$$

where $N$ is the number of photons counted. For purposes of accuracy $N$ must therefore be sufficiently high. As a result, either the counting time $\theta$ or the emerging intensity $I$ must therefore be sufficiently high. Consequently, if the incident intensity $I_0$ is fixed, the mass absorption coefficient must be decreased, and therefore the energy increased.

Constraints resulting from the contrast and the photon emission fluctuations therefore have opposite effects on the choice of the photon energy and of the source intensity. In this respect, X-rays have a beam intensity $10^3$-$10^4$ times higher than $\gamma$-rays.

The photon beam must be monochromatic for the exponential absorption law [Eq. (1)] to be applicable. In this respect, $\gamma$-ray emission is more monochromatic than X-ray bremsstrahlung emission. However, filters or electronic discriminators can be used to select or detect only those photons with energy in a certain band.

The stability of the source in time is another important parameter of measurement accuracy. In this respect, $\gamma$-rays are to be preferred to X-rays because of the rather long half-lives of the principal sources of $\gamma$-rays. However, certain methods can be used to avoid the inconvenience of X-ray generator drifts and/or fluctuations (Nyer, 1969; du Bousquet, 1969; Jones, 1973; Lahey, 1977; Solésio, 1978).

## 1.6 Radiation Detection

The emerging intensity $I$ is measured with a scintillator coupled to a photomultiplier supplied with correctly stabilized high voltage. Generally, a pretime method is used in which the pulses are counted during a predetermined time. Xenon ionization chambers seem to be more stable than photomultipliers, but further studies are needed to confirm this possibility.

## 1.7 Examples

1. *Steady-State Steam-Water Flow at High Pressure* Martin (1969, 1972) measured the line void fraction at high pressure (up to 140 bar) in a rectangular channel (53 × 2.8 mm), simulating a subchannel of a nuclear reactor plate-type fuel element. The test section consisted of two vertical plates heated by Joulean effect and was enclosed in a thick casing drilled with holes that allowed a horizontal X-ray beam to pass through the two 0.5-mm-thick lateral sides (Fig. 3). This X-ray beam, 2 mm in height and 0.05 mm in width, scanned the test cross section in a movement parallel to the heated plates. The line void fraction $\overline{R_{G1}}$ of the upward steam-water flow was measured every 0.025 mm with a position accuracy of about 0.005 mm despite the severe experimental conditions (that is, for a fluid pressure of 140 bar, a fluid temperature of 335°C, and an ambient temperature of 50°C). Typical distributions of the line void fraction $\overline{R_{G1}}$ are given in Fig. 3.

2. *Liquid Film Thickness Measurements* The thickness of a wavy liquid film flowing down an inclined plane was measured by Solésio (1978) by means of an

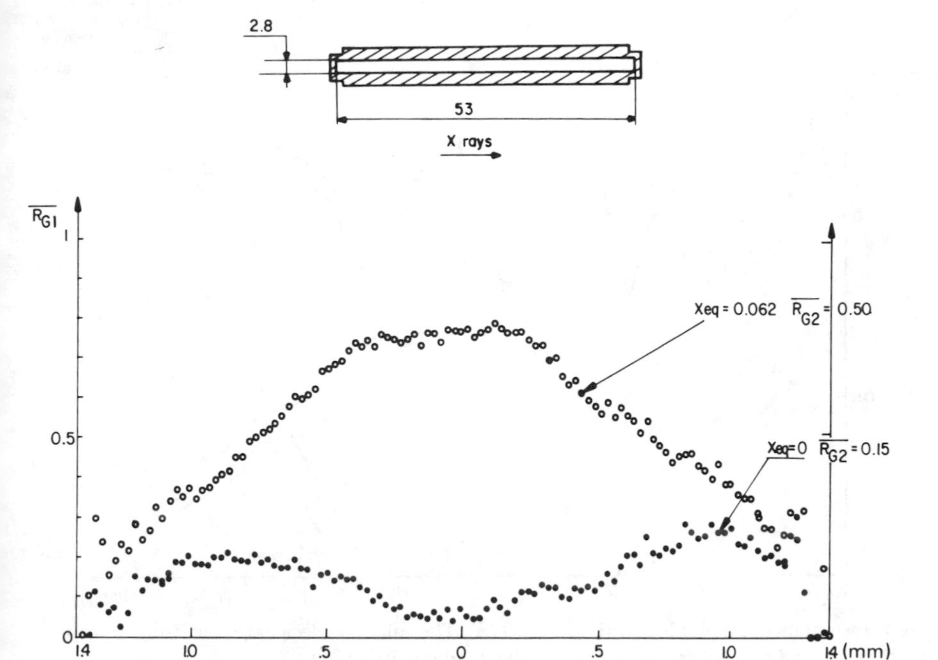


**Figure 3** Line void fractions $\overline{R_{G_1}}$ in steady-state steam–water flow at high pressure (Martin, 1969, 1972). Pressure, 80 bar; heat flux density, 110 W/cm²; mass velocity, 220 g/cm² s.

X-ray absorption technique (Fig. 4). The film thickness, whose averaged value is approximately 1 mm, is determined each millisecond with a precision finer than 0.050 mm. Figure 4 shows the film thickness as related to time when 35-Hz waves are generated at the free surface of the liquid film. The solid line represents the signal obtained by the X-ray absorption technique, whereas the dashed line represents the signal obtained by a conductivity probe that measures the liquid conductance between two electrodes mounted flush with the wall.

# 2 ELECTROMAGNETIC FLOWMETERS

## 2.1 Use in Single-Phase Flow

The theoretical and practical aspects of the use of electromagnetic flowmeters in single-phase flow are described in several papers or articles (Shercliff, 1962; Turner, 1968; Treenhaus, 1972; Bevir, 1974; Demagny, 1976). Whatever the conducting fluid used, the electromagnetic flowmeter gives a signal proportional to the area-averaged liquid velocity,

$$e = KBd \langle w \rangle \tag{11}$$

where $e$ = induced emf
$K$ = calibration constant

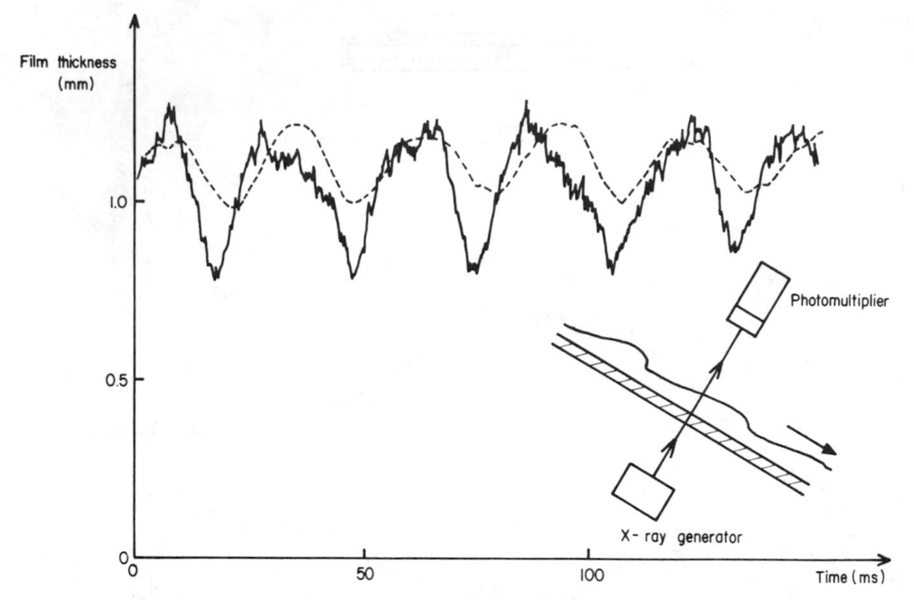


**Figure 4** Instantaneous liquid height (Solésio, 1978). The solid line designates the X-ray absorption technique, and the dashed line designates the conductivity probe.

$B$ = magnetic induction
$d$ = tube diameter
$\langle\!\langle w \rangle\!\rangle$ = area-averaged fluid velocity in the cross section containing the electrodes

The most commonly used flowmeter is the transverse flowmeter. The electrodes are placed diametrically opposite in a plane perpendicular to the plane formed by the magnetic induction and the flow axis. The magnetic induction is either constant or alternating. Constant-induction flowmeters generally use a permanent magnet, and their utilization is limited to liquid metals (Thatcher, 1972). Alternating-induction flowmeters are used when the liquid is a poor conductor. In this case, the generation of a magnetic induction sufficiently high to generate a satisfactory induced emf would require too powerful direct currents. The use of an alternating field also avoids polarization of the electrodes in the case of electrolytes.

## 2.2 Use in Steady-State Two-Phase Flow

Although several attempts have been made to establish a complete theory on electromagnetic flowmeters in two-phase flow (Alad'yev et al., 1971; Fitremann, 1972a, 1972b), their use is generally based on the assumption that they measure the area-averaged velocity of the continuous conducting phase in contact with the electrodes (Charlety, 1971). Hence, in dispersed annular flow, the electromagnetic flowmeter gives a signal proportional to the area-averaged velocity of the liquid film without taking into account the presence of liquid drops in the gas core.

Usually one flowmeter is placed in a single-phase zone of the flow (superscript $'$), while a second flowmeter is placed in the two-phase zone under consideration (superscript ″). We have

$$e' = K'B'd \lessdot w'_L \gtrdot_2 \tag{12}$$

$$e'' = K''B''d < w''_L >_2 \tag{13}$$

$$M'_L = \rho_L \lessdot w'_L \gtrdot_2 A \tag{14}$$

$$M'' = \rho_L < w''_L >_2 A(1 - R_{G2}) \tag{15}$$

where $M_L$ = liquid mass flow rate
$\rho_L$ = liquid density (assumed to be a constant)
$A$ = constant area of the tube cross section
$R_{G2}$ = area fraction

1. *Two-component steady-state two-phase flow* (Heinemann et al., 1963; Hori et al., 1966; Milliot et al., 1967). For this case

$$M''_L = M'_L \tag{16}$$

Hence

$$R_{G2} = 1 - \frac{\lessdot w'_L \gtrdot_2}{< w''_L >_2} = 1 - \frac{K''B''e'}{K'B'e''} \tag{17}$$

2. *One-component steady-state two-phase flow* (Heinemann, 1965; Lurie, 1965; Charlety, 1971). For this case

$$M''_L = (1 - x)M'_L \tag{18}$$

where $x$ is the quality. Hence

$$R_{G2} = 1 - (1 - x)\frac{\lessdot w'_L \gtrdot_2}{< w''_L >_2} = 1 - (1 - x)\frac{K''B''e'}{K'B'e''} \tag{19}$$

If $x$ is not negligible compared with unity, it must be either measured or computed with a model to obtain the void fraction. In return, if the void fraction is known by another method, the quality can be determined.

## 3 TURBINE FLOWMETERS

In single-phase flow one can define a flow coefficient $C$ and a rotation Reynolds number $N$ by the following relations:

$$C \triangleq \frac{Q}{nD^3} \tag{20}$$

$$N \triangleq \frac{nD^2}{\nu} \tag{21}$$

where $Q$ = volumetric flow rate
$n$ = rotational frequency
$D$ = pipe diameter
$\nu$ = kinematic viscosity of the fluid

Hochreiter (1958) has shown that the flow coefficient $C$ is a function of the rotation Reynolds number $N$. However, the flow coefficient appears to be a constant in a wide range of $N$, and here the rotational frequency $n$ is proportional to the volumetric flow rate. Introducing the mass velocity $G$ and the fluid density $\rho$, we thus obtain

$$n = K \frac{G}{\rho} \tag{22}$$

where $K$ is a dimensional calibration constant.

With respect to two-phase flow the same type of formula as Eq. (22) has been sought, relating the rotational frequency $n$, the total mass velocity $G$ of the mixture, and a two-phase density $\rho$ and using the dimensional calibration constant $K$, determined for single-phase liquid flow. So far the most successful attempt was suggested by Popper (1961) and tested by Rouhani (1964) and more recently by Frank et al. (1977). In Eq. (22) the two-phase mixture density is the so-called momentum, or impulse, density defined by

$$\frac{1}{\rho} \triangleq \frac{x^2}{R_{G2}\rho_G} + \frac{(1-x)^2}{(1-R_{G2})\rho_L} \tag{23}$$

where $x$ = quality
$R_{G2}$ = area void fraction
$\rho_G, \rho_L$ = gas and liquid densities, respectively

Frank et al. (1977) found good agreement between Eqs. (22) and (23) and the experimental results obtained in an argon-water loop where the void fraction $R_{G3}$ was measured by using quick-closing valves. In these experiments the pressure varied from 18 to 83 bar, the quality from 0.03 to 0.75, and the void fraction from 0.3 to 0.99. Another model for turbine flowmeters used in two-phase flow was developed by Aya (1975), but no better agreement has been obtained with the experimental results of Frank et al. (1977).

These models constitute exploratory investigations that must be supplemented by a more rigorous and detailed analysis of the turbine flowmeter in steady-state and transient two-phase flow. A step in this direction has been made recently by Kamath and Lahey (1977), but more results are still sorely needed. Finally, at high pressure and high temperature, technological problems remain to be solved, for example, the bearing behavior.

## 4 VARIABLE–PRESSURE–DROP METERS

When the total mass flow rate is known, orifice plates and venturi tubes can be used in two-phase flow to measure the gas mass quality, which is defined as the ratio of the gas mass flow rate to the total mass flow rate (NEL, 1966).

Among several interesting studies by different authors, Collins and Gacesa (1971) suggested the following correlation:

$$x = -\frac{R}{D_1/\epsilon - R} + D_2 \frac{\alpha d^2}{D_1/\epsilon - R}(10^8 y)^{1/2} + D_3 \left[\frac{\alpha d^2}{D_1/\epsilon - R}(10^8 y)^{1/2}\right]^2 \tag{24}$$

if

$$y \triangleq \frac{H\rho_G}{M^2} \tag{25}$$

$$R \triangleq \left(\frac{\rho_G}{\rho_L}\right)^{1/2} \tag{26}$$

where $x$ = quality
$H$ = differential water head (in; at 20°C)
$\rho_G, \rho_L$ = gas and liquid densities, respectively (lb/ft$^3$)
$M$ = total mass flow rate (lb/h)
$d$ = diameter of the orifice or of the throat (in)
$\epsilon$ = expansion factor
$\alpha$ = flow coefficient

The authors give the following values for orifice plates set in pipes of internal diameter greater than 50 mm:

$$D_1 = 1.275 \qquad D_2 = 3.066 \times 10^{-2} \qquad D_3 = 6.586 \times 10^{-4}$$

With venturi tubes placed in pipes of internal diameter greater than 65 mm, the authors find

$$D_1 = 11.8 \qquad D_2 = -9.99 \times 10^{-2} \qquad D_3 = 0.278$$

Equation (24) is valid for vertical, upward steam-water flow, at 67 bar, with a total mass flow rate between 1.9 and 12.6 kg/s and for steam quality between 0.05 and 0.90. Orifice plates and venturi tubes must comply with ASME standards (1959). The same correlation can be used for larger diameters (up to 200 mm), for lower pressures (1 bar), and for higher mass flow rates (up to 64 kg/s). The influence of the diameter is important, and the authors frequently recommend that the correlation should not be used for pipings with an internal diameter less than 50 mm for orifice plates and less than 65 mm for venturi tubes.

Zanker (1968) used orifice plates and venturi tubes to measure the total mass flow rate in horizontal air-water flows, with void fractions less than 0.05. For tubes with an internal diameter less than 5 cm, the total mass flow rate is equal to the mass flow rate calculated as if the water flowed alone, multiplied by the quantity

$$\frac{\rho_G}{(1-x)\rho_G + 0.5x\rho_L}$$

## 4.1 Correlations Valid Only for Orifice Plates

Kremlevskii and Dyudina (1972) used an orifice plate to measure the total mass flow rate for steam-water flows at 1-4 bar and for quality higher than 0.7. The total mass flow rate is equal to the mass flow rate calculated as if the steam flowed alone, multiplied by the quantity $1 + 0.56(1 - x)$.

Smith et al. (1977) evaluated the existing two-phase flow correlations for the flow of steam-water mixtures and two-component gas-liquid mixtures through ASME-code measuring orifices. They concluded that

1. For the flow of steam-water mixtures, the James correlation (1965–1966) gives the best results. This correlation reads

$$M = 0.61 \frac{YF}{\sqrt{1-\beta^4}} S_2 \sqrt{\frac{2\,\Delta p}{x^{1.5}(1/\rho_G - 1/\rho_L) + 1/\rho_L}} \tag{27}$$

where $M$ = total mass flow rate
$Y$ = gas expansion factor
$F$ = orifice thermal expansion factor
$\beta$ = ratio of orifice diameter to internal pipe diameter
$S_2$ = orifice cross-sectional area

The correlation was tested by Smith et al. (1977) for the parameter range given in Table 2.

2. For the flow of two-component gas-liquid mixtures the Murdock correlation (1962) gives the best results. This correlation reads

$$M = \frac{K_G Y_G F S_2 \sqrt{2\,\Delta p \rho_G}}{x + 1.26(1-x)(K_G Y_G / K_L)\sqrt{\rho_G/\rho_L}} \tag{28}$$

where $Y_G$ and $Y_L$ are the gas and liquid expansion factors. The flow coefficients $K_G$ and $K_L$ are treated as unknown parameters so that this correlation requires an iteration scheme. The correlation was tested by Smith et al. (1977) for the parameter range given in Table 3.

In a recent paper, Chisholm (1977) developed correlations for the pressure drop over sharp-edged orifices during the flow of incompressible two-phase mixtures. They read

$$\frac{\Delta p}{\Delta p_L} = 1 + \frac{C}{X} + \frac{1}{X^2} \tag{29}$$

$$X \triangleq \frac{1-x}{x}\left(\frac{\rho_G}{\rho_L}\right)^{1/2} \tag{30}$$

**Table 2 Range of parameters for which the James correlation was tested (steam–water flow)**

| Parameter | Minimum value | Maximum value |
|---|---|---|
| Internal pipe diameter (mm) | 63 | 200 |
| $\beta$ | 0.500 | 0.707 |
| Pressure (bar) | 16.8 | 77.2 |
| Quality | 0.06 | 0.95 |
| Mass velocity (kg/m² s) | 676 | 2514 |

**Table 3 Range of parameters for which the Murdock correlation was tested (two-component two-phase flow)**

| Parameter | Minimum value | Maximum value |
|---|---|---|
| Internal pipe diameter (mm) | 75 | 98 |
| $\beta$ | 0.26 | 0.43 |
| Pressure (bar) | 1 | 64 |
| Quality | 0.11 | 0.98 |
| Mass velocity (kg/m² s) | 46.5 | 1002 |

$$C = \left(\frac{\rho_L}{\rho_G}\right)^{1/4} + \left(\frac{\rho_G}{\rho_L}\right)^{1/4} \quad \text{for } X < 1 \tag{31}$$

$$C = \left(\frac{\rho}{\rho_G}\right)^{1/2} + \left(\frac{\rho_G}{\rho}\right)^{1/2} \quad \text{for } X > 1 \tag{32}$$

$$\frac{1}{\rho} \triangleq \frac{x}{\rho_G} + \frac{1-x}{\rho_L} \tag{33}$$

where $\Delta p_L$ is the pressure change over the orifice when liquid flows alone. The proposed correlation agrees with the experimental results of Collins and Gacesa (1971), among others, but disagrees with the data of James (1965–1966) and Murdock (1962).

## 4.2 Correlations Valid Only for Venturis

Harris (1967) suggested a formula for steam–water flows with quality less than 0.20, for pressures of 65.5 bar in horizontal tubes of internal diameter 105 mm, and for mass flow rates between 15.1 and 22.7 kg/s. The venturi tubes must comply with British standards. Harris gave the following formula:

$$10^2 x = 0.584 \frac{\Delta p - 17.0M^2}{M^2} - 2.44 \times 10^{-3} \frac{(\Delta p - 17.0M^2)^2}{M^2} \tag{34}$$

where the total mass flow rate is in $10^5$ pounds per hour and $\Delta p$ is in inches of water.

Del Tin and Panella (1972) measured the pressure drop in a venturi tube containing a steam–water flow at pressures from 5 to 2 bar. The total mass flow rate varied from 80 to 180 g/s, and the mass steam quality $x$ from 0 to 0.40. Tests were carried out in vertical and horizontal tubes. The authors plotted the following curves:

$$x = x(\phi_{L0}^2, p) \tag{35}$$

where $\phi_{L0}^2$ is the ratio of the pressure drop in two-phase flow to the pressure drop that would occur with the liquid flowing alone, at saturation temperature and with total mass flow rate.

Following the ideas developed by Collins and Gacesa (1971), Fouda (1975) and Fouda and Rhodes (1978) tried to correlate the pressure drop by using either a separated flow model or a homogeneous model. Although better agreement was obtained by taking into account the hydrostatic pressure head and the diameter effect, much work remains to be done to generalize the results.

At high pressure the more extensive results seem to have been obtained by Frank et al. (1977), who carried out an experimental program on an argon-water loop running between 18 and 83 bar. The results are fairly well correlated by the following equations:

$$\Delta p = K \frac{M^2}{\rho} \tag{36}$$

and

$$\frac{1}{\rho} = \frac{x^2}{R_G \rho_G} + \frac{(1-x)^2}{R_L \rho_L} \tag{37}$$

where $K$ is a dimensional constant determined in single-phase liquid flow. The correlation was tested for the parameter range given in Table 4.

## 5 LOCAL MEASUREMENTS

Two-phase flow modeling efforts require among other information local instantaneous measurements of phase density functions, liquid and gas velocities, interfacial passage frequencies, and liquid and vapor temperatures and their statistical characteristics (such as probability density functions and spectral densities). The purpose of these measurements is to obtain data regarding interfacial area densities and correlation coefficients between void and velocity, void and energy, and so on, and to verify hypotheses regarding the shape of void, velocity, and temperature profiles, the profile interrelations, and their statistical variations. A detailed review of transient and statistical measurement techniques for two-phase flows, including local techniques, can be found in Jones and Delhaye (1976).

### 5.1 Electrical Probes

The first requirement to be met when using an electrical probe in two-phase flow is that one phase has a significantly different electrical conductivity from the other.

**Table 4 Range of parameters for which Eqs. (36) and (37) were tested**

| Parameter | Minimum value | Maximum value |
|---|---|---|
| Argon–water pressure (bar) | 18 | 83 |
| Steam–water equivalent pressure (bar) | 40 | 140 |
| Quality | 0.03 | 0.75 |
| Void fraction | 0.30 | 0.99 |
| Mass flow rate (kg/s) | 0.10 | 0.80 |

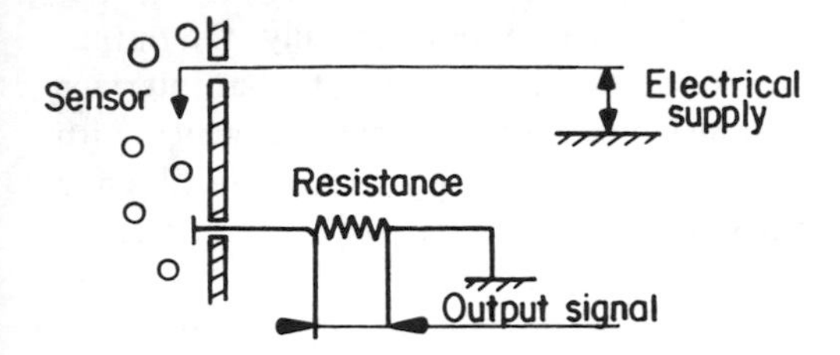


**Figure 5** Electrical diagram of a resistive probe.

Consequently, variations in conductance permit the measurement of the local void fraction and the arrival frequency of the bubbles at a given point in a continuous, conducting fluid. By using a double probe, a transit velocity can be measured, but one has to be very careful when giving a physical significance to this velocity (Galaup, 1975; Delhaye and Achard, 1977).

Figure 5 shows the classical electrical diagram of a resistive probe and Fig. 6 displays a typical probe geometry. Impedance changes caused by the passage of bubbles at the tip of the probe produce a fluctuation in the output signal. One of the principal features that differentiates the electrical circuits shown in Fig. 5 is the type of electrical supply. Direct-current supply requires low voltages to reduce electrochemical phenomena on the sensor. Resultant electronics may become troublesome and sensors may still sustain alterations because of electrochemical deposits at low flows. When an alternating-current supply is used, phase changes are detected by amplitude modulation of the alternating output signal. This technique has been used by several investigators to eliminate the electrochemical phenomena on the sensor. When high-speed flows are investigated, the required supply frequency can be very high, for example, 1 MHz, and much trouble occurs with the electronics. Galaup (1975) used a supply frequency lower than the frequency of the physical phenomenon, eliminating electrochemical effects and providing pseudo-direct-current operation in each half wave.

According to the way that the sensor is energized, the ideal output signal of a resistive probe is either a binary wave sequence or a sequence of bursts of

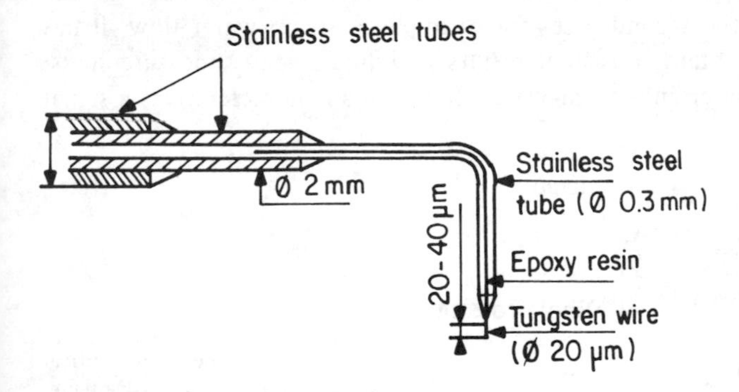


**Figure 6** Miniature probe geometry (Lecroart and Porte, 1971).

constant-amplitude oscillations separated by zero-voltage zones. Actually, the output signal is misshapen with respect to the ideal signal because of the interface deformations. The true signal is generally transformed into a binary sequence with the help of a trigger level. Galaup (1975) used a level adjustment based on a comparison between the integrated void profile and the line void fraction obtained with a $\gamma$-ray absorption method.

## 5.2 Optical Probes

An optical probe is sensitive to the change in the refractive index of the surrounding medium and is thus responsive to interfacial passages enabling measurements of local void fraction and of interface passage frequencies to be obtained even in a nonconducting fluid. By using two sensors and a cross-correlation method, one may obtain information on a transit velocity (Galaup, 1975).

A tiny optical sensor was proposed by Danel and Delhaye (1971) and developed by Galaup (1975). This probe consists of a single optical fiber, 40 $\mu$m in diameter. The overall configuration is shown in Fig. 7. The active element of the probe is obtained by bending the fiber into a U shape. The entire fiber, except the U-shaped bend, is protected inside a stainless steel tube, 2 mm in diameter. The active part of the probe, as shown in Fig. 8, has a characteristic size of 0.1 mm.

Signals are analyzed through an adjustable threshold that enables the signal to be transformed into a binary signal. Consequently, the local void fraction is a function of this threshold, which is adjusted and then held fixed during a traverse in order to obtain agreement between the profile average and a $\gamma$-ray measurement of the line void fraction. Experimental results for void profiles in Freon two-phase flow with phase change and in air-water flow are given by Galaup (1975).

## 5.3 Thermal Anemometers

It has been found that hot-wire or hot-film anemometry can be used in two-component two-phase flow or in one-component two-phase flow with phase change. In the first case–for example, air-water flow–it is possible to measure the local void fraction, the instantaneous velocity, and the turbulence intensity of the liquid phase. However, in the second case–for example, a steam-water flow–it has been so far impossible to obtain consistent results on liquid velocity measurements. Nevertheless, in both air-water and steam-water flow, the anemometer gives a signal

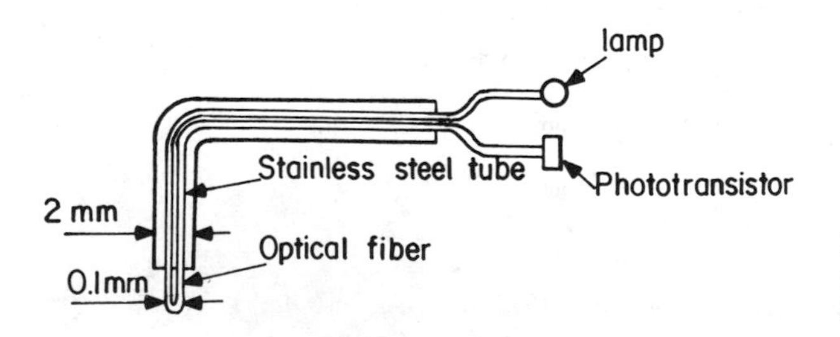


**Figure 7** U-shaped fiber optical sensor (Danel and Delhaye, 1971).

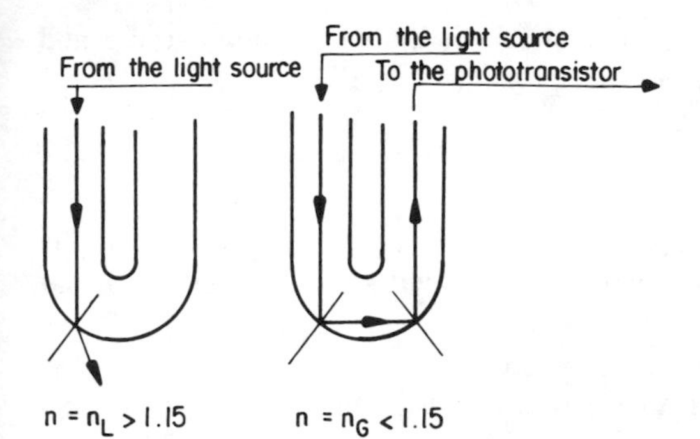


**Figure 8** Active part of the U-shaped fiber optical sensor (Danel and Delhaye, 1971).

characteristic of the flow pattern, although the signal is misshapen because of the interaction between the probe and the pierced interfaces (Bremhorst and Gilmore, 1976; Remke, 1978).

It is evident that if the gas and liquid signals could be separated, the turbulent structure of the liquid phase could be obtained. To achieve this separation, Delhaye (1969) and Galaup (1975) used the amplitude probability density function of the output signal (Fig. 9). In a first approximation, the local void fractions were calculated as the ratio of the hatched area to the total area. The separation line was set to ensure the identity between the averaged value of the local void fraction and the line void fraction measured by a $\gamma$-ray absorption technique. The liquid time-averaged velocity and the liquid turbulent intensity were calculated from the nonhatched area of the amplitude histogram (Fig. 9) and the calibration curve of the probe immersed in the liquid. The same method was used by Serizawa et al. (1975), Herringe and Davis (1976), and Remke (1976) for measuring the turbulent characteristics of air–water two-phase flow in a pipe.

A different signal processing was proposed by Resch (1975), Resch and Leutheusser (1972), and Resch et al. (1974) in a study of bubbly two-phase flow in a hydraulic jump. The analog signal from the anemometer is digitally analyzed according to a conditional sampling. Another technique was proposed by Jones (1973), who used a discriminator with a cutoff level depending on the local velocity.

For steam–water flow, Hsu et al. (1963) noted that the only reference temperature is the saturation temperature. If water velocity measurements are carried out, the probe temperature must not exceed saturation temperature by more than 5°C to avoid nucleate boiling on the sensor. Conversely, if only a high

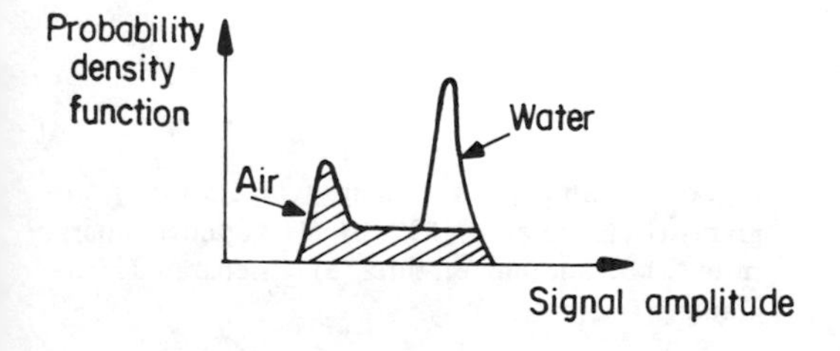


**Figure 9** Typical amplitude histogram of anemometer signal (Delhaye, 1969).

sensitivity to phase change is looked for, the superheat should range between 5 and 55°C, causing nucleate boiling to occur on the probe.

## 5.4 Phase-indicating Microthermocouple

Although the classical microthermocouple has contributed to a large extent to the understanding of the local structure of two-phase flow with change of phase, it has not provided any reliable statistical information on the distribution of the temperature between the liquid and the vapor phases.

The work done by Delhaye et al. (1973b) is based on the possibility of separating the liquid temperature from the steam temperature and of determining the local void fraction. An insulated 20-$\mu$m thermocouple was used both as a temperature sensor and as an electrical phase indicator by means of a Kohlrausch bridge that sensed the presence of the conducting liquid between the noninsulated junction and the ground. The phase signal was used to route the thermocouple signal to two separate subgroups of a multichannel analyzer. As a result, separate histograms of liquid and vapor temperatures can be obtained as shown in Fig. 10 for subcooled boiling.

## 5.5 Reliability of Local Measurements

Questions of reproducibility and the equivalence of one method compared with another always arise. It is encouraging that the results from three different

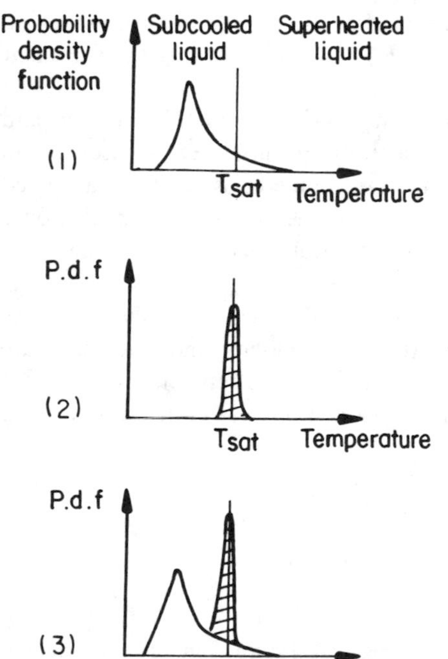


**Figure 10** Subcooled boiling temperature histograms (Delhaye et al., 1973b). (1) Liquid temperature. (2) Steam temperature. (3) Unseparated histogram.

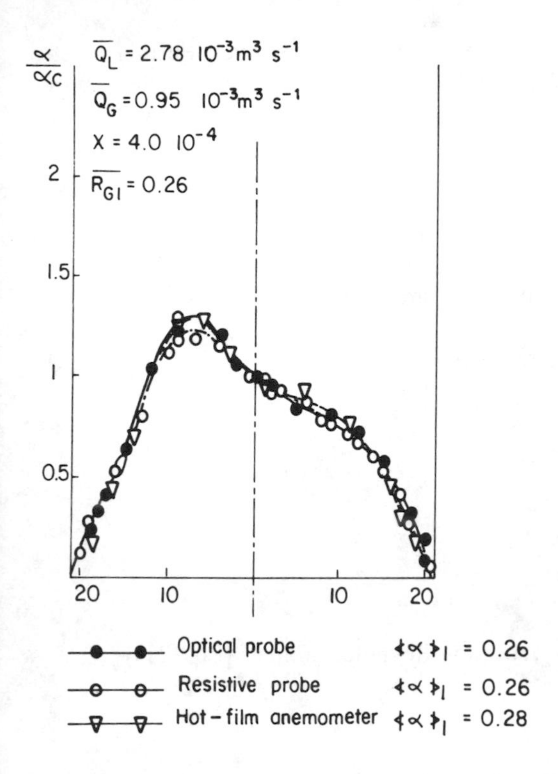


**Figure 11** Comparison of results of optical, resistive, and thermal probe techniques in an air-water flow (Galaup, 1975). $\bar{Q}$, volumetric flow rate; $x$, quality; $\overline{R_{G1}}$, time-averaged line void fraction measured with $\gamma$-rays; $\alpha$, local void fraction; $\alpha_c$, centerline local void fraction; and $\langle\alpha\rangle_1$, averaged void fraction measured on the profile.

instruments–optical probe, thermal anemometer, and resistive probe–all give comparable results as shown by Galaup (1975) (see Fig. 11). Nevertheless, more work is needed on the cross-correlation technique to obtain the gas velocity as well as the interface speed of displacement.

## 6 CONCLUSIONS

This review of measurement techniques in two-phase flow is far from exhaustive. Some methods were not presented because more work is needed to evaluate their reliability. Among others, we could mention the local isokinetic sampling (Schraub, 1969; Delhaye et al., 1973a), the subchannel isokinetic sampling (Bayoumi, 1976), or the Pitot tube technique (Delhaye et al., 1973a; Banerjee and Nguyen, 1977).

New techniques have been under development for a few years. Except for the use of drag bodies (Aya, 1975), which has been known for a long time in single-phase flow, most of these new techniques are highly sophisticated. Good examples are given by the neutron-scattering technique (Rousseau et al., 1978) and the nuclear magnetic resonance method, both used to measure the instantaneous area void fraction. Laser techniques were reviewed by Delhaye (1976) and seem to be promising for low or high void fraction two-phase flows (Ohba et al. 1976, 1977a, 1977b; Mahalingam et al., 1976).

Apart from the photon attenuation technique, which has been studied rather extensively, most two-phase flow measuring techniques need to be substantiated by a more rigorous analysis and more systematic experimental studies. Furthermore, it has been generally recognized that the validity of each technique must be assessed according to the existing flow pattern and the transient aspect of the flow.

## NOMENCLATURE

| | |
|---|---|
| $A$ | area |
| $B$ | magnetic induction |
| $C$ | flow coefficient [Eq. (20)] ; Chisholm parameter |
| $c$ | contrast |
| $D$ | diameter; Collins and Gacesa parameter |
| $d$ | distance; diameter |
| $e$ | thickness; emf |
| $F$ | orifice thermal expansion factor |
| $G$ | mass velocity |
| $H$ | differential water head |
| $I$ | intensity |
| $K$ | calibration constant |
| $M$ | mass flow rate |
| $N$ | number of photons [Eq. (7)] ; rotation Reynolds number [Eq. (21)] |
| $n$ | rotational frequency |
| $p$ | pressure |
| $Q$ | volumetric flow rate |
| $R$ | area fraction; tube radius; density ratio [Eq. (26)] |
| $S$ | cross-sectional area |
| $w$ | axial component of velocity vector |
| $X$ | Lockhart-Martinelli parameter |
| $x$ | quality |
| $Y$ | gas expansion factor |
| $y$ | Collins and Gacesa parameter [Eq. (25)] |
| $\alpha$ | local time fraction; flow coefficient |
| $\beta$ | ratio of orifice diameter to internal pipe diameter |
| $\epsilon$ | expansion factor |
| $\theta$ | time interval |
| $\mu$ | linear absorption coefficient |
| $\nu$ | kinematic viscosity |
| $\rho$ | density |
| $\phi_{L0}$ | Martinelli-Nelson parameter |

**Subscripts**

| | |
|---|---|
| $c$ | centerline |
| eq | equilibrium |
| $G$ | gas |
| $L$ | liquid |
| $P$ | wall |
| 0 | incident |

**Operators**

$\langle\ \rangle_1$ line-averaging operator
$\langle\ \rangle_2$ area-averaging operator

## REFERENCES


Alad'yev, I. T., Gavrilova, N. D., and Dodonov, L. D., Effective Electrical Conductivity of Two-Phase Gas–Liquid Metal Flows, *Heat Transfer Sov. Res.*, vol. 13, no. 4, p. 21, 1971.

ASME, *Fluid Meters–Their Theory and Application*, ASME Research Committee on Fluid Meters rept., 5th ed., ASME, New York, 1959.

Aya, I., A Model to Calculate Mass Flowrate and Other Quantities of Two-Phase Flow in a Pipe with a Densitometer, a Drag Disk, and a Turbine Meter, ORNL-TM-4759, 1975.

Banerjee, S. and Nguyen, D. M., Mass Velocity Measurement in Steam-Water Flow by Pitot Tubes, *AIChE J.*, vol. 23, no. 3, pp. 385–387, 1977.

Banerjee, S., Heidrick, T. R., and Saltvold, J. R., Measurement of Void Fraction and Mass Velocity in Transient Two-Phase Flow, in *Transient Two-Phase Flow, Proc. CSNI Specialists Meet., Aug. 3 and 4, 1976, Toronto*, eds. S. Banerjee and K. R. Weaver, vol. 2, pp. 789–832, AECL, 1978.

Bayoumi, M. A. A., Etude des Répartitions de Débit et d'Enthalpie dans les Sous-Canaux d'une Géométrie en Grappe des Réacteurs Nucléaires en Ecoulements Monophasique et Diphasique, Thèse de docteur ingénieur, Université Scientifique et Médicale de Grenoble el Institut National Polytechnique, Grenoble, 1976.

Benedict, R. P., *Fundamentals of Temperature, Pressure, and Flow Measurements*, Wiley, New York, 1969.

Bevir, M. K., Electromagnetic Flow Measurement, in *Techniques de Mesure dans les Ecoulements*, pp. 407–425, Eyrolles, Paris, 1974.

Bradshaw, P., *An Introduction to Turbulence and Its Measurement*, Pergamon, New York, 1971.

Bremhorst, K. and Gilmore, D. B., Response of Hot Wire Anemometer Probes to a Stream of Air Bubbles in a Water Flow, *J. Phys. E*, vol. 9, pp. 347–352, 1976.

Brockett, G. F. and Johnson, R. T., Single-Phase and Two-Phase Flow Measurement Techniques for Reactor Safety Studies, EPRI NP-195, 1976.

Charlety, P., Ebullition du Sodium en Convection Forcée, Thèse de docteur ingénieur, Faculté des Sciences, Université de Grenoble, 1971.

Chisholm, D., Two-Phase Flow through Sharp-edged Orifices, *J. Mech. Eng. Sci.*, vol. 19, no. 3, pp. 128–130, 1977.

Collins, D. B. and Gacesa, M., Measurement of Steam Quality in Two-Phase Upflow with Venturimeters and Orifice Plates, *J. Basic Eng.*, pp. 11–21, 1971.

Considine, D. M., ed., *Encyclopedia of Instrumentation and Control*, McGraw-Hill, New York, 1971.

Danel, F. and Delhaye, J. M., Sonde Optique pour Mesure du Taux de Présence Local en Ecoulement Diphasique, *Mesures-Régulation-Automatisme*, Aug.–Sept., pp. 99–101, 1971.

Delhaye, J. M., Hot-Film Anemometry, in *Two-Phase Flow Instrumentation*, eds. B. W. LeTourneau and A. E. Bergles, pp. 58–69, ASME, New York, 1969.

Delhaye, J. M., Two-Phase Flow Instrumentation and Laser Beams, in *The Accuracy of Flow Measurements by Laser Doppler Methods–Proceedings of the LDA Symposium, Copenhagen 1975*, eds. P. Buchhave, J. M. Delhaye, F. Durst, W. K. George, K. Refslund, and J. H. Whitelaw, Hemisphere, Washington, D.C., 1976.

Delhaye, J. M. and Achard, J. L., On the Use of Averaging Operators in Two-Phase Flow Modeling, in *Thermal and Hydraulic Aspects of Nuclear Reactor Safety*, vol. 1. *Light Water Reactors*, eds. O. C. Jones and S. G. Bankoff, pp. 289–332, ASME, New York, 1977.

Delhaye, J. M., Galaup, J. P., Réocreux, M., and Ricque, R., Métrologie des Ecoulements Diphasiques–Quelques Procédés, CEA-R-4457, 1973a.

Delhaye, J. M., Séméria, R., and Flamand, J. C., Void Fraction, Vapor and Liquid Temperatures: Local Measurements in Two-Phase Flow Using a Micro-thermocouple, *J. Heat Transfer*, vol. 95, no. 3, pp. 365–370, 1973b.

Del Tin, G. and Panella, B., Misura del Titolo di Una Miscela Bifase Mediante Venturimetro, *Energ. Elettr.*, no. 10, pp. 643–653, 1972.

Demagny, J. C., Mesures de Vitesses et de Débits par la Méthode Electromagnétique. Choix des Matériels Industriels, *Houille Blanche*, no. 5, pp. 361–367, 1976.

Doebelin, E. O., *Measurement Systems: Application and Design*, McGraw-Hill, New York, 1966.

Dowdell, R. B., ed., *Flow–Its Measurement and Control in Science and Industry*, vol. 1, Instrument Society of America, 1974.

du Bousquet, J. L., Etude du Mélange entre Deux Sous-canaux d'un Elément Combustible Nucléaire à Grappe, Thèse de docteur ingénieur, Faculté des Sciences, Université de Grenoble, 1969.

Fitremann, J. M., Théorie des Vélocimètres et Débitmètres Electromagnétiques en Ecoulement Diphasique, *C. R. Acad. Sci. Ser. C*, vol. 274, pp. 440–443, 1972a.

Fitremann, J. M., La Débitmétrie Electromagnétique Appliquée aux Emulsions, Société Hydrotechnique de France, XII° Journées de l'Hydraulique, Paris, question 4, rapport 6, 1972b.

Fouda, A. E., Two-Phase Flow Behaviour in Manifolds and Networks, Ph.D. thesis, Department of Chemical Engineering, University of Waterloo, Canada, 1975.

Fouda, A. E. and Rhodes, E., Total Mass Flow and Quality Measurement in Gas-Liquid Flow, in *Two-Phase Transport and Reactor Safety*, eds. T. N. Veziroğlu and S. Kakaç, vol. 4, pp. 1331–1355, Hemisphere, Washington, D.C., 1978.

Frank, R., Mazars, J., and Ricque, R., Determination of Mass Flowrate and Quality Using a Turbine Meter and a Venturi, in *Heat and Fluid Flow in Water Reactor Safety*, pp. 63–68, IME, London, 1977.

Galaup, J. P., Contribution à l'Etude des Méthodes de Mesure en Ecoulement Diphasique, Thèse de docteur ingénieur, Université Scientifique et Médicale de Grenoble et Institut National Polytechnique, Grenoble, 1975.

Gardner, R. P., Bean, R. H., and Ferrel, J. K., On the Gamma-Ray One-Shot-Collimator Measurement of Two-Phase Flow Void Fractions, *Nucl. Appl. Technol.*, vol. 8, pp. 88–94, 1970.

Hancox, W. T., Forrest, C. F., and Harms, A. A., Void Determination in Two-Phase Systems Employing Neutron Transmission, ASME paper 72-HT-2, 1972.

Harms, A. A. and Laratta, F. A., The Dynamic Bias in Radiation Interrogation of Two-Phase Flow, *Int. J. Heat Mass Transfer*, vol. 16, pp. 1459–1465, 1973.

Harris, D. M., Calibration of a Steam Quality Meter for Channel Power Measurement in the Prototype S.G.H.W. Reactor, Paper presented at European Two-Phase Heat Transfer Meeting, Bournemouth, United Kingdom, 1967.

Heinemann, J. B., Forced Convection Boiling Sodium Studies at Low Pressure, ANL-7100, pp. 189–194, 1965.

Heinemann, J. B., Marchaterre, J. F., and Mehta, S., Electromagnetic Flowmeters for Void Fraction Measurement in Two-Phase Metal Flow, *Rev. Sci. Instrum.*, vol. 34, no. 4, pp. 399–401, 1963.

Herringe, R. A. and Davis, M. R., Structural Development of Gas-Liquid Mixture Flows, *J. Fluid Mech.*, vol. 73, pt. 1, pp. 97–123, 1976.

Hewitt, G. F. and Lovegrove, P. C., Experimental Methods in Two-Phase Flow Studies, EPRI NP-118, 1976.

Hochreiter, H. M., Dimensionless Correlation of Coefficients of Turbine-Type Flowmeters, *Trans. ASME*, vol. 80, pp. 1363–1368, 1958.

Hori, M., Kobori, T., and Ouchi, Y., Method for Measuring Void Fraction by Electromagnetic Flowmeters, JAERI 1111, 1966.

Hsu, Y. Y., Simon, F. F., and Graham, R. W., Application of Hot-Wire Anemometry for Two-Phase Flow Measurements Such as Void Fraction and Slip Velocity, in *Multiphase Flow Symposium*, ed. N. J. Lipstein, pp. 26–34, ASME, New York, 1963.

James, R., Metering of Steam-Water Two-Phase Flow by Sharp-edged Orifices, *Proc. Inst. Mech. Eng.*, vol. 180, pt. 1, no. 23, pp. 549–566, 1965–1966.

Jones, O. C., Statistical Considerations in Heterogeneous, Two-Phase Flowing Systems, Ph.D. thesis, Rensselaer Polytechnic Institute, Troy, N.Y., 1973.

Jones, O. C. and Delhaye, J. M., Transient and Statistical Measurement Techniques for Two-Phase Flows: A Critical Review, *Int. J. Multiphase Flow*, vol. 3, pp. 89–116, 1976.

Jones, O. C. and Zuber, N., The Interrelation Between Void Fraction Fluctuations and Flow Patterns in Two-Phase Flow, *Int. J. Multiphase Flow*, vol. 2, pp. 273–306, 1975.

Kamath, P. S. and Lahey, R. T., A Turbine-Meter Evaluation Model for Two-Phase Transients (TEMPT), topical rept. prepared for EG & G Idaho Inc. by Department of Nuclear Engineering, Rensselaer Polytechnic Institute, Troy, N.Y., 1977.

Katys, G. P., *Continuous Measurement of Unsteady Flow*, Pergamon, New York, 1964.

Kremlevskii, P. P. and Dyudina, I. A., Measurements of Moist Vapor Discharge by Means of Gauging Diaphragms, *Meas. Tech.*, vol. 15, no. 5, pp. 741–744, 1972.

Lahey, R. T., Two-Phase Flow Phenomena in Nuclear Reactor Technology, Contract AT (49-24)-0301, Quarterly progress rept., March 1, 1977–May 31, 1977, prepared for U.S. NRC by Department of Nuclear Engineering, Rensselaer Polytechnic Institute, Troy, N.Y., 1977.

Laratta, F. A. and Harms, A. A., A Reduced Formula for the Dynamic Bias in Radiation Interrogation of Two-Phase Flow, *Int. J. Heat Mass Transfer*, vol. 17, p. 464, 1974.

Lecroart, H. and Porte, R., Electrical Probes for Study of Two-Phase Flow at High Velocity, *Int. Symp. on Two-Phase Systems, Haifa, Israel*, 1971.

Leonchik, B. I., Mayakin, V. P., and Lebedeva, P. D., Measurements in Dispersed Flows, AEC-TR-7600, 1976.

LeTourneau, B. W. and Bergles, A. E., eds., *Two-Phase Flow Instrumentation, Proc. 11th Natl. Heat Transfer Conf., Minneapolis, Aug. 3–6*, ASME, New York, 1969.

Levert, F. E. and Helminski, E., A Dual-Energy Method for Measuring Void Fractions in Flowing Mediums, *Nucl. Technol.*, vol. 19, pp. 58–60, 1973.

Lurie, H., Sodium Boiling Heat Transfer and Hydrodynamics, ANL-7100, pp. 549–571, 1965.

Mahalingam, R., Limaye, R. S., and Brink, J. A., Velocity Measurements in Two-Phase Bubble-Flow Regime with Laser-Doppler Anemometry, *AIChE J.*, vol. 22, no. 6, pp. 1152–1155, 1976.

Martin, R., Mesure du Taux de Vide à Haute-pression dans un Canal Chauffant, CEA-R-3781, 1969.

Martin, R., Measurements of the Local Void Fraction at High Pressure in a Heating Channel, *Nucl. Sci. Eng.*, vol. 48, pp. 125–138, 1972.

McMaster, W. H., Kerr Del Grande, N., Mallett, J. H., and Hubbell, J. H., Compilation of X-ray Cross Sections, Lawrence Radiation Laboratory rept. UCRL-50174, sec. 2, rev. 1, Calif., 1969.

Milliot, B., Lazarus, J., and Navarre, J. P., Mesure de la Fraction de Vide dans un Ecoulement Double-Phase NaK-Argon, EUR-3486f, 1967.

Moore, M. J., A Review of Instrumentation for Wet Steam, in *Two-Phase Steam Flow in Turbines and Separators*, eds. M. J. Moore and C. H. Sieverding, pp. 191–249, Hemisphere, Washington, D.C., 1976.

Murdock, J. W., Two-Phase Flow Measurements with Orifices, *J. Basic Eng.*, vol. 84, pp. 419–432, 1962.

NEL, Metering of Two-Phase Mixtures, Report of a meeting at NEL, Jan. 6, 1965, NEL-217, 1966.

Nyer, M., Etude des Phénomènes Thermiques et Hydrauliques Accompagnant une Excursion Rapide de Puissance sur un Canal Chauffant, CEA-R-3497, 1969.

Ohba, K., Kishimoto, I., and Ogasawara, M., Simultaneous Measurement of Local Liquid Velocity and Void Fraction in Bubbly Flows Using a Gas Laser, part I, Principle and Measuring Procedure, *Technol. Rep. Osaka Univ.*, vol. 26, no. 1328, pp. 547–556, 1976.

Ohba, K., Kishimoto, I., and Ogasawara, M., Simultaneous Measurement of Local Liquid Velocity and Void Fraction in Bubbly Flows Using a Gas Laser, Part II, Local Properties of Turbulent Bubbly Flow, *Technol. Rep. Osaka Univ.*, vol. 27, no. 1358, pp. 229–238, 1977a.

Ohba, K., Kishimoto, I., and Ogasawara, M., Simultaneous Measurement of Local Liquid Velocity and Void Fraction in Bubbly Flows Using a Gas Laser, Part III, Accuracy of Measurement, *Technol. Rep. Osaka Univ.*, vol. 27, no. 1383, pp. 475–483, 1977b.

Popper, G. F., *Proc. Boiling Water Reactor In-core Instrumentation Meet., Halden, May 15–18, 1961*, ed. R. W. Bowring, OECD Halden Reactor Project, HPR 16, 1961.

Remke, K., Ein Beitrag zur Anwendung der Heissfilmanemometrie auf die Turbulenzmessung in Gas-Flüssigkeitsströmungen, *ZAMM*, vol. 56, pp. 480–483, 1976.

Remke, K., Some Remarks on the Response of Hot-Wire and Hot-Film Probes to Passage through an Air-Water Interface, *J. Phys. E*, vol. 11, pp. 94–96, 1978.

Réocreux, M., Contribution à l'Etude des Débits Critiques en Ecoulement Diphasique Eau-Vapeur, Thèse de doctorat ès sciences, Université Scientifique et Médicale de Grenoble, 1974.

Resch, F. J., Phase Separation in Turbulent Two-Phase Flow, in *Turbulence in Liquids*, eds. G. K. Patterson and J. L. Zakin, *Proc. 3d Symp. Turbulence in Liquids, Sept. 1973*, pp. 243–249, Department of Chemical Engineering, University of Missouri-Rolla, 1975.

Resch, F. J. and Leutheusser, J. H., Le Ressaut Hydraulique: Mesures de Turbulence dans la Région Diphasique, *Houille Blanche*, no. 4, pp. 279–293, 1972.

Resch, F. J., Leutheusser, H. J., and Alemu, S., Bubbly Two-Phase Flow in Hydraulic Jump, *ASCE J. Hydraulics Div.*, vol. 100, no. HY1, proc. paper 10297, pp. 137–149, 1974.

Richards, B. E., ed., *Measurement of Unsteady Fluid Dynamic Phenomena*, Hemisphere, Washington, D.C., 1977.

Rouhani, Z., Application of the Turbine Type Flowmeters in the Measurement of Steam Quality and Void, *Symp. on In-core Instrumentation, Oslo, June 1964*, OECD Halden Reactor Project, USAEC-CONF-640607, 1964.

Rousseau, J. C., Czerny, J., and Riegel, B., Void Fraction Measurements during Blowdown by Neutron Absorption or Scattering Methods, in *Transient Two-Phase Flow, Proc. CSNI Specialists Meet., Aug. 3 and 4, 1976, Toronto*, eds. S. Banerjee and K. R. Weaver, vol. 2, pp. 890–904, AECL, 1978.

Schraub, F. A., Isokinetic Probes and Other Two-Phase Sampling Devices: A Survey, in *Two-Phase Flow Instrumentation*, eds. B. W. LeTourneau and A. E. Bergles, pp. 47–57, ASME, New York, 1969.

Serizawa, A., Kataoka, I., and Michiyoski, I., Turbulence Structure of Air-Water Bubbly Flow, I, Measuring Techniques, *Int. J. Multiphase Flow*, vol. 2, no. 3, pp. 221–223, 1975.

Shercliff, J. A., *The Theory of Electromagnetic Flow Measurement*, Cambridge University Press, Cambridge, England, 1962.

Smith, L. T., Murdock, J. W., and Applebaum, R. S., An Evaluation of Existing Two-Phase Flow Correlations for Use with ASME Sharp Edge Metering Orifices, *J. Eng. Power*, vol. 99, no. 3, pp. 343–347, 1977.

Solésio, J. N., Mesure de l'Epaisseur Locale Instantanée d'un Film Liquide Ruisselant sur une Paroi. Examen des Méthodes Existantes. Mise en Oeuvre d'une Technique Basée sur l'Absorption de Rayons X, CEA-R-4925, 1978.

Storm, E. and Israel, H. I., Photon Cross Sections from 1 keV to 100 MeV for Elements $Z = 1$ to $Z = 100$, *Nucl. Data Tables*, A7, pp. 565–681, 1970.

Stukenbroeker, G. L., Bonilla, C. F., and Peterson, R. W., The Use of Lead as a Shielding Material, *Nucl. Eng. Des.*, vol. 13, no. 1, pp. 3–145, 1970.

Thatcher, G., Electromagnetic Flowmeters for Liquid Metals, in *Modern Developments in Flow Measurement*, ed. C. G. Clayton, pp. 359–380, Peregrinus, London, 1972.

Treenhaus, R., Modern Developments and New Applications of Magnetic Flowmeters, *Modern Developments in Flow Measurement*, ed. C. G. Clayton, pp. 347–358, Peregrinus, London, 1972.

Turner, G. E., Liquid Metal Flow Measurement (Sodium). State-of-the-art study, LMEC-Memo-68-9, 1968.

Zanker, K. J., The Influence of Air on the Performance of Differential Pressure Water Flowmeters, Euromech Colloquium 7, Grenoble, 1968.

CHAPTER

# FIVE

# LOCAL INSTANTANEOUS EQUATIONS

**J. M. Delhaye**

In single-phase flow, local balance laws at point **x** are expressed in terms of partial differential equations if point **x** does not belong to a surface of discontinuity. If point **x** belongs to a surface of discontinuity, local balance laws are then formulated in terms of jump conditions that relate the values of the flow parameters on both sides of the surface of discontinuity (Truesdell and Toupin, 1960).

In two-phase flow interfaces can be considered as surfaces of discontinuity. Consequently, the balance laws for each phase are expressed in terms of partial differential equations; on the interface balance laws are formulated in terms of jump conditions.

Local instantaneous equations constitute the rational basis for almost all two-phase flow modeling procedures. They are used directly, for example, in the study of bubble dynamics or film flows, or in an averaged form, for example, in the study of pipe flows.

Jump conditions constitute a characteristic feature of two-phase flow analysis and provide relations between the phase-interaction terms that appear in the averaged equations (Delhaye, 1974; Ishii, 1975). Two kinds of jump conditions can be derived: primary and secondary. Primary jump conditions are directly derived from the integral laws written for the following quantities: mass, linear momentum, angular momentum, total energy, and entropy. Secondary jump conditions are established by combining the primary ones. They are written for the following quantities: mechanical energy, internal energy, enthalpy, and entropy.

The interfacial entropy source, obtained from entropy jump conditions, leads to interfacial boundary conditions. If the interface transfers are assumed reversible, it can thus be demonstrated that there is neither slip nor temperature jump across the interface and that a certain relation exists between the free enthalpies of each phase.

The derivation of the local instantaneous equations starts with the integral balance laws written for a fixed control volume containing both phases. These integral laws are then transformed by using the Leibniz rule and the Gauss theorems to obtain a sum of two volume integrals and a surface integral. The volume integrals lead to the local instantaneous partial differential equations valid in each phase, whereas the surface integral furnishes the local instantaneous jump conditions valid on the interface only.

# 1 MATHEMATICAL TOOLS

## 1.1 Speed of Displacement of a Geometric Surface (Truesdell and Toupin, 1960)

Consider a set of geometric surfaces defined by

$$\mathbf{r} = \mathbf{r}(u, v, t) \tag{1}$$

where $u$ and $v$ are the coordinates of a point on this surface. The velocity of the surface point $(u, v)$ is defined by

$$\mathbf{v}_i \triangleq \left(\frac{\partial \mathbf{r}}{\partial t}\right)_{u,v\ \mathrm{const}} \tag{2}$$

The surface equation may also be expressed by

$$f(x, y, z, t) = 0 \tag{3}$$

The normal unit vector $\mathbf{n}$ is related to the surface equation by

$$\mathbf{n} = \frac{\nabla f}{|\nabla f|} \tag{4}$$

The speed of displacement of the surface is defined as the scalar product of $\mathbf{v}_i$ and $\mathbf{n}$. We then deduce

$$\mathbf{v}_i \cdot \mathbf{n} = -\frac{\partial f/\partial t}{|\nabla f|} \tag{5}$$

Equation (5) shows that the projection of $\mathbf{v}_i$ on the normal vector depends only on the surface equation expressed by Eq. (3). Equation (5) is unique whereas Eq. (2) depends on the choice of parameters $u$ and $v$.

## 1.2 Leibniz Rule (Truesdell and Toupin, 1960)

Let us consider a geometric volume $\mathcal{V}(t)$ moving in space (Fig. 1). This volume is bounded by a closed surface $\mathcal{A}(t)$. At a given point belonging to this surface $\mathcal{A}(t)$, let

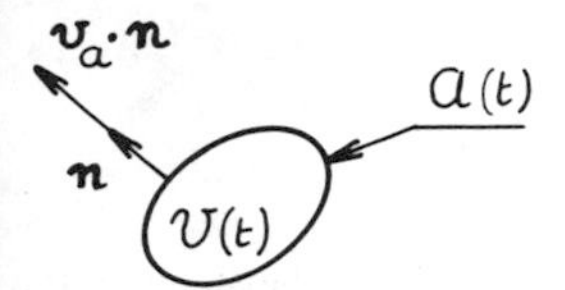


**Figure 1** Geometric volume used for the statement of the Leibniz rule.

**n** be the unit normal vector, outwardly directed. The speed of displacement of the surface at that point is denoted $\mathbf{v}_{\mathfrak{a}} \cdot \mathbf{n}$. The Leibniz rule enables the time rate of change of a volume integral to be transformed into the sum of a volume integral and a surface integral:

$$\frac{d}{dt}\int_{\mathfrak{V}(t)} f(x, y, z, t)\, d\mathfrak{V} = \int_{\mathfrak{V}(t)} \frac{\partial f}{\partial t}\, d\mathfrak{V} + \oint_{\mathfrak{a}(t)} f\mathbf{v} \cdot \mathbf{n}\, d\mathfrak{a} \tag{6}$$

Note that volume $\mathfrak{V}(t)$ is a geometric volume and not necessarily a material volume.

### 1.3 Gauss Theorems

Let us consider a geometric volume (Fig. 2). This volume, bounded by surface $\mathfrak{a}$, is material or not, moving or not. At a given point of surface $\mathfrak{a}$, the unit normal vector **n** is outwardly directly. Let **B** and **M** be some vector and tensor fields. The Gauss theorems enable a surface integral to be transformed into a volume integral according to the following relations:

$$\oint_{\mathfrak{a}} \mathbf{n} \cdot \mathbf{B}\, d\mathfrak{a} = \int_{\mathfrak{V}} \nabla \cdot \mathbf{B}\, d\mathfrak{V} \tag{7}$$

$$\oint_{\mathfrak{a}} \mathbf{n} \cdot \mathbf{M}\, d\mathfrak{a} = \int_{\mathfrak{V}} \nabla \cdot \mathbf{M}\, d\mathfrak{V} \tag{8}$$

## 2 INTEGRAL BALANCES

Let us consider a fixed control volume $\mathfrak{V}$ cut by an interface $\mathfrak{a}_i(t)$. This interface divides the control volume $\mathfrak{V}$ into subvolumes $\mathfrak{V}_1(t)$ and $\mathfrak{V}_2(t)$, respectively bounded by surfaces $\mathfrak{a}_1(t)$ and $\mathfrak{a}_i(t)$ and $\mathfrak{a}_2(t)$ and $\mathfrak{a}_i(t)$ (Fig. 3).

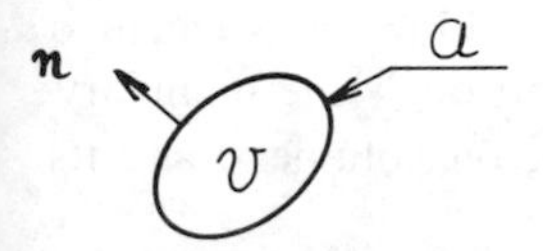


**Figure 2** Geometric volume used for the statement of the Gauss theorems.

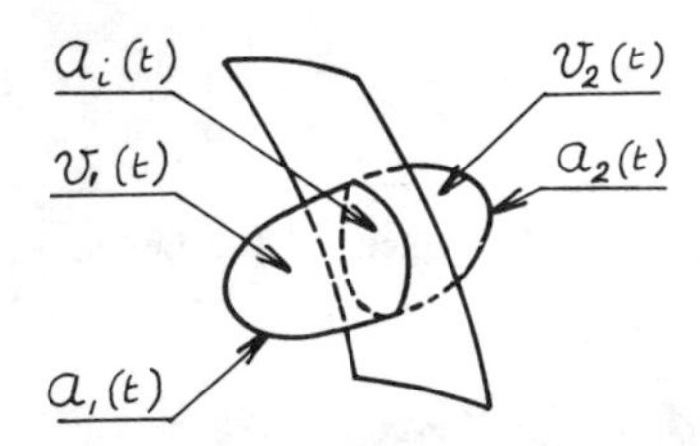


**Figure 3** Two-phase control volume used for the statements of the integral balances of mass, momenta, energy, and entropy.

## 2.1 Mass

The time rate of change of mass in the control volume $\mathcal{V}$ is equal to the net influx of mass into $\mathcal{V}$ through the boundary surface $\mathcal{A}$,

$$\frac{d}{dt}\int_{\mathcal{V}_1(t)} \rho_1 \, d\mathcal{V} + \frac{d}{dt}\int_{\mathcal{V}_2(t)} \rho_2 \, d\mathcal{V} = -\int_{\mathcal{A}_1(t)} \rho_1 \mathbf{v}_1 \cdot \mathbf{n}_1 \, d\mathcal{A} - \int_{\mathcal{A}_2(t)} \rho_2 \mathbf{v}_2 \cdot \mathbf{n}_2 \, d\mathcal{A} \tag{9}$$

where $\rho_k$ and $\mathbf{v}_k$ are the density and the velocity vector of phase $k (k = 1, 2)$.

## 2.2 Linear Momentum

The time rate of change of linear momentum in the control volume $\mathcal{V}$ is equal to the sum of (1) the net influx of momentum into $\mathcal{V}$ through the boundary surface $\mathcal{A}$ and (2) the resultant of the external forces acting on volume $\mathcal{V}$ and on its boundary $\mathcal{A}$. These forces are composed of volume forces, that is, gravity forces, and surface forces, that is, a stress tensor. For simplicity, we do not introduce surface tension here. As a result, the integral linear momentum balance reads

$$\begin{aligned}\frac{d}{dt}\int_{\mathcal{V}_1(t)} \rho_1 \mathbf{v}_1 \, d\mathcal{V} + \frac{d}{dt}\int_{\mathcal{V}_2(t)} \rho_2 \mathbf{v}_2 \, d\mathcal{V} &= -\int_{\mathcal{A}_1(t)} \rho_1 \mathbf{v}_1 (\mathbf{v}_1 \cdot \mathbf{n}_1) \, d\mathcal{A} \\ &- \int_{\mathcal{A}_2(t)} \rho_2 \mathbf{v}_2 (\mathbf{v}_2 \cdot \mathbf{n}_2) \, d\mathcal{A} + \int_{\mathcal{V}_1(t)} \rho_1 \mathbf{F} \, d\mathcal{V} + \int_{\mathcal{V}_2(t)} \rho_2 \mathbf{F} \, d\mathcal{V} \\ &+ \int_{\mathcal{A}_1(t)} \mathbf{n}_1 \cdot \mathbf{T}_1 \, d\mathcal{A} + \int_{\mathcal{A}_2(t)} \mathbf{n}_2 \cdot \mathbf{T}_2 \, d\mathcal{A}\end{aligned} \tag{10}$$

where $\mathbf{F}$ is the external force per unit of mass and $\mathbf{T}$ is the stress tensor.

## 2.3 Angular Momentum

The time rate of change of angular momentum in the control volume $\mathcal{V}$ is equal to the sum of (1) the net influx of angular momentum into $\mathcal{V}$ through the boundary surface $\mathcal{A}$ and (2) the resultant of the external torques acting on volume $\mathcal{V}$ and its boundary $\mathcal{A}$:

$$\frac{d}{dt}\int_{\mathcal{V}_1(t)} \mathbf{r}\times\rho_1\mathbf{v}_1\,d\mathcal{V} + \frac{d}{dt}\int_{\mathcal{V}_2(t)} \mathbf{r}\times\rho_2\mathbf{v}_2\,d\mathcal{V} = -\int_{\mathcal{A}_1(t)} \mathbf{r}\times\rho_1\mathbf{v}_1(\mathbf{v}_1\cdot\mathbf{n}_1)\,d\mathcal{A}$$

$$-\int_{\mathcal{A}_2(t)} \mathbf{r}\times\rho_2\mathbf{v}_2(\mathbf{v}_2\cdot\mathbf{n}_2)\,d\mathcal{A} + \int_{\mathcal{V}_1(t)} \mathbf{r}\times\rho_1\mathbf{F}\,d\mathcal{V} + \int_{\mathcal{V}_2(t)} \mathbf{r}\times\rho_2\mathbf{F}\,d\mathcal{V}$$

$$+\int_{\mathcal{A}_1(t)} \mathbf{r}\times(\mathbf{n}_1\cdot\mathbf{T}_1)\,d\mathcal{A} + \int_{\mathcal{A}_2(t)} \mathbf{r}\times(\mathbf{n}_2\cdot\mathbf{T}_2)\,d\mathcal{A} \qquad (11)$$

where $\mathbf{r}$ is the position vector.

## 2.4 Total Energy

The time rate of change of total energy (kinetic energy and internal energy) in the control volume $\mathcal{V}$ is equal to the sum of (1) the net influx of total energy into $\mathcal{V}$ through the boundary surface $\mathcal{A}$, (2) the power of the external forces acting on volume $\mathcal{V}$ and on its boundary $\mathcal{A}$, and (3) the heat flux entering volume $\mathcal{V}$ through $\mathcal{A}$:

$$\frac{d}{dt}\int_{\mathcal{V}_1(t)} \rho_1\left(\frac{1}{2}v_1^2 + u_1\right)d\mathcal{V} + \frac{d}{dt}\int_{\mathcal{V}_2(t)} \rho_2\left(\frac{1}{2}v_2^2 + u_2\right)d\mathcal{V}$$

$$= -\int_{\mathcal{A}_1(t)} \rho_1\left(\frac{1}{2}v_1^2 + u_1\right)\mathbf{v}_1\cdot\mathbf{n}_1\,d\mathcal{A} - \int_{\mathcal{A}_2(t)} \rho_2\left(\frac{1}{2}v_2^2 + u_2\right)\mathbf{v}_2\cdot\mathbf{n}_2\,d\mathcal{A}$$

$$+\int_{\mathcal{V}_1(t)} \rho_1\mathbf{F}\cdot\mathbf{v}_1\,d\mathcal{V} + \int_{\mathcal{V}_2(t)} \rho_2\mathbf{F}\cdot\mathbf{v}_2\,d\mathcal{V} + \int_{\mathcal{A}_1(t)} (\mathbf{n}_1\cdot\mathbf{T}_1)\cdot\mathbf{v}_1\,d\mathcal{A}$$

$$+\int_{\mathcal{A}_2(t)} (\mathbf{n}_2\cdot\mathbf{T}_2)\cdot\mathbf{v}_2\,d\mathcal{A} - \int_{\mathcal{A}_1(t)} \mathbf{q}_1\cdot\mathbf{n}_1\,d\mathcal{A} - \int_{\mathcal{A}_2(t)} \mathbf{q}_2\cdot\mathbf{n}_2\,d\mathcal{A} \qquad (12)$$

where $u$ is the internal energy per unit of mass and $\mathbf{q}$ is the heat flux.

## 2.5 Entropy

The time rate of change of entropy in the control volume $\mathcal{V}$ is equal to the sum of (1) the net influx of entropy into $\mathcal{V}$ through the boundary surface $\mathcal{A}$ owing to the mass flow, (2) the net influx of entropy into $\mathcal{V}$ owing to conduction through $\mathcal{A}$, and (3) the entropy source within volume $\mathcal{A}$:

$$\frac{d}{dt}\int_{\mathcal{V}_1(t)} \rho_1 s_1 \, d\mathcal{V} + \frac{d}{dt}\int_{\mathcal{V}_2(t)} \rho_2 s_2 \, d\mathcal{V} + \int_{\mathcal{A}_1(t)} \rho_1 s_1 \mathbf{v}_1 \cdot \mathbf{n}_1 \, d\mathcal{A}$$

$$+ \int_{\mathcal{A}_2(t)} \rho_2 s_2 \mathbf{v}_2 \cdot \mathbf{n}_2 \, d \; + \int_{\mathcal{A}_1(t)} \frac{1}{T_1}\mathbf{q}_1 \cdot \mathbf{n}_1 \, d\mathcal{A} + \int_{\mathcal{A}_2(t)} \frac{1}{T_2}\mathbf{q}_2 \cdot \mathbf{n}_2 \, d\mathcal{A}$$

$$= \int_{\mathcal{V}_1(t)} \Delta_1 \, d\mathcal{V} + \int_{\mathcal{V}_2(t)} \Delta_2 \, d\mathcal{V} + \int_{\mathcal{A}_i(t)} \Delta_i \, d\mathcal{A} \geqslant 0 \tag{13}$$

where the equality occurs when the evolution is reversible. In Eq. (13), $s_k$ is the entropy per unit mass, $T_k$ is the absolute temperature, $\Delta_k$ is the local entropy source per unit of volume and per unit of time within each phase, and $\Delta_i$ is the local entropy source per unit area of interface and per unit of time. Since Eq. (13) is verified, whatever $\mathcal{V}_1(t)$, $\mathcal{V}_2(t)$, and $\mathcal{A}_i(t)$ are, we then deduce that

$$\Delta_k \geqslant 0 \quad (k = 1, 2) \quad \Delta_i \geqslant 0 \tag{14}$$

## 2.6 Generalized Integral Balance

The integral balances of Eqs. (9)-(13) can be rewritten in the following condensed form:

$$\sum_{k=1,2} \frac{d}{dt}\int_{\mathcal{V}_k(t)} \rho_k \psi_k \, d\mathcal{V} = - \sum_{k=1,2} \int_{\mathcal{A}_k(t)} \rho_k \psi_k (\mathbf{v}_k \cdot \mathbf{n}_k) \, d\mathcal{A}$$

$$+ \sum_{k=1,2} \int_{\mathcal{V}_k(t)} \rho_k \phi_k \, d\mathcal{V} - \sum_{k=1,2} \int_{\mathcal{A}_k(t)} \mathbf{n}_k \cdot \mathbf{J}_k \, d\mathcal{A} + \int_{\mathcal{A}_i(t)} \phi_i \, d\mathcal{A} \tag{15}$$

For each balance law, the values of $\psi_k$, $\mathbf{J}_k$, and $\phi_k$ are given in Table 1.

**Table 1 Definitions of symbols used in the generalized integral balance [Eq. (15)]**

| Balance | $\psi_k$ | $\mathbf{J}_k$ | $\phi_k$ | $\phi_i$ |
|---|---|---|---|---|
| Mass | 1 | 0 | 0 | 0 |
| Linear momentum | $\mathbf{v}_k$ | $-\mathbf{T}_k$ | $\mathbf{F}$ | 0 |
| Angular momentum | $\mathbf{r} \times \mathbf{v}_k$ | $-\mathbf{T}_k \cdot \mathbf{R}^*$ | $\mathbf{r} \times \mathbf{F}$ | 0 |
| Total energy | $u_k + \frac{1}{2}v_k^2$ | $\mathbf{q}_k - \mathbf{T}_k \cdot \mathbf{v}_k$ | $\mathbf{F} \cdot \mathbf{v}_k$ | 0 |
| Entropy | $s_k$ | $\frac{1}{T_k}\mathbf{q}_k$ | $\frac{1}{\rho_k}\Delta_k$ | $\Delta_i$ |

*R is the antisymmetric tensor corresponding to vector r (Aris, 1962).

## 3 TRANSFORMATION OF THE INTEGRAL BALANCE

The integral balance law [Eq. (15)] is transformed by using the Leibniz rule [Eq. (6)] and the Gauss theorems [Eqs. (7) and (8)] into the following form:

$$\sum_{k=1,2} \int_{\mathcal{V}_k(t)} \left[ \frac{\partial}{\partial t} \rho_k \psi_k + \nabla \cdot (\rho_k \psi_k \mathbf{v}_k) + \nabla \cdot \mathbf{J}_k - \rho_k \phi_k \right] d\mathcal{V}$$
$$- \int_{\mathfrak{a}_i(t)} \left[ \sum_{k=1,2} \dot{m}_k \psi_k + \mathbf{n}_k \cdot \mathbf{J}_k + \phi_i \right] d\mathfrak{a} = 0 \tag{16}$$

with
$$\dot{m}_k \triangleq \rho_k (\mathbf{v}_k - \mathbf{v}_i) \cdot \mathbf{n}_k \tag{17}$$

Equation (16) has to be satisfied, whatever $\mathcal{V}_k(t)$ and $\mathfrak{a}_i(t)$ are. We thus deduce (1) the local instantaneous phase equations:

$$\frac{\partial}{\partial t} \rho_k \psi_k + \nabla \cdot (\rho_k \psi_k \mathbf{v}_k) + \nabla \cdot \mathbf{J}_k - \rho_k \phi_k = 0 \tag{18}$$

and (2) the local instantaneous jump condition:

$$\sum_{k=1,2} (\dot{m}_k \psi_k + \mathbf{n}_k \cdot \mathbf{J}_k + \phi_i) = 0 \tag{19}$$

For each local instantaneous balance law, the values of $\psi_k$, $\mathbf{J}_k$, and $\phi_k$ are given in Table 1.

## 4 PHASE EQUATIONS

### 4.1 Primary Equations

Mass:

$$\frac{\partial \rho_k}{\partial t} + \nabla \cdot (\rho_k \mathbf{v}_k) = 0 \tag{20}$$

Linear momentum:

$$\frac{\partial \rho_k \mathbf{v}_k}{\partial t} + \nabla \cdot (\rho_k \mathbf{v}_k \mathbf{v}_k) - \rho_k \mathbf{F} - \nabla \cdot \mathbf{T}_k = 0 \tag{21}$$

Angular momentum: This balance law does not supply any new local instantaneous equation; it simplifies to:

$$\mathbf{T}_k = \mathbf{T}_k^t \tag{22}$$

where $\mathbf{T}_k^t$ is the transposed form of tensor $\mathbf{T}_k$. The stress tensor is thus symmetric.

Total energy:

$$\frac{\partial}{\partial t}\left[\rho_k\left(\frac{1}{2}v_k^2+u_k\right)\right]+\nabla\cdot\left[\rho_k\left(\frac{1}{2}v_k^2+u_k\right)\mathbf{v}_k\right]$$
$$-\rho_k\mathbf{F}\cdot\mathbf{v}_k-\nabla\cdot(\mathbf{T}_k\cdot\mathbf{v}_k)+\nabla\cdot\mathbf{q}_k=0 \tag{23}$$

Entropy inequality:

$$\frac{\partial}{\partial t}(\rho_k s_k)+\nabla\cdot(\rho_k s_k\mathbf{v}_k)+\nabla\cdot\frac{\mathbf{q}_k}{T_k}=\Delta_k\geqslant 0 \tag{24}$$

## 4.2 Secondary Equations

Mechanical energy: Multiplying the momentum equation, [Eq. (21)] by $\mathbf{v}_k$, we obtain

$$\frac{\partial}{\partial t}\left(\frac{1}{2}\rho_k v_k^2\right)+\nabla\cdot\left(\frac{1}{2}\rho_k v_k^2\mathbf{v}_k\right)-\rho_k\mathbf{F}\cdot\mathbf{v}_k$$
$$-\nabla\cdot(\mathbf{T}_k\cdot\mathbf{v}_k)+\mathbf{T}_k:\nabla\mathbf{v}_k=0 \tag{25}$$

Internal energy: Subtracting the mechanical energy equation [Eq. (25)] from the total energy equation [Eq. (23)], we obtain

$$\frac{\partial}{\partial t}(\rho_k u_k)+\nabla\cdot(\rho_k u_k\mathbf{v}_k)+\nabla\cdot\mathbf{q}_k-\mathbf{T}_k:\nabla\mathbf{v}_k=0 \tag{26}$$

Enthalpy: We now introduce the enthalpy per unit of mass $i_k$ and the deviatoric stress tensor $\tau_k$, which gives

$$u_k=i_k-\frac{p_k}{\rho_k} \tag{27}$$

$$\mathbf{T}_k=-p_k\mathbf{U}+\tau_k \tag{28}$$

where $p_k$ is the pressure within phase $k$, and $\mathbf{U}$ is the unit tensor. Equation (26) then reads

$$\frac{\partial}{\partial t}(\rho_k i_k-p_k)+\nabla\cdot(\rho_k i_k\mathbf{v}_k)-\mathbf{v}_k\cdot\nabla p_k+\nabla\cdot\mathbf{q}_k-\tau_k:\nabla\mathbf{v}_k=0 \tag{29}$$

Entropy: First we recall the fundamental thermodynamic equation (Callen, 1960)

$$u_k=u_k(s_k,\rho_k) \tag{30}$$

and the definitions of pressure and temperature

$$p_k\triangleq\rho_k^2\left(\frac{\partial u_k}{\partial\rho_k}\right)_{s_k} \tag{31}$$

$$T_k\triangleq\left(\frac{\partial u_k}{\partial s_k}\right)_{\rho_k} \tag{32}$$

Equations (30)–(32) lead directly to the Gibbs equations

$$\frac{du_k}{dt} = T_k \frac{ds_k}{dt} - p_k \frac{d}{dt}\left(\frac{1}{\rho_k}\right) \tag{33}$$

Equation (33) is used to transform the internal energy equation [Eq. (26)] into an entropy equation. If we take into account the mass balance [Eq. (20)] and the definition of the Lagrangian derivative,

$$\frac{d}{dt} \triangleq \frac{\partial}{\partial t} + \mathbf{v}_k \cdot \nabla \tag{34}$$

we can transform the internal energy equation [Eq. (26)] into

$$\rho_k \frac{du_k}{dt} + \nabla \cdot \mathbf{q}_k - \mathbf{T}_k : \nabla \mathbf{v}_k = 0$$

or, with Eq. (28),

$$\rho_k \frac{du_k}{dt} + \nabla \cdot \mathbf{q}_k + p_k \nabla \cdot \mathbf{v}_k - \boldsymbol{\tau}_k : \nabla \mathbf{v}_k = 0 \tag{35}$$

On the other hand, the Gibbs equation can be rewritten by using Eqs. (20) and (34)

$$\frac{du_k}{dt} = T_k \frac{ds_k}{dt} - \frac{p_k}{\rho_k} \nabla \cdot \mathbf{v}_k \tag{36}$$

Combining Eqs. (35) and (36), we obtain

$$\rho_k \frac{ds_k}{dt} + \nabla \cdot \frac{\mathbf{q}_k}{T_k} - \mathbf{q}_k \cdot \nabla \frac{1}{T_k} - \frac{1}{T_k} \boldsymbol{\tau}_k : \nabla \mathbf{v}_k = 0 \tag{37}$$

## 4.3 Entropy Source

If the mass balance [Eq. (20)] and the Lagrangian derivative definition [Eq. (34)] are used, the entropy inequality of Eq. (24) becomes

$$\rho_k \frac{ds_k}{dt} + \nabla \cdot \frac{\mathbf{q}_k}{T_k} = \Delta_k \geqslant 0 \tag{38}$$

By comparing Eqs. (37) and (38), we obtain the expression for the entropy source:

$$\Delta_k = \mathbf{q}_k \cdot \nabla \frac{1}{T_k} + \frac{1}{T_k} \boldsymbol{\tau}_k : \nabla \mathbf{v}_k \geqslant 0 \tag{39}$$

# 5 PRIMARY JUMP CONDITIONS

## 5.1 Mass

$$\rho_1(\mathbf{v}_1 - \mathbf{v}_i) \cdot \mathbf{n}_1 + \rho_2(\mathbf{v}_2 - \mathbf{v}_i) \cdot \mathbf{n}_2 = 0 \tag{40}$$

or

$$\dot{m}_1 + \dot{m}_2 = 0 \tag{41}$$

with
$$\dot{m}_k \triangleq \rho_k(\mathbf{v}_k - \mathbf{v}_i) \cdot \mathbf{n}_k \tag{42}$$

If there is no mass transfer, we have

$$\dot{m}_1 \equiv \dot{m}_2 \equiv 0 \tag{43}$$

and consequently,

$$(\mathbf{v}_1 - \mathbf{v}_i) \cdot \mathbf{n}_1 = 0 \tag{44}$$

$$(\mathbf{v}_2 - \mathbf{v}_i) \cdot \mathbf{n}_2 = 0 \tag{45}$$

Eliminating the interface displacement velocity, we obtain

$$(\mathbf{v}_1 - \mathbf{v}_2) \cdot \mathbf{n}_1 = 0 \tag{46}$$

If $\mathbf{v}_1$ and $\mathbf{v}_2$ are resolved into their normal and tangential components,

$$\mathbf{v}_1 = (\mathbf{v}_1 \cdot \mathbf{n}_1)\mathbf{n}_1 + \mathbf{v}_1^t \tag{47}$$

$$\mathbf{v}_2 = (\mathbf{v}_2 \cdot \mathbf{n}_2)\mathbf{n}_2 + \mathbf{v}_2^t \tag{48}$$

If we admit no slip of one phase on the other at the interface (see Sec. 8), we have

$$\mathbf{v}_1^t \equiv \mathbf{v}_2^t \tag{49}$$

which, taking into account Eqs. (46)-(49), gives

$$\mathbf{v}_1 \equiv \mathbf{v}_2 \tag{50}$$

## 5.2 Linear Momentum

$$\dot{m}_1\mathbf{v}_1 + \dot{m}_2\mathbf{v}_2 - \mathbf{n}_1 \cdot \mathbf{T}_1 - \mathbf{n}_2 \cdot \mathbf{T}_2 = 0 \tag{51}$$

or, by taking into account the mass balance of Eq. (43),

$$\dot{m}_1(\mathbf{v}_1 - \mathbf{v}_2) - \mathbf{n}_1 \cdot \mathbf{T}_1 - \mathbf{n}_2 \cdot \mathbf{T}_2 = 0 \tag{52}$$

Inviscid phases: The stress tensor simplifies to the pressure term only. Therefore Eq. (52) becomes

$$\dot{m}_1(\mathbf{v}_1 - \mathbf{v}_2) + (p_1 - p_2)\mathbf{n}_1 = 0 \tag{53}$$

Separating the normal and tangential components of $\mathbf{v}_k$, we have

$$\dot{m}_1(\mathbf{v}_1^n - \mathbf{v}_2^n) + \dot{m}_1(\mathbf{v}_1^t - \mathbf{v}_2^t) + (p_1 - p_2)\mathbf{n}_1 = 0 \tag{54}$$

with
$$\mathbf{v}_k = \mathbf{v}_k^n + \mathbf{v}_k^t \tag{55}$$

By resolving Eq. (54) into the normal and tangential components, we deduce

$$\dot{m}_1(\mathbf{v}_1^n - \mathbf{v}_2^n) + (p_1 - p_2)\mathbf{n}_1 = 0 \tag{56}$$

$$\mathbf{v}_1^t = \mathbf{v}_2^t \tag{57}$$

We can therefore conclude that if there is mass transfer between two inviscid fluids, there is no slip at the interface.

Viscous phases: Consider a one-dimensional flow in which the interface is perpendicular to the flow direction (Fig. 4). The momentum jump condition [Eq. (52)] reads

$$\dot{m}_1(\mathbf{v}_1 - \mathbf{v}_2) + (p_1 - p_2)\mathbf{n}_1 + (\tau_2 - \tau_1) \cdot \mathbf{n}_1 = 0 \tag{58}$$

Multiplying scalarly by $\mathbf{n}_1$, we obtain

$$\dot{m}_1(\mathbf{v}_1 - \mathbf{v}_2) \cdot \mathbf{n}_1 + (p_1 - p_2) + [(\tau_2 - \tau_1) \cdot \mathbf{n}_1] \cdot \mathbf{n}_1 = 0 \tag{59}$$

Because the flow is assumed to be one-dimensional, the mass balance in phase $k$, assumed to be incompressible, is simplified to

$$\frac{du_k}{dx} \equiv 0 \tag{60}$$

where $u_k$ is the $x$ component of $\mathbf{v}_k$. Consequently,

$$\tau_k \equiv 0 \tag{61}$$

The momentum jump condition [Eq. (59)] is therefore simplified to

$$\dot{m}_1(\mathbf{v}_1 - \mathbf{v}_2) \cdot \mathbf{n}_1 + p_1 - p_2 = 0 \tag{62}$$

But because we have, from Eq. (42),

$$\left(\frac{1}{\rho_1} - \frac{1}{\rho_2}\right) \dot{m}_1 = (\mathbf{v}_1 - \mathbf{v}_2) \cdot \mathbf{n}_1 \tag{63}$$

Eq. (62) leads to

$$p_1 - p_2 = \frac{\rho_1 - \rho_2}{\rho_1 \rho_2} \dot{m}_1^2 \tag{64}$$

The pressure is therefore higher in the denser fluid, whatever the transfer sense ($\dot{m}_1 \gtrless 0$).

Two-dimensional case with surface tension: If surface tension is taken into account, the momentum jump condition is written (Delhaye, 1974)

$$\dot{m}_1 \mathbf{v}_1 + \dot{m}_2 \mathbf{v}_2 - \mathbf{n}_1 \cdot \mathbf{T}_1 - \mathbf{n}_2 \cdot \mathbf{T}_2 + \frac{d\sigma}{dl} \tau - \frac{\sigma}{R} \mathbf{n}_1 = 0 \tag{65}$$

where $R$ = radius of curvature
$\tau$ = tangential unit vector
$l$ = curvilinear abscissa along the interface (Fig. 5)

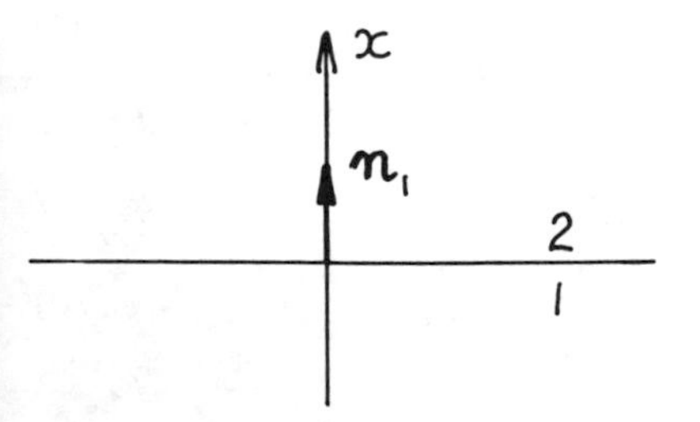


**Figure 4** Plane interface separating phase 1 and phase 2.

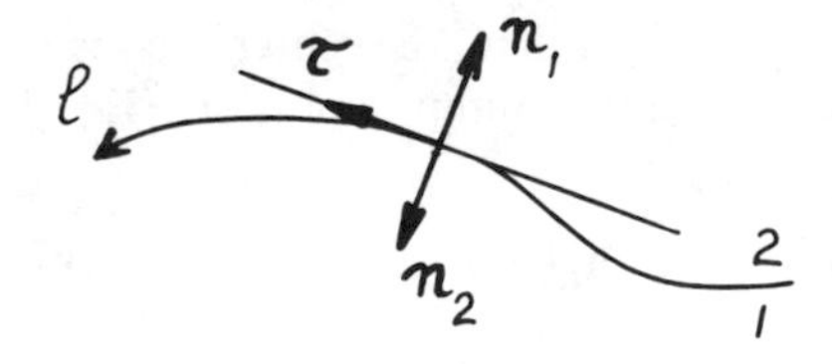


**Figure 5** Cylindrical interface separating phase 1 and phase 2.

The fifth term in Eq. (65) expresses the mechanical effects resulting from surface-tension gradients and is responsible for the Marangoni effect (Kenning, 1968; Levich and Krylov, 1969). If there is no mass transfer and the phases are inviscid, Eq. (65) becomes

$$(p_1 - p_2)\mathbf{n}_1 + \frac{d\sigma}{dl}\tau - \frac{\sigma}{R}\mathbf{n}_1 = 0 \tag{66}$$

This relation implies

$$\frac{d\sigma}{dl} = 0 \tag{67}$$

which is not the general case. Consequently, the hypothesis of inviscid phases is not consistent with the presence of surface-tension gradients. If the phases are viscous, the surface-tension gradients will necessarily induce fluid motions. If the surface tension is constant, the momentum jump condition [Eq. (66)] simplifies to

$$p_1 - p_2 = \frac{\sigma}{R} \tag{68}$$

which is the well-known Laplace law.

## 5.3 Angular Momentum

No new information is afforded by the angular-momentum jump condition.

## 5.4 Total Energy

$$\dot{m}_1(u_1 + \tfrac{1}{2}v_1^2) + \dot{m}_2(u_2 + \tfrac{1}{2}v_2^2) + \mathbf{q}_1 \cdot \mathbf{n}_1 + \mathbf{q}_2 \cdot \mathbf{n}_2 - (\mathbf{n}_1 \cdot \mathbf{T}_1) \cdot \mathbf{v}_1 - (\mathbf{n}_2 \cdot \mathbf{T}_2) \cdot \mathbf{v}_2 = 0 \tag{69}$$

## 5.5 Entropy Inequality

$$\Delta_i = -\dot{m}_1 s_1 - \dot{m}_2 s_2 - \frac{1}{T_1}\mathbf{q}_1 \cdot \mathbf{n}_1 - \frac{1}{T_2}\mathbf{q}_2 \cdot \mathbf{n}_2 \geqslant 0 \tag{70}$$

# 6 SECONDARY JUMP CONDITIONS

## 6.1 Mechanical Energy

A mechanical-energy jump condition can be derived by multiplying the momentum jump condition [Eq. (51)] by a velocity, arbitrarily chosen, whose normal component is the speed of displacement of the interface. We thus have

$$\mathbf{v}_p = (\mathbf{v}_i \cdot \mathbf{n}_k)\mathbf{n}_k + \mathbf{v}^t \tag{71}$$

where $\mathbf{v}^t$ is the arbitrary tangential component of $\mathbf{v}_p$. Equation (51) becomes

$$\dot{m}_1 \mathbf{v}_1 \cdot \mathbf{v}_p + \dot{m}_2 \mathbf{v}_2 \cdot \mathbf{v}_p - (\mathbf{n}_1 \cdot \mathbf{T}_1) \cdot \mathbf{v}_p - (\mathbf{n}_2 \cdot \mathbf{T}_2) \cdot \mathbf{v}_p = 0 \tag{72}$$

## 6.2 Internal Energy

Subtracting the mechanical-energy jump condition [Eq. (72)] from the total energy jump condition [Eq. (69)], we obtain

$$\begin{aligned} &\dot{m}_1 [u_1 + \tfrac{1}{2}(\mathbf{v}_1 - \mathbf{v}_p)^2] + \dot{m}_2 [u_2 + \tfrac{1}{2}(\mathbf{v}_2 - \mathbf{v}_p)^2] + \mathbf{q}_1 \cdot \mathbf{n}_1 + \mathbf{q}_2 \cdot \mathbf{n}_2 \\ &\quad - (\mathbf{n}_1 \cdot \mathbf{T}_1) \cdot (\mathbf{v}_1 - \mathbf{v}_p) - (\mathbf{n}_2 \cdot \mathbf{T}_2) \cdot (\mathbf{v}_2 - \mathbf{v}_p) = 0 \end{aligned} \tag{73}$$

Unlike what occurs in the phase equation derivation, the kinetic energy of each phase is not eliminated. In fact, the method that we are using here enables the interface kinetic energy to be eliminated when the interfacial thermodynamic properties are taken into account (Delhaye, 1974).

## 6.3 Enthalpy

From the definition of $\dot{m}_k$ given in Eq. (42) we can derive the following identity:

$$\mathbf{v}_k - \mathbf{v}_p = \frac{\dot{m}_k}{\rho_k} \mathbf{n}_k + \mathbf{v}_k^t - \mathbf{v}^t \tag{74}$$

where $\mathbf{v}_k^t$ is the tangential component of $\mathbf{v}_k$. By taking into account the definitions of the enthalpy [Eq. (27)] and of the stress tensor [Eq. (28)], Eq. (73), combined with Eq. (74), can be transformed into

$$\begin{aligned} &\dot{m}_1 \left[ i_1 + \frac{1}{2}(\mathbf{v}_1 - \mathbf{v}_p)^2 - \frac{1}{\rho_1}(\boldsymbol{\tau}_1 \cdot \mathbf{n}_1) \cdot \mathbf{n}_1 \right] + \dot{m}_2 \left[ i_2 + \frac{1}{2}(\mathbf{v}_2 - \mathbf{v}_p)^2 \right. \\ &\quad \left. - \frac{1}{\rho_2}(\boldsymbol{\tau}_2 \cdot \mathbf{n}_2) \cdot \mathbf{n}_2 \right] + \mathbf{q}_1 \cdot \mathbf{n}_1 + \mathbf{q}_2 \cdot \mathbf{n}_2 - (\boldsymbol{\tau}_1 \cdot \mathbf{n}_1) \cdot (\mathbf{v}_1^t - \mathbf{v}^t) \\ &\quad - (\boldsymbol{\tau}_2 \cdot \mathbf{n}_2) \cdot (\mathbf{v}_2^t - \mathbf{v}^t) = 0 \end{aligned} \tag{75}$$

## 6.4 Entropy

To transform the enthalpy jump condition [Eq. (75)] into an entropy jump condition, we have to introduce the free enthalpies of $g_k$ of each phase:

$$g_k \triangleq i_k - T_k s_k \tag{76}$$

Equation (75) then becomes

$$\dot{m}_1 \left[ T_1 s_1 + g_1 + \frac{1}{2}(\mathbf{v}_1 - \mathbf{v}_p)^2 - \frac{1}{\rho_1}(\tau_1 \cdot \mathbf{n}_1) \cdot \mathbf{n}_1 \right] + \dot{m}_2 \left[ T_2 s_2 + g_2 + \frac{1}{2}(\mathbf{v}_2 - \mathbf{v}_p)^2 - \frac{1}{\rho_2}(\tau_2 \cdot \mathbf{n}_2) \cdot \mathbf{n}_2 \right] + \mathbf{q}_1 \cdot \mathbf{n}_1 + \mathbf{q}_2 \cdot \mathbf{n}_2$$
$$- (\tau_1 \cdot \mathbf{n}_1) \cdot (\mathbf{v}_1^t - \mathbf{v}^t) - (\tau_2 \cdot \mathbf{n}_2) \cdot (\mathbf{v}_2^t - \mathbf{v}^t) = 0 \tag{77}$$

# 7 INTERFACIAL ENTROPY SOURCE

To combine the entropy inequality [Eq. (70)] with the entropy equation [Eq. (77)], we have to introduce an arbitrary temperature $T_i$, which appears to be the interface temperature when the interface thermodynamic properties are taken into account (Delhaye, 1974; Ishii, 1975). Combining these two equations gives the following expression for the entropy source:

$$\Delta_i = \sum_{k=1,2} \left\{ \frac{\dot{m}_k}{T_i} \left[ g_k + \frac{1}{2}(\mathbf{v}_k - \mathbf{v}_p)^2 - \frac{1}{\rho_k}(\tau_k \cdot \mathbf{n}_k) \cdot \mathbf{n}_k \right] + (\mathbf{q}_k \cdot \mathbf{n}_k + \dot{m}_k s_k T_k) \left( \frac{1}{T_i} - \frac{1}{T_k} \right) - \frac{1}{T_i}(\tau_k \cdot \mathbf{n}_k) \cdot (\mathbf{v}_k^t - \mathbf{v}^t) \right\} \geqslant 0 \tag{78}$$

# 8 INTERFACIAL BOUNDARY CONDITIONS

If we assume that the interface transfers are reversible, the entropy source $\Delta_i$ must be equal to zero whatever the mass flux, the viscous stress tensors, and the heat fluxes are. We then deduce the following interfacial boundary conditions:

1. Thermal boundary condition

$$T_1 \equiv T_2 \equiv T_i \tag{79}$$

2. Mechanical boundary condition

$$\mathbf{v}_1^t \equiv \mathbf{v}_2^t \equiv \mathbf{v}^t \tag{80}$$

3. Phase change boundary condition

$$g_1 - g_2 = \left[ \frac{1}{2}(\mathbf{v}_2 - \mathbf{v}_p)^2 - \frac{1}{2}(\mathbf{v}_1 - \mathbf{v}_p)^2 \right] - \left[ \frac{1}{\rho_2}(\tau_2 \cdot \mathbf{n}_2) \cdot \mathbf{n}_2 - \frac{1}{\rho_1}(\tau_1 \cdot \mathbf{n}_1) \cdot \mathbf{n}_1 \right] \tag{81}$$

which can be written by using Eqs. (74) and (80) as

$$g_1 - g_2 = \frac{1}{2}\dot{m}_k\left(\frac{1}{\rho_2^2} - \frac{1}{\rho_1^2}\right) - \left[\frac{1}{\rho_2}(\tau_2 \cdot \mathbf{n}_2)\cdot\mathbf{n}_2 - \frac{1}{\rho_1}(\tau_1 \cdot \mathbf{n}_1)\cdot\mathbf{n}_1\right] \tag{82}$$

# 9 EXAMPLE 1: DYNAMICS OF VAPOR BUBBLES

In this section we will search for the set of local instantaneous equations governing the dynamics of a spherical vapor bubble. For simplicity we make use of the following hypotheses:

1. On bubble motion:
   H1. No gravity effect
   H2. Spherical symmetry
2. On liquid behavior:
   H3. Single-component fluid
   H4. Newtonian fluid
   H5. Constant viscosity $\mu_L$
   H6. Fluid obeying Fourier's law
   H7. Constant thermal conductivity $k_L$
3. On vapor behavior:
   H8. Single-component fluid
   H9. Newtonian fluid
   H10. Constant viscosity $\mu_V$
   H11. Fluid obeying Fourier's law
   H12. Constant thermal conductivity $k_V$
4. On interface behavior:
   H13. No interfacial mass
   H14. No interfacial viscosity
   H15. Constant surface tension $\sigma$

The local instantaneous equations are written in spherical coordinates (Bird et al., 1960) taking into account the hypothesis of spherical symmetry (H2). They express the mass, momentum, and energy balances for the liquid phase, the vapor phase, and the interface. Finally, we will derive a special form of the liquid momentum equation called the Rayleigh equation. The hypotheses used will be given for each equation.

## 9.1 Mass Balance

1. Liquid phase: H3, single-component fluid

$$\frac{\partial \rho_L}{\partial t} + \frac{1}{r^2}\frac{\partial}{\partial r}(\rho_L r^2 w_L) = 0 \tag{83}$$

2. Vapor phase: H8, single-component fluid

$$\frac{\partial \rho_V}{\partial t} + \frac{1}{r^2} \frac{\partial}{\partial r} (\rho_V r^2 w_V) = 0 \tag{84}$$

3. Interface: H13, no interfacial mass
The jump condition [Eq. (40)] yields

$$\rho_{Vi}(w_{Vi} - \dot{R}) = \rho_{Li}(w_{Li} - \dot{R}) \tag{85}$$

where the overdot denotes a time derivative. Note that if $\rho_{Vi} \ll \rho_{Li}$, we obtain,

$$w_{Li} \cong \dot{R} \tag{86}$$

## 9.2 Momentum Balance

1. Liquid phase:

H1. No gravity effect
H3. Single-component fluid
H4. Newtonian fluid
H5. Constant viscosity $\mu_L$

$$\rho_L \left( \frac{\partial w_L}{\partial t} + w_L \frac{\partial w_L}{\partial r} \right) = -\frac{\partial p_L}{\partial r} + \frac{4}{3} \mu_L \left( \frac{\partial^2 w_L}{\partial r^2} + \frac{2}{r} \frac{\partial w_L}{\partial r} - \frac{2 w_L}{r^2} \right) \tag{87}$$

Note that for an incompressible liquid ($\rho_L$ = constant) the viscous term cancels out because of the liquid mass balance [Eq. (83)].

2. Vapor phase:

H1. No gravity effect
H8. Single-component fluid
H9. Newtonian fluid
H10. Constant viscosity $\mu_V$

$$\rho_V \left( \frac{\partial w_V}{\partial t} + w_V \frac{\partial w_V}{\partial r} \right) = -\frac{\partial p_V}{\partial r} + \frac{4}{3} \mu_V \left( \frac{\partial^2 w_V}{\partial r^2} + \frac{2}{r} \frac{\partial w_V}{\partial r} - \frac{2 w_V}{r^2} \right) \tag{88}$$

3. Interface:

H4. Newtonian liquid phase
H9. Newtonian vapor phase
H13. No interfacial mass
H14. No interfacial viscosity

When surface tension is taken into account, the jump condition [Eq. (51)] can be generalized (Delhaye, 1974) and leads to

$$p_{Vi} - p_{Li} = \frac{2\sigma}{R} - \rho_{Vi}(w_{Vi} - \dot{R})w_{Vi} - \rho_{Li}(\dot{R} - w_{Li})w_{Li}$$
$$+ \frac{4}{3}\mu_V \left( \frac{\partial w_V}{\partial r}\bigg|_i - \frac{w_{Vi}}{R} \right) - \frac{4}{3}\mu_L \left( \frac{\partial w_L}{\partial r}\bigg|_i - \frac{w_{Li}}{R} \right) \quad (89)$$

## 9.3 Internal Energy Balance:

1. Liquid phase:

H3. Single-component fluid
H4. Newtonian fluid
H5. Constant viscosity $\mu_L$
H6. Fluid obeying Fourier's law
H7. Constant thermal conductivity $k_L$

$$\rho_L c_L^v \left( \frac{\partial T_L}{\partial t} + w_L \frac{\partial T_L}{\partial r} \right) = \frac{k_L}{r^2} \frac{\partial}{\partial r} \left( r^2 \frac{\partial T_L}{\partial r} \right)$$
$$- T_L \left( \frac{\partial p_L}{\partial T_L} \right)_{\rho_L} \frac{1}{r^2} \frac{\partial}{\partial r} (w_L r^2) + \frac{4}{3} \mu_L \left( \frac{\partial w_L}{\partial r} - \frac{w_L}{r} \right)^2 \quad (90)$$

In Eq. (90) $c_L^v$ denotes the specific heat at constant volume, and the last term of the right-hand side represents the viscous dissipation. Note that for an incompressible liquid ($\rho_L$ = constant) the second term of the right-hand side drops out.

2. Vapor phase:

H8. Single-component fluid
H9. Newtonian fluid
H10. Constant viscosity
H11. Fluid obeying Fourier's law
H12. Constant thermal conductivity $k_V$

$$\rho_V c_V^v \left( \frac{\partial T_V}{\partial t} + w_V \frac{\partial T_V}{\partial r} \right) = \frac{k_V}{r^2} \frac{\partial}{\partial r} \left( r^2 \frac{\partial T_V}{\partial r} \right)$$
$$- T_V \left( \frac{\partial p_V}{\partial T_V} \right)_{\rho_V} \frac{1}{r^2} \frac{\partial}{\partial r} (w_V r^2) + \frac{4}{3} \mu_V \left( \frac{\partial w_V}{\partial r} - \frac{w_V}{r} \right)^2 \quad (91)$$

In Eq. (91) $c_V^v$ denotes the specific heat at constant volume and the last term of the right-hand side represents the viscous dissipation.

3. Interface:

H4. Newtonian liquid phase
H6. Liquid phase obeying Fourier's law
H9. Newtonian vapor phase
H11. Vapor phase obeying Fourier's law
H13. No interfacial mass

H14. No interfacial viscosity
H15. Constant surface tension

When surface tension is taken into account, the enthalpy jump condition [Eq. (75)] can be generalized (Delhaye, 1974). Nevertheless when surface tension is constant, this generalized equation simplifies to Eq. (75), which yields

$$k_L \left.\frac{\partial T_L}{\partial r}\right|_i - k_V \left.\frac{\partial T_V}{\partial r}\right|_i = \rho_{Vi}(\dot{R} - w_{Vi})\left[i_{Vi} + \frac{1}{2}(w_{Vi} - \dot{R})^2 - \frac{4}{3}\frac{\mu_V}{\rho_{Vi}}\left(\left.\frac{\partial w_V}{\partial r}\right|_i - \frac{w_{Vi}}{r}\right)\right] - \rho_{Li}(\dot{R} - w_{Li})\left[i_{Li} + \frac{1}{2}(w_{Li} - \dot{R})^2 - \frac{4}{3}\frac{\mu_L}{\rho_{Li}}\left(\left.\frac{\partial w_L}{\partial r}\right|_i - \frac{w_{Li}}{r}\right)\right] \tag{92}$$

## 9.4 Rayleigh Equation

In the studies of vapor bubble dynamics a special form of the liquid momentum equation, the Rayleigh equation, is often used.

In addition to the hypotheses required to ensure the validity of the liquid continuity equation [Eq. (83)], the mass jump condition [Eq. (85)], and the liquid momentum equation [Eq. (87)], we assume that

H16. The liquid is incompressible ($\rho_L$ = constant)
H17. The vapor density $\rho_V$ is negligible with respect to the liquid density $\rho_L$ ($\rho_V \ll \rho_L$)

When hypothesis H16 is taken into account, the liquid continuity equation [Eq. (83)] can be directly integrated, which leads to

$$w_L = \frac{A(t)}{r^2} \tag{93}$$

As a result, the liquid momentum equation [Eq. (87)] takes the following form, whatever the liquid viscosity:

$$\rho_L \left(\frac{\partial w_L}{\partial t} + w_L \frac{\partial w_L}{\partial r}\right) = -\frac{\partial p_L}{\partial r} \tag{94}$$

Taking into account Eq. (93) and integrating from $r = R$ to $r$ = infinity, we find

$$\dot{w}_{Li}R + 2\dot{R}w_{Li} - \frac{1}{2}w_{Li}^2 = \frac{p_{Li} - p_{L\infty}}{\rho_L} \tag{95}$$

where the overdot denotes a time derivative.

Finally, hypothesis H17 enables the mass jump condition to be written as Eq. (86). Consequently, Eq. (95) is rewritten as

$$\ddot{R}R + \frac{3}{2}\dot{R}^2 = \frac{p_{Li} - p_{L\infty}}{\rho_L} \tag{96}$$

which is called the Rayleigh equation.

# 10 EXAMPLE 2: HYDRODYNAMICS OF LIQUID FILMS

In this section we will derive the mass and momentum jump conditions at the free surface of an isothermal, viscous liquid film flowing down an inclined plane. Because we will deal only with values at the interface, we will omit the subscript $i$. For simplicity we use the following hypotheses:

1. On film motion:
   H1. Two-dimensional flow
2. On liquid behavior:
   H2. Single-component fluid
   H3. Newtonian fluid
3. On gas behavior:
   H4. The gas phase intervenes only by its pressure $p_G$, which is assumed constant
4. On interface behavior:
   H5. No interfacial mass
   H6. No interfacial viscosity
   H7. Constant surface tension $\sigma$
   H8. No phase change at the interface

The local instantaneous jump conditions are written in rectangular coordinates (Bird et al., 1960). The $x$ axis is along the plane; the $y$ axis is perpendicular to it. The origin of the coordinates is at the wall.

## 10.1 Mass Balance

H5. No interfacial mass
H8. No phase change at the interface

Equation (43) yields

$$\dot{m}_L \triangleq \rho_L(\mathbf{v}_L - \mathbf{v}_i) \cdot \mathbf{n}_L \equiv 0 \tag{97}$$

Let $y = \eta(x, t)$ be the thickness of the liquid film. Equations (4) and (5) lead to

$$\mathbf{n}_L = \begin{Bmatrix} \dfrac{-\partial\eta/\partial x}{[1 + (\partial\eta/\partial x)^2]^{1/2}} \\ \dfrac{1}{[1 + (\partial\eta/\partial x)^2]^{1/2}} \end{Bmatrix} \tag{98}$$

$$\mathbf{v}_i \cdot \mathbf{n}_L = \frac{\partial \eta / \partial t}{[1 + (\partial \eta / \partial x)^2]^{1/2}} \tag{99}$$

Consequently, Eq. (97) reads

$$v_L = \frac{\partial \eta}{\partial t} + u_L \frac{\partial \eta}{\partial x} \tag{100}$$

where $u_L$ and $v_L$ are the $x$ and $y$ components of the velocity vector $\mathbf{v}_L$. Equation (100) is often called the kinematic condition.

## 10.2 Momentum Balance

H3. Newtonian liquid phase
H4. The gas phase intervenes only by its pressure $p_G$, which is assumed constant
H5. No interfacial mass
H6. No interfacial viscosity
H7. Constant surface tension $\sigma$
H8. No phase change at the interface

In these conditions the two-dimensional momentum jump condition reads

$$-\mathbf{n}_G \cdot \mathbf{T}_G - \mathbf{n}_L \cdot \mathbf{T}_L = -\frac{\partial^2 \eta / \partial x^2}{[1 + (\partial \eta / \partial x)^2]^{3/2}} \sigma \mathbf{n}_L \tag{101}$$

where

$$\mathbf{T}_G = \begin{bmatrix} -p_G & 0 \\ 0 & -p_G \end{bmatrix} \qquad \mathbf{T}_L = \begin{bmatrix} -p_L + 2\mu_L \dfrac{\partial u_L}{\partial x} & \mu_L \left( \dfrac{\partial u_L}{\partial y} + \dfrac{\partial v_L}{\partial x} \right) \\ \mu_L \left( \dfrac{\partial u_L}{\partial y} + \dfrac{\partial v_L}{\partial x} \right) & -p_L + 2\mu_L \dfrac{\partial v_L}{\partial y} \end{bmatrix}$$

1. Normal projection of the momentum jump condition. Taking the scalar product of Eq. (101) by $\mathbf{n}_L$, we obtain

$$p_L - p_G = -\sigma \frac{\partial^2 \eta / \partial x^2}{[1 + (\partial \eta / \partial x)^2]^{3/2}} + 2\mu_L \frac{\partial v_L / \partial y - (\partial u_L / \partial y + \partial v_L / \partial x)(\partial \eta / \partial x) + (\partial u_L / \partial x)(\partial \eta / \partial x)^2}{1 + (\partial \eta / \partial x)^2} \tag{102}$$

2. Tangential projection of the momentum jump condition. Taking the scalar product of Eq. (101) by a unit vector tangent to the interface, we obtain

$$2 \frac{\partial \eta}{\partial x} \left( \frac{\partial u_L}{\partial x} - \frac{\partial v_L}{\partial y} \right) - \left( \frac{\partial u_L}{\partial y} + \frac{\partial v_L}{\partial x} \right) \left[ 1 - \left( \frac{\partial \eta}{\partial x} \right)^2 \right] = 0 \tag{103}$$

## NOMENCLATURE

| | |
|---|---|
| $A$ | function |
| $\mathbf{\mathfrak{a}}$ | surface |
| $\mathbf{B}$ | vector |
| $c^v$ | specific heat at constant volume |
| $\mathbf{F}$ | external force per unit of mass |
| $f$ | function |
| $g$ | free enthalpy per unit of mass |
| $i$ | enthalpy per unit of mass |
| $\mathbf{J}$ | flux term |
| $k$ | thermal conductivity |
| $l$ | line coordinate |
| $\mathbf{M}$ | tensor |
| $\dot{m}$ | mass transfer per unit area of interface and per unit of time [Eq. (17)] |
| $\mathbf{n}$ | normal unit vector |
| $p$ | pressure [Eq. (31)] |
| $\mathbf{q}$ | heat flux |
| $R$ | radius of curvature |
| $\mathbf{R}$ | antisymmetric tensor |
| $r$ | radial coordinate |
| $\mathbf{r}$ | position vector |
| $s$ | entropy per unit of mass |
| $T$ | temperature [Eq. (32)] |
| $\mathbf{T}$ | stress tensor |
| $t$ | time |
| $\mathbf{U}$ | unit tensor |
| $u$ | surface coordinate; internal energy per unit of mass; $x$ component of the velocity vector |
| $\mathcal{V}$ | volume |
| $v$ | surface coordinate; $y$ component of the velocity vector |
| $\mathbf{v}$ | velocity |
| $w$ | radial component of the velocity vector |
| $\Delta$ | entropy source per unit of volume and per unit of time |
| $\eta$ | film thickness |
| $\mu$ | viscosity |
| $\rho$ | density |
| $\sigma$ | surface tension |
| $\boldsymbol{\tau}$ | tangential unit vector |
| $\boldsymbol{\tau}$ | viscous stress tensor |
| $\phi$ | source term |
| $\psi$ | specific quantity |

**Subscripts**

$a$ surface
$G$ gas
$i$ interface
$k$ phase index
$L$ liquid
$p$ interface velocity
$V$ vapor

**Superscripts**

$n$ normal
$t$ tangential, transposed
$v$ at constant volume

## REFERENCES


Aris, R., *Vectors, Tensors and the Basic Equations of Fluid Mechanics*, Prentice Hall, Englewood Cliffs, N.J., 1962.

Bird, R. B., Stewart, W. E., and Lightfoot, E. N., *Transport Phenomena*, Wiley, New York, 1960.

Callen, H. B., *Thermodynamics*, Wiley, New York, 1960.

Delhaye, J. M., Jump Conditions and Entropy Sources in Two-Phase Systems, Local Instant Formulation, *Int. J. Multiphase Flow*, vol. 1, pp. 395–409, 1974.

Ishii, M., *Thermo-fluid Dynamic Theory of Two-Phase Flow*, Eyrolles, Paris, 1975.

Kenning, D. B., Two-Phase Flow with Nonuniform Surface Tension, *Appl. Mech. Rev.*, vol. 21, no. 11, pp. 1101–1111, 1968.

Levich, V. G. and Krylov, V. S., Surface-Tension-Driven Phenomena, *Ann. Rev. Fluid Mech.*, vol. 1, pp. 293–316, 1969.

Truesdell, C. A. and Toupin, R. A., The Classical Field Theories, in *Encyclopedia of Physics*, ed. S. Flugge, vol. III/1, *Principles of Classical Mechanics and Field Theory*, Springer-Verlag, Germany, pp. 226–858, 1960.

CHAPTER

# SIX

# INTERFACIAL EQUILIBRIUM AND NUCLEATION

**M. Giot**

This chapter presents fundamental ideas concerning basic nucleation phenomena. A more extensive approach would have required a much larger chapter. Several important aspects, such as cavitation, vapor explosions, and even condensation, are not treated.

At the beginning of this chapter we would like to point out the following definitions: Nucleation is called *homogeneous* when the second phase is generated completely within a preexistent phase. The nucleation is *heterogeneous* when it occurs at an interface between the preexistent phase and another phase.

Three types of superheat in boiling phenomena should be distinguished:

Wall superheat, which is the difference between wall temperature and saturation temperature of the fluid
Bulk superheat, which is the difference between liquid bulk temperature and saturation temperature
Local superheat, which is the difference between a local liquid temperature and saturation temperature

# 1 MECHANICAL EQUILIBRIUM

## 1.1 Equilibrium Condition of an Interface

According to Young's model, an interface between isothermal and nonmiscible fluids may be considered as a membrane uniformly stretched and without thickness. Surface tension is a consequence of this model: it is the modulus $\sigma$ of the force acting on a unit length of each closed contour drawn on the interface. This force ensures static equilibrium of the portion of the membrane located inside this contour.

Consider a portion of interface defined by elements $dl_1$ and $dl_2$ of orthogonal lines (Fig. 1). Pressure of the bulk fluid on the concave side near the interface is denoted $p_1$, pressure on the convex side, $p_2$; $R_1$ and $R_2$ are the radii of curvature corresponding to $dl_1$ and $dl_2$, respectively. The resultant of the forces acting on the contour is normal to the interface and directed toward the concave side. Its modulus is

$$|d\mathbf{F}| = \sigma(d\alpha_1\, dl_2 + d\alpha_2\, dl_1)$$

and the equilibrium condition is

$$\sigma(d\alpha_1\, dl_2 + d\alpha_2\, dl_1) - p_1\, dl_1\, dl_2 + p_2\, dl_1\, dl_2 = 0$$

or

$$p_1 - p_2 = \sigma\left(\frac{d\alpha_1}{dl_1} + \frac{d\alpha_2}{dl_2}\right) = \sigma\left(\frac{1}{R_1} + \frac{1}{R_2}\right) \tag{1}$$

This relation is called the Young-Laplace equation. One sees that $p_1 > p_2$; the pressure always has a higher value on the concave side. Equation (1) does not depend on the selected net of orthogonal lines because a property of the surfaces is that at a given point the sum of the curvatures of two orthogonal sections, called the mean curvature $C$, is constant.

$$p_1 - p_2 = \sigma C \tag{2}$$

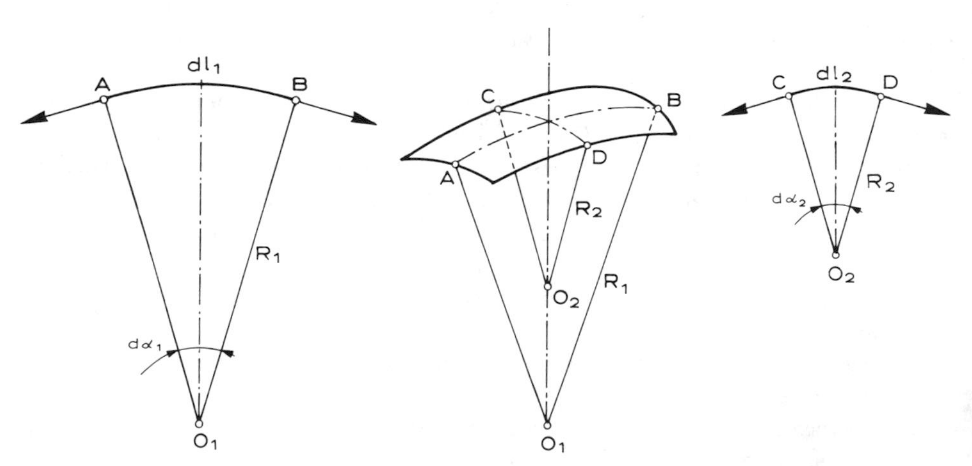


**Figure 1** Young's model of interface.

In the gravity field (acceleration $g$), where a level of reference (index 0) crossing the interface is defined and where the $0z$ axis is along the upwardly directed vertical, if densities $\rho_1$ and $\rho_2$ are constant,

$$p_1 = p_{1,0} - \rho_1 g z \qquad p_2 = p_{2,0} - \rho_2 g z$$

and consequently

$$p_1 - p_2 = (p_{1,0} - p_{2,0}) + (\rho_2 - \rho_1) g z = \sigma C \tag{3}$$

Let

$$\Delta\rho = |\rho_2 - \rho_1|$$

The capillary constant, defined as

$$a \triangleq \sqrt{\frac{2\sigma}{\Delta\rho g}} \tag{4}$$

is expressed in meters and is related to a nondimensional number called the Eötvos number, or Bond number.

$$\mathrm{Eo} \triangleq \frac{\Delta\rho g D^2}{\sigma} = \frac{2D^2}{a}$$

where $D$ is a characteristic dimension of the interface, for example, a spherical equivalent diameter of a bubble or a droplet. This nondimensional parameter expresses the ratio of buoyancy forces to capillary forces.

If the gravity term of Eq. (3) is negligible, then at each point of the interface

$$C = \frac{1}{R_1} + \frac{1}{R_2} = \frac{p_{1,0} - p_{2,0}}{\sigma} \triangleq C_0 = \text{const}$$

Thus the interface is spherical ($R_1 = R_2 = R = 2/C$), and

$$p_1 - p_2 = \frac{2\sigma}{R}$$

If

$$z^* \triangleq \frac{z}{D}$$

Eq. (3) may be written in a nondimensional form:

$$C - C_0 = \frac{1}{D}\,\mathrm{Eo}\, z^* \tag{5}$$

Approximation of spherical interfaces is verified in the following examples, which correspond to small values of the Eötvos number:

1. Large capillary constant. This condition occurs when the surface tension has a large value or the system consists of two nonmiscible fluids having roughly the same densities, or when both situations occur; this condition also occurs in a liquid-vapor system when the pressure is near the critical value.
2. Small dimensions of the dispersed particles.

To clarify statement 2, consider the sessile drop. Integration of Eq. (3) yields the results shown in Fig. 2, where the nondimensional height ($h/a$) and volume ($V_0/a^3$) of the droplets are compared with the nondimensional half-width ($x_{max}/a$). For small droplets

$$\frac{h}{a} \simeq 2\,\frac{x_{max}}{a}$$

whereas for large droplets, one finds (see, for example, Wachters, 1965):

$$h \simeq a\sqrt{2}$$

## 1.2 Equilibrium Condition of a Contact Line

**1.2.1 Contact line for three fluid phases** Consider at point $M$ of a contact line between three fluid phases a section normal to this line (Fig. 3). If there is no tension along this line, the equilibrium condition of the capillary forces is expressed by the law of Neumann's triangle:

$$\boldsymbol{\sigma}_{12} + \boldsymbol{\sigma}_{23} + \boldsymbol{\sigma}_{31} = 0 \tag{6}$$

where $\boldsymbol{\sigma}_{ij}$ denotes the surface forces per unit length between phases $i$ and $j$ at point $M$.

In particular, along the contact line $C$ between a liquid lens $L'$, a liquid substrate $L''$, and gas $G$, the law of Neumann's triangle [Eq. (6)] yields

$$\sigma^2_{L'L''} = \sigma^2_{L'G} + \sigma^2_{L''G} - 2\sigma_{L'G}\sigma_{L''G}\cos\beta$$

where $\beta$ is the Neumann angle defined in Fig. 4.

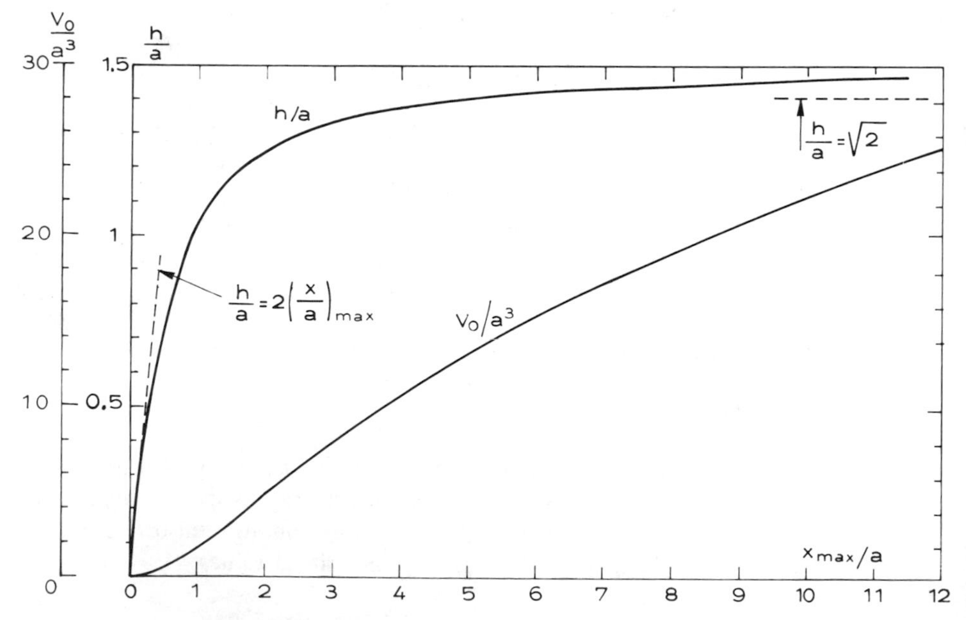


**Figure 2** Height and volume of sessile droplets compared with their half-width.

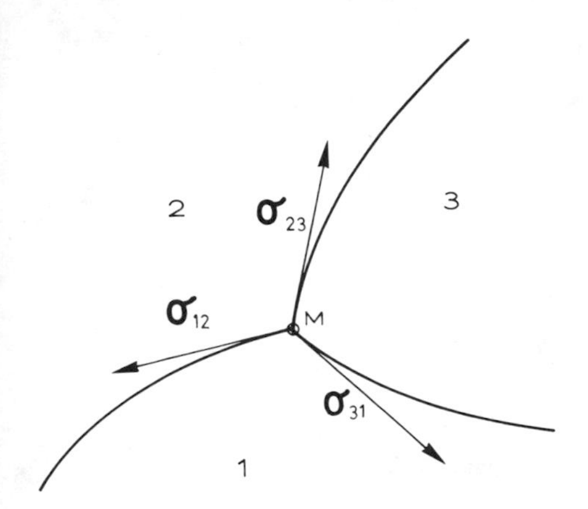


**Figure 3** Equilibrium of a contact line.

**1.2.2 Contact line along solid wall** Consider a contact line located along a solid wall, where the other two phases are a liquid and a gas (Fig. 5). Young-Dupré's equilibrium condition is

$$\sigma_{SL} + \sigma_{LG} \cos \theta = \sigma_{SG} \tag{7}$$

where the contact angle $\theta$ is defined in Fig. 5. Equation (7) has given rise to much controversy in the literature because it expresses only the equilibrium of forces along a plane that is tangential to the solid wall, but the equilibrium cannot be justified in the normal direction without a reaction $f$ per unit length of the contact line exerted by the solid on the liquid.

However, Chappuis (1974) recently proved the existence of this reaction, and Goodrich (1969) derived the Young-Dupré equation in a simple geometric situation from a principle of minimum energy.

**1.2.3 Contact line coinciding with solid surface discontinuity** When the contact line coincides with a discontinuity of the solid surface, the equilibrium value of the contact angle measured from the solid-liquid interface is increased with respect to a continuous solid surface. Oliver et al. (1977) verified experimentally that this increase is well within the limits calculated by Gibbs, as will now be shown. Therefore, consider a solid surface with a sharp edge of angle $\alpha$ (Fig. 6). The

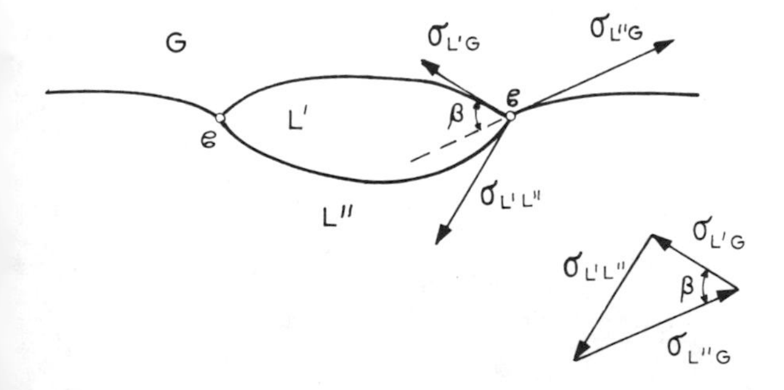


**Figure 4** Neumann angle.

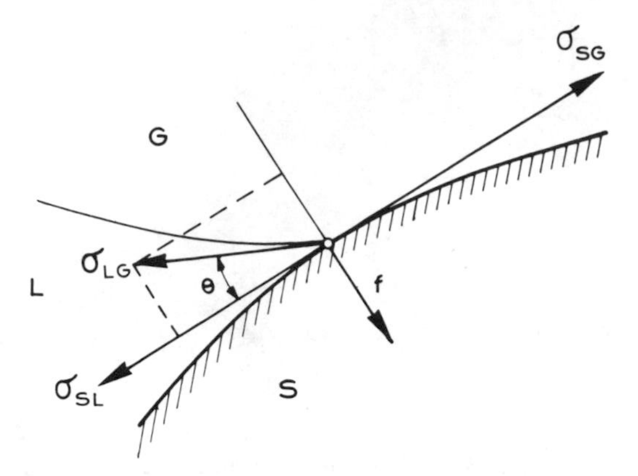


Figure 5 Contact angle.

equilibrium of forces acting on the contact line can be expressed by the following relationships, written along the directions of the two sides of the sharp edge:

$$\sigma_{SL} + \sigma_{SG} \cos \alpha + \sigma_{LG} \cos \theta = 0$$

$$\sigma_{SG} + \sigma_{SL} \cos \alpha - \sigma_{LG} \cos [\theta - (\pi - \alpha)] = 0$$

or

$$\begin{aligned} \sigma_{SG} \cos (\pi - \alpha) - \sigma_{SL} &= \sigma_{LG} \cos \theta \\ \sigma_{SG} - \sigma_{SL} \cos (\pi - \alpha) &= \sigma_{LG} \cos [\theta - (\pi - \alpha)] \end{aligned} \tag{8}$$

If $\theta_0$ is the value of the contact angle when $\alpha = \pi$, that is, in the absence of a sharp edge, then the Young-Dupré equation yields

$$\sigma_{SG} - \sigma_{SL} = \sigma_{LG} \cos \theta_0$$

and, comparing with Eq. (8),

$$\theta_0 \leqslant \theta \leqslant \theta_0 + (\pi - \alpha)$$

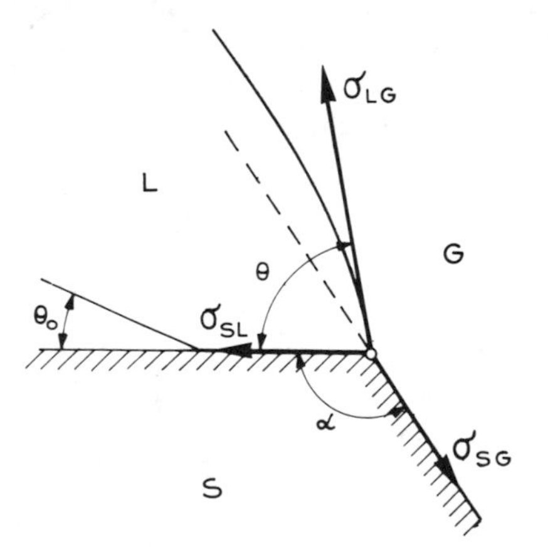


Figure 6 Contact line on a sharp edge.

The upper value $\theta_c \triangleq \theta_0 + (\pi - \alpha)$ is obtained when the contact line jumps over the edge during the forward movement of the liquid.

Since the interfacial tensions result from interactions between the molecules of the adjacent phases, in the absence of any solid-liquid attraction, the liquid does not wet the wall, $\theta = \pi$, and the Young-Dupré equation becomes

$$\sigma_{SL} = \sigma_{SG} + \sigma_{LG}$$

When solid-liquid intermolecular attraction exists, the liquid spreads over the wall up to the point where an angle $\theta$ $(0 \leqslant \theta \leqslant \pi)$ can exist that satisfies the Young-Dupré equilibrium condition. For this reason, the parameter $\sigma_{LG} \cos \theta$ is called adhesion.

Calculation of the contact angle by the Young-Dupre equation implicitly assumes that the solid surface is smooth, homogeneous, isotropic, and nondeformable and that the fluids are pure.

1. *Smooth surface.* When the surface is rough, the real solid-liquid contact area $\mathfrak{a}$ is larger than the apparently smooth area $\mathfrak{a}'$. In 1936 Wenzel proposed the following relationship between the real contact angle $\theta_{\text{real}}$ and the apparent one $\theta$:

$$\frac{\cos \theta}{\cos \theta_{\text{real}}} = \frac{\mathfrak{a}}{\mathfrak{a}'}$$

This expression has been established in a simple geometric case from energetic considerations by Johnson and Dettre (1964).

2. *Homogeneous and isotropic surface.* If the solid surface is composed of various elements, the contact-angle value may be the result of an average. An analysis of the effects of macroscopic and microscopic heterogeneities is given in Adamson and Ling (1964). The presence of surface heterogeneities is most often claimed as the cause of contact-angle hysteresis.

3. *Pure fluids.* Schwartz (1975) pointed out that dust and chemical reactions, like oxidation, at the liquid-vapor interface can modify the rheological properties of the interfacial layer and make it rigid.

Contact-angle hysteresis is an effect of difference between the model and a practical situation with real surfaces and real fluids.

## 2 WORK OF THE CAPILLARY FORCES

### 2.1 Gas-Liquid and Liquid-Liquid Systems

The Young-Laplace equation

$$p_1 - p_2 = \sigma C$$

implies the definition of a particular surface of the interfacial layer, called the surface of tension. Let us consider a section of the two fluid phases and the interfacial layer, this section containing the $0z$ axis perpendicular to the layer (Fig. 7). We will denote by $t$ the module of the tension force per unit area exerted by the portion of the fluid located to the left of the plane perpendicular to the figure and containing $0z$, on the portion located to the right.

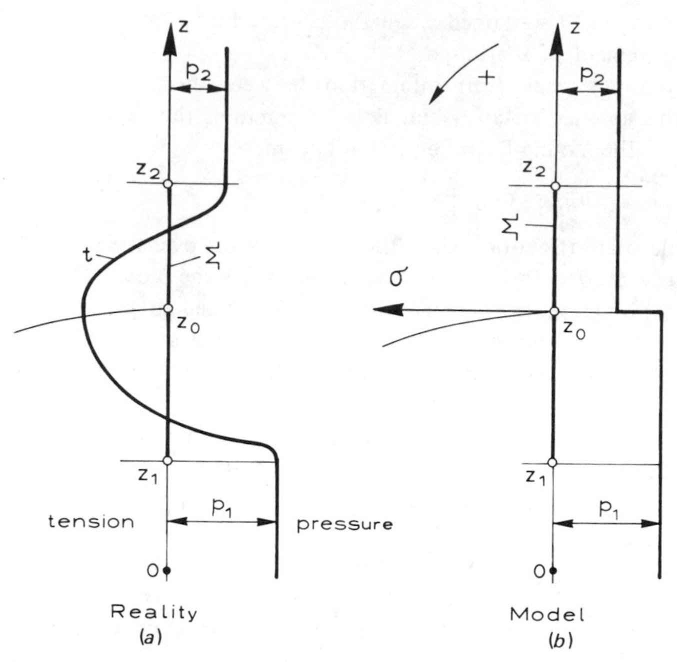


**Figure 7** Surface of tension.

In reality (Fig. 7*a*),

$$t = -p_1 \qquad \text{for } z < z_1$$

$$t = -p_2 \qquad \text{for } z > z_2$$

According to the model involving the surface of tension (Fig. 7*b*),

$$t = -p_1 \qquad \text{for } z < z_0$$

$$t = -p_2 \qquad \text{for } z > z_0$$

Let $\Sigma$ be the surface containing $0z$, perpendicular to the figure, comprised between $z = z_1$ and $z = z_2$, and having unit width. The equality of the projections on plane $z = z_0$ of the forces acting on $\Sigma$ in the model and in reality yields

$$\sigma - p_1(z_0 - z_1) - p_2(z_2 - z_0) = \int_{z_1}^{z_2} t\, dz$$

or

$$\sigma = \int_{z_1}^{z_0} (t + p_1)\, dz + \int_{z_0}^{z_2} (t + p_2)\, dz$$

We define the excess tension $t^*$ by setting

$$t^* = t + p_1 \qquad \text{for } z < z_0$$

$$t^* = t + p_2 \qquad \text{for } z > z_0$$

Let us denote by $M_0$, $M_1$, and $M_2$, respectively, the moments of zero, first, and second order of the excess tension $t^*$ with reference to point 0:

$$M_0 \triangleq \int_{z_1}^{z_2} t^* \, dz \qquad M_1 \triangleq \int_{z_1}^{z_2} t^* z \, dz \qquad M_2 \triangleq \int_{z_1}^{z_2} t^* z^2 \, dz$$

We have

$$\sigma = M_0$$

The equality of the resultants of the moments of forces acting on $\Sigma$ in the model and in reality gives

$$\sigma z_0 - \int_{z_1}^{z_0} p_1 z \, dz - \int_{z_0}^{z_2} p_2 z \, dz = \int_{z_1}^{z_0} tz \, dz$$

or

$$\sigma z_0 = \int_{z_1}^{z_0} (t + p_1) z \, dz + \int_{z_0}^{z_2} (t + p_2) z \, dz = \int_{z_1}^{z_2} t^* z \, dz$$

This result gives the position of the surface of tension:

$$z_0 = \frac{M_1}{M_0}$$

Calculating the works of forces acting on the interfacial layer for a deformation of this layer resulting in a tangential deformation of the surface of tension ($R_1 R_2 = \text{const}$) and a normal deformation of the surface of tension ($d\alpha_1 \, d\alpha_2 = \text{const}$), Dennery and Guénot (1976) obtained

$$dW = \sigma \, d\mathfrak{a} + M_2 \frac{d\mathfrak{a}}{R_1 R_2} \tag{9}$$

where $dW$ is the elementary work accompanying the variation $d\mathfrak{a}$ of the surface of tension area, and where $M_2$ is calculated by putting $z_0 = 0$ (origin of the $0z$ axis on the surface of tension). The classical expression

$$dW = \sigma \, d\mathfrak{a} \tag{10}$$

is related only to the normal deformations of the interface ($d\alpha_1 \, d\alpha_2 = \text{const}$) and coresponds to the work of the internal forces. The numerical importance of the correction found by Dennery and Guénot still remains unknown.

## 2.2 Cohesion, Adhesion, and Spreading

The surface tension $\sigma$ appears in Eq. (10) as the work needed to create a unit area of interface. In the absence of any field of external forces, the work of separation of a liquid column (Fig. 8A) having a unit cross-sectional area, into two portions (Fig. 8B) is

$$W_C \triangleq 2\sigma_{LG} \tag{11}$$

$W_C$ is called work of cohesion.

If system A in Fig. 9 consists of two liquid phases $L'$ and $L''$, the work of separation is given by

$$W_A \triangleq \sigma_{L'G} + \sigma_{L''G} - \sigma_{L'L''} \tag{12}$$

$W_A$ is called work of adhesion. It plays an important role in the theory of emulsions (Becher, 1965).

Let us consider the spreading of a droplet of liquid $L'$ on liquid $L''$ (Fig. 10). The work of formation of a contact area $\mathfrak{a}_{L'L''}$ by the spreading is

$$\sigma_{L'L''}\,\mathfrak{a}_{L'L''} + \sigma_{L'G}\,\mathfrak{a}_{L'L''} - \sigma_{L''G}\,\mathfrak{a}_{L'L''}$$

that is, per unit area,

$$\sigma_{L'L''} + \sigma_{L'G} - \sigma_{L''G} = W_C(L') - W_A$$

We will call work of spreading $W_E$ (often called spreading coefficient) the difference between work of adhesion and work of cohesion of the spreading liquid:

$$W_E \triangleq W_A - W_C(L') = \sigma_{L''G} - \sigma_{L'L''} - \sigma_{L'G} \tag{13}$$

When this work is positive, liquid $L'$ spreads on liquid $L''$. In the opposite case, lenses are formed. Indeed, the equilibrium condition on the contact surface is

$$\sigma_{L''G} = \sigma_{L'L''} + \sigma_{L'G} \cos\theta$$

Thus, $W_E > 0$ means $\cos\theta > 1$, that is, spreading, and $W_E < 0$ means $\cos\theta < 1$, that is, lenses.

It must be emphasized that in solid-liquid-vapor contact lines, the Young-Dupré equation requires the preliminary definition of the surface energies $\sigma_{SG}$ and $\sigma_{SL}$. Concerning $\sigma_{SG}$, one must distinguish between the value $\sigma_S$ related to the

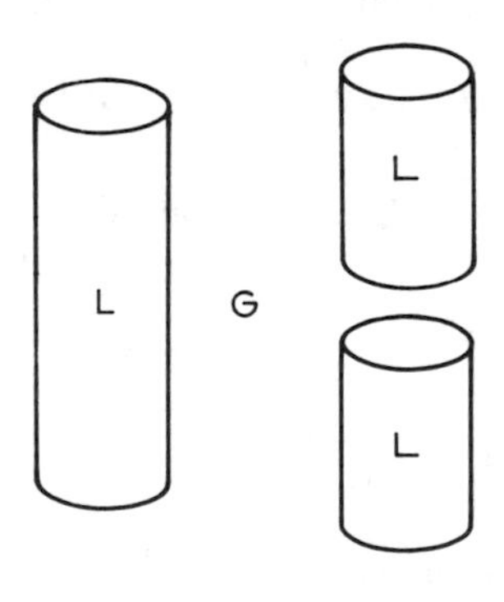


**Figure 8** Cohesion.

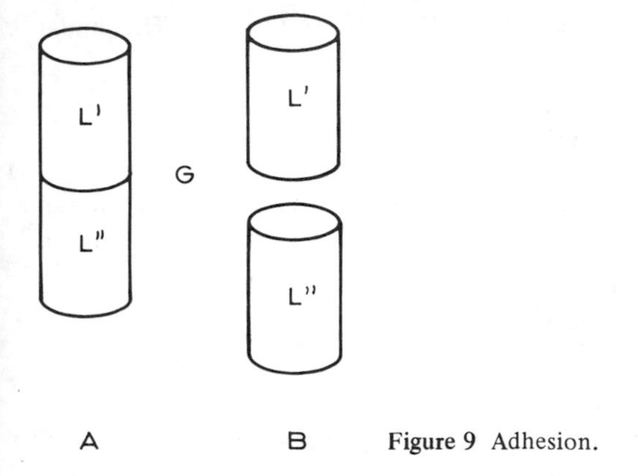


**Figure 9** Adhesion.

solid-void interface and the value $\sigma_{SV}$ of the solid surface near the liquid, that is, at places where a film of adsorbed vapor exists and increases the contact angle if $\theta < \pi/2$ (see Zisman, 1964).

# 3 THERMODYNAMIC EQUILIBRIUM

## 3.1 The Gibbs Model

**3.1.1 The model** Consider a two-phase system (subscripts 1 and 2) occupying a volume $V$ and separated by interfaces having a total area $\mathfrak{a}$. Further consider that $m$ components coexist in the volume $V$. Let us denote by $c_{i,1}$ and $c_{i,2}$ the molar concentrations of component $i$ ($i = 1, \ldots, m$) in each phase. The Gibbs model consists of assuming that the interfacial layers have no thickness and that the molar concentrations $c_{i,1}$ and $c_{i,2}$ are uniform up to the separating surfaces between the phases. Volume $V$ is then divided into two volumes $V_1$ and $V_2$, where the numbers of moles of component $i$ are, respectively,

$$n_{i,1} = c_{i,1} V_1 \qquad \text{and} \qquad n_{i,2} = c_{i,2} V_2$$

If $n_i$ denotes the total number of moles of component $i$ in volume $V$, a consequence of the Gibbs model is that there exists a number $n_{i,\sigma}$ of moles in excess:

$$n_{i,\sigma} = n_i - (n_{i,1} + n_{i,2})$$

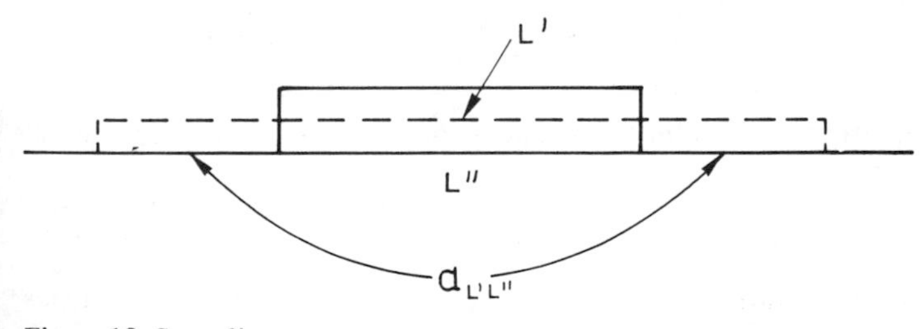


**Figure 10** Spreading.

These $n_{i,\sigma}$ moles are considered as being adsorbed on the interfaces. The molar adsorption of component $i$ is defined by

$$\gamma_i \triangleq \frac{n_{i,\sigma}}{\mathfrak{a}}$$

As material properties are given to the interface, an internal energy $U_\sigma$ and an entropy $S_\sigma$ of the interface exist; for the system consisting of two phases and an interface,

$$U = U_1 + U_2 + U_\sigma$$

$$S = S_1 + S_2 + S_\sigma$$

A similar decomposition may be written for the Helmholtz free energy $F$ at thermal equilibrium of the system:

$$F \triangleq U - TS = (U_1 - TS_1) + (U_2 - TS_2) + (U_\sigma - TS_\sigma) = F_1 + F_2 + F_\sigma$$

The positions of the separation surfaces between the two phases in the Gibbs model must still be defined. To satisfy the mechanical equilibrium requirements, these surfaces will coincide with the surface of tension. If necessary, thermodynamic variables invariant with the position of the interface may be used. This question is treated in Adamson (1967) and Defay et al. (1966).

**3.1.2 Internal energy equation** From the first law of thermostatics we can write the internal energy variation rate:

$$\dot{U} = Q - p_1 \dot{V}_1 - p_2 \dot{V}_2 + \sigma \dot{\mathfrak{a}} \tag{14}$$

where $Q$ is the heat power received by the system and the other three terms of the right-hand side are the powers of internal forces.

**3.1.3 Entropy balance** The second law of thermostatics gives the expression of the entropy variation rate:

$$\dot{S} = \frac{Q}{T} + \Delta \tag{15}$$

where the entropy source $\Delta$ owing to irreversibilities is positive or zero:

$$\Delta \geqslant 0$$

**3.1.4 Fundamental equation and definitions** The supposedly known fundamental equation of the system,

$$U = U(S, V_1, V_2, \mathfrak{a}, n_{i,1}, \ldots, n_{i,2}, \ldots, n_{i,\sigma}, \ldots) \tag{16}$$

is a relation among extensive variables that allows the following intensive quantities to be defined:

Temperature:

$$T \triangleq \frac{\partial U}{\partial S}$$

Pressure:

$$p_1 \triangleq -\frac{\partial U}{\partial V_1} \qquad p_2 \triangleq -\frac{\partial U}{\partial V_2}$$

Surface tension:

$$\sigma \triangleq \frac{\partial U}{\partial \mathfrak{a}}$$

Chemical potential:

$$\mu_{i,1}^* \triangleq \frac{\partial U}{\partial n_{i,1}} \qquad \mu_{i,2}^* \triangleq \frac{\partial U}{\partial n_{i,2}} \qquad \mu_{i,\sigma}^* \triangleq \frac{\partial U}{\partial n_{i,\sigma}}$$

The adsorption affinities will be

$$A_{i,1} \triangleq \mu_{i,1}^* - \mu_{i,\sigma}^*$$

$$A_{i,2} \triangleq \mu_{i,2}^* - \mu_{i,\sigma}^*$$

Mass transfer of any component from one phase to the other takes place through the interface; consequently this transfer may be considered as consisting of two reactions, namely, adsorption and desorption, characterized by distinct velocities, $\dot{\xi}_{i,1}$ and $\dot{\xi}_{i,2}$. In the absence of chemical reactions

$$\dot{\xi}_{i,1} \triangleq -\dot{n}_{i,1} \qquad \text{and} \qquad \dot{\xi}_{i,2} \triangleq -\dot{n}_{i,2}$$

Since we have

$$\dot{n}_i = 0 = \dot{n}_{i,1} + \dot{n}_{i,2} + \dot{n}_{i,\sigma}$$

we find

$$\dot{n}_{i,\sigma} = \dot{\xi}_{i,1} + \dot{\xi}_{i,2}$$

Further, if $r$ chemical reactions can take place and are characterized by velocities $\dot{\xi}_\rho$ ($\rho = 1, \ldots, r$) and stoichiometric coefficients $\nu_{i,1}^\rho$, $\nu_{i,2}^\rho$, and $\nu_{i,\sigma}^\rho$,

$$\dot{n}_{i,1} = -\dot{\xi}_{i,1} + \sum_\rho \nu_{i,1}^\rho \dot{\xi}_\rho$$

$$\dot{n}_{i,2} = -\dot{\xi}_{i,2} + \sum_\rho \nu_{i,2}^\rho \dot{\xi}_\rho \tag{17}$$

$$\dot{n}_{i,\sigma} = \dot{\xi}_{i,1} + \dot{\xi}_{i,2} + \sum_\rho \nu_{i,\sigma}^\rho \dot{\xi}_\rho$$

Taking the derivative of the fundamental equation with respect to time yields

$$\dot{U} = \frac{\partial U}{\partial S}\dot{S} + \frac{\partial U}{\partial V_1}\dot{V}_1 + \frac{\partial U}{\partial V_2}\dot{V}_2 + \frac{\partial U}{\partial \alpha}\dot{\alpha} + \sum_i \frac{\partial U}{\partial n_{i,1}}\dot{n}_{i,1} + \sum_i \frac{\partial U}{\partial n_{i,2}}\dot{n}_{i,2} + \sum_i \frac{\partial U}{\partial n_{i,\sigma}}\dot{n}_{i,\sigma} \tag{18}$$

Using the definitions mentioned above gives, in the absence of chemical reactions,

$$T\dot{S} = \dot{U} + p_1\dot{V}_1 + p_2\dot{V}_2 - \sigma\dot{\alpha} + \sum_i A_{i,1}\dot{\xi}_{i,1} + \sum_i A_{i,2}\dot{\xi}_{i,2} \tag{19}$$

When chemical reactions occur,

$$\sum_\rho A_\rho \dot{\xi}_\rho$$

must be added to the right-hand side; $A_\rho$ is the affinity of the $\rho$th chemical reaction.

**3.1.5 Entropy equation** If $\dot{U}$ of Eq. (19) is replaced by Eq. (14) (given by the first law of thermostatics),

$$T\dot{S} = Q + \sum_i A_{i,1}\dot{\xi}_{i,1} + \sum_i A_{i,2}\dot{\xi}_{i,2} \tag{20}$$

**3.1.6 Expression of the entropy source** Comparing the entropy equation [Eq. (20)] with the entropy balance [Eq. (21)] yields De Donder's inequality for two-phase systems without chemical reactions:

$$\Delta = \frac{1}{T}\left(\sum_i A_{i,1}\dot{\xi}_{i,1} + \sum_i A_{i,2}\dot{\xi}_{i,2}\right) \geqslant 0 \tag{21}$$

At the phase change equilibrium

$$A_{i,1} = 0 \quad \text{and} \quad A_{i,2} = 0$$

which correspond to the equality of the chemical potentials:

$$\mu^*_{i,1} = \mu^*_{i,2} = \mu^*_{i,\sigma} \tag{22}$$

**3.1.7 Helmholtz free energy** The Helmholtz free energy is classically defined as

$$F \triangleq U - TS$$

This definition implies that both phases and the interface have the same temperature $T$. If the derivative of this relationship is taken with respect to time,

$$\dot{F} = \dot{U} - T\dot{S} - S\dot{T}$$

and $\dot{U} - T\dot{S}$ is eliminated by using Eq. (19), then

$$\dot{F} = -p_1 \dot{V}_1 - p_2 \dot{V}_2 + \sigma \dot{\mathfrak{a}} - S\dot{T} - \sum_i A_{i,1} \dot{\xi}_{i,1} - \sum_i A_{i,2} \dot{\xi}_{i,2}$$

From this expression of $\dot{F}$, one can deduce in particular

$$\sigma = \left(\frac{\partial F}{\partial \mathfrak{a}}\right)_{T, V_1, V_2, \xi_i}$$

$$A_{i,1} = -\left(\frac{\partial F}{\partial \xi_{i,1}}\right)_{T, V_1, V_2, \mathfrak{a}, \xi_{i,2}}$$

$$A_{i,2} = -\left(\frac{\partial F}{\partial \xi_{i,2}}\right)_{T, V_1, V_2, \mathfrak{a}, \xi_{i,1}}$$

**3.1.8 Phase variables** As for the internal energy, the free energy of an isothermal two-phase system has three components:

$$F = F_1 + F_2 + F_\sigma$$

where $F_1$ and $F_2$ depend only on the phase quantities

$$F_1 = F_1(T, S_1, V_1, n_{i,1}, \ldots)$$

$$F_2 = F_2(T, S_2, V_2, n_{i,2}, \ldots)$$

whereas $F_\sigma$ depends not only on the interfacial parameters but also on the phase quantities

$$F_\sigma = F_\sigma(T, S_1, S_2, S_\sigma, V_1, V_2, \mathfrak{a}, n_{i,1} \ldots, n_{i,2}, \ldots, n_{i,\sigma}, \ldots)$$

Consequently, we have

$$\begin{aligned}
\frac{\partial F}{\partial n_{i,1}} &= \frac{\partial F_1}{\partial n_{i,1}} + \frac{\partial F_\sigma}{\partial n_{i,1}} = \mu^*_{i,1} = \mu_{i,1} + \frac{\partial F_\sigma}{\partial n_{i,1}} \\
\frac{\partial F}{\partial n_{i,2}} &= \frac{\partial F_2}{\partial n_{i,2}} + \frac{\partial F_\sigma}{\partial n_{i,2}} = \mu^*_{i,2} = \mu_{i,2} + \frac{\partial F_\sigma}{\partial n_{i,2}} \\
\frac{\partial F}{\partial n_{i,\sigma}} &= \frac{\partial F_\sigma}{\partial n_{i,\sigma}} = \mu^*_{i,\sigma} = \mu_{i,\sigma}
\end{aligned} \tag{23}$$

where $\mu_{i,1}$ and $\mu_{i,2}$ are the usual single-phase chemical potentials. Defay et al. (1966) give the expressions of the partial derivatives appearing on the right-hand side of Eqs. (23) as functions of the intensive quantities $c_{i,1}$, $c_{i,2}$, and $f_\sigma$, where $f_\sigma$ is the interface free energy per unit area:

$$f_\sigma \triangleq \frac{F_\sigma}{\mathcal{A}}$$

One finds

$$\frac{\partial F_\sigma}{\partial n_{i,1}} = \frac{\mathcal{A}}{V_1} \frac{\partial f_\sigma}{\partial c_{i,1}} \qquad \frac{\partial F_\sigma}{\partial n_{i,2}} = \frac{\mathcal{A}}{V_2} \frac{\partial f_\sigma}{\partial c_{i,2}}$$

If $\epsilon_{i,1}$ and $\epsilon_{i,2}$ are the cross chemical potentials,

$$\epsilon_{i,1} \triangleq \frac{\partial f_\sigma}{\partial c_{i,1}} \qquad \epsilon_{i,2} \triangleq \frac{\partial f_\sigma}{\partial c_{i,2}}$$

then at equilibrium, the equalities given in Eq. (22) may be written

$$\mu_{i,1} + \frac{\mathcal{A}}{V_1} \epsilon_{i,1} = \mu_{i,2} + \frac{\mathcal{A}}{V_2} \epsilon_{i,2} = \mu_{i,\sigma} \tag{24}$$

These relationships remaining true whatever the values of $V_1$ and $V_2$, they imply

$$\epsilon_{i,1} = 0 \qquad \text{and} \qquad \epsilon_{i,2} = 0$$

and hence

$$\mu_{i,1} = \mu_{i,2} = \mu_{i,\sigma} \tag{25}$$

## 3.2 Influence of Interfacial Curvature on the Equilibrium State of a Single Bubble or Droplet of a Single-Component Fluid

Let us consider a single vapor bubble in its generating liquid at uniform temperature; if we neglect the gravity term, the interface is spherical (see Sec. 1.1), and the mechanical equilibrium condition at rest is

$$p_G - p_L = \frac{2\sigma}{R} \tag{26}$$

Similarly, for a single liquid droplet in its generating steam at uniform temperature:

$$p_L - p_G = \frac{2\sigma}{R} \tag{27}$$

As the liquid and vapor phases are single components, we may write

$$\begin{aligned} \dot{G}_L &= V_L \dot{p}_L - S_L \dot{T}_L + \mu_L \dot{n}_L \\ \dot{G}_G &= V_G \dot{p}_G - S_G \dot{T}_G + \mu_G \dot{n}_G \end{aligned} \tag{28}$$

where $G_k$, $V_k$, and $S_k$ $(k = L, G)$ are quantities relative to the total mass of the corresponding phase. Further, since

$$\dot{G}_k = n_k \dot{\mu}_k + \mu_k \dot{n}_k$$

Eqs. (28) become

$$\dot{\mu}_L = v_L \dot{p}_L - s_L \dot{T}_L \qquad \dot{\mu}_G = v_G \dot{p}_G - s_G \dot{T}_G \tag{29}$$

which are the Gibbs-Duhem equations of each phase, where $\mu_k$, $v_k$, and $s_k$ are molar quantities. At equilibrium, and with $T_L = T_G$, Eqs. (25) and (29) yield

$$(s_G - s_L)\dot{T} = v_G \dot{p}_G - v_L \dot{p}_L \tag{30}$$

The pressures $p_G$ and $p_L$, and temperature $T$ appearing in Eqs. (26)-(30) are equilibrium values for liquid-vapor systems, where both phases are separated by a spherical interface (radius $R$); they can differ significantly from the saturation values ($R = \infty$) listed in tables for systems where both phases are separated by a plane interface. In the sequel, saturation values will be denoted by means of the subscript $\infty$, for example, $p_\infty$ and $T_\infty$. Figure 11 illustrates the two cases: the metastable state of liquid in equilibrium with a small bubble (Fig. 11a), and the metastable state of vapor in equilibrium with a small droplet (Fig. 11b). Temperature-entropy diagrams have been used for this representation. The differences $T - T_\infty$, $p_L - p_\infty$, and $p_G - p_\infty$ will be calculated later in this chapter.

**3.2.1 Equilibrium pressure calculations** Let us consider isothermal expansion $AB$ (Fig. 11a), resulting in the formation of a bubble, with radius $R$, from saturated liquid at saturation pressure $p_\infty(T)$. Differentiating Eq. (26) with respect to time, then eliminating $\dot{p}_L$ by using Eq. (30) (where $\dot{T} = 0$), and finally multiplying both members by $dt$ yield

$$d\left(\frac{2\sigma}{R}\right) = dp_G - \frac{v_G}{v_L}\, dp_L$$

Recalling that $v_G/v_L = \rho_L/\rho_G$, where $\rho_L/\rho_G$ is the density ratio, and assuming that the vapor phase is an ideal gas (constant $\mathcal{R}$), we obtain

$$d\left(\frac{2\sigma}{R}\right) = dp_G - \rho_L \mathcal{R} T \frac{dp_G}{p_G}$$

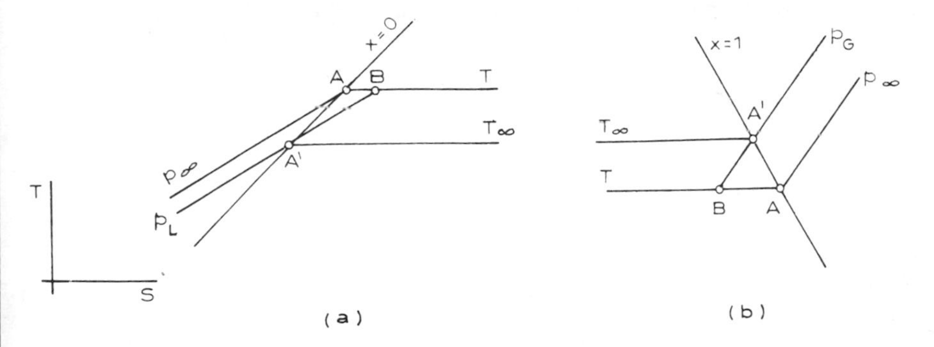


**Figure 11** Equilibrium at the incipience of the second phase.

Integration within the previously defined limits $A - B$ of the isothermal expansion, gives, assuming $\rho_L$ = const,

$$\frac{2\sigma}{R} = (p_G - p_\infty) + \rho_L \mathcal{R} T \ln \frac{p_\infty}{p_G} \tag{31}$$

In consideration of the small value of the relative pressure difference,

$$x \triangleq \frac{p_\infty - p_G}{p_G}$$

the expansion of $\ln (p_\infty / p_G)$ may be limited to the first order and reads

$$p_G - p_\infty \simeq \frac{\rho_G}{\rho_G - \rho_L} \frac{2\sigma}{R} \tag{32}$$

and hence

$$p_L - p_\infty = (p_L - p_G) + (p_G - p_\infty) \simeq \frac{\rho_L}{\rho_G - \rho_L} \frac{2\sigma}{R} \tag{33}$$

These two expressions, Eqs. (32) and (33), show that at equilibrium the pressure field around and inside a small bubble (Fig. 12) is such that

$$p_L(T) \leqslant p_G(T) \leqslant p_\infty(T)$$

At low pressure, $\rho_G \ll \rho_L$ and

$$p_\infty - p_L \simeq \frac{2\sigma}{R} = p_G - p_L$$

When, at constant temperature $T$, $p_L$ is decreased, the equilibrium radius decreases to a minimum value $R_{\min}$:

$$R_{\min} = \lim_{p_L = 0} R = \frac{2\sigma}{p_\infty}$$

Figure 13 shows the equilibrium radius of the steam bubbles plotted as a function of the pressure of the liquid phase for three constant temperatures of the liquid phase. The saturation pressures $T_\infty(p_L)$ are also indicated along the abscissa.

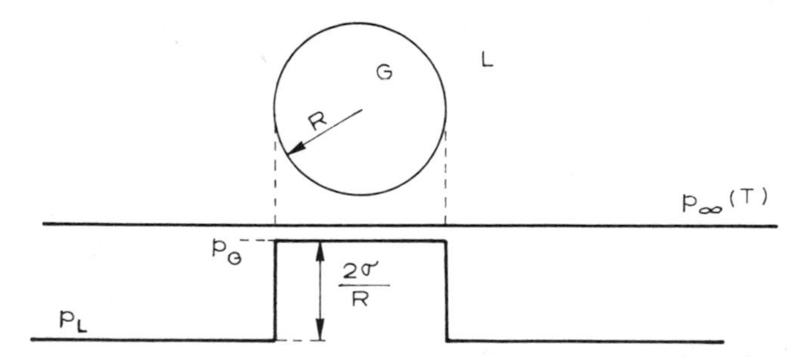


**Figure 12** Pressure field around and inside a spherical bubble.

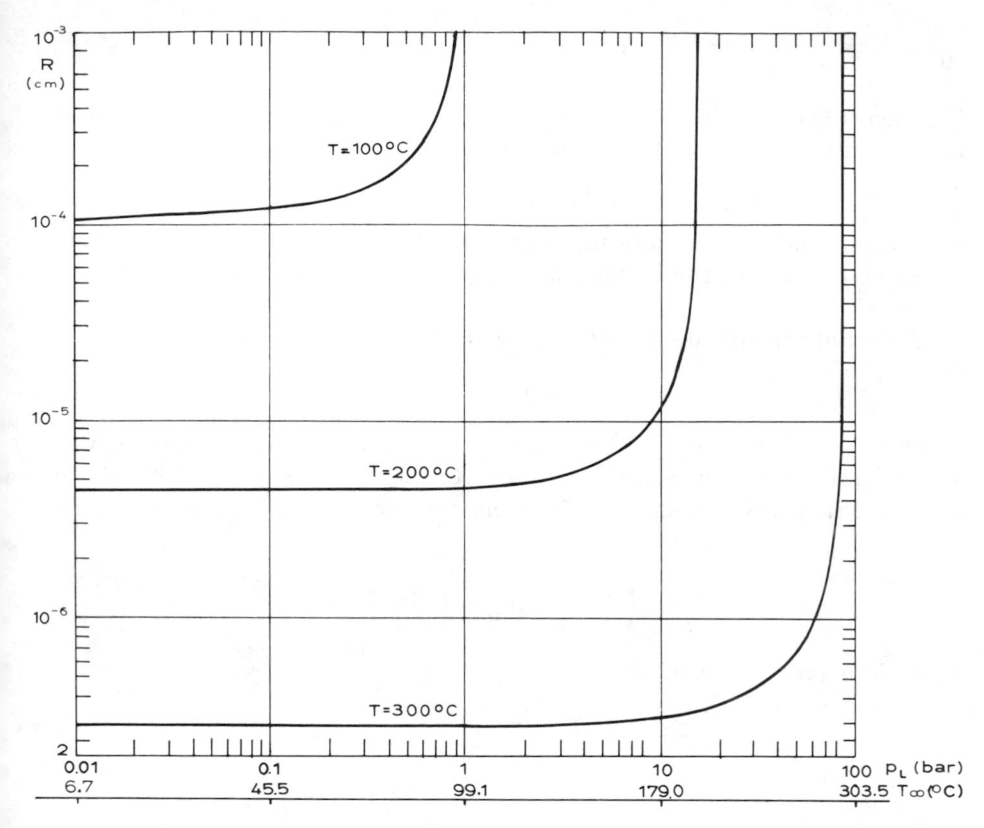


**Figure 13** Equilibrium radius of steam bubbles as a function of pressure of the liquid phase for three constant temperatures.

Consider now the isothermal compression *AB* (Fig. 11b) resulting in the formation of a droplet with radius $R$, from saturated vapor at saturation pressure $p_\infty(T)$. Differentiating Eq. (27) with respect to time, then eliminating $\dot{p}_L$ by using Eq. (30) (where $\dot{T} = 0$), and finally multiplying both members by $dt$ yield

$$d\left(\frac{2\sigma}{R}\right) = \rho_L \mathcal{R} T \frac{dp_G}{p_G} - dp_G$$

where the vapor phase is assumed to behave like an ideal gas. For constant $\rho_L$, integration of this equation within the above-defined limits *A-B* of the isothermal compression gives:

$$\frac{2\sigma}{R} = -\rho_L \mathcal{R} T \ln \frac{p_\infty}{p_G} - (p_G - p_\infty) \tag{34}$$

A series expansion limited to the first order yields

$$p_G - p_\infty \simeq \frac{\rho_G}{\rho_L - \rho_G} \frac{2\sigma}{R} \tag{35}$$

and
$$p_L - p_\infty \simeq \frac{\rho_L}{\rho_L - \rho_G} \frac{2\sigma}{R} \tag{36}$$

Equations (35) and (36) show that, at equilibrium, the pressure field around and inside a small droplet (Fig. 14) is such that

$$p_\infty(T) \leqslant p_G(T) \leqslant p_L(T)$$

Let us point out that for both the bubble and the droplet, pressure $p_G$ is near to saturation pressure for temperatures far enough from the critical temperature.

**3.2.2 Heat of vaporization** The heat of vaporization $\mathfrak{L}$ may be defined as

$$\mathfrak{L} \triangleq T(s_G - s_L) \tag{37}$$

where $s_G(p_G, T)$ and $s_L(p_L, T)$ are entropies per unit mass (or mole). Pressures $p_G$ and $p_L$ are functions of radius $R$ according to Eqs. (32) and (33) for bubbles and to Eqs. (35) and (36) for droplets. Differentiating Eq. (37) yields, for an isothermal path,

$$\frac{d\mathfrak{L}}{T} = \left(\frac{\partial s_G}{\partial p_G}\right)_T dp_G - \left(\frac{\partial s_L}{\partial p_L}\right)_T dp_L$$

Using the Maxwell relationships,

$$\left(\frac{\partial s_k}{\partial p_k}\right)_T = \frac{1}{\rho_k^2}\left(\frac{\partial \rho_k}{\partial T}\right)_{p_k} \qquad (k = L, G)$$

one finds

$$\frac{d\mathfrak{L}}{T} = \frac{1}{\rho_G^2}\left(\frac{\partial \rho_G}{\partial T}\right)_{p_G} dp_G - \frac{1}{\rho_L^2}\left(\frac{\partial \rho_L}{\partial T}\right)_{p_L} dp_L$$

Eliminating $dp_G$ and $dp_L$ by using Eqs. (26) and (30) gives, for the isothermal vaporization,

$$\frac{d\mathfrak{L}}{T} = \frac{1}{\rho_G - \rho_L}\left[\frac{1}{\rho_G}\left(\frac{\partial \rho_G}{\partial T}\right)_{p_G} - \frac{1}{\rho_L}\left(\frac{\partial \rho_L}{\partial T}\right)_{p_L}\right] d\left(\frac{2\sigma}{R}\right) \tag{38}$$

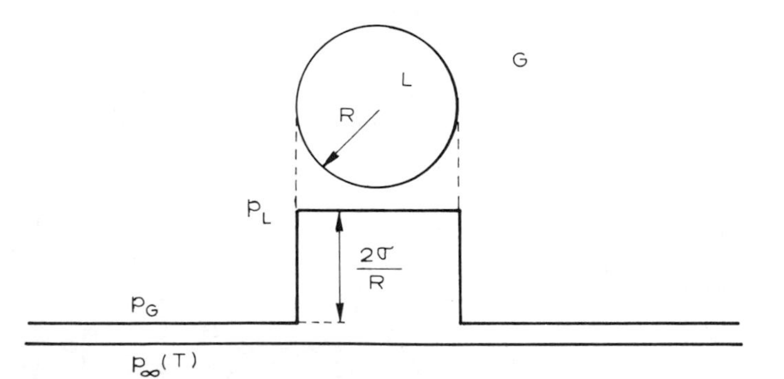


**Figure 14** Pressure field around and inside a spherical droplet.

Similarly, eliminating $dp_G$ and $dp_L$ by using Eqs. (27) and (30) gives an expression having the opposite sign for the isothermal condensation.

If it is assumed that $\rho_L$ = const and that vapor behaves like an ideal gas, Eq. (38) becomes

$$|d\mathcal{L}| = \frac{1}{\rho_L} d\left(\frac{2\sigma}{R}\right)$$

Integrating, one obtains

$$|\Delta\mathcal{L}| = |\mathcal{L}(p_G, p_L, R) - \mathcal{L}(p_\infty)| = \frac{2\sigma}{R\rho_L} \tag{39}$$

Numerically the difference $|\Delta\mathcal{L}|$ is found to be very small with respect to $\mathcal{L}$: for example, for water at 20°C and $R = 10^{-2}$ μm

$$\frac{|\Delta\mathcal{L}|}{\mathcal{L}(p_\infty)} = 6 \times 10^{-3}$$

**3.2.3 Equilibrium temperature calculation** Let us consider isobaric heating $A'B$ (Fig. 11a) of the liquid phase ($\dot{p}_L = 0$), resulting in the formation of a bubble of radius $R$, from saturated liquid at temperature $T_\infty(p_L)$. Equation (30) multiplied by $dt$ may now be written

$$\frac{\mathcal{L}}{T} dT = \frac{1}{\rho_G} dp_G$$

and, as for the ideal gas,

$$\rho_G = \frac{p_G}{\mathcal{R}T} = \frac{1}{\mathcal{R}T}\left(p_L + \frac{2\sigma}{R}\right)$$

Hence
$$\mathcal{L}\frac{dT}{T^2} = \mathcal{R}\frac{d(p_L + 2\sigma/R)}{p_L + 2\sigma/R}$$

If it is assumed that heat of vaporization and surface tension are constant, integration yields

$$\mathcal{L}\left(\frac{1}{T_\infty} - \frac{1}{T}\right) = \mathcal{R} \ln \frac{p_L + 2\sigma/R}{p_L}$$

or
$$\frac{1}{T_\infty} - \frac{1}{T} = \frac{\mathcal{R}}{\mathcal{L}} \ln\left(1 + \frac{2\sigma}{Rp_L}\right) \tag{40}$$

This expression shows that $T > T_\infty$, as indicated in Fig. 11a; for sufficiently large values of $R$, Eq. (40) is approximated by

$$\Delta T_{\text{sat}} \triangleq T - T_\infty \simeq \frac{\mathcal{R}T_\infty^2}{\mathcal{L}} \frac{2\sigma}{Rp_L} \tag{41}$$

The so-called superheat, calculated for water by using Eq. (40), has been plotted against the radius $R$ in Fig. 15. It will be noted that superheat reaches high values for small radii.

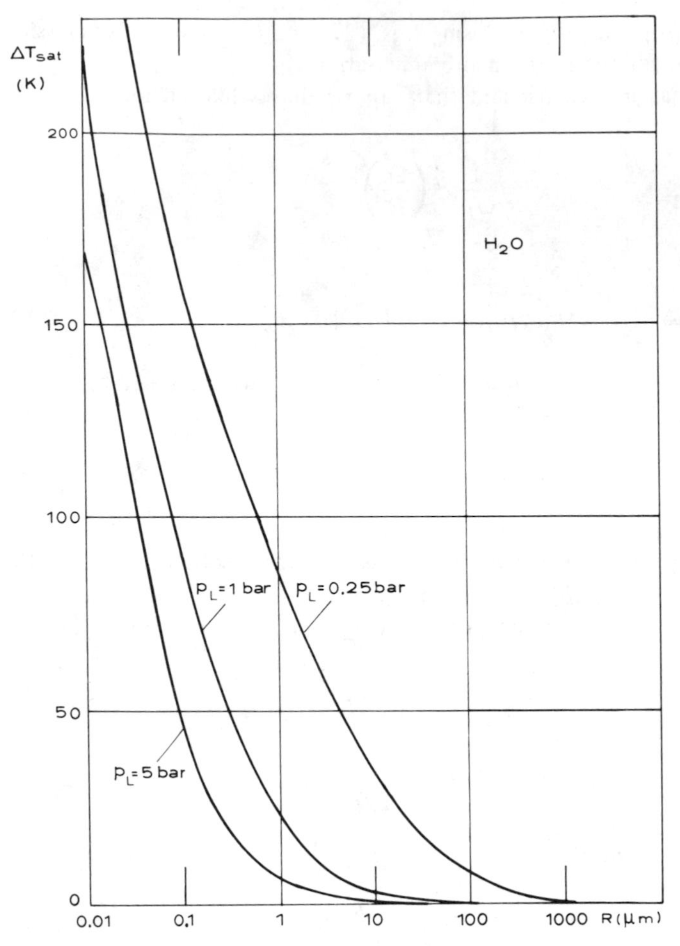


**Figure 15** Water equilibrium superheats.

Consider now the isobaric cooling $A'B$ (Fig. 11b) of the vapor phase ($\dot{p}_G = 0$), resulting in the formation of a droplet of radius $R$, from saturated vapor at temperature $T_\infty(p_G)$. Equation (30) multiplied by $dt$ may now be written

$$\frac{\mathfrak{L}}{T}\,dT = -\frac{1}{\rho_L}\,dp_L$$

and since

$$dp_L = d\left(\frac{2\sigma}{R}\right)$$

one obtains

$$\frac{dT}{T} = -\frac{1}{\rho_L \mathfrak{L}}\,d\left(\frac{2\sigma}{R}\right)$$

Assuming constant values for heat of condensation and surface tension, integration yields

$$\ln \frac{T}{T_\infty} = -\frac{2\sigma}{R\rho_L \mathfrak{L}} \tag{42}$$

Equation (42) shows that $T < T_\infty$, as indicated in Fig. 11b; one obtains, approximately,

$$\Delta T_{\text{sat}} \triangleq T - T_\infty \simeq -\frac{2\sigma}{R} \frac{T_\infty}{\rho_L \mathfrak{L}} \tag{43}$$

The results of calculations made for steam are presented in Fig. 16 and may be compared with those given in Fig. 15. Superheats (measured as temperature differences) required for the metastable equilibrium of droplets are much smaller than those required for bubbles. Moreover, the pressure influence is insignificant for droplets, whereas the pressure affects the liquid superheat in bubbles.

## 3.3 Influence of Interfacial Curvature on the Equilibrium State of a Single Bubble or Droplet of a Dilute Binary Mixture

Let us denote the solvent by subscript 1 and the solute by subscript 2; the numbers of moles of solute per mole of solvent in the gas and liquid phases are, respectively,

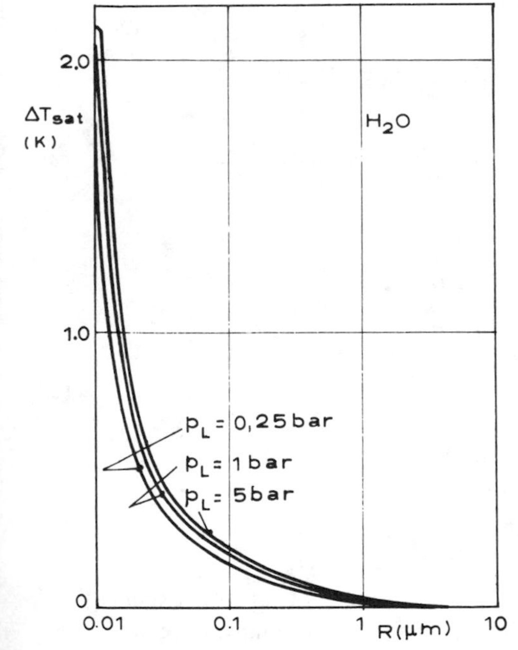


**Figure 16** Steam metastable equilibrium.

$c_G$ and $c_L$. The value of $c_L$ at saturation is $c_L^{\text{sat}}$. The analysis developed below is according to Ward et al. (1970).

In the liquid phase, the molar chemical potential $\mu_{1,L}$ of the solvent at pressure $p_L$ and temperature $T$ of the liquid mixture is

$$\mu_{1,L} = \mu_{1,L}^{\circ}(p_L, T) + \mathcal{R}T \ln \frac{1}{1+c_L} \simeq \mu_{1,L}^{\circ}(p_L, T) - \mathcal{R}Tc_L \tag{44}$$

where $\mu_{1,L}^{\circ}$ is the molar chemical potential of the solvent as a single component; similarly,

$$\mu_{2,L} = \mu_{2,L}^{\circ}(p_L, T) + \mathcal{R}T \ln \frac{c_L}{1+c_L^{\text{sat}}} \simeq \mu_{2,L}^{\circ}(p_L, T) + \mathcal{R}T \ln \frac{c_L}{c_L^{\text{sat}}} \tag{45}$$

If the saturation pressure of the solvent at temperature $T$ is $p_\infty$, and $\rho_L = \text{const}$, Eq. (44) may be written

$$\mu_{1,L} \simeq \mu_{1,L}^{\circ}(p_\infty, T) + \frac{1}{\rho_{1,L}}(p_L - p_\infty) - \mathcal{R}Tc_L \tag{46}$$

If $a_1$ and $a_2$ are the activities of the two components in the gas mixture, then

For the solvent

$$\mu_{1,G} = \mu_{1,G}^{\circ}(p_G, T) + \mathcal{R}T \ln \frac{a_1}{1+c_G}$$

or

$$\mu_{1,G} = \mu_{1,G}^{\circ}(p_\infty, T) + \mathcal{R}T \ln \frac{p_G}{p_\infty} + \mathcal{R}T \ln \frac{a_1}{1+c_G} \tag{47}$$

For the solute

$$\mu_{2,G} = \mu_{2,G}^{\circ}(p_G, T) + \mathcal{R}T \ln \frac{a_2 c_G}{1+c_G}$$

or

$$\mu_{2,G} = \mu_{2,G}^{\circ}(p_L, T) + \mathcal{R}T \ln \frac{p_G}{p_L} + \mathcal{R}T \ln \frac{a_2 c_G}{1+c_G} \tag{48}$$

The three equilibrium conditions are

$$\mu_{1,L} = \mu_{1,G} \tag{49}$$

$$\mu_{2,L} = \mu_{2,G} \tag{50}$$

$$p_G - p_L = \pm \frac{2\sigma}{R} \tag{51}$$

By using Eqs. (46) and (47), the equilibrium condition [Eq. (49)] becomes

$$\frac{a_1}{1+c_G}\frac{p_G}{p_\infty} \simeq \exp\left[\frac{p_L - p_\infty}{\rho_{1,L}\mathcal{R}T} - c_L\right] \tag{52}$$

As the binary solution is dilute ($c_L \ll 1$), the argument of the exponential is negligible if the temperature is small with respect to critical temperature, and

$$a_1 \frac{p_G}{p_\infty} \simeq 1 + c_G \tag{53}$$

From Eqs. (45) and (48), the equilibrium condition [Eq. (50)] becomes

$$\frac{c_L}{c_L^{\text{sat}}} = \frac{a_2 c_G}{1 + c_G} \frac{p_G}{p_L} \tag{54}$$

Eliminating $p_G$ by combining Eqs. (53) and (54) gives

$$c_G = \frac{c_L}{c_L^{\text{sat}}} \frac{a_1}{a_2} \frac{p_L}{p_\infty} \tag{55}$$

Further, eliminating $p_G$ by combining Eqs. (51) and (53) gives

$$p_\infty \frac{1 + c_G}{a_1} - p_L = \pm \frac{2\sigma}{R} \tag{56}$$

Replacing $c_G$ in this expression by its value given by Eq. (55) yields

$$\frac{1}{a_1} p_\infty - \frac{a_2 c_L^{\text{sat}} - c_L}{a_2 c_L^{\text{sat}}} p_L = \pm \frac{2\sigma}{R} \tag{57}$$

If the gas–vapor mixture behaves as an ideal gas, $a_1 = a_2 = 1$, then Eq. (57) reduces to

$$p_\infty - \frac{c_L^{\text{sat}} - c_L}{c_L^{\text{sat}}} p_L = \pm \frac{2\sigma}{R} \tag{58}$$

This expression must be compared with Eqs. (33) and (36) in which $\rho_G \ll \rho_L$. For a single bubble in its generating liquid, Eq. (56) may be rewritten

$$R = \frac{2\sigma}{p_\infty - [(c_L^{\text{sat}} - c_L)/c_L] p_L} \tag{59}$$

In Fig. 17 we have plotted the equilibrium radius as related to temperature for four different assumptions concerning $c_L$, namely $c_L = 0$ (pure component), $c_L = 0.5c_L^{\text{sat}}$, $c_L = c_L^{\text{sat}}$, and $c_L = 2c_L^{\text{sat}}$. The presence of dissolved gas in a bubble can significantly decrease the value of the equilibrium radius. This observation might help to explain the degassing-boiling nucleation.

Let us consider, for example, a weak binary mixture, where solubility of the dissolved gas decreases when temperature increases. Point A (Fig. 17) corresponds to an initial state of the mixture. By isobaric heating, $c_L^{\text{sat}}$ decreases, and consequently $c_L$ progressively reaches $c_L^{\text{sat}}$ as indicated by the dashed line in the figure. It is even possible that $c_L^{\text{sat}} > c_L$ before reaching the solvent saturation temperature (point B). Nucleation starts when the temperature is high enough and the equilibrium radius small enough to ensure a sufficiently large nucleation probability. According to the point where nucleation occurs along the path AB, and if the processes are slow (equilibrium), the bubbles contain more or less gas during the nucleation initiation.

Solubility of the sodium–argon mixtures increases with the temperature; consequently, the path AB for these mixtures is located to the left of the vertical of A.

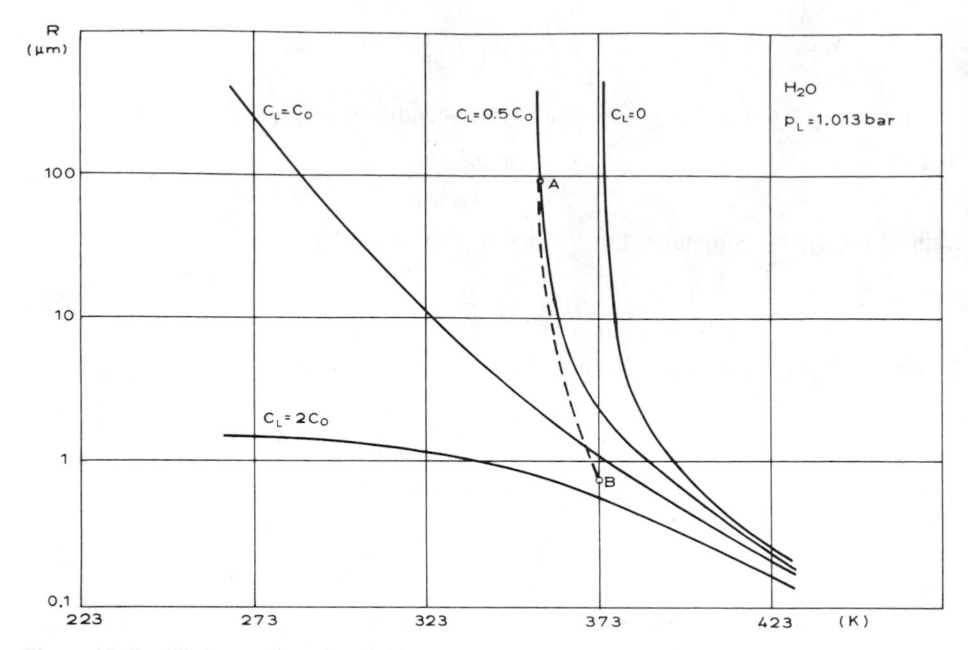


**Figure 17** Equilibrium radius of a bubble consisting of a vapor–gas mixture.

## 3.4 Stability of the Equilibrium

It will be proved below that the equilibrium of a single nucleus whose conditions are described above is metastable.

Let us therefore consider the infinitesimal variation of free energy along an isothermal path associated with a phase change (single component):

$$dF = -p_L\,dV_L - p_G\,dV_G + \sigma\,d\mathfrak{a} + \mu_L\,dn_L + \mu_G\,dn_G + \mu_\sigma\,dn_\sigma$$

Because $F$ is a linear and homogeneous function with respect to the extensive variables, we have

$$F = -p_L V_L - p_G V_G + \sigma \mathfrak{a} + \mu_L n_L + \mu_G n_G + \mu_\sigma n_\sigma \tag{60}$$

**3.4.1 Vaporization process with constant pressure** In a vaporization process, pressure $p_L$ may be assumed constant and the Gibbs free enthalpy defined as

$$G \triangleq F + p_L(V_L + V_G) \tag{61}$$

Hence, through Eq. (60),

$$G = (p_L - p_G)V_G + \sigma\,\mathfrak{a} + \mu_L n_L + \mu_G n_G + \mu_\sigma n_\sigma \tag{62}$$

For liquid at the same temperature $T$ and pressure $p_L$ but without a vapor nucleus, the free enthalpy is

$$G_0 = (n_L + n_G + n_\sigma)\mu_L^\circ \tag{63}$$

Since the number of molecules of the nucleus is small compared with the number of molecules of liquid, we assume

$$\mu_L \simeq \mu_L^\circ$$

The free-enthalpy change resulting from the formation of the nucleus is then obtained by subtracting Eq. (63) from Eq. (62):

$$(\Delta G)_{T,p_L} \triangleq G - G_0 \simeq (p_L - p_G)V_G + \sigma \mathfrak{a} + (\mu_G - \mu_L)n_G + (\mu_\sigma - \mu_L)n_\sigma \quad (64)$$

Since, near equilibrium (radius $R_c$),

$$p_L - p_G \simeq -\frac{2\sigma}{R_c} \qquad \mu_G - \mu_L \cong 0$$

and

$$\mu_\sigma - \mu_L \cong 0$$

Eq. (64) may be rewritten

$$(\Delta G)_{T,p_L} \simeq -\frac{2\sigma}{R_c}\left(\frac{4}{3}\pi R^3\right) + 4\pi R^2 \sigma$$

with

$$R = R_c + \epsilon$$

or

$$(\Delta G)_{T,p_L} \simeq 4\pi R^2 \sigma \left(1 - \frac{2}{3}\frac{R}{R_c}\right) \quad (65)$$

This relationship is illustrated by Fig. 18, where the free energy of formation of a nucleus reaches a maximum value when the radius of the nucleus reaches its equilibrium value. This maximum is

$$\Delta G_c = \tfrac{4}{3}\pi R_c^2 \sigma \quad (66)$$

The corresponding equilibrium is thus unstable, which means that any nucleus created with a radius larger than the equilibrium radius would grow and that any nucleus created with a radius smaller than that at equilibrium would collapse. The equilibrium radius is therefore called the *critical radius.*

**3.4.2 Vaporization process with constant total volume** If, instead of having considered an isothermal path with constant pressure $p_L$, we had considered an

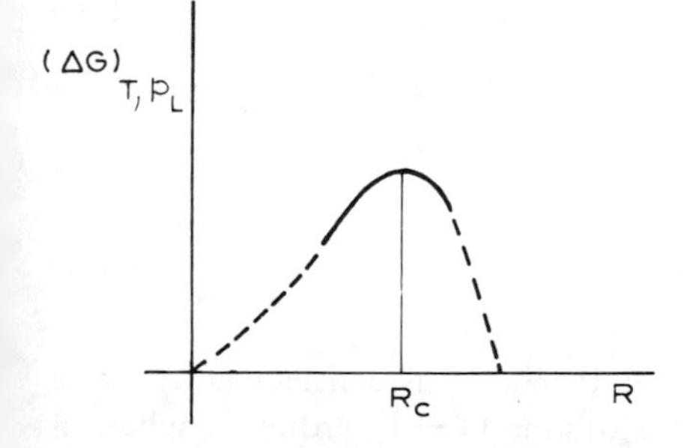


**Figure 18** The behavior of the Gibbs function in the neighborhood of the critical radius.

isothermal path with constant total volume $V_L + V_G$, the free energy of the liquid in the absence of vapor bubbles would have been

$$F_0 = -p_L^\circ(V_L + V_G) + (n_L + n_G + n_\sigma)\mu_L^\circ \tag{67}$$

where $p_L^\circ$ is the corresponding pressure of the liquid in this state, and, with the approximations

$$p_L^\circ \cong p_L \quad \text{and} \quad \mu_L^\circ \cong \mu_L$$

we would have obtained, by subtracting Eq. (67) from Eq. (60),

$$(\Delta F)_{T,V} \triangleq F - F_0 = (p_L - p_G)V_G + \sigma \mathfrak{a} + (\mu_G - \mu_L)n_G + (\mu_\sigma - \mu_L)n_\sigma \tag{68}$$

At equilibrium

$$\Delta F_c = \tfrac{4}{3}\pi R_c^2 \sigma \tag{69}$$

The free energy of formation of the nucleus also exhibits a maximum for $R = R_c$.

**3.4.3 Condensation process** In a condensation process for the creation of a liquid nucleus, pressure $p_G$ of the vapor phase (pure component) will be assumed constant. The Gibbs free enthalpy may then be defined as

$$G \triangleq F + p_G(V_L + V_G) \tag{70}$$

and Eq. (60) becomes

$$G = (p_G - p_L)V_L + \sigma \mathfrak{a} + \mu_L n_L + \mu_G n_G + \mu_\sigma n_\sigma \tag{71}$$

The free enthalpy of vapor at the same temperature and at the same pressure $p_G$, but without a liquid nucleus, is

$$G_0 = (n_L + n_G + n_\sigma)\mu_G^\circ \tag{72}$$

If

$$\mu_G^\circ = \mu_G$$

is assumed, the free-enthalpy change resulting from formation of the nucleus is obtained by subtraction of Eq. (72) from Eq. (71):

$$(\Delta G)_{T,p_G} \triangleq G - G_0 \cong (p_G - p_L)V_L + \sigma \mathfrak{a} + (\mu_L - \mu_G)n_L + (\mu_\sigma - \mu_G)n_\sigma \tag{73}$$

At equilibrium

$$\Delta G_c = \tfrac{4}{3}\pi R_c^2 \sigma \tag{74}$$

An identical expression is obtained for the free energy of formation of the critical nucleus at given temperature and volume.

## 3.5 Equilibrium of a Nucleus on a Solid Spherical Particle

Following a line of argument given by Plesset in 1969, let us consider (Fig. 19) a spherical vapor nucleus of radius $R$ occupying a volume $V' - V_S$ around a spherical

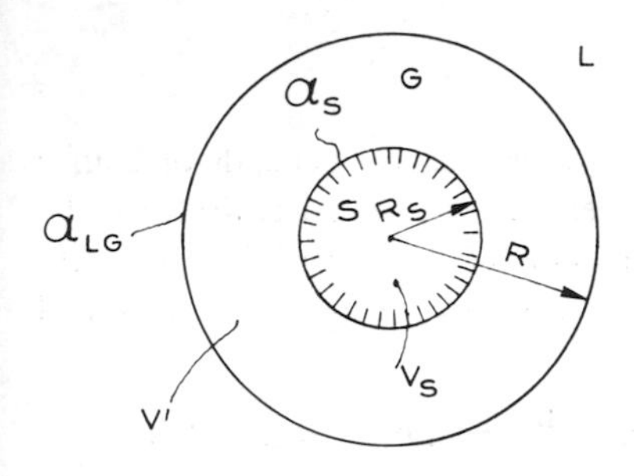


**Figure 19** Geometry of a vapor nucleus on a solid spherical particle.

particle of radius $R_S$ and volume $V_S$. In the absence of the vapor nucleus, the liquid volume around the particle will be $V + V' - V_S$. The free energy of the three-phase system defined in Fig. 19 is given by

$$F = -p_L V - \left(p_L + \frac{2\sigma_{LG}}{R}\right)(V' - V_S) - \left(p_L + \frac{2\sigma_{LG}}{R} + \frac{2\sigma_{GS}}{R_S}\right) V_S$$
$$+ \sigma_{LG}\, a_{LG} + \sigma_{GS}\, a_S + \mu_L n_L + \mu_G n_G + \mu_\sigma n_\sigma \tag{75}$$

In the absence of the vapor nucleus, the free energy is

$$F_0 = -p_L(V + V' - V_s) - \left(p_L + \frac{2\sigma_{LS}}{R_S}\right) V_S + \sigma_{LS}\, a_S + (n_L + n_G + n_\sigma)\mu_L \tag{76}$$

The free energy of formation of the vapor nucleus is thus

$$(\Delta F)_{T,V} \triangleq F - F_0 = \sigma_{LG}\left(a_{LG} - \frac{2V'}{R}\right) + (\sigma_{GS} - \sigma_{LS})\left(a_S - \frac{2V_S}{R_S}\right)$$
$$+ (\mu_G - \mu_L)n_G + (\mu_\sigma - \mu_L)n_\sigma \tag{77}$$

At equilibrium,

$$\Delta F_c = \tfrac{4}{3}\pi R^2 \sigma_{LG} + \tfrac{4}{3}\pi R_S^2(\sigma_{GS} - \sigma_{LS}) \tag{78}$$

Depending on the sign of the difference $(\sigma_{GS} - \sigma_{LS})$ between surface energies per unit area, the presence of a solid particle can increase or decrease the free energy of formation. According to the Young-Dupré relationship,

$$\sigma_{GS} - \sigma_{LS} = \sigma_{LG} \cos\theta$$

and Eq. (78) may be written

$$\Delta F_c = \frac{4}{3}\pi R^2 \sigma_{LG}\left(1 + \frac{R_S^2}{R^2}\cos\theta\right) \tag{79}$$

A decrease of the free energy of formation is obtained when the liquid does not wet the wall ($\pi/2 < \theta < \pi$).

A similar calculation holds for the case of a liquid nucleus on a solid spherical particle. The expression of the free energy of formation is then

$$\Delta F_c = \frac{4}{3}\pi R^2 \sigma_{LG}\left(1 - \frac{R_S^2}{R^2}\cos\theta\right) \tag{80}$$

## 3.6 Equilibrium of a Nucleus in a Wall Cavity or on a Flat Solid Surface

The free energy of formation of a vapor nucleus in a wall cavity depends of course on the geometry of the cavity. In the example given below, conical cavities will be considered; the symbols denoting the geometric parameters are defined in Fig. 20.

Let us consider system II of Fig. 21. Its boundaries are defined by the dashed lines; its free energy is

$$F_{\mathrm{II}} = -p_L V_L - p_G V_G + \sigma_{LG}\, \mathfrak{a}_{LG}^{\mathrm{II}} + \sigma_{LS}\, \mathfrak{a}_{LS}^{\mathrm{II}} + \sigma_{GS} \mathfrak{a}_{GS}^{\mathrm{II}} + \mu_L n_L + \mu_G n_G + \mu_\sigma n_\sigma \tag{81}$$

If free enthalpy is defined by Eq. (61) for a system where pressure $p_L$ is kept constant, one has

$$G_{\mathrm{II}} = -\frac{2\sigma_{LG}}{R} V_G + \sigma_{LG}\mathfrak{a}_{LG}^{\mathrm{II}} + \sigma_{LS}\mathfrak{a}_{LS}^{\mathrm{II}} + \sigma_{GS}\mathfrak{a}_{GS}^{\mathrm{II}} + \mu_L n_L + \mu_G n_G + \mu_\sigma n_\sigma \tag{82}$$

For system I, $V_G = 0$, and the free enthalpy is

$$G_{\mathrm{I}} = \sigma_{LS}\mathfrak{a}_{LS}^{\mathrm{I}} + \mu_{L,0}(n_L + n_G + n_\sigma) \tag{83}$$

Consequently, the free enthalpy of formation is given by

$$(\Delta G)_{T,p_L} \simeq -\frac{2\sigma_{LG}}{R} V_G + \sigma_{LG}\mathfrak{a}_{LG}^{\mathrm{II}} + \sigma_{LS}(\mathfrak{a}_{LS}^{\mathrm{II}} - \mathfrak{a}_{LS}^{\mathrm{I}}) + \sigma_{GS}\mathfrak{a}_{GS}^{\mathrm{II}} \tag{84}$$

and since

$$\mathfrak{a}_{LS}^{\mathrm{I}} = \mathfrak{a}_{LS}^{\mathrm{II}} + \mathfrak{a}_{GS}^{\mathrm{II}}$$

then

$$(\Delta G)_{T,p_L} \simeq -\frac{2\sigma_{LG}}{R} V_G + \sigma_{LG}\mathfrak{a}_{LG}^{\mathrm{II}} + (\sigma_{GS} - \sigma_{LS})\mathfrak{a}_{GS}^{\mathrm{II}}$$

and, using the Young-Dupré relationship,

$$(\Delta G)_{T,p_L} \simeq \sigma_{LG}\left(\mathfrak{a}_{LG}^{\mathrm{II}} - \frac{2}{R} V_G + \mathfrak{a}_{GS}^{\mathrm{II}} \cos\theta\right) \tag{85}$$

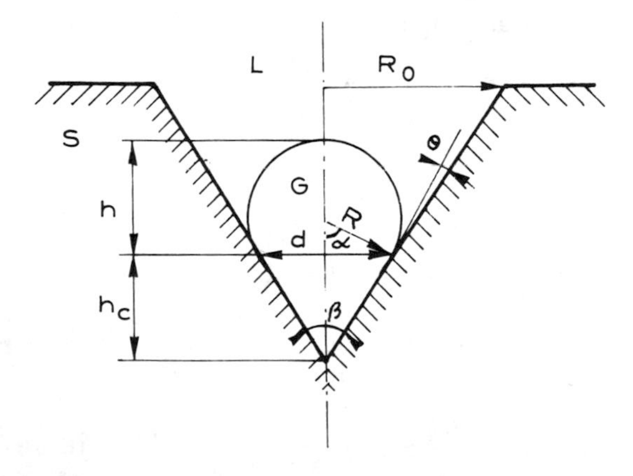


**Figure 20** Geometry of a vapor nucleus in a conical wall cavity.

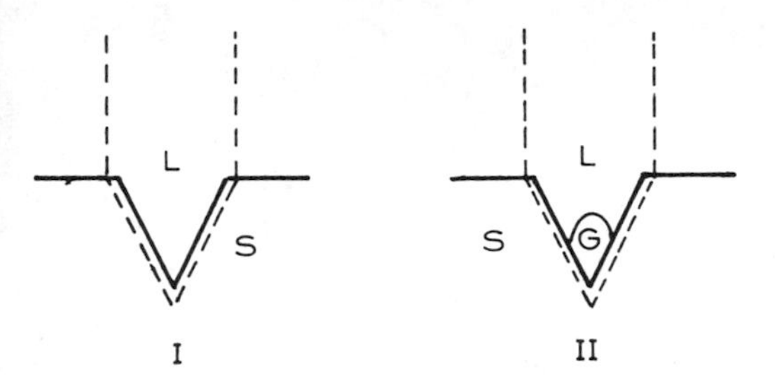


**Figure 21** Formation of a vapor nucleus in a cavity.

Introducing into this expression the geometric relationship

$$V_G = \frac{1}{3}\pi R^2 h_c \sin^2\alpha + \pi h\left(\frac{d^2}{8}+\frac{h^2}{6}\right) = \frac{1}{3}\pi R^3 \cot\frac{\beta}{2}\cos^2\left(\theta-\frac{\beta}{2}\right)$$

$$+\frac{1}{3}\pi R^3\left[2-3\sin\left(\theta-\frac{\beta}{2}\right)-\sin^3\left(\theta-\frac{\beta}{2}\right)\right]$$

$$\mathfrak{a}_{LG}^{\mathrm{II}} = \pi\left(h^2+\frac{d^2}{4}\right) = 2\pi R^2\left[1-\sin\left(\theta-\frac{\beta}{2}\right)\right]$$

$$\mathfrak{a}_{GS}^{\mathrm{I}} = \pi R^2\cos^2\left(\theta-\frac{\beta}{2}\right)\left(\sin\frac{\beta}{2}\right)^{-1}$$

one finds

$$(\Delta G)_{T,p_L} \simeq \frac{1}{3}\pi R^2\sigma_{LG}\left[2-2\sin^3\left(\theta-\frac{\beta}{2}\right)-2\cot\frac{\beta}{2}\cos^3\left(\theta-\frac{\beta}{2}\right)\right.$$

$$\left.+3\cos\theta\cos^2\left(\theta-\frac{\beta}{2}\right)\left(\sin\frac{\beta}{2}\right)^{-1}\right] \tag{86}$$

The term in brackets in Eq. (86) is a function of both the angle of the cavity and the contact angle. Radius $R$ is related to the radius of cavity $R_0$ by the relationship

$$R = R_0\frac{\cot\beta/2}{1-\sin(\theta-\beta/2)+\cos(\theta-\beta/2)\cot\beta/2}$$

For example, if $\theta = 0°$, $\beta/2 = 30°$, and $R_0 = 1$ $\mu$m, one obtains $R = 0.58$ $\mu$m, $(\Delta G)_{T,p_L} \simeq \frac{4.5}{3}\pi R^2\sigma_{LG}$.

A particular example of Eq. (86) is obtained when setting $\beta/2 = \pi/2$. This yields

$$(\Delta G)_{T,p_L} = \tfrac{1}{3}\pi R^2\sigma_{LG}(2+3\cos\theta-\cos^3\theta) \tag{87}$$

This result is the expression of the free enthalpy of formation of a nucleus at equilibrium on a flat horizontal surface (Fig. 22). For $\theta = 0$, Eq. (87) becomes

$$(\Delta G)_{T,p_L} = \tfrac{4}{3}\pi R^2\sigma_{LG}$$

Results for some other cavity shapes are presented in Kottowski (1971).

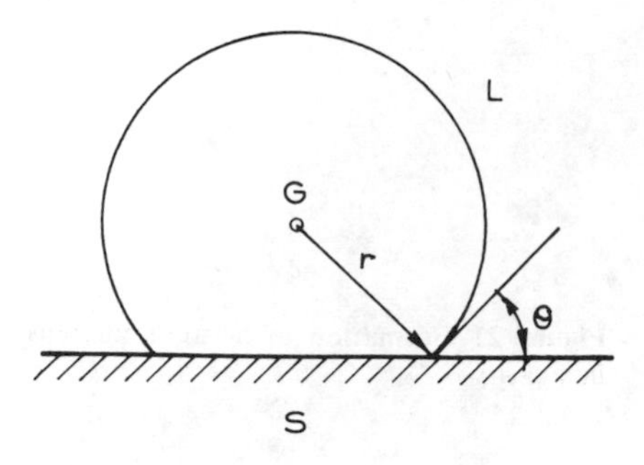


**Figure 22** Spherical nucleus on a flat horizontal surface.

# 4 HOMOGENEOUS NUCLEATION

Recent reviews on this subject are given in Cole (1974), Skripov (1974), Blander and Katz (1975), and Mathieu (1976).

## 4.1 Thermodynamic Limit of Liquid Superheat

The thermodynamic limit of liquid superheat is, in a pressure-volume diagram, the curve of the minima of the isotherms. This curve is the limit between a metastable and an unstable region in the $p$-$V$ diagram.

To determine this curve, one needs an accurate equation of state for the fluid. If the van der Waals equation, written in terms of reduced variables ($p_r$, $V_r$, $T_r$), is used,

$$\left(p_r + \frac{3}{V_r^2}\right)\left(V_r - \frac{1}{3}\right) = \frac{8}{3} T_r$$

the condition

$$\left(\frac{\partial p_r}{\partial V_r}\right)_{T_r} = 0$$

is

$$T_r = \frac{9}{4} \frac{(V_r - \frac{1}{3})^2}{V_r^3}$$

Calculating $T_r$ related to $p_r$ gives the van der Waals curve in Fig. 23. The experimental points lie between the theoretical results obtained by the van der Waals and Berthelot equations.

## 4.2 Kinetic Limits of Liquid Superheat

According to the kinetic theory, clusters are formed as density fluctuations in the liquid. The growth or decay of clusters is a consequence of evaporation or condensation of molecules at the gas-liquid interface, in a step by step, or bimolecular process. The rate of formation of clusters containing $x$ molecules is given by

$$J = N\left(\frac{2\sigma}{\pi m B}\right)^{1/2} \exp\left(-\frac{\Delta G}{kT}\right) \tag{88}$$

where $N$ = number density of the liquid
$m$ = molecular mass
$B \simeq \frac{2}{3}$ (except for cavitation, where $B \simeq 1$)

If one expresses $\Delta G$ by Eq. (66), where $R_c^2$ is eliminated by Eq. (33), then

$$J = N\left(\frac{2\sigma}{\pi m B}\right)^{1/2} \exp\left[-\frac{16\pi\sigma^3\rho_L^2}{3kT(\rho_L-\rho_G)^2(p_\infty - p_L)^2}\right] \tag{89}$$

The exponential term is strongly dependent on temperature, whereas the prefactor brings a relatively insignificant change in the kinetic limit of superheat.

The basic theory of homogeneous nucleation presented by Döring and Volmer has been developed by Zel'dovich and Kagan. The latter (see Skripov, 1974) formulated the boundary conditions of growing bubbles in more detail, taking into account the influence of heat transfer, inertia, and viscosity on the bubble nucleation rate. However, these factors seem to have little influence on the nucleation rate because the exponential factor, always the same, varies by about three or four orders of magnitude per degree near the limit of superheat of pure substances.

It is often recommended to consider as a superheat limit the temperature at which a given nucleation rate, predicted, for example, by Eq. (89), is obtained. For example, Cole (1974) considered it reasonable to assume a nucleation rate of one vapor nucleus per second per cubic centimeter of liquid, whereas Blander and Katz (1975) obtained accurate results when assuming $J = 10^{-6}$ bubbles per cubic centimeter per second.

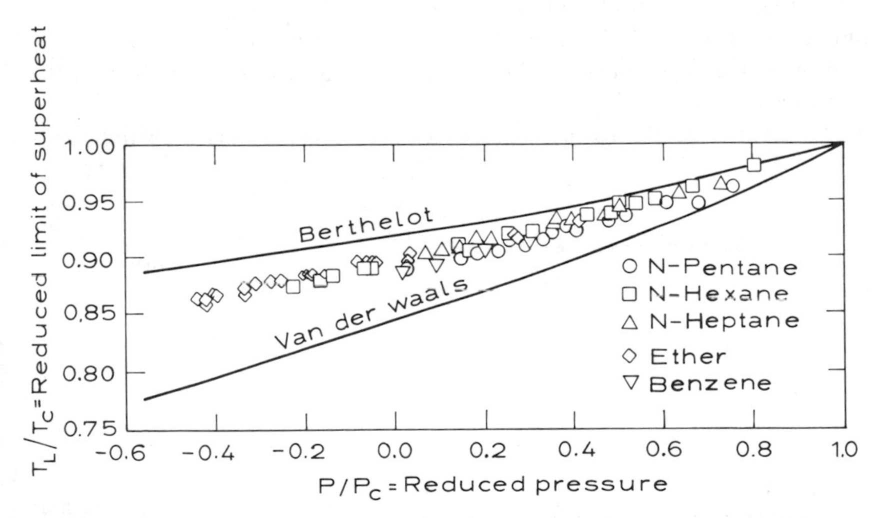


**Figure 23** Reduced limits of liquid superheat as functions of reduced pressure (Blander and Katz, 1975).

# 5 HETEROGENEOUS NUCLEATION

The decrease of the critical radius in several examples where nucleation takes place at interfaces has been examined in Sec. 3. A considerable number of experimental investigations have been devoted to this subject, mostly for liquid metals, in relation to the safety of fast-breeder nuclear reactors, because superheat can reach very high values for these fluids (see, for example, Fig. 24).

Many parameters are supposed to affect the incipient boiling superheat; for example,

1. Gas content of liquid
2. Gas pockets in the wall cavities
3. Oxide impurity in the liquid
4. Pressure–temperature history
5. Heating surface condition
6. Velocity of the liquid
7. Nuclear radiation
8. Heat flux

Let us summarize some results:

1. *Gas Content of the Liquid.* Inert gas bubbles are likely to form on the wall of the cold part of the sodium-cooled reactor loops because solubility of an inert gas decreases with decreasing temperatures in liquid metals. According to Kottowski et al. (1970), for most of the superheat data reported, the transported gas effect overshadows other effects. Inert gas bubble dynamics is the object of a model proposed by Holtz and Singer (1969).

2. *Gas Pockets in the Wall Cavities.* If wall cavities contain entrapped gas with radii greater than the critical radius, nucleation will occur as soon as the temperature exceeds saturation temperature. The gas will then be progressively carried away with the vapor bubbles, the duration of this process depending on the shape of the cavities.

3. *Oxide Impurity in the Liquid.* Experiments at several laboratories indicate a

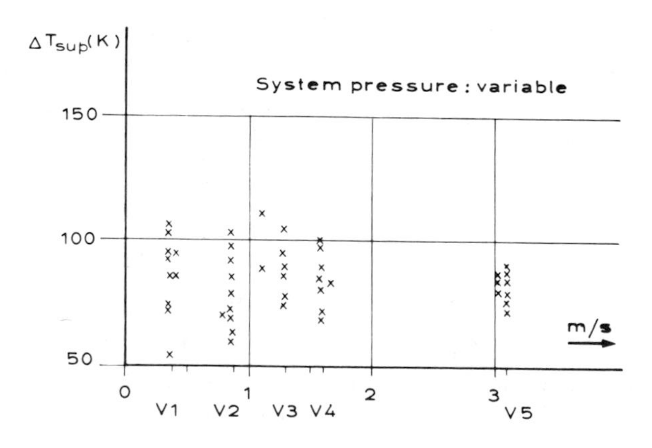


**Figure 24** Sodium bulk superheat compared with velocity (Kottowski et al., 1974).

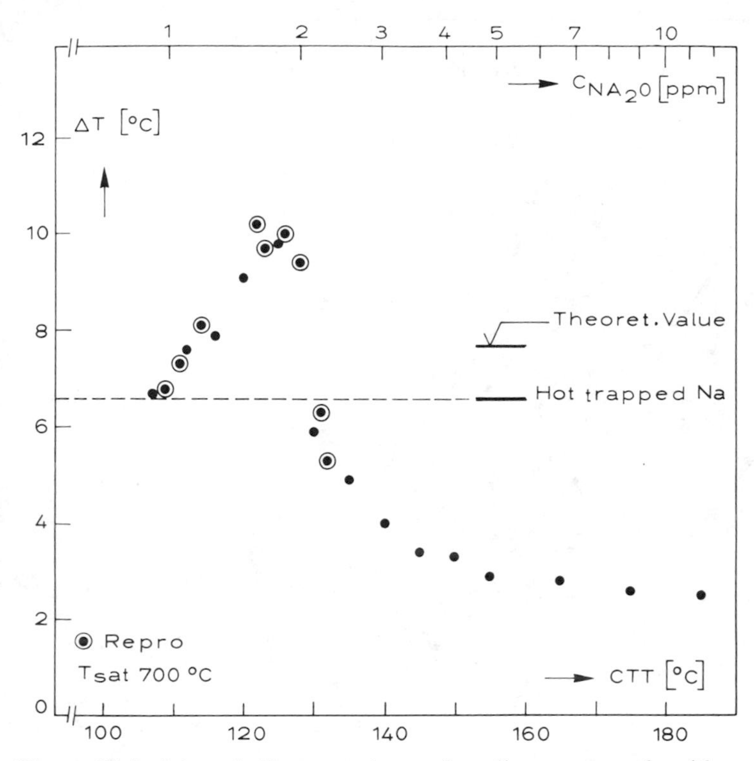


**Figure 25** Incipient boiling superheat of sodium compared with cold-trap temperature (Schultheiss, 1971).

decrease of superheat when oxygen concentration is increased. For example, consider Fig. 25 (Schultheiss, 1971), showing the incipient boiling superheat of sodium compared with sodium oxide concentration, which is changed by cold-trap temperature variation.

4. *Pressure-Temperature History.* Holtz and Singer (1969) showed that the preboiling of an experimental apparatus determines the size of the available nucleation sites and has greater influence than the surface roughness on the superheat. In Fig. 26, $p_{Ai}$ denotes the initial inert gas partial pressure. Dwyer (1976) extensively reported the theoretical considerations and experimental results in this field as well as results relative to the other topics of this section.

5. *Heating Surface Condition.* According to Peppler and Kottowski (1977), laboratory-scale experiments on artificial cavities of known size and shape show good agreement between superheat prediction and measurement, whereas measurements obtained on normal or mechanically treated surfaces are less definite.

6. *Velocity of the Liquid.* Since the studies by Pezzilli et al. (1970) and Logan et al. (1970), many investigations have been carried out to verify and explain the trend of bulk superheat to decrease while velocity is increased. To determine whether turbulent pressure fluctuations at the wall could explain the observed decrease in bulk superheat, Bankoff (1971) developed a random-walk theory. He

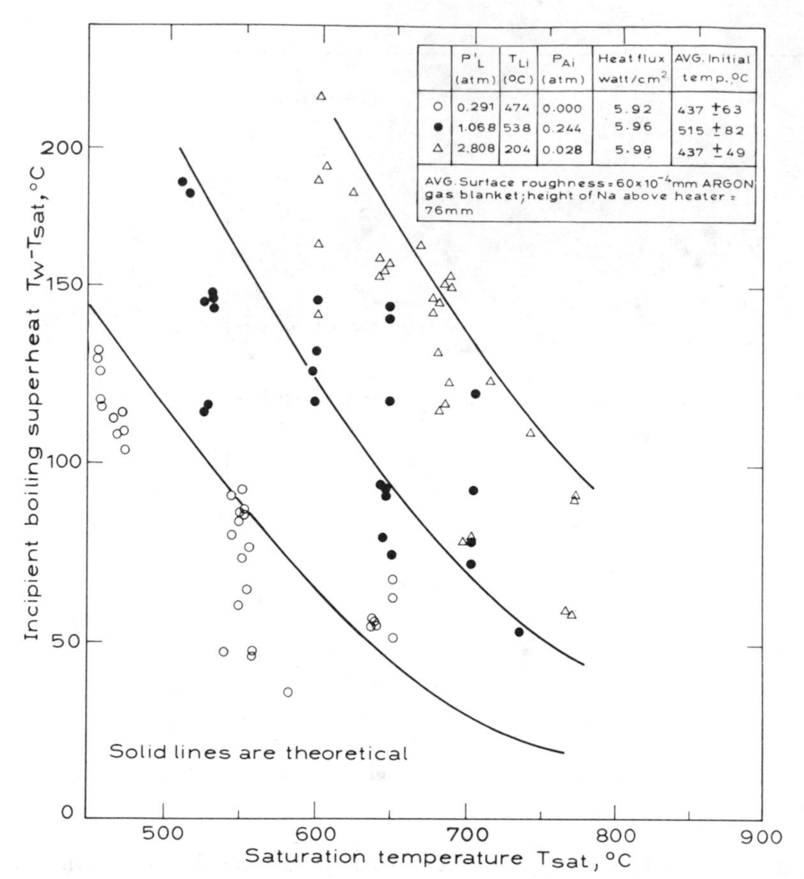


**Figure 26** Influence of preboiling history on incipient boiling superheat of sodium (Holtz and Singer, 1969).

determined the expected waiting time between two successive bubbles in a circulating liquid model system as a function of wall superheat and bulk velocity. For waiting times in the range of 0.033–0.1 s, which correspond to the usual bubbling frequencies, and turbulent intensities of 0.05–0.3, the predicted velocities are almost an order of magnitude greater than the experimental ones.

7. *Nuclear Radiation.* Holtz and Singer (1969) did not find any significant effect of thermal neutron fluxes on sodium superheat. However, they agreed that these results cannot be extrapolated for fast-breeder reactors. Working with water, acetone, and benzene, El-Nagdy and Harris (1970) found for each liquid and for each neutron energy in the range 2.45–14.1 MeV that there exists a superheat below which the liquid is insensitive to radiation.

8. *Heat Flux.* Hsu (1962) was the first to publish a theoretical relation for predicting the effect of wall heat flux on superheat. Bergles and Rohsenow (1964) extended Hsu's concept. The following assumptions were made:

1. The nuclei forming on the wall cavities have a truncated spherical shape.
2. Mechanical equilibrium is obtained during growth of the nuclei.
3. A nucleus grows if the liquid temperature at a distance $y'$ from the wall surface equal to the height of the nucleus is larger than the superheat temperature required for the equilibrium of the nucleus.
4. The presence of a small bubble does not affect the temperature profile of the liquid.
5. There is no disturbance from other nuclei.

Because the height $y'$ is proportional to the radius $R$ of the bubble, which in turn is proportional to the cavity radius $R_c$,

$$y' = c_1 R$$

and Eq. (40) may be written

$$\Delta T_{\text{sat}} = \frac{\mathcal{R} T_\infty^2}{\mathcal{L}} \ln \left(1 + \frac{2c_1 \sigma}{p_L y'}\right) \tag{90}$$

In the viscous sublayer, the heat flux density is given by

$$\dot{q} = -k \left.\frac{\partial T}{\partial y}\right|_{y=0}$$

and the temperature profile by

$$T(y) = T_w - \frac{\dot{q}}{k} y \tag{91}$$

where $T_w$ is the wall temperature (Fig. 27). For $y = y'$, Eqs. (90) and (91) give

$$\left.\frac{dT}{dy}\right|_{y=y'} = -\frac{\mathcal{R} T_\infty^2}{\mathcal{L}} \left(1 + \frac{2C_1 \sigma}{p_L y'^2}\right) = -\frac{\dot{q}}{k}$$

or with

$$S \triangleq \frac{T_\infty^2 \sigma}{p_L}$$

$$y' = -\frac{C_1 \sigma}{p_L} + \sqrt{\left(\frac{C_1 \sigma}{p_L}\right)^2 + \frac{2kSC_1}{\dot{q}}}$$

The corresponding wall superheat is

$$\Delta T_{\text{sat}} = T_w - T \simeq \frac{qy'}{k} + \frac{2C_1 S}{y'}$$

or

$$\Delta T_{\text{sat}} \simeq \left(\frac{8C_1 Sq}{k}\right)^{1/2}$$

The influence of the cavity size distribution on the wall superheat can easily be explained by using this model.

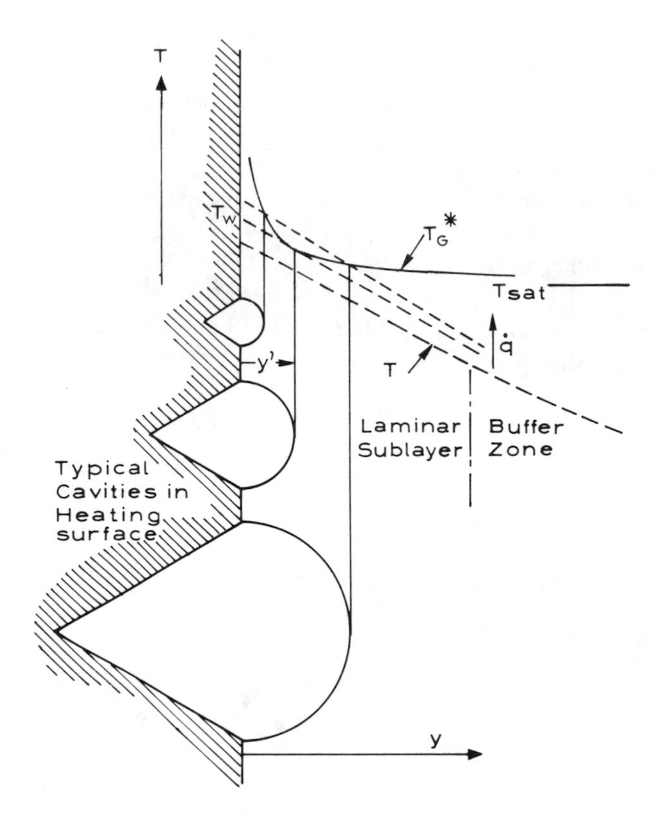


**Figure 27** Boiling nucleation on a heated surface.

## NOMENCLATURE

| | |
|---|---|
| $A$ | area |
| $A_i$ | affinity |
| $a$ | capillary constant [Eq. (4)] |
| $a_1, a_2$ | activities |
| $B$ | constant [Eq. (88)] |
| $C$ | mean curvature |
| $c_G, c_L$ | moles of solute per mole of solvent in the gas or in the liquid phase |
| $c_i$ | molar concentration |
| $D$ | characteristic dimension |
| Eo | Eötvos number |
| $F$ | Helmholtz free energy |
| $\mathbf{F}$ | force |
| $f_\sigma$ | interface free energy per unit area |
| $G$ | Gibbs free enthalpy [Eq. (61)] |
| $g$ | gravity constant |
| $h$ | height |
| $J$ | rate of nucleation per unit volume |

| | |
|---|---|
| $k$ | Boltzmann constant |
| $k$ | thermal conductivity |
| $\mathcal{L}$ | heat of vaporization |
| $l$ | length |
| $M_0, M_1, M_2$ | moments of the excess tension |
| $m$ | molecular mass |
| $N$ | number density |
| $n_i$ | number of moles |
| $p$ | pressure |
| $p_{Ai}$ | initial inert gas partial pressure |
| $Q$ | heat power |
| $\dot{q}$ | heat flux density |
| $R$ | radius of curvature |
| $\mathcal{R}$ | ideal gas constant |
| $S$ | entropy |
| $s$ | molar entropy |
| $T$ | temperature |
| $t$ | modulus of the tension force per unit area |
| $U$ | internal energy |
| $V$ | volume |
| $v$ | molar volume |
| $W$ | work |
| $W_A$ | work of adhesion |
| $W_E$ | work of spreading |
| $W_C$ | work of cohesion |
| $x_{max}$ | half-width of a droplet |
| $y$ | distance from a wall |
| $z$ | vertical coordinate |
| $\alpha$ | angle |
| $\beta$ | Neumann angle |
| $\gamma_i$ | molar adsorption |
| $\Delta$ | entropy source |
| $\Delta T_{sat}$ | superheat |
| $\epsilon_i$ | cross chemical potential |
| $\theta$ | contact angle |
| $\mu_i$ | chemical potential |
| $\nu_i^\rho$ | stoichiometric coefficient |
| $\dot{\xi}_i$ | velocity of a reaction |
| $\rho$ | density |
| $\Sigma$ | surface in Fig. 7 |
| $\sigma$ | surface tension |
| $\sigma_{ij}$ | capillary force per unit length between phases *i* and *j* |

**Subscripts**

| | |
|---|---|
| 0 | reference level |
| 1 | concave side |

| | |
|---|---|
| 2 | convex side |
| $\infty$ | saturation (infinite radius) |
| $c$ | critical |
| $G$ | gas phase |
| $L$ | liquid phase |
| $r$ | reduced variable |
| $S$ | solid phase |
| sup | superheat |
| $w$ | wall |
| $\sigma$ | interface |

**Superscripts**

| | |
|---|---|
| * | nondimensional or excess |
| $\cdot$ | time derivative |
| $\circ$ | single component |

## REFERENCES


Adamson, A. W., *Physical Chemistry of Surfaces*, 2d ed., Wiley-Interscience, New York, 1967.

Adamson, A. W. and Ling, I., The Status of Contact Angle as a Thermodynamic Property, *Adv. Chem. Ser.*, vol. 43, pp. 57–73, 1964.

Bankoff, S. G., A Random-Walk Theory for the Inception of Bubble Growth in Flowing Liquid Metals at a Heated Wall, Paper presented at 4th Int. Seminar on Heat Transfer in Liquid Metal, Trogir, Yugoslavia, 1971.

Becher, P., *Emulsions: Theory and Practice*, 2d ed., Reinhold, New York, 1965.

Bergles, A. E. and Rohsenow, N. M., The Determination of Forced-Convection Surface Boiling Heat Transfer, *J. Heat Transfer*, vol. 86, pp. 365–372, 1964.

Blander, M. and Katz, J. L., Bubble Nucleation in Liquids, *AIChE J.*, vol. 21, no. 5, pp. 833–848, 1975.

Chappuis, J., Contribution à l'Etude du Mouillage, Application aux Problèmes de Lubrification, Thèse de doctorat ès sciences, Université Claude Bernard, Lyon, 1974.

Cole, R., Boiling Nucleation, *Adv. Heat Transfer*, vol. 10, pp. 85–166, 1974.

Defay, R., Prigogine, I., Bellemans, A., and Everett, D. H., *Surface Tension and Adsorption*, Longmans, London, 1966.

Dennery, F. and Guénot, R., Considérations sur les Phénomènes Interfaciaux, Conférence à la Société Française des Thermiciens, Paris, 1976.

Dwyer, O. E., *Boiling Liquid Metal Heat Transfer*, American Nuclear Society, Hinsdale, Ill., 1976.

El-Nagdy, M. M. and Harris, M. J., Experimental Study of Radiation-induced Boiling in Superheated Liquids, University of Manchester, England, 1970.

Goodrich, F. C., The Thermodynamics of Fluid Interfaces, in *Surface and Colloid Science*, vol. 1, ed. E. Matijevic, pp. 1–37, Wiley-Interscience, New York, 1969.

Holtz, R. E. and Singer, R. M., A Study of the Incipient Boiling of Sodium, ANL-7608, 1969.

Hsu, Y. Y., On the Size Range of Active Nucleation Cavities on a Heating Surface, *J. Heat Transfer*, vol. 84, p. 207, 1962.

Johnson, R. E., Jr. and Dettre, R. H., Contact Angle Hysteresis, *Adv. Chem. Ser.*, vol. 43, pp. 112–135, 1964.

Kottowski, H. M., The Mechanism of Nucleation, Superheating and Reducing Effects on the

Activation Energy of Nucleation, Paper presented at 4th Int. Seminar on Heat Transfer in Liquid Metal, Trogir, Yugoslavia, 1971.

Kottowski, H., Grass, G., Birke, A., and Lazarus, J., Prevention of Superheat by Injection of Argon Gas Bubbles into Sodium, private communication, 1970.

Kottowski, H., Mol, M., and Warnsing, R., Experimental Investigation of the Influence of Velocity Effects on Local and Bulk Superheat, private communication, 1974.

Logan, A. D., Baroczy, C. J., Landoni, J. A., and Morewitz, H. A., Effects of Velocity, Oxide Level and Flow Transients on Boiling Initiation in Sodium, *Proc. Symp. Liquid Metal Heat Transfer Fluid Dynamics*, p. 116, ASME, New York, 1970.

Mathieu, P., Condensation de Vapeurs dans des Ecoulements Supersoniques en Tuyères, Thèse de doctorat, Université de Liège, 1976.

Oliver, J. F., Huh, C., and Mason, S. G., Resistance to Spreading of Liquids by Sharp Edges, *J. Colloid Interface Sci.*, vol. 59, no. 3, pp. 568–581, 1977.

Peppler, W. and Kottowski, H., Survey on Basic Sodium Boiling Phenomena, private communication, 1977.

Pezzilli, M., Sacco, A., Scarano, G., Tomasetti, G., and Pinchera, G. C., The Nemi Code for Sodium Boiling and Its Experimental Basis, *Proc. Symp. Liquid Metal Heat Transfer Fluid Dynamics*, p. 153, ASME, New York, 1970.

Plesset, M. S., The Tensile Strength of Liquids, in *Cavitation State of Knowledge*, eds. J. M. Robertson and G. F. Wislicenus, *Proc. ASME Fluids Engineering and Applied Mechanics Conf., Evanston, Ill., June 16-18, 1969*, pp. 15-25, 1969.

Schultheiss, G. F., Aspects of Liquid Metal Superheat and the Effects on Dynamic Boiling, Paper presented at 4th Int. Seminar on Heat Transfer in Liquid Metal, Trogir, Yugoslavia, 1971.

Schwartz, A. M., The Dynamics of Contact Angle Phenomena, *Adv. Colloid Interface Sci.*, vol. 4, pp. 349–374, 1975.

Skripov, V. P., *Metastable Liquids*, Wiley, New York, 1974.

Wachters, De Warmteoverdracht van een Hete Wand naar Druppels in de Sferoidale Toestand, Doctoral thesis, Technische Hogeschool Delft, 1965.

Ward, C. A., Balakrishnan, A., and Hooper, F. C., On the Thermodynamics of Nucleation in Weak Gas-Liquid Solutions, *J. Basic Eng.*, vol. 92, no. 4, pp. 695–704, 1970.

Zisman, W. A., Relation of Equilibrium Contact Angle to Liquid and Solid Constitution, *Adv. Chem. Ser.*, vol. 43, pp. 1–51, 1964.

CHAPTER

# SEVEN

# INSTANTANEOUS SPACE-AVERAGED EQUATIONS

**J. M. Delhaye**

Space-averaged equations for two-phase flow were derived by Delhaye (1968) and Vernier and Delhaye (1968). These calculations were developed following an article by Birkhoff (1964) concerning the errors to be avoided when setting up such equations for single-phase flow. The most thorough study of these equations and their application, within the frame of a two-fluid model, to a problem of liquid film hydrodynamics, is to be found in Kocamustafaogullari's thesis (1971).

## 1 MATHEMATICAL TOOLS

### 1.1 Limiting Forms of the Leibniz and Gauss Theorems for Volume $\mathcal{V}_k$

Given a fixed tube, axis $Oz$ (unit vector $\mathbf{n}_z$) in which volume $\mathcal{V}_k^*$ is cut by two cross-sectional planes located a distance $Z$ apart over areas $\mathcal{A}_{k1}$ and $\mathcal{A}_{k2}$ (Fig. 1). Let $\mathcal{V}_k$ be the volume limited by $\mathcal{A}_{k1}$, $\mathcal{A}_{k2}$ and the portions $\mathcal{A}_i$ and $\mathcal{A}_{kw}$ of interface and wall enclosed between the two cross-sectional planes. The unit vector normal to the interface and directed away from phase $k$ is denoted by $\mathbf{n}_k$. The cross-sectional planes limiting the volume $\mathcal{V}_k$ are not necessarily fixed and their speeds of displacement are denoted by $-\mathbf{v}_{\mathcal{A}_{k1}} \cdot \mathbf{n}_z$ and $\mathbf{v}_{\mathcal{A}_{k2}} \cdot \mathbf{n}_z$.

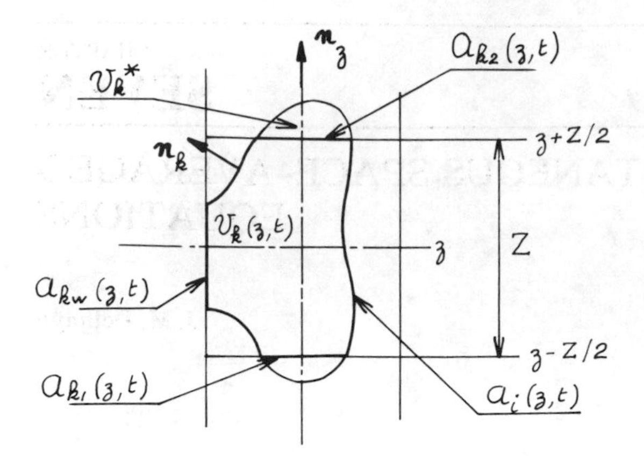


**Figure 1** Control volume used for the special forms of the Leibniz rule [Eq. (1)] and of the Gauss theorems [Eqs. (2)–(4)].

**1.1.1 Leibniz rule** The Leibniz rule applied to volume $\mathcal{V}_k$ leads to

$$\frac{\partial}{\partial t}\int_{\mathcal{V}_k(z,t)} f(x,y,z,t)\,d\mathcal{V} = \int_{\mathcal{V}_k(z,t)} \frac{\partial f}{\partial t}\,d\mathcal{V} + \int_{\mathcal{a}_i(z,t)} f\mathbf{v}_i \cdot \mathbf{n}_k\,d\mathcal{a}$$
$$- \int_{\mathcal{a}_{k1}(z,t)} f\mathbf{v}_{\mathcal{a}_{k1}} \cdot \mathbf{n}_z\,d\mathcal{a} + \int_{\mathcal{a}_{k2}(z,t)} f\mathbf{v}_{\mathcal{a}_{k2}} \cdot \mathbf{n}_z\,d\mathcal{a} \tag{1}$$

where $\mathbf{v}_i \cdot \mathbf{n}_k$ is the speed of displacement of the interface $\mathcal{a}_i$.

**1.1.2 Gauss theorems** The Gauss theorem applied to volume $\mathcal{V}_k$ leads to

$$\int_{\mathcal{V}_k(z,t)} \nabla \cdot \mathbf{B}\,d\mathcal{V} = \int_{\mathcal{a}_i(z,t)} \mathbf{n}_k \cdot \mathbf{B}\,d\mathcal{a} + \int_{\mathcal{a}_{kw}(z,t)} \mathbf{n}_k \cdot \mathbf{B}\,d\mathcal{a}$$
$$- \int_{\mathcal{a}_{k1}(z,t)} \mathbf{n}_z \cdot \mathbf{B}\,d\mathcal{a} + \int_{\mathcal{a}_{k2}(z,t)} \mathbf{n}_z \cdot \mathbf{B}\,d\mathcal{a} \tag{2}$$

which yields

$$\int_{\mathcal{V}_k(z,t)} \nabla \cdot \mathbf{B}\,d\mathcal{V} = \int_{\mathcal{a}_i(z,t)} \mathbf{n}_k \cdot \mathbf{B}\,d\mathcal{a} + \int_{\mathcal{a}_{kw}(z,t)} \mathbf{n}_k \cdot \mathbf{B}\,d\mathcal{a}$$
$$+ \frac{\partial}{\partial z}\int_{\mathcal{V}_k(z,t)} B_z\,d\mathcal{V} \tag{3}$$

where $B_z$ is the $z$ component of $\mathbf{B}$.

For a tensor field we have

$$\int_{\mathcal{V}_k(z,t)} \nabla \cdot \mathbf{M}\, d\mathcal{V} = \int_{\mathcal{a}_i(z,t)} \mathbf{n}_k \cdot \mathbf{M}\, d\mathcal{a} + \int_{\mathcal{a}_{kw}(z,t)} \mathbf{n}_k \cdot \mathbf{M}\, d\mathcal{a}$$

$$+ \frac{\partial}{\partial z} \int_{\mathcal{V}_k(z,t)} \mathbf{n}_z \cdot \mathbf{M}\, d\mathcal{V} \tag{4}$$

## 1.2 Limiting Forms of the Leibniz and Gauss Theorems for Area $\mathcal{a}_k$

The following is given: a tube, axis $0z$ (unit vector $\mathbf{n}_z$), in which a volume $\mathcal{V}_k$ is limited by a boundary $\mathcal{a}_i$ and cut by a fixed cross-sectional plane over area $\mathcal{a}_k$ (Fig. 2). The unit vector normal to the interface and directed away from phase $k$ is denoted by $\mathbf{n}_k$. The intersection of interface $\mathcal{a}_i$ with the cross-sectional plane is denoted by $\mathcal{C}$. The unit vector normal to $\mathcal{C}$, located in the cross-sectional plane and directed away from phase $k$ is denoted by $\mathbf{n}_{k\mathcal{C}}$.

### 1.2.1 Leibniz rule

$$\frac{\partial}{\partial t} \int_{\mathcal{a}_k(z,t)} f(x, y, z, t)\, d\mathcal{a} = \int_{\mathcal{a}_k(z,t)} \frac{\partial f}{\partial t}\, d\mathcal{a} + \int_{\mathcal{C}(z,t)} f\mathbf{v}_i \cdot \mathbf{n}_k \frac{d\mathcal{C}}{\mathbf{n}_k \cdot \mathbf{n}_{k\mathcal{C}}} \tag{5}$$

**1.2.2 Gauss theorems** For vector fields we have

$$\int_{\mathcal{a}_k(z,t)} \nabla \cdot \mathbf{B}\, d\mathcal{a} = \frac{\partial}{\partial z} \int_{k(z,t)} B_z\, d\mathcal{a} + \int_{\mathcal{C}(z,t)} \mathbf{n}_k \cdot \mathbf{B} \frac{d\mathcal{C}}{\mathbf{n}_k \cdot \mathbf{n}_{k\mathcal{C}}} \tag{6}$$

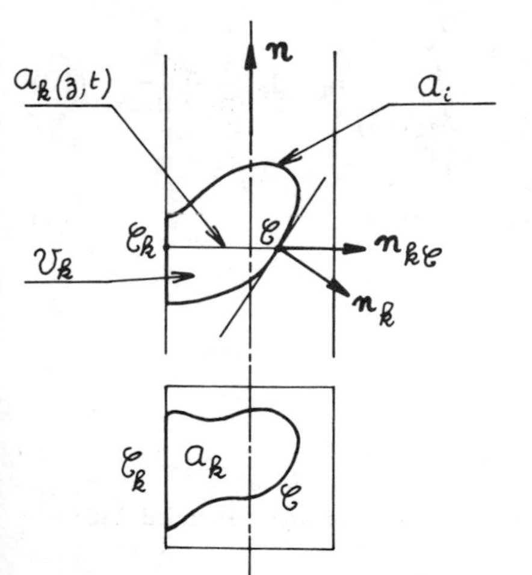


**Figure 2** Notations used in the Leibniz and Gauss theorems written for an area.

If we take $\mathbf{B} = \mathbf{n}_z$ we obtain

$$\frac{\partial \mathcal{A}_k(z, t)}{\partial z} = -\int_{\mathcal{C}(z,t)} \mathbf{n}_k \cdot \mathbf{n}_z \frac{d\mathcal{C}}{\mathbf{n}_k \cdot \mathbf{n}_{k\mathcal{C}}} \tag{7}$$

For tensor fields

$$\int_{\mathcal{A}_k(z,t)} \nabla \cdot \mathbf{M} \, d\mathcal{A} = \frac{\partial}{\partial z} \int_{\mathcal{A}_k(z,t)} \mathbf{n}_z \cdot \mathbf{M} \, d\mathcal{A} + \int_{\mathcal{C}(z,t)} \mathbf{n}_k \cdot \mathbf{M} \frac{d\mathcal{C}}{\mathbf{n}_k \cdot \mathbf{n}_{k\mathcal{C}}} \tag{8}$$

## 2 INSTANTANEOUS AREA-AVERAGED EQUATIONS

We will average the local instantaneous balance equation for phase $k$ over the cross-sectional area occupied by phase $k$ (Fig. 3). The local, instantaneous balance equation is integrated over the area $\mathcal{A}_k(z, t)$ limited by the boundaries $\mathcal{C}(z, t)$ with the other phase and $\mathcal{C}_k(z, t)$ with the pipe wall:

$$\int_{\mathcal{A}_k(z,t)} \frac{\partial}{\partial t} \rho_k \psi_k \, d\mathcal{A} + \int_{\mathcal{A}_k(z,t)} \nabla \cdot (\rho_k \psi_k \mathbf{v}_k) \, d\mathcal{A} + \int_{\mathcal{A}_k(z,t)} \nabla \cdot \mathbf{J}_k \, d\mathcal{A}$$

$$- \int_{\mathcal{A}_k(z,t)} \rho_k \phi_k \, d\mathcal{A} = 0 \tag{9}$$

Applying the limiting forms of the Leibniz rule [Eq. (5)] and of the Gauss theorems [Eqs. (6) and (8)] we obtain, because the pipe wall is supposed to be fixed and impermeable,

$$\frac{\partial}{\partial t} \mathcal{A}_k \langle \rho_k \psi_k \rangle_2 + \frac{\partial}{\partial z} \mathcal{A}_k \langle \mathbf{n}_z \cdot (\rho_k \psi_k \mathbf{v}_k) \rangle_2 + \frac{\partial}{\partial z} \mathcal{A}_k \langle \mathbf{n}_z \cdot \mathbf{J}_k \rangle_2 - \mathcal{A}_k \langle \rho_k \phi_k \rangle_2$$

$$= -\int_{\mathcal{C}(z,t)} (\dot{m}_k \psi_k + \mathbf{n}_k \cdot \mathbf{J}_k) \frac{d\mathcal{C}}{\mathbf{n}_k \cdot \mathbf{n}_{k\mathcal{C}}} - \int_{\mathcal{C}_k(z,t)} \mathbf{n}_k \cdot \mathbf{J}_k \frac{d\mathcal{C}}{\mathbf{n}_k \cdot \mathbf{n}_{k\mathcal{C}}} \tag{10}$$

where

$$\langle f_k \rangle_2 \triangleq \frac{1}{\mathcal{A}_k} \int_{\mathcal{A}_k(z,t)} f_k(x, y, z, t) \, d\mathcal{A} \tag{11}$$

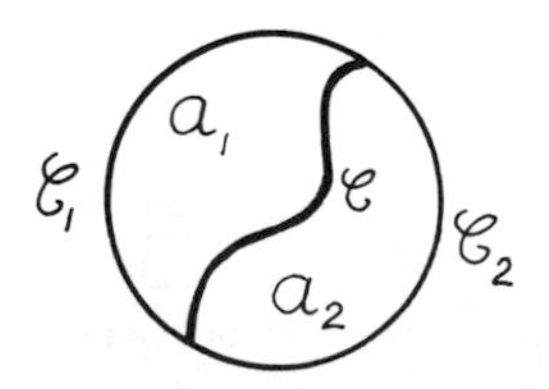


**Figure 3** Pipe cross section showing the areas occupied by each phase at a given time.

$$\dot{m}_k \triangleq \rho_k(\mathbf{v}_k - \mathbf{v}_i) \cdot \mathbf{n}_k \tag{12}$$

By means of Table 1 given in Chap. 5, Eq. (10) gives the instantaneous area-averaged equations for the balance of mass, momentum, total energy, and entropy.

## 2.1 Balance of Mass

$$\frac{\partial}{\partial t} \mathfrak{a}_k \langle \rho_k \rangle_2 + \frac{\partial}{\partial z} \mathfrak{a}_k \langle \rho_k w_k \rangle_2 = -\int_{\mathfrak{c}(z,t)} \dot{m}_k \frac{d\mathfrak{c}}{\mathbf{n}_k \cdot \mathbf{n}_{k\mathfrak{c}}} \tag{13}$$

## 2.2 Balance of Momentum

Splitting the stress tensor $\mathbf{T}_k$ into the pressure term and the viscous stress tensor, we have

$$\frac{\partial}{\partial t} \mathfrak{a}_k \langle \rho_k \mathbf{v}_k \rangle_2 + \frac{\partial}{\partial z} \mathfrak{a}_k \langle \rho_k w_k \mathbf{v}_k \rangle_2 - \mathfrak{a}_k \langle \rho_k \mathbf{F} \rangle_2 + \frac{\partial}{\partial z} \mathfrak{a}_k \langle p_k \mathbf{n}_z \rangle_2$$

$$- \frac{\partial}{\partial z} \mathfrak{a}_k \langle \mathbf{n}_z \cdot \tau_k \rangle_2 = - \int_{\mathfrak{c}(z,t)} (\dot{m}_k \mathbf{v}_k - \mathbf{n}_k \cdot \mathbf{T}_k) \frac{d\mathfrak{c}}{\mathbf{n}_k \cdot \mathbf{n}_{k\mathfrak{c}}}$$

$$+ \int_{\mathfrak{c}_k(z,t)} \mathbf{n}_k \cdot \mathbf{T}_k \frac{d\mathfrak{c}}{\mathbf{n}_k \cdot \mathbf{n}_{k\mathfrak{c}}} \tag{14}$$

If we project along the tube axis we have

$$\frac{\partial}{\partial t} \mathfrak{a}_k \langle \rho_k w_k \rangle_2 + \frac{\partial}{\partial z} \mathfrak{a}_k \langle \rho_k w_k^2 \rangle_2 - \mathfrak{a}_k \langle \rho_k F_z \rangle_2 + \frac{\partial}{\partial z} \mathfrak{a}_k \langle p_k \rangle_2 - \frac{\partial}{\partial z} \mathfrak{a}_k \langle (\mathbf{n}_z \cdot \tau_k) \cdot \mathbf{n}_z \rangle_2$$

$$= - \int_{\mathfrak{c}(z,t)} \mathbf{n}_z \cdot (\dot{m}_k \mathbf{v}_k - \mathbf{n}_k \cdot \mathbf{T}_k) \frac{d\mathfrak{c}}{\mathbf{n}_k \cdot \mathbf{n}_{k\mathfrak{c}}} + \int_{\mathfrak{c}_k(z,t)} \mathbf{n}_z \cdot (\mathbf{n}_k \cdot \mathbf{T}_k) \frac{d\mathfrak{c}}{\mathbf{n}_k \cdot \mathbf{n}_{k\mathfrak{c}}} \tag{15}$$

where $w_k$ is the $z$ component of the velocity vector $\mathbf{v}_k$. If we assume that the pressure $p_k$ is constant along $\mathfrak{c}$ and $\mathfrak{c}_k$ and equal to the averaged pressure $\langle p_k \rangle_2$ over $\mathfrak{a}_k$, then by using Eq. (7), Eq. (15) can be written

$$\frac{\partial}{\partial t} \mathfrak{a}_k \langle \rho_k w_k \rangle_2 + \frac{\partial}{\partial z} \mathfrak{a}_k \langle \rho_k w_k^2 \rangle_2 - \mathfrak{a}_k \langle \rho_k F_z \rangle_2 + \mathfrak{a}_k \frac{\partial p_k}{\partial z} - \frac{\partial}{\partial z} \mathfrak{a}_k \langle (\mathbf{n}_z \cdot \tau_k) \cdot \mathbf{n}_z \rangle_2$$

$$= - \int_{\mathfrak{c}(z,t)} \mathbf{n}_z \cdot (\dot{m}_k \mathbf{v}_k - \mathbf{n}_k \cdot \tau_k) \frac{d\mathfrak{c}}{\mathbf{n}_k \cdot \mathbf{n}_{k\mathfrak{c}}} + \int_{\mathfrak{c}_k(z,t)} \mathbf{n}_z \cdot (\mathbf{n}_k \cdot \tau_k) \frac{d\mathfrak{c}}{\mathbf{n}_k \cdot \mathbf{n}_{k\mathfrak{c}}} \tag{16}$$

## 2.3 Balance of Total Energy

If we introduce the enthalpy $i_k$ per unit of mass, we obtain

$$\frac{\partial}{\partial t}\,\alpha_k\langle\rho_k\left(\frac{1}{2}v_k^2+u_k\right)\rangle_2+\frac{\partial}{\partial z}\,\alpha_k\langle\rho_k\left(\frac{1}{2}v_k^2+i_k\right)w_k\rangle_2-\alpha_k\langle\rho_k\mathbf{F}\cdot\mathbf{v}_k\rangle_2$$
$$-\frac{\partial}{\partial z}\,\alpha_k\langle(\tau_k\cdot\mathbf{v}_k)\cdot\mathbf{n}_z\rangle_2+\frac{\partial}{\partial z}\,\alpha_k\langle\mathbf{q}_k\cdot\mathbf{n}_z\rangle_2$$
$$=-\int_{\mathcal{C}(z,t)}\left[\dot{m}_k\left(\frac{1}{2}v_k^2+u_k\right)-(\mathbf{T}_k\cdot\mathbf{v}_k)\cdot\mathbf{n}_k+\mathbf{q}_k\cdot\mathbf{n}_k\right]\frac{d\mathcal{C}}{\mathbf{n}_k\cdot\mathbf{n}_{k\mathcal{C}}}$$
$$-\int_{\mathcal{C}_k(z,t)}\mathbf{q}_k\cdot\mathbf{n}_k\,\frac{d\mathcal{C}}{\mathbf{n}_k\cdot\mathbf{n}_{k\mathcal{C}}}\qquad(17)$$

## 2.4 Entropy Inequality

$$\frac{\partial}{\partial t}\,\alpha_k\langle\rho_k s_k\rangle_2+\frac{\partial}{\partial z}\,\alpha_k\langle\rho_k s_k w_k\rangle_2+\frac{\partial}{\partial z}\,\alpha_k\langle\frac{\mathbf{q}_k}{T_k}\cdot\mathbf{n}_z\rangle_2$$
$$+\int_{\mathcal{C}(z,t)}\left(\dot{m}_k s_k+\frac{\mathbf{q}_k}{T_k}\cdot\mathbf{n}_k\right)\frac{d\mathcal{C}}{\mathbf{n}_k\cdot\mathbf{n}_{k\mathcal{C}}}+\int_{\mathcal{C}_k(z,t)}\frac{\mathbf{q}_k}{T_k}\cdot\mathbf{n}_k\,\frac{d\mathcal{C}}{\mathbf{n}_k\cdot\mathbf{n}_{k\mathcal{C}}}$$
$$=\alpha_k\langle\Delta_k\rangle_2\geqslant 0\qquad(18)$$

# 3 INSTANTANEOUS VOLUME-AVERAGED EQUATIONS

Transient two-phase flow modeling sometimes requires the solution of a set of partial differential equations written in terms of instantaneous area averages. This set of partial differential equations is usually solved by using one of the following techniques:

1. A finite-difference method involving a discretization of the partial differential equations over finite-sized meshes and leading to a *distributed* parameter model
2. A profile method involving an integration over a finite length and leading to a *lumped* parameter model

In essence, both resolution techniques deal with volume-averaged quantities. It is the purpose of this section to give the instantaneous volume-averaged equations with which the equations of both models must finally reconcile.

By assuming no mass flow through the wall, the integral balance over $\mathcal{V}_k(z, t)$ reads

$$\frac{\partial}{\partial t}\int_{\mathcal{V}_k(z,t)} \rho_k \psi_k \, d\mathcal{V} = -\int_{\mathcal{A}_i(z,t)} \rho_k \psi_k (\mathbf{v}_k - \mathbf{v}_i) \cdot \mathbf{n}_k \, d\mathcal{A}$$

$$- \int_{\mathcal{A}_{k1}(z,t)} \rho_k \psi_k (\mathbf{v}_k - \mathbf{v}_{\mathcal{A}_{k1}}) \cdot \mathbf{n}_z \, d\mathcal{A} - \int_{\mathcal{A}_{k2}(z,t)} \rho_k \psi_k (\mathbf{v}_k - \mathbf{v}_{\mathcal{A}_{k2}}) \cdot \mathbf{n}_z \, d\mathcal{A}$$

$$- \int_{\mathcal{A}_i(z,t)} \mathbf{n}_k \cdot \mathbf{J}_k \, d\mathcal{A} - \int_{\mathcal{A}_{kw}(z,t)} \mathbf{n}_k \cdot \mathbf{J}_k \, d\mathcal{A} + \int_{\mathcal{A}_{k1}(z,t)} \mathbf{n}_z \cdot \mathbf{J}_k \, d\mathcal{A}$$

$$- \int_{\mathcal{A}_{k2}(z,t)} \mathbf{n}_z \cdot \mathbf{J}_k \, d\mathcal{A} + \int_{\mathcal{V}_k(z,t)} \rho_k \phi_k \, d\mathcal{V} \tag{19}$$

By taking account of the following definitions,

$$\langle f_k \rangle_3 \triangleq \frac{1}{\mathcal{V}_k} \int_{\mathcal{V}_k(z,t)} f_k \, d\mathcal{V} \tag{20}$$

$$\dot{m}_k \triangleq \rho_k (\mathbf{v}_k - \mathbf{v}_i) \cdot \mathbf{n}_k \tag{21}$$

the integral balance [Eq. (19)] can be written as

$$\underbrace{\frac{\partial}{\partial t} \mathcal{V}_k \langle \rho_k \psi_k \rangle_3}_{①} - \underbrace{\mathcal{V}_k \langle \rho_k \phi_k \rangle_3}_{②}$$

$$\left.\begin{aligned} = \int_{\mathcal{A}_{k1}(z,t)} \mathbf{n}_z \cdot [\rho_k \psi_k (\mathbf{v}_k - \mathbf{v}_{\mathcal{A}_{k1}}) + \mathbf{J}_k] \, d\mathcal{A} \\ - \int_{\mathcal{A}_{k2}(z,t)} \mathbf{n}_z \cdot [\rho_k \psi_k (\mathbf{v}_k - \mathbf{v}_{\mathcal{A}_{k2}}) + \mathbf{J}_k] \, d\mathcal{A} \end{aligned}\right| ③$$

$$\underbrace{- \int_{\mathcal{A}_i(z,t)} (\dot{m}_k \psi_k + \mathbf{n}_k \cdot \mathbf{J}_k) \, d\mathcal{A}}_{④} \underbrace{- \int_{\mathcal{A}_{kw}(z,t)} \mathbf{n}_k \cdot \mathbf{J}_k \, d\mathcal{A}}_{⑤} \tag{22}$$

The physical significance of Eq. (22) is straightforward: Term (1) is the storage of quantity $\psi_k$ inside volume $\mathcal{V}_k$, term (2) is the source of $\psi_k$ inside volume $\mathcal{V}_k$, terms (3) are the fluxes of $\psi_k$ across the cross-sectional planes. Terms (4) and (5) are the fluxes of $\psi_k$ across the interface and the wall. They are of a particular importance because they are directly connected to the flow pattern through the interfacial area $\mathcal{A}_i(z, t)$ and the wall area $\mathcal{A}_{kw}(z, t)$ in contact with phase $k$.

An equivalent form of Eq. (22) is

$$\frac{\partial}{\partial t} \mathcal{V}_k \langle \rho_k \psi_k \rangle_3 + \frac{\partial}{\partial z} \mathcal{V}_k \langle \rho_k \psi_k w_k + \mathbf{n}_z \cdot \mathbf{J}_k \rangle_3 - \mathcal{V}_k \langle \rho_k \phi_k \rangle_3$$

$$= \int_{\mathcal{A}_{k2}(z,t)} \rho_k \psi_k \mathbf{v}_{\mathcal{A}_{k2}} \cdot \mathbf{n}_z \, d\mathcal{A} - \int_{\mathcal{A}_{k1}(z,t)} \rho_k \psi_k \mathbf{v}_{\mathcal{A}_{k1}} \cdot \mathbf{n}_z \, d\mathcal{A}$$

$$- \int_{\mathcal{A}_i(z,t)} (\dot{m}_k \psi_k + \mathbf{n}_k \cdot \mathbf{J}_k) \, d\mathcal{A} - \int_{\mathcal{A}_{kw}(z,t)} \mathbf{n}_k \cdot \mathbf{J}_k \, d\mathcal{A} \tag{23}$$

where the integrals over $\mathcal{A}_{k1}$ and $\mathcal{A}_{k2}$ vanish if the cross-sectional planes are fixed.

By means of Table 1 given in Chap. 5, Eq. (22) or (23) gives the instantaneous volume-averaged equations for the balances of mass, momentum, total energy, and entropy.

## 4 STRATIFIED FLOW IN A HORIZONTAL CHANNEL

In this section we will look for the instantaneous area-averaged equations governing the isothermal stratified flow of two inviscid, incompressible fluids in a horizontal channel. The calculation will concentrate on the interface behavior in the absence of phase change. For the sake of simplicity we will assume that the flow is two-dimensional and that the interface has no mass, no viscosity, and no surface tension. Figure 4 shows the system geometry. The channel width is denoted by $H$ whereas $h$ and $H-h$ denote the thicknesses of fluids 1 and 2, respectively.

In the following we will write the area-averaged equations directly, ignoring completely the two-dimensional details of the flow. Some characteristic features of two-phase flow modeling will appear and will be discussed in relation to Chap. 10.

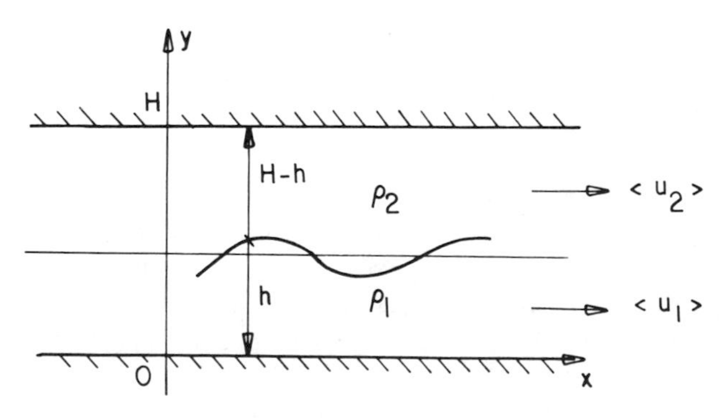


**Figure 4** Two-dimensional stratified flow.

### 4.1 Mass Balance

H1. Incompressible fluids
H2. No phase change at the interface

Equation (13) simplifies to the following relations:

For fluid 1:

$$\frac{\partial}{\partial t} h\rho_1 + \frac{\partial}{\partial x} h\rho_1 \langle u_1 \rangle = 0 \tag{24}$$

For fluid 2:

$$\frac{\partial}{\partial t}(H-h)\rho_2 + \frac{\partial}{\partial x}(H-h)\rho_2 \langle u_2 \rangle = 0 \tag{25}$$

### 4.2 Momentum Balance along the $x$ Axis

H1. Incompressible fluids
H2. No phase change at the interface
H3. Horizontal flow
H4. Inviscid fluids

Equation (15) together with Eq. (7) yields

For fluid 1:

$$\frac{\partial}{\partial t} h\rho_1 \langle u_1 \rangle + \frac{\partial}{\partial x} h\rho_1 \langle u_1^2 \rangle + \frac{\partial}{\partial x} h \langle p_1 \rangle = p_{1i} \frac{\partial h}{\partial x} \tag{26}$$

For fluid 2:

$$\frac{\partial}{\partial t}(H-h)\rho_2 \langle u_2 \rangle + \frac{\partial}{\partial x}(H-h)\rho_2 \langle u_2^2 \rangle + \frac{\partial}{\partial x}(H-h)\langle p_2 \rangle = -p_{2i} \frac{\partial h}{\partial x} \tag{27}$$

### 4.3 Momentum Jump Condition

H2. No phase change at the interface
H4. Inviscid fluids
H5. No surface tension

The jump condition [Eq. (51)] of Chap. 5 simplifies to the following relation:

$$p_{1i} = p_{2i} \triangleq p_i \tag{28}$$

### 4.4 Two-Phase Flow Modeling

Considering Eqs. (24)–(28) we realize that, given $H$, $\rho_1$, and $\rho_2$ we have four partial differential equations [Eqs. (24)–(27)], five dependent variables (i.e., $h$, $\langle u_1 \rangle$, $\langle u_2 \rangle$, $\langle p_1 \rangle$, and $\langle p_2 \rangle$), and three supplementary variables (i.e., $\langle u_1^2 \rangle$, $\langle u_2^2 \rangle$, and $p_i$).

**4.4.1 Space correlation coefficients** One can define a space correlation coefficient $C_k$ by the following relation:

$$C_k \triangleq \frac{\langle u_k^2 \rangle}{\langle u_k \rangle^2} = C_k(x, t) \tag{29}$$

Generally speaking, $C_k$ should be expressed as a functional of the dependent variables. However, because of the absence of information concerning this correlation coefficient, we will assume that

$$C_1 = C_2 = 1 \tag{30}$$

**4.4.2 Mechanical law** We recall here that in the method based on the area-averaged equations, we completely ignore the two-dimensional equations that govern the flow. As a consequence, we have to postulate an equation that relates the pressure $p_i$ at the interface to the dependent variables, knowing that we have also to reduce the number of these dependent variables by one. Physical intuition leads us to assume that we have a hydrostatic pressure distribution over the height of the channel:

$$\langle p_1 \rangle = p_i + \rho_1 g \frac{h}{2} \tag{31}$$

$$\langle p_2 \rangle = p_i - \rho_2 g \frac{H-h}{2} \tag{32}$$

As a result, we postulate the following mechanical law:

$$\langle p_2 \rangle = \langle p_1 \rangle - (\rho_1 - \rho_2) g \frac{h}{2} - \rho_2 g \frac{H}{2} \tag{33}$$

**4.4.3 Final system** Thanks to assumptions (30) and (33) the four balance equations [Eqs. (24)–(27)] can be written under the following matrix form:

$$A \frac{\partial \mathbf{X}}{\partial t} + B \frac{\partial \mathbf{X}}{\partial x} = 0 \tag{34}$$

with matrices $A$ and $B$ and the solution vector $\mathbf{X}$ given by

$$A \triangleq \begin{bmatrix} \rho_1 & 0 & 0 & 0 \\ -\rho_2 & 0 & 0 & 0 \\ \rho_1 \langle u_1 \rangle & h\rho_1 & 0 & 0 \\ -\rho_2 \langle u_2 \rangle & 0 & (H-h)\rho_2 & 0 \end{bmatrix} \tag{35}$$

$$B \triangleq \begin{bmatrix} \rho_1\langle u_1\rangle & h\rho_1 & 0 & 0 \\ -\rho_2\langle u_2\rangle & 0 & (H-h)\rho_2 & 0 \\ \rho_1\langle u_1\rangle^2 + \dfrac{\rho_1 gh}{2} & 2h\rho_1\langle u_1\rangle & 0 & h \\ -\rho_2\langle u_2\rangle^2 + \rho_2 g(H-h) - \dfrac{\rho_1 g(H-h)}{2} & 0 & 2(H-h)\rho_2\langle u_2\rangle & H-h \end{bmatrix} \tag{36}$$

$$\mathbf{X} \triangleq [h \quad \langle u_1\rangle \quad \langle u_2\rangle \quad \langle p_1\rangle] \tag{37}$$

**4.4.4 Interface stability** The characteristic equation reads

$$\det(B - cA) = 0 \tag{38}$$

where $c$ is the characteristic speed. After some calculation we obtain the following relation:

$$\rho_1(\langle u_1\rangle - c)^2(H-h) + \rho_2(\langle u_2\rangle - c)^2 h - gh(H-h)(\rho_1 - \rho_2) = 0 \tag{39}$$

which is fully consistent with the gravity long-wave theory (Milne-Thomson, 1955).

Stability is ensured if the characteristic speeds are real; i.e., if we have

$$(\langle u_1\rangle - \langle u_2\rangle)^2 < \frac{g(\rho_1 - \rho_2)[\rho_1(H-h) + \rho_2 h]}{\rho_1\rho_2} \tag{40}$$

This condition means that the restoring effect caused by the gravity should be high enough to overcome the destabilizing effect caused by the slip velocity.

If we assume a constant pressure over the channel height instead of a hydrostatic distribution, the gravity term no longer appears in the characteristic equation [Eq. (39)]. As a result the solution is always unstable, which is physically unrealistic.

## NOMENCLATURE

| | |
|---|---|
| $\mathfrak{a}$ | area |
| $A$ | matrix [Eq. (35)] |
| $B$ | matrix [Eq. (36)] |
| $\mathbf{B}$ | vector |
| $c$ | characteristic speed |
| $C$ | space correlation coefficient |
| $\mathfrak{c}$ | line |
| $f$ | function |
| $\mathbf{F}$ | external force per unit of mass |
| $h$ | height |

| | |
|---|---|
| $H$ | channel height |
| $i$ | enthalpy per unit of mass |
| $\mathbf{J}$ | flux |
| $L$ | segment |
| $\dot{m}$ | mass transfer per unit area of interface and per unit of time [Eq. (12)] |
| $\mathbf{M}$ | tensor |
| $\mathbf{n}$ | unit normal vector |
| $p$ | pressure |
| $\mathbf{q}$ | heat flux |
| $s$ | entropy per unit of mass |
| $t$ | time |
| $\mathbf{T}$ | total stress tensor |
| $u$ | internal energy per unit of mass; $x$ component of the velocity vector |
| $\mathbf{v}$ | velocity vector |
| $\mathcal{V}$ | volume |
| $w$ | $z$ component of the velocity vector |
| $\mathbf{X}$ | solution vector [Eq. (37)] |
| $Z$ | distance |
| $\rho$ | density |
| $\tau$ | deviatoric stress tensor |
| $\phi$ | source term |
| $\psi$ | specific quantity |

**Subscripts**

| | |
|---|---|
| $i$ | interface |
| $k$ | phase index |
| $w$ | wall |

**Operators**

| | |
|---|---|
| $\langle\ \rangle_2$ | area-averaging operator over $\mathcal{A}_k$ [Eq. (11)] |
| $\langle\ \rangle_3$ | volume-averaging operator over $\mathcal{V}_k$ [Eq. (20)] |

## REFERENCES

Birkhoff, G., Averaged-Conservation Laws in Pipes, *J. Math. Anal. Appl.*, vol. 8, pp. 66–77, 1964.

Delhaye, J. M., Equations Fondamentales des Ecoulements Diphasiques, CEA-R-3429, 1968.

Kocamustafaogullari, G., Thermo-Fluid Dynamics of Separated Two-Phase Flow, Ph.D. thesis, School of Mechanical Engineering, Georgia Institute of Technology, Atlanta, Ga., 1971.

Milne-Thomson, L. M., *Theoretical Hydrodynamics*, Macmillan, New York, 1955.

Vernier, Ph. and Delhaye, J. M., General Two-Phase Flow Equations Applied to the Thermohydrodynamics of Boiling Water Nuclear Reactors, *Energie Primaire*, vol. 4, no. 1–2, pp. 5–46, 1968.

CHAPTER

# EIGHT

# LOCAL TIME-AVERAGED EQUATIONS

J. M. Delhaye

The use of the time-averaged local variables in two-phase flows was proposed by Teletov (1958) and reconsidered by Vernier and Delhaye (1968). Ishii (1975), in chap. 3 of his book, traces the history of the subject in detail and compares the different types of averages used.

The method used here to establish time-averaged equations differs from that used by Ishii. The interfaces we consider are density discontinuity surfaces. They have no thickness, unlike the interfaces considered by Ishii. Nevertheless, we find the same results but by using simpler calculations.

For the solution of two-dimensional or three-dimensional transient problems, the local instantaneous equations can be time-averaged over a time interval $[t - T/2; t + T/2]$. As for single-phase turbulent flow, this time interval $[T]$ must be carefully chosen, large enough compared with the turbulence fluctuations, and small enough compared with the overall flow fluctuations. This is not always possible, and a thorough discussion of this complex problem is to be found in the works of Delhaye and Achard (1977, 1978).

After recalling a few theorems on the derivatives of piecewise continuous functions, we will define a single time-averaging operator and a double time-averaging operator.

# 1 MATHEMATICAL TOOLS

If we consider a given point in a two-phase flow, phase $k$ passes this point intermittently, and a function $f_k$ associated with phase $k$ will have the appearance shown in Fig. 1. It will be a piecewise continuous function.

If we consider the time interval $[t - T/2; t + T/2]$, let $[T_k]$ be the subset of residence time intervals of phase $k$ belonging to the interval $[T]$ and $T_k$ the cumulated residence time of phase $k$ in the interval $[T]$.

## 1.1 Limiting Form of the Leibniz Theorem

$$\int_{[T_k]} \frac{\partial f_k}{\partial t} \, dt = \frac{\partial}{\partial t} \int_{[T_k]} f_k \, dt - \sum_{\substack{\text{disc} \\ \in [T]}} \frac{1}{|\mathbf{v}_i \cdot \mathbf{n}_k|} f_k \mathbf{v}_i \cdot \mathbf{n}_k \tag{1}$$

If $f_k \equiv 1$, Eq. (1) reads

$$\frac{\partial \alpha_k}{\partial t} = \frac{1}{T} \sum_{\substack{\text{disc} \\ \in [T]}} \frac{\mathbf{v}_i \cdot \mathbf{n}_k}{|\mathbf{v}_i \cdot \mathbf{n}_k|} \tag{2}$$

with $\alpha_k$ the residence time fraction of phase $k$, defined by

$$\alpha_k \triangleq \frac{T_k}{T} \tag{3}$$

## 1.2 Limiting Forms of the Gauss Theorems

$$\int_{[T_k(x)]} \nabla \cdot \mathbf{B}_k(\mathbf{x}, t) \, dt = \nabla \cdot \int_{[T_k(x)]} \mathbf{B}_k(\mathbf{x}, t) \, dt + \sum_{\substack{\text{disc} \\ \in [T]}} \frac{1}{|\mathbf{v}_i \cdot \mathbf{n}_k|} \mathbf{n}_k \cdot \mathbf{B}_k(\mathbf{x}, t) \tag{4}$$

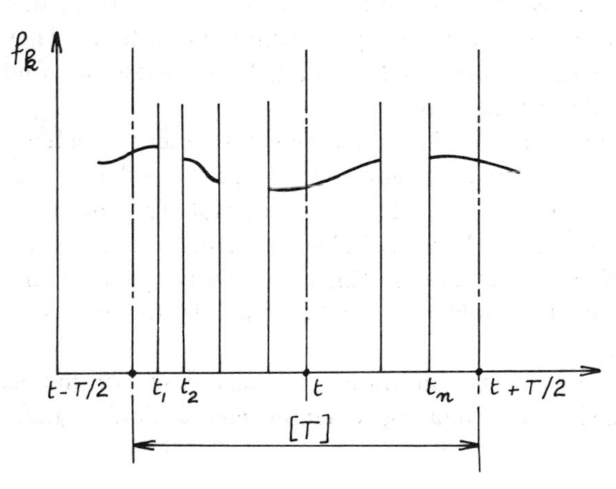


**Figure 1** Time evolution of function $f_k$ associated with phase $k$.

The vector $\mathbf{B}_k$ can be replaced by a tensor $\mathbf{M}_k$. A relation identical to Eq. (4) is obtained. In particular, if

$$\mathbf{M}_k = \mathbf{U} \tag{5}$$

we have

$$\nabla\alpha_k = -\frac{1}{T}\sum_{\substack{\text{disc}\\ \in[T]}} \frac{\mathbf{n}_k}{|\mathbf{v}_i \cdot \mathbf{n}_k|} \tag{6}$$

## 2 SINGLE TIME-AVERAGED EQUATIONS

The local instantaneous balance law is integrated over the time interval $[T_k]$:

$$\int_{[T_k]} \frac{\partial}{\partial t}\rho_k\psi_k\,dt + \int_{[T_k]} \nabla\cdot(\rho_k\psi_k\mathbf{v}_k)\,dt + \int_{[T_k]} \nabla\cdot\mathbf{J}_k\,dt$$
$$-\int_{[T_k]} \rho_k\phi_k\,dt = 0 \tag{7}$$

By taking into account the limiting forms of the Leibniz and Gauss theorems [Eqs. (1) and (4)], Eq. (7) becomes

$$\frac{\partial}{\partial t}\alpha_k\overline{\rho_k\psi_k}^X + \nabla\cdot\alpha_k\overline{\rho_k\psi_k\mathbf{v}_k}^X + \nabla\cdot\alpha_k\overline{\mathbf{J}_k}^X - \alpha_k\overline{\rho_k\phi_k}^X$$
$$= -\sum_j l_j^{-1}(\dot{m}_k\,\psi_k + \mathbf{J}_k\cdot\mathbf{n}_k)_j \tag{8}$$

where

$$\overline{f_k}^X \triangleq \frac{1}{T_k}\int_{[T_k]} f_k\,dt \tag{9}$$

$$l_j \triangleq T|\mathbf{v}_i\cdot\mathbf{n}_k|_j \tag{10}$$

$j$ denoting the $j$th interface passing through $\mathbf{x}$ during the time interval $[T]$.

### 2.1 Remarks on the Definition of $\overline{f_k}^X$

A phase density function $X_k(\mathbf{x}, t)$ is defined by

$$X_k(\mathbf{x},t) \triangleq \begin{cases} 1 & \text{if point } \mathbf{x} \text{ pertains to phase } k \\ 0 & \text{if point } \mathbf{x} \text{ does not pertain to phase } k \end{cases} \tag{11}$$

The time fraction $\alpha_k$ defined by Eq. (3) is then equal to

$$\alpha_k(\mathbf{x}, t) \triangleq \frac{T_k}{T} = \frac{1}{T} \int_{[T]} X_k(\mathbf{x}, t)\, dt \triangleq \overline{X_k}(\mathbf{x}, t) \tag{12}$$

On the other hand $\overline{f_k}^X$, defined by Eq. (9) is equal to

$$\overline{f_k}^X = \frac{(1/T) \int_{[T]} X_k f_k \, dt}{(1/T) \int_{[T]} X_k \, dt} \triangleq \frac{\overline{X_k f_k}}{\overline{X_k}} \tag{13}$$

Hence, we can see that the time average of $f_k$ over the time interval $[T_k]$ is the $X_k$-weighted average of $f_k$ over $[T]$. This assumption justifies the notation $\overline{f_k}^X$.

## 2.2 Primary Balance Equations

From Table 1 in Chap. 5, Eq. (8) gives the local time-averaged equations for the balances of mass, momentum, total energy, and entropy.

Mass:

$$\frac{\partial}{\partial t} \alpha_k \overline{\rho_k}^X + \nabla \cdot \alpha_k \overline{\rho_k \mathbf{v}_k}^X = - \sum_j l_j^{-1} \dot{m}_{k_j} \tag{14}$$

Momentum:

$$\frac{\partial}{\partial t} \alpha_k \overline{\rho_k \mathbf{v}_k}^X + \nabla \cdot \alpha_k \overline{\rho_k \mathbf{v}_k \mathbf{v}_k}^X - \nabla \cdot \alpha_k \overline{\mathbf{T}_k}^X - \alpha_k \overline{\rho_k \mathbf{F}}^X$$

$$= - \sum_j l_j^{-1} (\dot{m}_k \mathbf{v}_k - \mathbf{T}_k \cdot \mathbf{n}_k)_j \tag{15}$$

Total energy:

$$\frac{\partial}{\partial t} \alpha_k \overline{\rho_k \left( u_k + \frac{1}{2} v_k^2 \right)}^X + \nabla \cdot \alpha_k \overline{\rho_k \left( u_k + \frac{1}{2} v_k^2 \right) \mathbf{v}_k}^X$$

$$+ \nabla \cdot \alpha_k \overline{\mathbf{T}_k \cdot \mathbf{v}_k}^X + \nabla \cdot \alpha_k \overline{\mathbf{q}_k}^X - \alpha_k \overline{\rho_k \mathbf{F} \cdot \mathbf{v}_k}^X$$

$$= - \sum_j l_j^{-1} \left[ \dot{m}_k \left( u_k + \frac{1}{2} v_k^2 \right) - (\mathbf{T}_k \cdot \mathbf{v}_k) \cdot \mathbf{n}_k + \mathbf{q}_k \cdot \mathbf{n}_k \right]_j \tag{16}$$

Entropy:

$$\frac{\partial}{\partial t} \alpha_k \overline{\rho_k s_k}^X + \nabla \cdot \alpha_k \overline{\rho_k s_k \mathbf{v}_k}^X + \nabla \cdot \alpha_k \overline{\frac{1}{T_k} \mathbf{q}_k}^X$$

$$+ \sum_j l_j^{-1} \left( \dot{m}_k s_k + \frac{1}{T_k} \mathbf{q}_k \cdot \mathbf{n}_k \right)_j = \alpha_k \overline{\Delta_k}^X \geqslant 0 \tag{17}$$

## 3 COMMENTS ON THE SINGLE TIME-AVERAGING OPERATORS

In Eqs. (12) and (13) we used the single time-averaging operator defined by

$$\bar{g}(\mathbf{x}, t) \triangleq \frac{1}{T} \int_{[T]} g(\mathbf{x}, \eta)\, d\eta \tag{18}$$

In two-phase flow the $g$ function is of the type

$$g(\mathbf{x}, t) = X_k f_k \tag{19}$$

and can be expanded into a Fourier series. The message $g(\mathbf{x}, t)$ can be split into two parts:

1. The signal $g_s(\mathbf{x}, t)$, the sum of the expansion terms whose angular frequencies are lower than an arbitrary cutoff frequency
2. The noise $g_n(\mathbf{x}, t)$, the sum of the remaining terms

Hence

$$g(\mathbf{x}, t) = g_s(\mathbf{x}, t) + g_n(\mathbf{x}, t) \tag{20}$$

The time-averaging operator [Eq. (18)] is expected to low-pass filter the message $g(\mathbf{x}, t)$ on which it is acting:

$$\bar{g}(\mathbf{x}, t) \simeq g_s(\mathbf{x}, t) \tag{21}$$

Because the first time derivatives of the time-averaged variables appear in the single time-averaged balance equation [Eq. (8)], it would be worthwhile to have also

$$\frac{\partial}{\partial t}\bar{g}(\mathbf{x}, t) \cong \frac{\partial}{\partial t} g_s(\mathbf{x}, t) \tag{22}$$

Delhaye and Achard (1977, 1978) showed that the single time-averaging operator [Eq. (18)] does not always fulfill conditions of Eqs. (21) and (22). Furthermore, the first time derivative of $\bar{g}$ is discontinuous. For these reasons, Delhaye and Achard (1977, 1978) introduced a double time-averaging operator, which will be defined in Sec. 4.

## 4 DOUBLE TIME-AVERAGED EQUATIONS

Delhaye and Achard (1978) showed that the condition of Eq. (22) on the time derivative is satisfied by the following average:

$$\overline{\overline{f_k}}^X \triangleq \frac{1}{T} \int_{t-T/2}^{t+T/2} \overline{f_k}^x \, d\tau = \frac{1}{T} \int_{t-T/2}^{t+T/2} \left( \frac{1}{T_k} \int_{[T_k]} f_k \, d\eta \right) d\tau \tag{23}$$

Unfortunately, the limiting forms of the Leibniz and Gauss theorems cannot be reduced to a simple form, and it seems better to introduce the following double time-averaging operator.

$$\overline{\overline{f_k}}^X \triangleq \frac{(1/T)\int_{t-T/2}^{t+T/2}\left[(1/T)\int_{\tau-T/2}^{\tau+T/2} X_k f_k \, d\eta\right] d\tau}{(1/T)\int_{t-T/2}^{t+T/2}\left[(1/T)\int_{\tau-T/2}^{\tau+T/2} X_k \, d\eta\right] d\tau} \triangleq \frac{\overline{\overline{X_k f_k}}}{\overline{\overline{X_k}}} \tag{24}$$

Delhaye and Achard (1977) showed that the conditions of Eqs. (21) and (22) are fulfilled by Eq. (24) under less restrictive conditions than the single time-averaging operator [Eq. (13)] and that the time derivative of $\overline{\overline{f_k}}^X$ has a physical significance and is continuous. Figure 2 shows the results obtained by applying the single time-averaging operator [Eq. (13)] and the double time-averaging operator [Eq. (24)] to a noisy square wave (Delhaye and Achard, 1977).

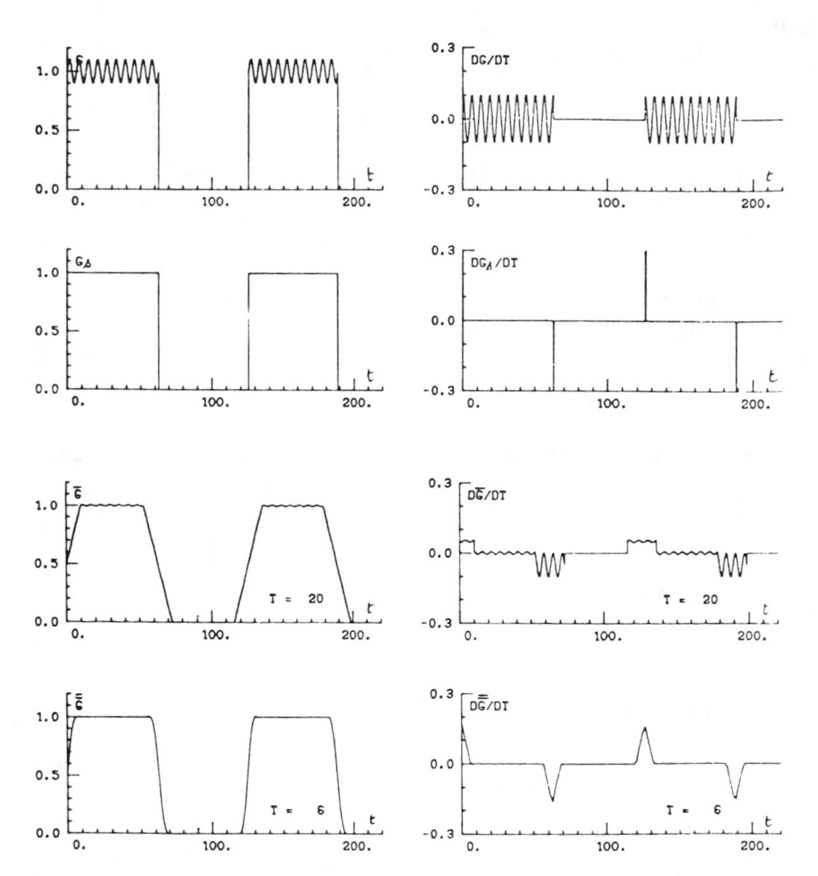


**Figure 2** Graphs of the original message $g(t)$, the signal $g_s(t)$, the message single time average $\bar{g}(t)$, the message double time-average $\bar{\bar{g}}(t)$, and their time derivatives (Delhaye and Achard, 1978).

## 4.1 Limiting Forms of the Leibniz and Gauss Theorems (Delhaye and Achard, 1978)

To derive these forms, new quantities must be defined. When the interval $[\tau - T/2;\ \tau + T/2]$ is displaced over the base interval $[t - T/2;\ t + T/2]$, it gains or loses interfaces at times denoted by $\tau^{+}(t)$ and $\tau^{-}(t)$.

By definition,

$$\Upsilon^{\pm}(t) \triangleq \begin{cases} \tau^{\pm}(t) - \left(t - \dfrac{T}{2}\right) & \text{if } \eta < t \\ \left(t + \dfrac{T}{2}\right) - \tau^{\pm}(t) & \text{if } \eta > t \end{cases} \tag{25}$$

$$L_j \triangleq \frac{T}{\Upsilon_j}\, l_j \tag{26}$$

with $l_j$ defined by Eq. (10).

Leibniz rule:

$$\overline{\overline{X_k \frac{\partial f_k}{\partial t}}} = \frac{\partial}{\partial t}\, \overline{\overline{X_k f_k}} - \sum_j L_j^{-1} (f_k \mathbf{v}_i \cdot \mathbf{n}_k)_j \tag{27}$$

Gauss theorem:

$$\overline{\overline{X_k \nabla \cdot \mathbf{B}_k}} = \nabla \cdot \overline{\overline{X_k \mathbf{B}_k}} + \sum_j L_j^{-1} (\mathbf{n}_k \cdot \mathbf{B}_k)_j \tag{28}$$

## 4.2 Balance Equation

If Eqs. (27) and (28) are taken into account, the local instantaneous balance equation is transformed into the following double time-averaged equation:

$$\frac{\partial}{\partial t}\, \beta_k \overline{\overline{\rho_k \psi_k}}^{X} + \nabla \cdot \beta_k \overline{\overline{\rho_k \psi_k \mathbf{v}_k}}^{X} + \nabla \cdot \beta_k \overline{\overline{\mathbf{J}_k}}^{X} - \beta_k \overline{\overline{\rho_k \phi_k}}^{X}$$

$$= - \sum_j L_j^{-1} (\dot{m}_k \psi_k + \mathbf{J}_k \cdot \mathbf{n}_k)_j \tag{29}$$

where $\beta_k \triangleq \overline{\overline{X_k}}$

Comparing the double time-averaged equation [Eq. (29)] with the single time-averaged equation [Eq. (8)], we conclude that these two equations are formally identical if $\alpha_k, \overline{f_k}^{X}$, and $l_j$ are replaced by $\beta_k, \overline{\overline{f_k}}^{X}$, and $L_j$.

## NOMENCLATURE

**B** vector
**F** external force per unit of mass

| | |
|---|---|
| $f$ | arbitrary function |
| $g$ | arbitrary function |
| $\mathbf{J}$ | flux term |
| $L$ | specific length [Eq. (26)] |
| $l$ | specific length [Eq. (10)] |
| $\mathbf{M}$ | tensor |
| $\dot{m}$ | mass transfer per unit area of interface and per unit of time |
| $\mathbf{n}$ | unit normal vector |
| $\mathbf{q}$ | heat flux |
| $s$ | entropy per unit of mass |
| $T$ | time interval |
| $\mathbf{T}$ | total stress tensor |
| $t$ | time |
| $u$ | internal energy per unit of mass |
| $\mathbf{U}$ | unit tensor |
| $\mathbf{v}$ | velocity |
| $X$ | phase density function [Eq. (11)] |
| $\mathbf{x}$ | position vector |
| $\alpha$ | time fraction [Eq. (3)] |
| $\beta$ | time fraction [Eq. (30)] |
| $\Delta$ | entropy source per unit of volume and per unit of time |
| $\eta$ | time dummy variable |
| $\rho$ | density |
| $\tau$ | time dummy variable |
| $\Upsilon$ | time interval [Eq. (25)] |
| $\phi$ | source term |
| $\psi$ | specific quantity |

**Subscripts**

| | |
|---|---|
| $i$ | interface |
| $k$ | phase index |
| $n$ | noise |
| $s$ | signal |

**Operators**

| | |
|---|---|
| $\overline{\phantom{x}}$ | single time-averaging operator [Eq. (13)] |
| $\overline{\phantom{x}}^{X}$ | $X$-weighted single time-averaging operator [Eq. (9)] |
| $\overline{\overline{\phantom{x}}}$ | double time-averaging operator [Eq. (24)] |
| $\overline{\overline{\phantom{x}}}^{X}$ | $X$-weighted double time-averaging operator [Eq. (24)] |

## REFERENCES

Delhaye, J. M. and Achard, J. L., On the Use of Averaging Operators in Two-Phase Flow Modeling, Modeling, in *Thermal and Hydraulic Aspects of Nuclear Reactor Safety*, vol. 1: *Light Water Reactors*, eds. O. C. Jones and S. G. Bankoff, pp. 289–332, ASME, New York, 1977.

Delhaye, J. M. and Achard, J. L., On the Averaging Operators Introduced in Two-Phase Flow Modeling, in *Transient Two-Phase Flow, Proc. CSNI Specialists Meet., Aug. 3 and 4, 1976, Toronto,* eds. S. Banerjee and K. R. Weaver, vol. 1, pp. 5–84, AECL, 1978.

Ishii, M., *Thermo-Fluid Dynamic Theory of Two-Phase Flow*, Eyrolles, Paris, 1975.

Teletov, S. G., Two-Phase Flow Hydrodynamics. 1. Hydrodynamics and Energy Equations, *Bull. Moscow Univ.*, vol. 2, 1958 (in Russian).

Vernier, Ph. and Delhaye, J. M., General Two-Phase Flow Equations Applied to the Thermohydrodynamics of Boiling Water Nuclear Reactors, *Energie Primaire*, vol. 4, no. 1–2, pp. 5–46, 1968.

CHAPTER

# NINE

# COMPOSITE-AVERAGED EQUATIONS

**J. M. Delhaye**

The importance of composite, that is, space-time- or time-space-averaged, equations is considerable because all the practical problems of two-phase flow in channels are now dealt with using these equations.

The composite-averaged equations can be obtained

1. By averaging the time-averaged local equations over a channel cross-sectional area or over a slice
2. By averaging over a time interval, the instantaneous equations averaged over the cross-sectional areas or slices occupied by each phase

To verify the equivalence of the results obtained with one or the other of these methods, we will first establish or recall theorems concerning the commutativity of averaging operators. We will also introduce local and integral specific areas because these quantities play an important part in mass, momentum, and energy interface transfers. We will conclude by demonstrating the identity of the space-time- and time-space-averaging operators. For the sake of simplicity, derivations will be carried out with the single time averages. Nevertheless, all the results are valid for the double time averages.

## 1 COMMUTATIVITY OF THE AVERAGING OPERATORS

If we consider any scalar, vector, or tensor function associated with phase $k$, the aim of our calculation is to find the group of variables on which a permutation of the time- and space-averaging operators can be carried out.

We have, by means of the averaging operator definitions,

$$\sphericalangle \alpha_k \overline{f_k}^X \gg_2 = \sphericalangle \overline{X_k f_k} \gg_2 \triangleq \frac{1}{\mathcal{A}} \int_{\mathcal{A}} \left( \frac{1}{T} \int_{[T]} X_k f_k \, dt \right) d\mathcal{A} \tag{1}$$

Reversing the order of integration yields

$$\sphericalangle \alpha_k \overline{f_k}^X \gg_2 = \frac{1}{T} \int_{[T]} \left( \frac{1}{\mathcal{A}} \int_{\mathcal{A}} X_k f_k \, d\mathcal{A} \right) dt \tag{2}$$

$$= \frac{1}{T} \int_{[T]} \left( \frac{1}{\mathcal{A}_k} \int_{\mathcal{A}_k} f_k \, d\mathcal{A} \right) dt \tag{3}$$

As a consequence we obtain the fundamental relation (Vernier and Delhaye, 1968)

$$\sphericalangle \alpha_k \overline{f_k}^X \gg_2 \equiv \overline{R_{k_2} \langle f_k \rangle_2} \tag{4}$$

where $R_{k_2}$ is the area fraction of phase $k$ in the cross section defined by

$$R_{k_2} \triangleq \frac{\mathcal{A}_k}{\mathcal{A}} \tag{5}$$

Equation (4) plays an essential part in two-phase flow modeling. Generally, hypotheses are more conveniently formulated on local time-averaged quantities (such as $\alpha_k$ or $\overline{f_k}^X$) than on instantaneous area-averaged quantities (such as $R_{k_2}$ or $\langle f_k \rangle_2$), depending on the flow pattern, for example, bubbly flow or slug flow. A particular case for Eq. (4) is obtained by taking

$$f_k \equiv 1$$

which leads to

$$\sphericalangle \alpha_k \gg_2 \equiv \overline{R_{k_2}} \tag{6}$$

Obviously, Eqs. (4) and (6) established for areas are also valid for segments and volumes. More generally, we have

$$\sphericalangle \alpha_k \overline{f_k}^X \gg_n \equiv \overline{R_{k_n} \langle f_k \rangle_n} \tag{7}$$

where $n = 1$ for segments
$n = 2$ for areas
$n = 3$ for volumes

If Eq. (7) is written for segments ($n = 1$), probe measurements of local void fraction can be tallied with radiation attenuation measurements of segment void fractions.

## 2 LOCAL AND INTEGRAL SPECIFIC AREAS

### 2.1 Fundamental Identity

If we have a pipe and a fixed control volume $\mathcal{V}$ limited by the pipe wall and the cross sections $\mathcal{A}'$ and $\mathcal{A}''$ located a distance $Z$ apart (Fig. 1), the area of the moving interfaces contained in volume $\mathcal{V}$ is denoted by $\mathcal{A}_i(t)$.

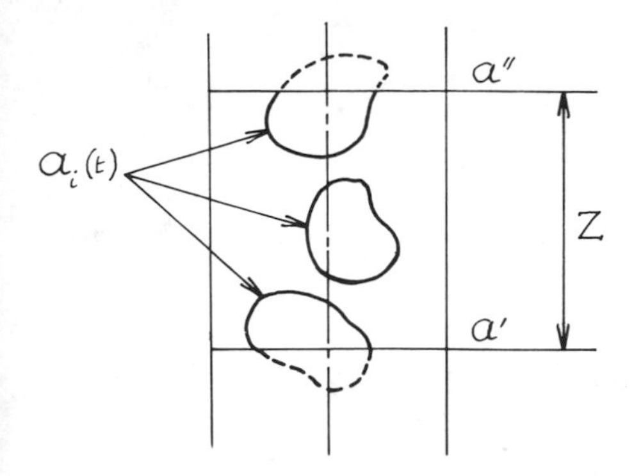


**Figure 1** Control volume used for the fundamental identity Eq. (8).

For any arbitrary continuous vector field $\mathbf{B}_k(\mathbf{x}, t)$, the following identity is satisfied (Delhaye, 1976; Delhaye and Achard, 1978):

$$\int_{\mathcal{V}} \left[ \sum_j l_j^{-1} (\mathbf{B}_k \cdot \mathbf{n}_k)_j \right] d\mathcal{V} \equiv \overline{\int_{\mathfrak{a}_i(t)} \mathbf{B}_k \cdot \mathbf{n}_k \, d\mathfrak{a}} \tag{8}$$

with

$$l_j \triangleq T |\mathbf{v}_i \cdot \mathbf{n}_k|_j \tag{9}$$

where $T$ = integration period

$\mathbf{v}_i \cdot \mathbf{n}_k$ = speed of displacement of $\mathfrak{a}_i(t)$

subscript $j$ = $j$th interface passing through $\mathbf{x}$ during the time interval $[T]$

If the vector field $\mathbf{B}_k$ is chosen such that on the interfaces

$$\mathbf{B}_k \equiv \mathbf{n}_k \tag{10}$$

Eq. (8) leads to a relation that connects the interfacial area to the speed of displacement of the interfaces:

$$\int_{\mathcal{V}} \left( \sum_j l_j^{-1} \right) d\mathcal{V} \equiv \overline{\mathfrak{a}_i(t)} \tag{11}$$

## 2.2 Specific Areas

The following definitions can then be set:

1. The local specific area $\gamma(\mathbf{x})$ is a local quantity defined over the time interval $[T]$ by the following relation:

$$\gamma(\mathbf{x}) \triangleq \sum_j l_j^{-1} \tag{12}$$

2. The integral specific area $\Gamma_3(t)$ is an instantaneous quantity defined on the volume $\mathcal{V}$ by the following relation:

$$\Gamma_3(t) \triangleq \frac{\mathfrak{a}_i(t)}{\mathcal{V}} \tag{13}$$

Equation (11) can then be written

$$\nless \gamma(\mathbf{x}) \ngtr_3 \equiv \overline{\Gamma_3(t)} \tag{14}$$

Identities (11) and (14) are fundamental for interfacial area measurements. They provide the link between the integral specific area, which can be measured by chemical methods, and the local specific area, which could be determined by probe techniques.

## 2.3 Limiting Form of the Fundamental Identity

If $Z \to 0$, Eq. (8) becomes

$$\nless \sum_j l_j^{-1} (\mathbf{B}_k \cdot \mathbf{n}_k)_j \ngtr_2 \equiv \frac{1}{\mathcal{A}} \overline{\int_{\mathcal{C}} \mathbf{B}_k \cdot \mathbf{n}_k \frac{d\mathcal{C}}{\mathbf{n}_k \cdot \mathbf{n}_{k\mathcal{C}}}} \tag{15}$$

which will appear as the link between the interaction terms occurring in the space–time- or time–space-averaged equations. In Eq. (15), $\mathbf{n}_{k\mathcal{C}}$ denotes the unit vector normal to $\mathcal{C}$, located in the cross-sectional plane and directed away from phase $k$. Curve $\mathcal{C}$ is the intersection of interface $\mathcal{A}_i$ with the cross-sectional plane.

# 3 SPACE–TIME- OR TIME–SPACE-AVERAGED EQUATIONS

If we time average over $[T]$ the instantaneous area-averaged balance equation, we obtain

$$\frac{\partial}{\partial t} \overline{\mathcal{A}_k \langle \rho_k \psi_k \rangle_2} + \frac{\partial}{\partial z} \overline{\mathcal{A}_k \langle \mathbf{n}_z \cdot (\rho_k \psi_k \mathbf{v}_k) \rangle_2} + \frac{\partial}{\partial z} \overline{\mathcal{A}_k \langle \mathbf{n}_z \cdot \mathbf{J}_k \rangle_2} - \overline{\mathcal{A}_k \langle \rho_k \phi_k \rangle_2}$$
$$= - \overline{\int_{\mathcal{C}(z,t)} (\dot{m}_k \psi_k + \mathbf{n}_k \cdot \mathbf{J}_k) \frac{d\mathcal{C}}{\mathbf{n}_k \cdot \mathbf{n}_{k\mathcal{C}}}} - \overline{\int_{\mathcal{C}_k(z,t)} \mathbf{n}_k \cdot \mathbf{J}_k \frac{d\mathcal{C}}{\mathbf{n}_k \cdot \mathbf{n}_{k\mathcal{C}}}} \tag{16}$$

If we area average the local time-averaged balance equation over the total cross-sectional area, we obtain

$$\frac{\partial}{\partial t} \nless \alpha_k \overline{\rho_k \psi_k}^X \ngtr_2 + \frac{\partial}{\partial z} \nless \alpha_k \overline{\mathbf{n}_z \cdot (\rho_k \psi_k \mathbf{v}_k)}^X \ngtr_2 + \frac{\partial}{\partial z} \nless \alpha_k \mathbf{n}_z \cdot \overline{\mathbf{J}_k}^X \ngtr_2$$
$$- \nless \alpha_k \overline{\rho_k \phi_k}^X \ngtr_2 = - \nless \sum_j l_j^{-1} (\dot{m}_k \psi_k + \mathbf{J}_k \cdot \mathbf{n}_k)_j \ngtr_2$$
$$- \int_{\mathcal{C}_1 + \mathcal{C}_2} \alpha_k \mathbf{n}_k \cdot \overline{\mathbf{J}_k}^X \frac{d\mathcal{C}}{\mathbf{n}_k \cdot \mathbf{n}_{k\mathcal{C}}} \tag{17}$$

As a result of Eqs. (4) and (15), Eqs. (16) and (17) are identical. Similar equations to Eqs. (16) and (17) can be written for volume-averaged quantities.

## NOMENCLATURE

| | |
|---|---|
| $\mathcal{A}$ | area |
| $\mathbf{B}$ | vector |
| $\mathcal{C}$ | line |
| $\mathcal{D}$ | domain |
| $f$ | arbitrary function |
| $\mathbf{J}$ | flux term |
| $l$ | specific length |
| $\dot{m}$ | mass transfer per unit area of interface and per unit of time |
| $\mathbf{n}$ | unit normal vector |
| $R$ | area fraction [Eq. (5)] |
| $t$ | time |
| $T$ | time interval |
| $\mathbf{v}$ | velocity |
| $\mathcal{V}$ | volume |
| $\mathbf{x}$ | position vector |
| $X$ | phase density function |
| $z$ | axial coordinate |
| $Z$ | vertical distance |
| $\alpha$ | time fraction |
| $\gamma$ | local specific area [Eq. (12)] |
| $\Gamma$ | integral specific area [Eq. (13)] |
| $\rho$ | density |
| $\phi$ | source term |
| $\psi$ | specific quantity |

**Susbscripts**

| | |
|---|---|
| $i$ | interface |
| $j$ | interface number |
| $k$ | phase index |
| $n$ | dimension index |

**Operators**

| | |
|---|---|
| $\langle\ \rangle$ | space-averaging operator over $\mathcal{D}_k$ |
| $\nless\ \ngtr$ | space-averaging operator over $\mathcal{D}$ |
| $\overline{\phantom{x}}$ | single time-averaging operator over $[T]$ |
| $\overline{\phantom{x}}^{X}$ | single time-averaging operator over $[T_k]$ |

## REFERENCES

Delhaye, J. M., Sur les Surfaces Volumiques Locale et Intégrale en Ecoulement Diphasique, *C. R. Acad. Sci., Ser. A*, vol. 283, pp. 243–246, 1976.

Delhaye, J. M. and Achard, J. L., On the Averaging Operators Introduced in Two-Phase Flow Modeling, in *Transient Two-Phase Flow, Proc. CSNI Specialists Meet., Aug. 3 and 4, 1976, Toronto*, eds. S. Banerjee and K. R. Weaver, vol. 1, pp. 5–84, AECL, 1978.

Vernier, Ph. and Delhaye, J. M., General Two-Phase Flow Equations Applied to the Thermohydrodynamics of Boiling Water Nuclear Reactors, *Energie Primaire*, vol. 4, no. 1–2, pp. 5–46, 1968.

CHAPTER

# TEN

# TWO-PHASE FLOW MODELING

**J. M. Delhaye**

## 1 TWO-FLUID MODEL

By its very nature the two-fluid model is the only model consistent with the balance laws written for each phase and the interfaces. This model can be written using either time averages (Ishii, 1975) or space averages (Kocamustafaogullari, 1971). If the two-fluid model is written using time averages, it is theoretically possible to solve three-dimensional problems of transient two-phase flow, whereas if it is written using space averages, the two-fluid model can deal only with transient flow problems with a single space variable.

First a list of hypotheses commonly accepted to simplify the equations is given. Next, the different forms of the six-equation system are recalled with particular emphasis on the difference equations introduced by Bouré (1975). These equations display the dynamic and thermodynamic nonequilibria, that is, the slip velocity and the differences between the enthalpies of each phase and the corresponding saturation enthalpies.

### 1.1 Simplified Balance Equations

The two averaging operators, that is, over a cross-sectional area and over a time interval, give averages of products that can be expressed as a function of the product of averages by correlation coefficients. For instance,

$$\lessdot fg \gtrdot = A \lessdot f \gtrdot \lessdot g \gtrdot$$

$$\overline{fg}^X = B\overline{f}^X\overline{g}^X$$

The following hypotheses are generally admitted (Vernier and Delhaye, 1968; Kocamustafaogullari, 1971; Bouré and Réocreux, 1972; Réocreux, 1972):

1. The space correlation coefficients are all equal to 1.
2. The time correlation coefficients are all equal to 1.
3. The equation of state valid for local quantities applies to averaged quantities.
4. Longitudinal conduction terms in each phase, as well as their derivatives, are negligible.
5. The phase viscous stress derivatives and the power of these viscous stresses are negligible.
6. In vertical flow, the pressure is constant over a cross section.

Moreover, if the flow is assumed to be symmetrical with respect to a straight line, the vectorial momentum equation is reduced to its projection on the symmetry axis. On the basis of these hypotheses, simplified phase and mixture equations are obtained as follows:

Mass:
Phase equation ($k = 1, 2$)

$$\frac{\partial}{\partial t}(R_k\rho_k) + \frac{\partial}{\partial z}(R_k\rho_k w_k) = -\frac{1}{\mathcal{A}}\overline{\int_{\mathcal{C}(z,t)} \dot{m}_k \frac{d\mathcal{C}}{\mathbf{n}_k \cdot \mathbf{n}_{k\mathcal{C}}}} \tag{1}$$

Mixture equation

$$\frac{\partial}{\partial t}(R_1\rho_1 + R_2\rho_2) + \frac{\partial}{\partial z}(R_1\rho_1 w_1 + R_2\rho_2 w_2) = 0 \tag{2}$$

Momentum:
Phase equation ($k = 1, 2$)

$$\begin{aligned}\frac{\partial}{\partial t}(R_k\rho_k w_k) &+ \frac{\partial}{\partial z}(R_k\rho_k w_k^2) - R_k\rho_k F_z + R_k\frac{\partial p}{\partial z} \\ &= -\frac{1}{\mathcal{A}}\overline{\int_{\mathcal{C}(z,t)} \mathbf{n}_z \cdot (\dot{m}_k \mathbf{v}_k - \mathbf{n}_k \cdot \tau_k)\frac{d\mathcal{C}}{\mathbf{n}_k \cdot \mathbf{n}_{k\mathcal{C}}}} \\ &\quad + \frac{1}{\mathcal{A}}\overline{\int_{\mathcal{C}_k(z,t)} \mathbf{n}_z \cdot (\mathbf{n}_k \cdot \tau_k)\frac{d\mathcal{C}}{\mathbf{n}_k \cdot \mathbf{n}_{k\mathcal{C}}}}\end{aligned} \tag{3}$$

Mixture equation

$$\begin{aligned}\frac{\partial}{\partial t}(R_1\rho_1 w_1 + R_2\rho_2 w_2) &+ \frac{\partial}{\partial z}(R_1\rho_1 w_1^2 + R_2\rho_2 w_2^2) - (R_1\rho_1 + R_2\rho_2)F_z + \frac{\partial p}{\partial z} \\ &= \frac{1}{\mathcal{A}}\sum_{k=1,2}\overline{\int_{\mathcal{C}_k(z,t)} \mathbf{n}_z \cdot (\mathbf{n}_k \cdot \tau_k)\frac{d\mathcal{C}}{\mathbf{n}_k \cdot \mathbf{n}_{k\mathcal{C}}}}\end{aligned} \tag{4}$$

Total energy:
Phase equation ($k = 1, 2$)

$$\frac{\partial}{\partial t}\left[R_k \rho_k \left(\frac{1}{2} v_k^2 + i_k\right)\right] + \frac{\partial}{\partial z}\left[R_k \rho_k \left(\frac{1}{2} v_k^2 + i_k\right) w_k\right] - R_k \frac{\partial p}{\partial t} - R_k \rho_k \mathbf{F} \cdot \mathbf{v}_k$$

$$= -\frac{1}{\mathfrak{a}} \overline{\int_{\mathfrak{C}(z,t)} \left[\dot{m}_k \left(\frac{1}{2} v_k^2 + i_k\right) - (\tau_k \cdot \mathbf{v}_k) \cdot \mathbf{n}_k + \mathbf{q}_k \cdot \mathbf{n}_k\right] \frac{d\mathfrak{C}}{\mathbf{n}_k \cdot \mathbf{n}_{k\mathfrak{C}}}}$$

$$- \frac{1}{\mathfrak{a}} \overline{\int_{\mathfrak{C}_k(z,t)} \mathbf{q}_k \cdot \mathbf{n}_k \frac{d\mathfrak{C}}{\mathbf{n}_k \cdot \mathbf{n}_{k\mathfrak{C}}}} \tag{5}$$

Mixture equation

$$\frac{\partial}{\partial t}\left[R_1 \rho_1 \left(\frac{1}{2} v_1^2 + u_1\right) + R_2 \rho_2 \left(\frac{1}{2} v_2^2 + u_2\right)\right]$$

$$+ \frac{\partial}{\partial z}\left[R_1 \rho_1 \left(\frac{1}{2} v_1^2 + i_1\right) w_1 + R_2 \rho_2 \left(\frac{1}{2} v_2^2 + i_2\right) w_2\right]$$

$$- (R_1 \rho_1 \mathbf{F} \cdot \mathbf{v}_1 + R_2 \rho_2 \mathbf{F} \cdot \mathbf{v}_2) = -\frac{1}{\mathfrak{a}} \sum_{k=1,2} \overline{\int_{\mathfrak{C}_k(z,t)} \mathbf{q}_k \cdot \mathbf{n}_k \frac{d\mathfrak{C}}{\mathbf{n}_k \cdot \mathbf{n}_{k\mathfrak{C}}}} \tag{6}$$

## 1.2 Constitutive Laws

To solve the six phase equations [Eqs. (1), (3), and (5)] written for each phase ($k = 1, 2$), seven constitutive laws must be known, that is,

1. At the interface, the mass, momentum, and energy transfer laws
2. At the wall, the friction and heat transfer laws for each phase

Note that hypothesis 6 in Sec. 1.1 introduced an additional law of mechanical behavior, that is, the constancy of the pressure over the channel cross section. The implications of this assumption, as well as the nature of this mechanical law were examined in detail by Bouré (1977, 1978a). The reader is also referred to Sec. 4 of Chap. 7.

Knowledge of the constitutive laws permits the calculation of the flow parameters, such as the phase velocities and the phase enthalpies. As a result, slip velocity and thermodynamic nonequilibria for each phase can be computed. Consequently, the knowledge of the constitutive laws forms the crux of two-phase flow modeling. To reduce the general form of the constitutive laws Bouré (1978a) proposed restrictions based on fundamental principles, such as the indifference to Galilean changes of frame and to some changes of origins, the second law of thermodynamics, and the assumption of local thermodynamic equilibrium.

## 1.3 Equivalent Systems

The simplified system, composed of the six phase equations [Eqs. (1), (3), and (5)] can be put in the following matrix form:

$$A \frac{\partial \mathbf{X}}{\partial t} + B \frac{\partial \mathbf{X}}{\partial z} = \mathbf{C} \tag{7}$$

where $A$ and $B$ are matrices, with $\mathbf{X}$, the solution vector having as components the following dependent variables:

$$R_2 \quad p \quad w_1 \quad w_2 \quad i_1 \quad i_2$$

and $\mathbf{C}$ is a vector without partial derivatives.

**1.3.1 First equivalent system** This simplified system has two disadvantages:

1. Matrices $A$ and $B$ are not simple.
2. Matrices $A$ and $B$ are generally singular for $R_k = 0$, which gives an ill-conditioned system for void fractions close to 0 or 1 and consequently entails difficulties in the subsequent numerical resolution. This disadvantage can be eliminated by dividing the two sides of the momentum and energy equations by $R_k$ and by ensuring that the corresponding right-hand sides remain finite when $R_k \to 0$.

To simplify $A$ and $B$ matrices by introducing zeros, Bouré (1975) proposed certain combinations to transform the original system [Eq. (7)] into a first equivalent system. This form, suitable for numerical calculation of the solution, does not, however, present in a simple way the particular models, that is, without slip or nonequilibria or without both, as particular solutions of the two-fluid model. Bouré's solution was to combine the equation so as to introduce difference variables representing the dynamic and thermal nonequilibria.

**1.3.2 Second equivalent system** The difference variables are defined in the following way:

Dynamic nonequilibrium:

$$\Delta w \triangleq w_2 - w_1 \tag{8}$$

Thermal nonequilibria:

$$\Delta i_2 \triangleq i_2 - i_2^{\text{sat}} \tag{9}$$

$$\Delta i_1 \triangleq i_1 - i_1^{\text{sat}} \tag{10}$$

The mixture variables are

1. The void fraction $R_2$
2. The pressure $p$
3. The superficial velocity of the mixture defined as

$$J \triangleq R_1 w_1 + R_2 w_2 \tag{11}$$

The six-equation system proposed by Bouré (1975) comprises

1. Three difference equations
2. Three mixture balance equations

These equations are written as a function of the six following dependent variables:

$$\Delta w \quad \Delta i_1 \quad \Delta i_2 \quad R_2 \quad p \quad J$$

**1.3.3 Third equivalent system** To simplify the system matrices by introducing zeros, Bouré (1975) combined the three mixture equations and kept the difference equations. He then obtained a six-equation system that seems to have the most suitable form for numerical applications. The complete equations of this system are given in Bouré's paper. The author used this system to reconstitute the usual particular models from the six-equation model. He indicated the restrictions that must be applied to the constitutive laws of a model with less than six equations to make it equivalent to a six-equation model.

## 2 TWO-PHASE FLOW MODELING

Although the two-fluid model is the most satisfactory in theory, it has been thought difficult to use because seven constitutive laws are required. To reduce this degree of complexity, several authors have hypothesized particular flow evolutions, expressed by nonequilibrium algebraic laws, such as,

A slip correlation, using

1. A slip ratio

$$\gamma \triangleq \frac{w_2}{w_1} = \text{function (thermodynamic equality, pressure)} \tag{12}$$

2. A drift velocity (e.g., Zuber and Findlay, 1965)

$$w_{2j} \triangleq w_2 - j = \text{function (flow pattern, physical properties)} \tag{13}$$

3. A void fraction (e.g., Martinelli and Nelson, 1948)

$$R_2 = \text{function (thermodynamic quality, pressure)} \tag{14}$$

Laws of thermal nonequilibrium

1. Vapor at saturation temperature

$$\Delta i_2 \triangleq i_2 - i_2^{\text{sat}}(p) \equiv 0 \tag{15}$$

2. Liquid at saturation temperature

$$\Delta i_1 \triangleq i_1 - i_1^{\mathrm{sat}}(p) \equiv 0 \tag{16}$$

With this method, the number of partial differential equations in the system can be reduced, along with the number of constitutive laws. In return, the mathematical nature of the system is changed radically. The different cases are summarized in Table 1 (Bouré, 1975).

# 3 SINGLE-FLUID MODEL

In the single-fluid model the idea is to replace the two-phase fluid by an equivalent compressible single-phase fluid. The physical properties of this single-phase fluid, such as its density and viscosity, as well as the flow parameters, for example, the velocity and temperature, must then be defined as functions of the properties of each of the phases. If one of the phases is finely dispersed, momentum and energy transfers will be sufficiently rapid for the average velocities and the average temperatures of the two phases to be equal, that is,

$$w_1 \equiv w_2 \triangleq w \tag{17}$$

$$T_1 \equiv T_2 \triangleq T \tag{18}$$

If the temperature $T$ is assumed to be the saturation temperature, the flow is said to be described by the homogeneous equilibrium model.

## 3.1 Expression of the Density

If there is no rapid variation of flow parameters such as the pressure, and if thermodynamic nonequilibrium is not a great influence, one can assume the following expression for the density:

$$\rho = R_1 \rho_1 + R_2 \rho_2 \tag{19}$$

with

$$R_k = \frac{Q_k}{Q_1 + Q_2} \triangleq \beta_k \qquad (k = 1, 2) \tag{20}$$

In Eq. (20) $Q_k$ is the volumetric flow rate of phase $k$ and $\beta_k$ is the volumetric quality of phase $k$.

The density can also be expressed as a function of the mass quality $x$ because Eqs. (19) and (20) can be rewritten as

$$\frac{1}{\rho} = \frac{1-x}{\rho_1} + \frac{x}{\rho_2} \tag{21}$$

with

$$x \triangleq \frac{M_2}{M_1 + M_2} \tag{22}$$

where $M_k$ is the mass flow rate of phase $k$.

**Table 1 Two-phase flow modeling**

| Imposed restrictions | | Remaining dependent variables | Balance equations written in practice | Constitutive laws needed | |
|---|---|---|---|---|---|
| Number | Type | | | Number | Type |
| 3 | $\Delta w$ $\Delta i_1$ $\Delta i_2$ (e.g. homogeneous model) | $R_2$ $p$ $w_1$ | 3 mixture balance equations | 2 | Mixture wall friction<br>Mixture wall heat flux |
| 2 | $\Delta w$ $\Delta i_2$ (or $\Delta i_1$) (e.g. diffusion models) | $R_2$ $p$ $w_1$ $\Delta i_1$ (or $\Delta i_2$) | 3 mixture balance equations<br>+ 1 phase balance equation (mass) | 3 | Mixture wall friction<br>Mixture wall heat flux<br>Mass interaction term |
| | | | 3 mixture balance equations<br>+ 1 phase balance equation (energy) | 4 | Mixture wall friction<br>Wall heat flux for each phase<br>Energy interaction term |
| | $\Delta i_1$ $\Delta i_2$ (e.g. thermal equilibrium) | $R_2$ $p$ $w_1$ $w_2$ | 3 mixture balance equations<br>+ 1 phase balance equation (momentum) | 4 | Wall friction for each phase<br>Mixture wall heat flux<br>Momentum interaction term |
| 1 | $\Delta w$ | $R_2$ $p$ $w_1$ $\Delta i_1$ $\Delta i_2$ | 3 mixture balance equations<br>+ 1 phase balance equation (mass)<br>+ 1 phase balance equation (energy) | 5 | Mixture wall friction<br>Wall heat flux for each phase<br>Mass, energy interaction terms |
| | $\Delta i_2$ (or $\Delta i_1$) | $R_2$ $p$ $w_1$ $w_2$ $\Delta i_1$ (or $\Delta i_2$) | 3 mixture balance equations<br>+ 1 phase balance equation (mass)<br>+ 1 phase balance equation (momentum) | 5 | Wall friction for each phase<br>Mixture wall heat flux<br>Mass, momentum interaction terms |
| | | | 3 mixture balance equations<br>+ 1 phase balance equation (momentum)<br>+ 1 phase balance equation (energy) | 6 | Wall friction for each phase<br>Wall heat flux for each phase<br>Momentum, energy interaction terms |
| 0 | | $R_2$ $p$ $w_1$ $w_2$ $\Delta i_1$ $\Delta i_2$ | 3 mixture balance equations<br>+ 3 phase balance equations<br>(or 6 phase balance equations) | 7 | Wall friction for each phase<br>Wall heat flux for each phase<br>Mass, momentum, energy interaction terms |

Note that choosing Eq. (19) for the definition of the density and introducing this density into the equations of single-phase flow would be the same as writing the simplified mixture equations and assuming equal velocities and equal temperatures [Eqs. (17) and (18)].

When nonequilibrium effects occur in the flow, one can assume that the density is given by a relaxation-type equation. Relaxation models of that type are currently under development (Bouré, 1978b).

The homogeneous model has been frequently used to study problems in oil extraction, steam generation, and refrigeration. The higher the pressures and velocities, the more realistic the homogeneous model.

## 3.2 Balance Equations for the Homogeneous Model

A convenient means to obtain equations of single-phase flow of a compressible fluid is to start from the phase equations of the simplified system described earlier. As there is only one phase, the integrals on $\mathcal{C}$ disappear, and $R_k \equiv 1$.

Mass balance:

$$\frac{\partial \rho}{\partial t} + \frac{\partial \rho w}{\partial z} = 0 \tag{23}$$

In steady-state flow we have

$$\rho w \triangleq G = \text{const} \tag{24}$$

Momentum balance:
Taking into account the mass balance we obtain

$$\rho \frac{\partial w}{\partial t} + \rho w \frac{\partial w}{\partial z} = -\frac{\partial p}{\partial z} + \rho F_z - \frac{\mathcal{P}}{\mathcal{A}} \tau_W \tag{25}$$

where $\mathcal{P}$ is the total perimeter of the cross section and $\tau_W$ the wall shear stress. For a tube of circular cross section of diameter $D$ we obtain

$$\frac{\mathcal{P}}{\mathcal{A}} = \frac{4}{D} \tag{26}$$

In steady-state flow, Eq. (25) can be rewritten as

$$\frac{dp}{dz} = -\rho w \frac{dw}{dz} + \rho F_z - \frac{\mathcal{P}}{\mathcal{A}} \tau_W \tag{27}$$

which means that the total pressure drop is composed of an acceleration pressure drop, a hydrostatic pressure drop, and a frictional pressure drop. Equation (27) is often written

$$\frac{dp}{dz} = \left(\frac{dp}{dz}\right)_A + \left(\frac{dp}{dz}\right)_G + \left(\frac{dp}{dz}\right)_F \tag{28}$$

The evaluation of the frictional pressure drop will be dealt with in Chap. 11.

Total energy balance:

Assuming that the heat flux at the wall is applied on the whole perimeter of the cross section, we have,

$$\frac{\partial}{\partial t}\left[\rho\left(\frac{1}{2}w^2+u\right)\right]+\frac{\partial}{\partial z}\left[\rho w\left(\frac{1}{2}w^2+i\right)\right]=\rho w F_z+\frac{\mathcal{P}}{\mathcal{A}}\varphi \tag{29}$$

where $\varphi$ is the wall heat flux density.

## 4 ZUBER AND FINDLAY MODEL

The Zuber and Findlay (1965) model is simple and enables the slip between the phases as well as the two-dimensional pattern of the flow to be considered. Unlike the homogeneous model, the Zuber and Findlay model can handle countercurrent flow and offer better results at low velocities. The basic development of this model appeared about a decade ago in a series of three papers (Zuber and Findlay, 1965; Zuber et al., 1967; Kroeger and Zuber, 1968).

### 4.1 The Zuber and Findlay Void Equation

Given a two-phase mixture flowing in a pipe, the following local quantities can be defined:

$$j_f \triangleq (1-\alpha)v_f \tag{30}$$

$$j_g \triangleq \alpha v_g \tag{31}$$

$$j \triangleq j_g + j_f \tag{32}$$

$$v_{gj} \triangleq v_g - j = (1-\alpha)(v_g - v_f) \tag{33}$$

where $\alpha$ denotes the gas local time fraction, that is, the local void fraction, and $v_k$ is the component, along the axis of the pipe, of the local time-averaged velocity.

We have locally

$$\alpha v_{gj} = \alpha v_g - \alpha j = j_g - \alpha j \tag{34}$$

Averaging this equation over the total cross section of the pipe, we obtain the Zuber and Findlay void equation:

$$\langle\!\langle \alpha \rangle\!\rangle = \frac{\langle\!\langle j_g \rangle\!\rangle}{C_0 \langle\!\langle j \rangle\!\rangle + \tilde{v}_{gj}} = \frac{J_g}{C_0 J + \tilde{v}_{gj}} \tag{35}$$

where $J_g$ and $J$ are the gas and mixture superficial velocities. In this equation two quantities appear: (1) the Zuber and Findlay distribution parameter $C_0$ defined by

$$C_0 \triangleq \frac{\langle\!\langle \alpha j \rangle\!\rangle}{\langle\!\langle \alpha \rangle\!\rangle \langle\!\langle j \rangle\!\rangle} \tag{36}$$

which accounts for the shape of the $\alpha$ and $j$ profiles, and (2) the Zuber and Findlay drift velocity $\widetilde{v}_{gj}$ defined by

$$\widetilde{v}_{gj} \triangleq \frac{\langle\!\langle \alpha v_{gj} \rangle\!\rangle}{\langle\!\langle \alpha \rangle\!\rangle} \tag{37}$$

which accounts for the relative velocity between the phases.

Note that no hypothesis has been formulated to establish the void equation [Eq. (35)]. To calculate $\langle\!\langle \alpha \rangle\!\rangle$ from this equation it is sufficient to know $J_g$, $J$, that is, the volumetric flow rates of each phase and the cross-sectional area, the distribution parameter $C_0$, and the drift velocity $\widetilde{v}_{gj}$.

## 4.2 Values of the Distribution Parameter and of the Drift Velocity

Zuber and Findlay realized that $C_0$ and $\widetilde{v}_{gj}$ are only functions of the flow pattern, and they recommended the values given in Table 2. Recently Ishii et al. (1976) proposed a new equation for the vapor drift velocity in annular flow.

For high pressure steam–water flows, $C_0$ and $\widetilde{v}_{gj}$ are given by Eqs. (38) and (39), which are valid whatever the flow pattern,

$$C_0 = 1.13 \tag{38}$$

$$\widetilde{v}_{gj} = 1.41 \left( \frac{\sigma g \, \Delta\rho}{\rho_f^2} \right)^{1/4} \tag{39}$$

Figure 1 gives the drift velocity as a function of pressure for steam–water flows.

**Table 2 Zuber and Findlay distribution parameter and drift velocity**

| Type of flow | Distribution parameter | Churn-turbulent flow or unknown bubble diameter |
|---|---|---|
| Bubbly | Circular cross section: $p_R \triangleq \dfrac{p}{p_c}$ | |
| | $D > 5$ cm $\quad C_0 = 1 - 0.5p_R$ | |
| | $D < 5$ cm $\quad p_R < 0.5 \quad C_0 = 1.2$ | $\widetilde{v}_{gj} = 1.41 \left( \dfrac{\sigma g \, \Delta p}{\rho_f^2} \right)^{1/4}$ |
| | $D < 5$ cm $\quad p_R < 0.5 \quad C_0 = 1.4 - 0.4p_R$ | |
| | Rectangular cross section: $C_0 = 1.4 - 0.4p_R$ | |
| Slug | $C_0 = 1.2$ | $\widetilde{v}_{gj} = 0.35 \left( \dfrac{g \, \Delta\rho D}{\rho_f} \right)^{1/2}$ |
| Annular | $C_0 = 1.0$ | $\widetilde{v}_{gj} = 23 \left( \dfrac{\mu_f j_f}{\rho_g D} \right)^{1/2} \dfrac{\Delta\rho}{\rho_f}$ |

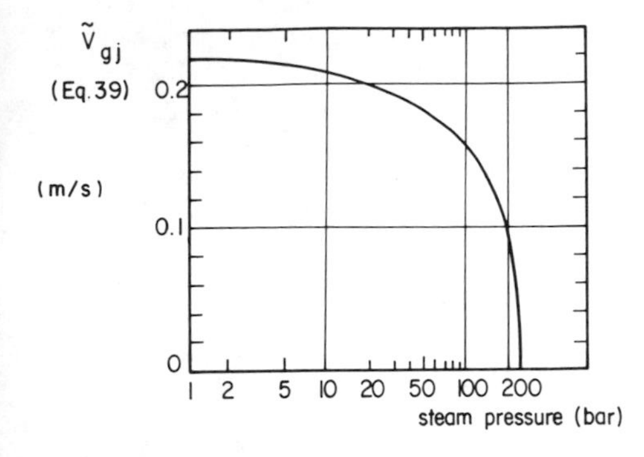


**Figure 1** Variation of the drift velocity with the pressure in a steam–water flow.

## 4.3 Zuber and Findlay Diagram

The void equation [Eq. (35)] can be transformed into

$$\frac{J_g}{\langle\!\langle \alpha \rangle\!\rangle} = C_0 J + \tilde{v}_{gj} \tag{40}$$

If $C_0$ and $\tilde{v}_{gj}$ are assumed constant for a given flow pattern, the quantity $J_g/\langle\!\langle \alpha \rangle\!\rangle$ is then a linear function of the mixture superficial velocity $J$. Figure 2, called the Zuber and Findlay diagram, represents the quantity $J_g/\langle\!\langle \alpha \rangle\!\rangle$ as a function of $J$.

Figure 3 shows some experimental results for a bubbly air–water mixture flowing out of a converging–diverging nozzle (Delhaye and Jacquemin, 1971).

## 4.4 Other Forms of the Void Equation

The void equation [Eq. (35)] can be written in terms of the volumetric quality $\beta$. One finds

$$\langle\!\langle \alpha \rangle\!\rangle = \frac{\beta}{C_0 + \tilde{v}_{gj}/J} \tag{41}$$

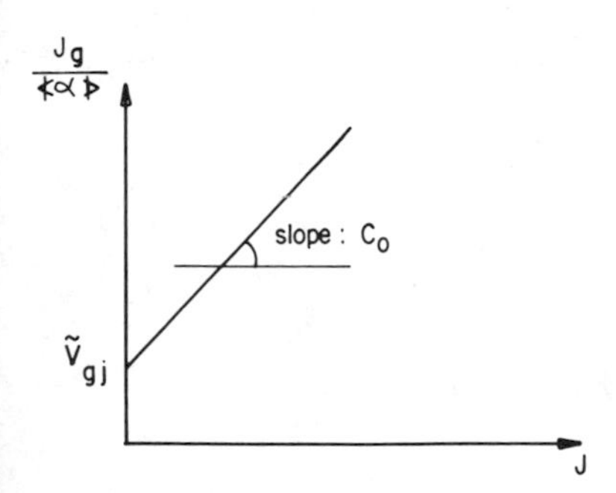


**Figure 2** Zuber and Findlay diagram.

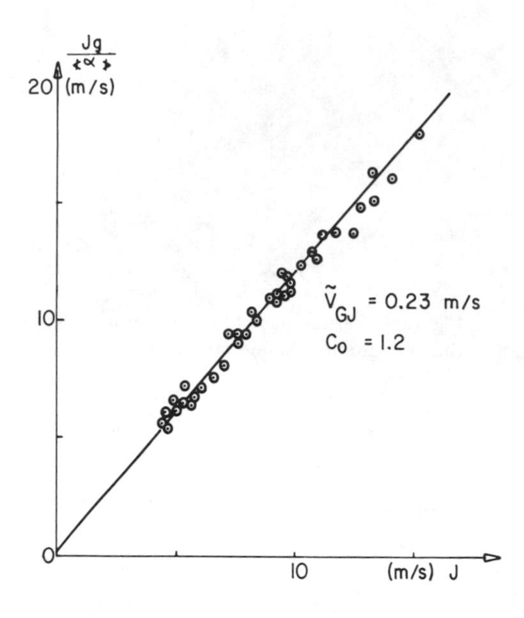


**Figure 3** Bubbly air–water flow experimental results (Delhaye and Jacquemin, 1971).

If it is assumed that the drift velocity is small compared to the mixture superficial velocity, Eq. (41) now reads

$$\langle\!\langle \alpha \rangle\!\rangle = \frac{\beta}{C_0} \triangleq K\beta \tag{42}$$

where $K$ is the so-called Armand parameter.

The void equation [Eq. (35)] can also be written in terms of the gas quality $x$ and of the mass velocity $G$. One finds

$$\langle\!\langle \alpha \rangle\!\rangle = \frac{x\rho_f G}{C_0\,[x\rho_f + (1-x)\rho_g]\,G + \tilde{v}_{gj}\rho_g\rho_f} \tag{43}$$

We thus see that the mass velocity enters the Zuber and Findlay void equation. As a result the calculated void fraction will depend on the mass velocity for a given quality and pressure. Such is not the case with the Lockhart-Martinelli or Martinelli-Nelson empirical correlations, which will be given in Chap. 11.

## 5 WALLIS MODEL

Wallis (1969) introduced the following quantity:

$$j_{21} \triangleq \alpha(v_2 - j) = \alpha(1-\alpha)(v_2 - v_1) \tag{44}$$

which can also be written in the form

$$j_{21} = (1-\alpha)j_2 - \alpha j_1 \tag{45}$$

The Wallis model does not take into account the distribution effect but only the relative velocity between the phases. Consequently the quantities entering Eqs. (44) and (45) can be considered either as local variables or area-averaged variables.

The quantity $j_{21}$ is a function of the void fraction $\alpha$ and the relative velocity $(v_2 - v_1)$. If there is no wall shear stress, this velocity depends only on the void fraction $\alpha$ and the physical properties of the system:

$$j_{21} = j_{21}(\alpha, \text{physical properties of the system}) \tag{46}$$

Often, the relationship in Eq. (46) can be written

$$j_{21} = v_\infty \alpha (1 - \alpha)^n \tag{47}$$

where $v_\infty$ is the terminal speed of a rising bubble in an infinite medium (Wallis, 1974).

The Wallis diagram (Fig. 4) is a plot of $j_{21}$ related to $\alpha$. Equation (46) is represented by the curve whereas Eq. (45) is represented by a straight line whose ordinate is $j_{21} = j_2$ for $\alpha = 0$ and $j_{21} = -j_1$ for $\alpha = 1$.

## 5.1 Cocurrent Upward Flow

In cocurrent upward flow $j_1$ and $j_2$ are positive and the straight line [Eq. (45)] intersects the curve [Eq. (46)] at a single point that gives the value of the void fraction $\alpha$.

## 5.2 Countercurrent Flow

The gas is flowing upward ($j_2 > 0$) and the liquid downward ($j_1 < 0$). If $|j_1| < |(j_1)_3|$ (Fig. 4) the straight line cuts the curve in two points that correspond to a loose-bed and a dense-packed-bed pattern in the same column. If the liquid flow rate is increased, the straight line becomes tangent to the curve when $|j_1| = |(j_1)_3|$. This corresponds to the flooding point. When the liquid flow rate is

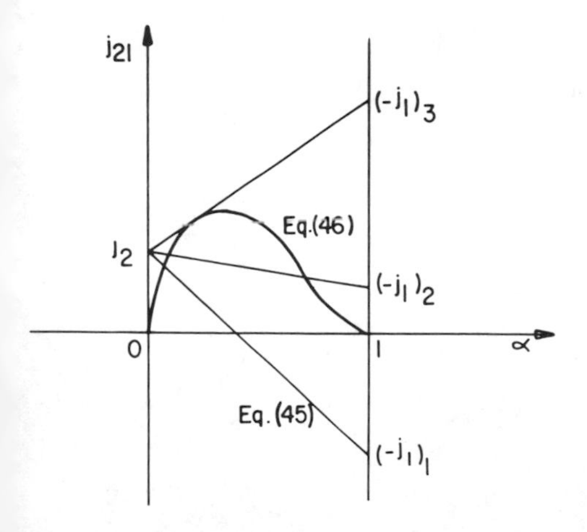


**Figure 4** Wallis diagram.

further increased, there is no solution pertaining to the countercurrent flow pattern. The flow becomes cocurrent downward.

## NOMENCLATURE

| | |
|---|---|
| $\mathfrak{a}$ | area |
| $A$ | correlation coefficient; matrix |
| $B$ | correlation coefficient; matrix |
| $\mathbf{C}$ | vector |
| $C_0$ | distribution parameter [Eq. (36)] |
| $\mathfrak{e}$ | line |
| $D$ | diameter |
| $f$ | arbitrary function |
| $\mathbf{F}$ | external force per unit of mass |
| $g$ | arbitrary function; acceleration owing to gravity |
| $G$ | mass velocity |
| $i$ | enthalpy per unit of mass |
| $j$ | local volumetric flux [Eqs. (30) and (31)] |
| $J$ | mixture superficial velocity [Eq. (11)] |
| $\mathbf{J}$ | flux term |
| $K$ | Armand parameter [Eq. (42)] |
| $\dot{m}$ | mass transfer per unit area of interface and per unit of time |
| $M$ | mass flow rate |
| $\mathbf{n}$ | unit normal vector |
| $p$ | pressure |
| $\mathcal{P}$ | perimeter |
| $\mathbf{q}$ | heat flux |
| $Q$ | volumetric flow rate |
| $R$ | area fraction |
| $t$ | time |
| $T$ | temperature |
| $u$ | internal energy |
| $v$ | $z$ component of the velocity vector |
| $\mathbf{v}$ | velocity |
| $w$ | $z$ component of the velocity vector |
| $x$ | quality [Eq. (22)] |
| $\mathbf{X}$ | solution vector |
| $z$ | axial coordinate |
| $\alpha$ | time fraction |
| $\beta$ | volumetric quality [Eq. (20)] |
| $\gamma$ | slip ratio [Eq. (12)] |
| $\mu$ | viscosity |
| $\rho$ | density |
| $\sigma$ | surface tension |

$\tau$ shear stress
$\boldsymbol{\tau}$ viscous stress tensor
$\varphi$ heat flux density

**Subscripts**

$A$ acceleration
$c$ critical pressure
$f$ liquid
$F$ friction
$g$ gas
$G$ gravity
$i$ interface
$j$ mixture
$k$ phase index
$R$ reduced pressure
$W$ wall
1 denser phase
2 lighter phase

**Superscript**

sat saturation conditions

**Operators**

$\langle\!\langle\ \rangle\!\rangle$ area-averaging operator over $\mathfrak{a}$
$\overline{\quad}^{x}$ single time-averaging over $[T_k]$

## REFERENCES


Bouré, J. A., Mathematical Modeling and the Two-Phase Constitutive Equations, European Two-Phase Flow Group Meet., Haifa, Israel, 1975.

Bouré, J. A., The Critical Flow Phenomenon with Reference to Two-Phase Flow and Nuclear Reactor Systems, in *Thermal and Hydraulic Aspects of Nuclear Reactor Safety*, vol. 1. *Light Water Reactors*, eds. O. C. Jones and S. G. Bankoff, pp. 195–216, ASME, New York, 1977.

Bouré, J. A., Les Lois Constitutives des Modèles d'Ecoulements Diphasiques Monodimensionnels à Deux Fluides. Formes Envisageables. Restrictions Résultant d'Axiomes Fondamentaux, CEA-R-4915, 1978a.

Bouré, J. A., Mathematical Modeling of Two-Phase Flows. Its Bases and Problems. A Review, in *Transient Two-Phase Flow, Proc. CSNI Specialists Meet., Aug. 3 and 4, 1976, Toronto*, eds. S. Banerjee and K. R. Weaver, vol. 1, pp. 85–111, AECL, 1978b.

Bouré, J. A. and Réocreux, M., General Equations of Two-Phase Flow. Application to Critical Flows and to Non-steady Flows, Fourth All-Union Heat and Mass Transfer Conf., Minsk, U.S.S.R., 1972.

Delhaye, J. M. and Jacquemin, J. P., Experimental Study of a Two-Phase Air-Water Flow in a Converging-Diverging Nozzle, CENG note TT 382, 1971.

Ishii, M., *Thermo-Fluid Dynamic Theory of Two-Phase Flow*, Eyrolles, Paris, 1975.

Ishii, M., Chawla, T. C., and Zuber, N., Constitutive Equation for Vapor Drift Velocity in Two-Phase Annular Flow, *AIChE J.*, vol. 22, no. 2, pp. 283–289, 1976.

Kocamustafaogullari, G., Thermo-Fluid Dynamics of Separated Two-Phase Flow, Ph.D. thesis, School of Mechanical Engineering, Georgia Institute of Technology, Atlanta, Ga., 1971.

Kroeger, P. G. and Zuber, N., Average Volumetric Concentration in Two-Phase Flow through Rectangular Channels, *J. Heat Transfer*, vol. 90, pp. 491–493, 1968.

Martinelli, A. C. and Nelson, D. B., Prediction of Pressure Drop during Forced Circulation Boiling of Water, *Trans. ASME*, vol. 70, pp. 695–702, 1948.

Réocreux, M., Applications des Equations Générales des Ecoulements Diphasiques à l'Etude des Débits Critiques, CENG rapport TT 109, 1972.

Vernier, Ph. and Delhaye, J. M., General Two-Phase Flow Equations Applied to the Thermohydrodynamics of Boiling Water Nuclear Reactors, *Energie Primaire*, vol. 4, no. 1–2, pp. 5–46, 1968.

Wallis, G. B., *One-dimensional Two-Phase Flow*, McGraw-Hill, New York, 1969.

Wallis, G. B., The Terminal Speed of Single Drops or Bubbles in an Infinite Medium, *Int. J. Multiphase Flow*, vol. 1, pp. 491–511, 1974.

Zuber, N. and Findlay, J. A., Average Volumetric Concentration in Two-Phase Flow Systems, *J. Heat Transfer*, vol. 87, pp. 453–468, 1965.

Zuber, N., Staub, F. W., Bijwaard, G., and Kroeger, P. G., Steady State and Transient Void Fraction in Two-Phase Flow Systems, GEAP 5417, 1967.

CHAPTER

# ELEVEN

# FRICTION FACTORS IN SINGLE CHANNELS

M. Giot

## 1 MOMENTUM BALANCE

The simplified momentum mixture equation, in steady-state conditions, reads

$$\frac{d}{dz}(R_1\rho_1 w_1^2 + R_2\rho_2 w_2^2) - (R_1\rho_1 + R_2\rho_2)F_z + \frac{dp}{dz} = \frac{1}{\mathfrak{a}} \sum_{k=1,2} \int_{\mathcal{C}_k(z,t)} \mathbf{n}_z$$

$$\cdot (\mathbf{n}_k \cdot \tau_k) \frac{d\mathcal{C}}{\mathbf{n}_k \cdot \mathbf{n}_{k\mathcal{C}}}$$

If $\qquad 1 \equiv L \quad$ and $\quad 2 \equiv G$

and with

$$R_2 \equiv \alpha \qquad R_1 \equiv 1 - \alpha \qquad F_z \equiv -g\cos\theta$$

and

$$\frac{1}{\mathfrak{a}} \sum_{k=1,2} \int_{\mathcal{C}_k(z,t)} \cdots \triangleq -\frac{\mathcal{P}}{\mathfrak{a}}\tau_w$$

the equation may be written

$$\frac{dp}{dz} = -[\alpha\rho_G + (1-\alpha)\rho_L]g\cos\theta - \frac{d}{dz}[\alpha\rho_G w_G^2 + (1-\alpha)\rho_L w_L^2] - \frac{\mathcal{P}}{\mathcal{A}}\tau_w \tag{1}$$

which means that the total pressure drop is composed, as in single-phase flow, of a gravity term, an acceleration term, and a friction term:

$$\frac{dp}{dz} = \left(\frac{dp}{dz}\right)_g + \left(\frac{dp}{dz}\right)_a + \left(\frac{dp}{dz}\right)_f$$

When one considers horizontal two-component two-phase flows in tubes, the hydrostatic pressure drop is zero, and the acceleration pressure drop is negligible. When one considers horizontal two-phase flows with phase change, the acceleration pressure drop can be important.

Generally, evaluation of the total pressure drop involves not only the friction term, but also the other two terms. All three must be calculated within the framework of the same model, using appropriate correlations or semiempirical formulas.

## 2 HOMOGENEOUS FLOW MODEL

The central assumption of the homogeneous flow model is that the liquid- and gas-phase velocities are identical:

$$w_L = w_G$$

The relationship between void fraction and quality is then

$$\frac{1-\alpha}{\alpha} = \frac{1-x}{x}\frac{\rho_G}{\rho_L} \tag{2}$$

Using Eq. (2) to eliminate $\alpha$ from Eq. (1) yields

$$\frac{dp}{dz} = -\rho_m g\cos\theta - G^2\frac{d}{dz}\left(\frac{1}{\rho_m}\right) - \frac{\mathcal{P}}{\mathcal{A}}\tau_w \tag{3}$$

where $G$ is the mass velocity and the mixture density $\rho_m$ is given by

$$\frac{1}{\rho_m} = \frac{1}{\alpha\rho_G + (1-\alpha)\rho_L} = \frac{x}{\rho_G} + \frac{1-x}{\rho_L} \tag{4}$$

The friction factor $f$ is defined by

$$\tau_w = f\frac{1}{2}\frac{G^2}{\rho_m} \tag{5}$$

Integrating Eq. (3) along a length $L$ of a cylindrical pipe gives

$$-\Delta p = g\cos\theta\int_0^L \rho_m\,dz + G^2\int_0^L \frac{d}{dz}\left(\frac{1}{\rho_m}\right)dz + \frac{4}{D}\frac{G^2}{2}\int_0^L f\frac{1}{\rho_m}\,dz \tag{6}$$

To calculate the integrals, the quality distribution along the pipe is needed.

1. If the quality is constant, and if the drop is given by

$$-\Delta p = \rho_m g L \cos\theta + \frac{4L}{D} f \frac{1}{2} \frac{G}{\rho_m}$$

where $\rho_m$ is evaluated at the mean pressure between the inlet and the

2. For a heated channel, a simplified energy balance leads to a linea of quality with length:

$$\frac{dx}{dz} = \frac{x_{out}}{L}$$

If $x(0) = 0$, one finds

$$-\Delta p \cong \rho_L g L \cos\theta \, \frac{\ln(\rho_L/\rho_{out})}{\rho_L/\rho_{out} - 1} + \frac{G^2}{\rho_L}\left(\frac{\rho_L}{\rho_{out}} - 1\right) + \frac{4L}{D} f \frac{G^2}{2\rho_L}\left(\frac{\rho_L}{\rho_{out}} + 1\right) \quad (8)$$

All parameters in Eqs. (7) and (8) are known except for the friction factor $f$. Several methods are proposed to calculate the friction factor. They generally rely on the use of a Reynolds number Re and the Moody chart. The Reynolds number is defined by

$$\text{Re} \triangleq \frac{GD}{\mu}$$

where $\mu$ is a mixture viscosity defined arbitrarily; the following formulas may be found in the literature:

$\mu = \mu_L$ (mainly for low-quality flow)
$\mu = \mu_G$ (mainly for high-quality flow)
$1/\mu = x/\mu_G + (1 - x)/\mu_L$ (Isbin et al., 1957)
$\mu = x\mu_G + (1 - x)\mu_L$ (Cicchitti et al., 1960)
$\mu = \lambda\mu_L + (1 - \lambda)\mu_G$ (Dukler et al., 1964b)

where $\lambda$ is a liquid fraction based on the volumetric flow rates:

$$\lambda \triangleq \frac{Q_L}{Q_L + Q_G}$$

Occasionally, for example, when the two-phase pressure drop in high-pressure steam-generating tubes has to be predicted, the homogeneous model gives good results. Whitcutt and Chojnowski (1973) presented results consisting of 486 data points obtained from a boiling-water test facility, covering a nominal range of mass fluxes from 400 to 3500 kg/m$^2$ s, pressures from 110 to 205 bar, and exit steam qualities from 0 to 80%. Choosing $\mu = \mu_L$, they found that 95% of the pressure drop measurements were predicted within ±9% both by the homogeneous model and by Thom's model (see Sec. 3.2). Andeen and Griffith (1968) measured the momentum flux through a section of pipe with steam–water and air–water two-phase flow. They reported that the homogeneous model correlated well with the data because of compensating errors in the assumptions.

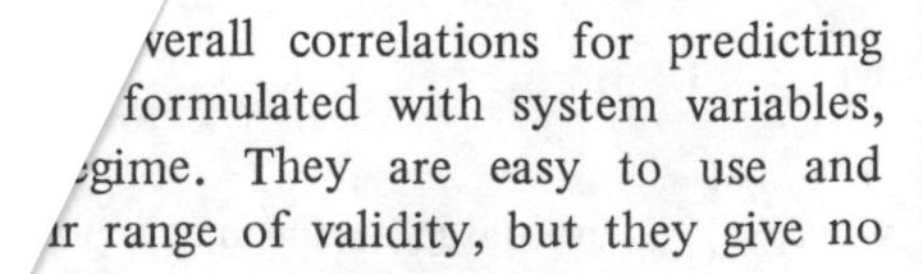

...verall correlations for predicting ... formulated with system variables, ...gime. They are easy to use and ...r range of validity, but they give no

...the American Institute of Chemical Engineers ...Martinelli defined the following variables:

$$\ldots \frac{\ldots dz}{\ldots dz)_G} \qquad \Phi_L^2 \triangleq \frac{dp/dz}{(dp/dz)_L}$$

$$\ldots^2 \triangleq \frac{\Phi_G^2}{\Phi_L^2} = \frac{(dp/dz)_L}{(dp/dz)_G}$$

where all the pressure gradients are frictional pressure gradients, $(dp/dz)_L$ and $(dp/dz)_G$ being evaluated, respectively, for the liquid and the gas phase flowing alone in the same tube and with the same flow rate ($M_L$ and $M_G$) as in the two-phase flow.

The authors considered four possible flow combinations:

1. Turbulent liquid and turbulent gas (subscript *tt*)
2. Laminar liquid and turbulent gas (subscript *lt*)
3. Turbulent liquid and laminar gas (subscript *tl*)
4. Laminar liquid and laminar gas (subscript *ll*)

Assuming gas and liquid turbulent flows, one obtains, according to the Blasius law,

$$\left(\frac{dp}{dz}\right)_L = 0.3164 \left(\frac{\mu_L}{G_L D}\right)^{0.25} \frac{G_L^2}{2\rho_L}$$

$$\left(\frac{dp}{dz}\right)_G = 0.3164 \left(\frac{\mu_G}{G_G D}\right)^{0.25} \frac{G_G^2}{2\rho_G}$$

We deduce

$$X_{tt}^2 = \left(\frac{\mu_L}{\mu_G}\right)^{0.25} \left(\frac{1-x}{x}\right)^{1.75} \frac{\rho_G}{\rho_L}$$

Lockhart and Martinelli plotted $\Phi_{Gtt}$ and $\Phi_{Ltt}$ as well as the void fraction $\alpha$ ($\equiv R_G$) compared with $X_{tt}$ (Fig. 1). A similar procedure was followed for the other three combinations (*lt*, *tl*, and *ll*). The curves are approximated by the following relationship:

$$\Phi_L^2 = 1 + \frac{C}{X} + \frac{1}{X^2} \qquad \text{and} \qquad \Phi_G^2 = 1 + CX + X^2 \tag{9}$$



...TS
...length is small enough, the pressure
...outlet. (7)
...variation

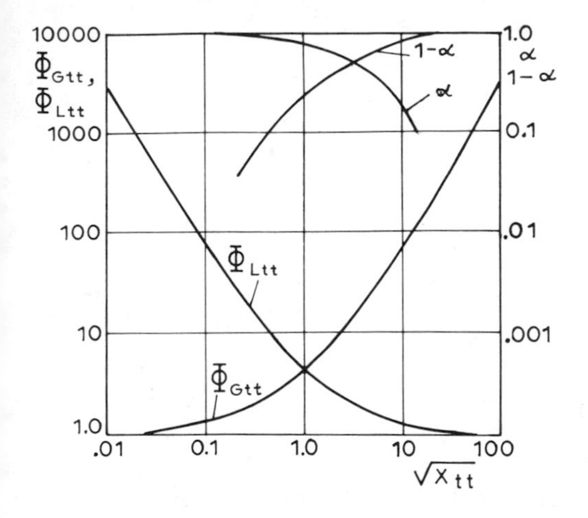


**Figure 1** Basic plot for frictional pressure gradients predictions (Lockhart and Martinelli, 1947).

$$1 - \alpha = \frac{X}{\sqrt{X^2 + 20X + 1}}$$

where $C$ has the following values:

Liquid turbulent, gas turbulent $C = 20$
Liquid laminar, gas turbulent $C = 12$
Liquid turbulent, gas laminar $C = 10$
Liquid laminar, gas laminar $C = 5$

Air-water horizontal flow data near to atmospheric pressure were used to determine these curves. Several authors plotted $\Phi^2_{Ltt}$ against $(1 - \alpha)$ and found

$$\Phi^2_{Ltt} = (1 - \alpha)^{-n}$$

Lottes and Flinn (1956) adopted $n = 2$, whereas later works such as those of Katsuhara (1958) and Richardson (1958), chose $n = 1.75$. Using the value of 1.75, one obtains

$$\left(\frac{dp}{dz}\right)_f = (1 - \alpha)^{-1.75} \left(\frac{dp}{dz}\right)_L = 0.3164(\text{Re}_L)^{-0.25} \frac{\rho_L w_L^2}{2} \tag{10}$$

where $$\text{Re}_L = \frac{\rho_L w_L D}{\mu_L}$$

since $$G_L = (1 - \alpha)\rho_L w_L$$

## 3.2 Martinelli-Nelson Method

Martinelli and Nelson adapted the previous method to steam-water flows. They introduced the ratio of frictional two-phase pressure drop to the frictional pressure drop of the liquid flowing alone in the tube with a flow rate equal to the total flow rate of the two-phase flow:

$$\Phi_{L0}^2 \triangleq \frac{dp/dz}{(dp/dz)_{L0}}$$

The relationship between $\Phi_L^2$ and $\Phi_{L0}^2$ is

$$\Phi_{L0}^2 = (1-x)^{1.75}\Phi_L^2$$

Since, at the critical pressure,

$$\mu_L = \mu_G \qquad \rho_L = \rho_G \qquad \Phi_{L0}^2 = 1 \qquad X_{tt}^2 = \left(\frac{1-x}{x}\right)^{1.75}$$

and consequently,

$$\Phi_{Ltt}^2 = \left(\frac{1 + X_{tt}^{1.14}}{X_{tt}^{1.14}}\right)^{1.75}$$

This expression is used to plot $\Phi_{Ltt}$ against $X_{tt}$ at the critical pressure in the Lockhart-Martinelli diagram (Fig. 2). Curves for pressures between atmospheric and critical pressures are obtained by interpolation.

The Martinelli-Nelson method consists of representing $\Phi_{L0}^2$ and the void fraction as a function of quality (Figs. 3 and 4). Each curve relates to a different pressure.

Further, to simplify the use of the method, $\Phi_{L0}^2$ was integrated and compared with quality at the outlet of the tubes (Martinelli and Nelson, 1948). Let us consider Eq. (1) again. By integration, we obtain if $x = 0$ for $z = 0$, and $dx/dz = x_{\text{out}}/L$:

$$-\Delta p = g\cos\theta \int_0^L [\alpha\rho_G + (1-\alpha)\rho_L]\, dz + G^2\left[\left(\frac{x^2}{\alpha\rho_G}\right)_{\text{out}} + \left(\frac{(1-x)^2}{(1-\alpha)\rho_L}\right)_{\text{out}} - \frac{1}{\rho_L}\right]$$

$$+ \frac{4L}{D} f_{L0} \frac{G^2}{2\rho_L}\left(\frac{1}{x_{\text{out}}}\int_0^{x_{\text{out}}} \Phi_{L0}^2\, dx\right) \tag{11}$$

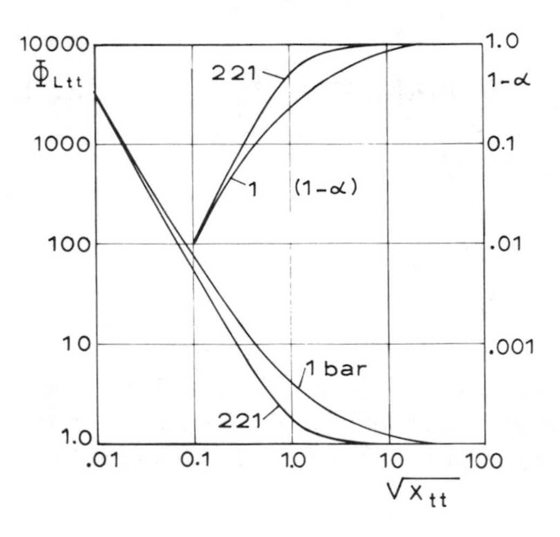


**Figure 2** $\Phi_{Ltt}$ and liquid volumetric concentration plotted as a function of Martinelli parameter at the atmospheric and critical pressures for steam–water flows (Martinelli and Nelson, 1948).

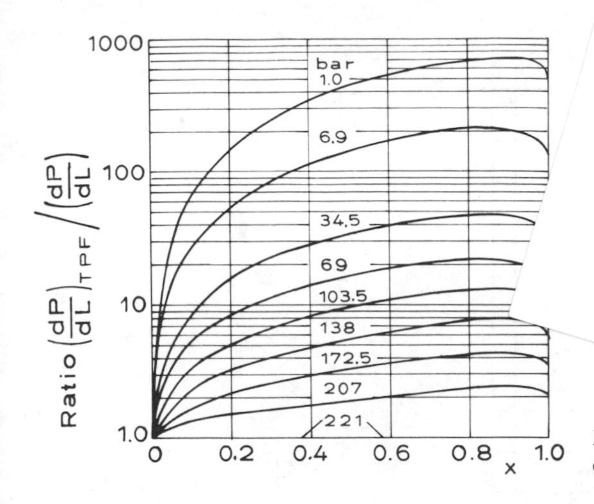


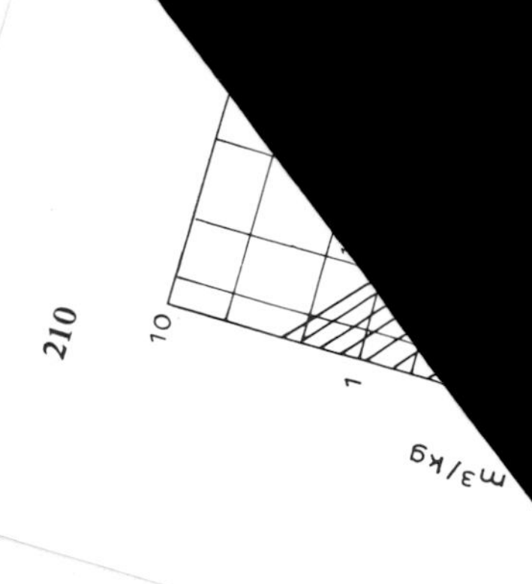


**Figure 3** $\Phi^2_{L0}$ plotted as a functio... quality (Martinelli and Nelson, 1948).

In this expression, $f_{L0}$ is the friction factor of the liquid flow when the liquid mass velocity is $G$. Martinelli and Nelson plotted

$$r_2 \triangleq \left(\frac{x}{\alpha\rho_G}\right)_{\text{out}} + \left(\frac{(1-x)^2}{(1-\alpha)\rho_L}\right)_{\text{out}} - \frac{1}{\rho_L}$$

and

$$\frac{\Delta p_{TPf}}{\Delta p_0} \triangleq \frac{1}{x_{\text{out}}} \int_0^{x_{\text{out}}} \Phi^2_{L0}\, dx$$

against pressure for various outlet qualities (Figs. 5 and 6).

Thom (1970) who correlated a large number of points, especially at high pressures, recommended use of the Martinelli-Nelson method with values of the multipliers $r_2$, $r_3$, and $r_4$ given, respectively, in Figs. 7, 8, and 9. The multipliers are defined by

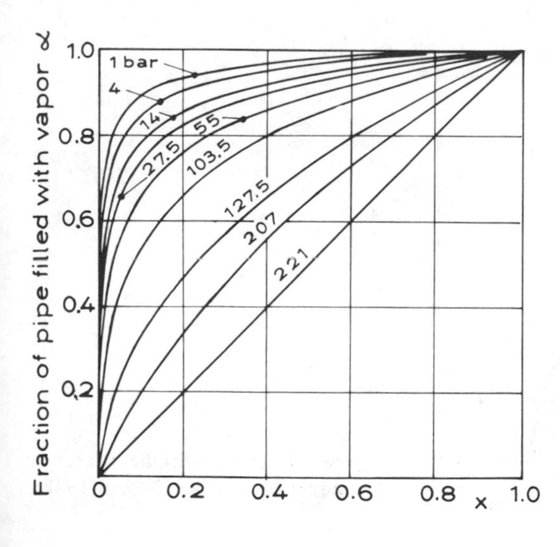


**Figure 4** Volumetric gas concentration plotted as a function of quality (Martinelli and Nelson, 1948).

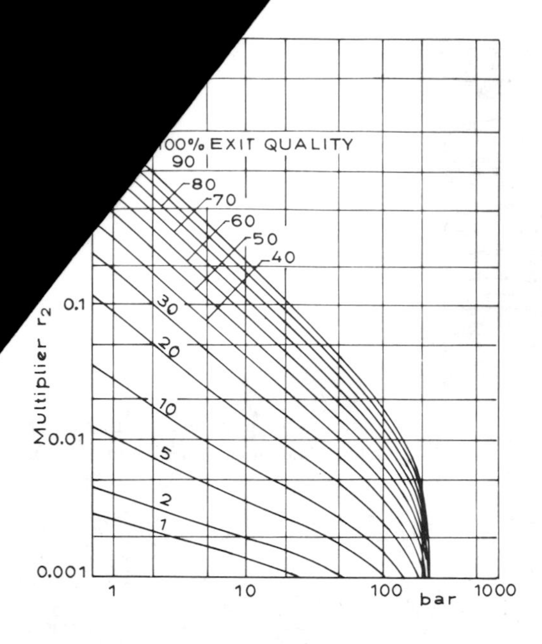


**Figure 5** Acceleration multiplier related to pressure (Martinelli and Nelson, 1948).

1. Acceleration multiplier:

$$r_2 \triangleq \left(\frac{x}{\alpha\rho_G}\right)_{\text{out}} + \left[\frac{(1-x)^2}{(1-\alpha)\rho_L}\right]_{\text{out}} - \frac{1}{\rho_L}$$

2. Friction multiplier:

$$r_3 \triangleq \frac{1}{x_{\text{out}}} \int_0^{x_{\text{out}}} \Phi_{L0}^2 \, dx$$

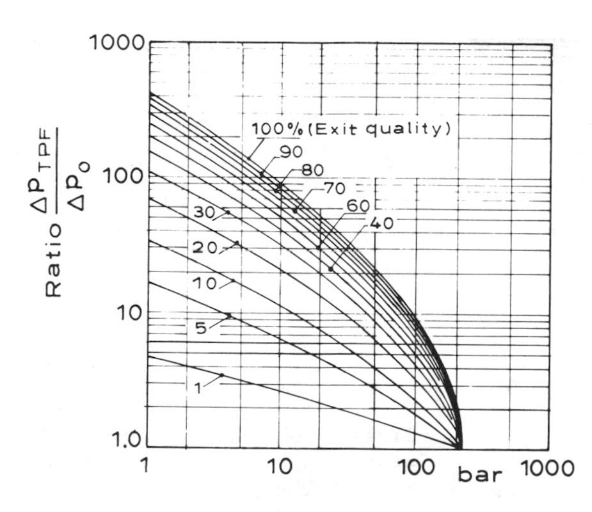


**Figure 6** Friction multiplier related to pressure (Martinelli and Nelson, 1948).

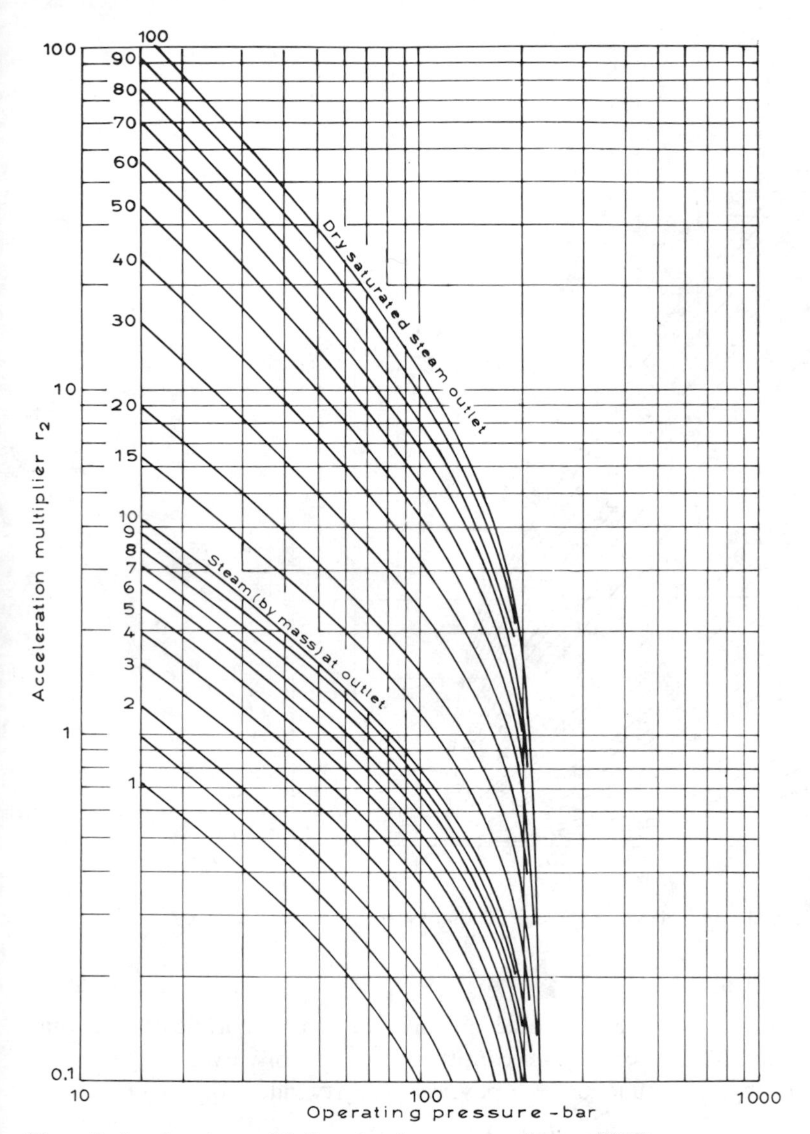


**Figure 7** Acceleration multiplier related to pressure (Thom, 1970).

3. Gravitation multiplier:

$$r_4 \triangleq \frac{1}{x_{\text{out}}} \int_0^{x_{\text{out}}} [\alpha\rho_G + (1-\alpha)\rho_L] \; dx$$

The integrated momentum balance equation is then

$$-\Delta p = r_4 Lg \cos\theta + r_2 G^2 + \frac{4L}{D} f_{L0} r_3 \frac{G^2}{2\rho_L} \tag{12}$$

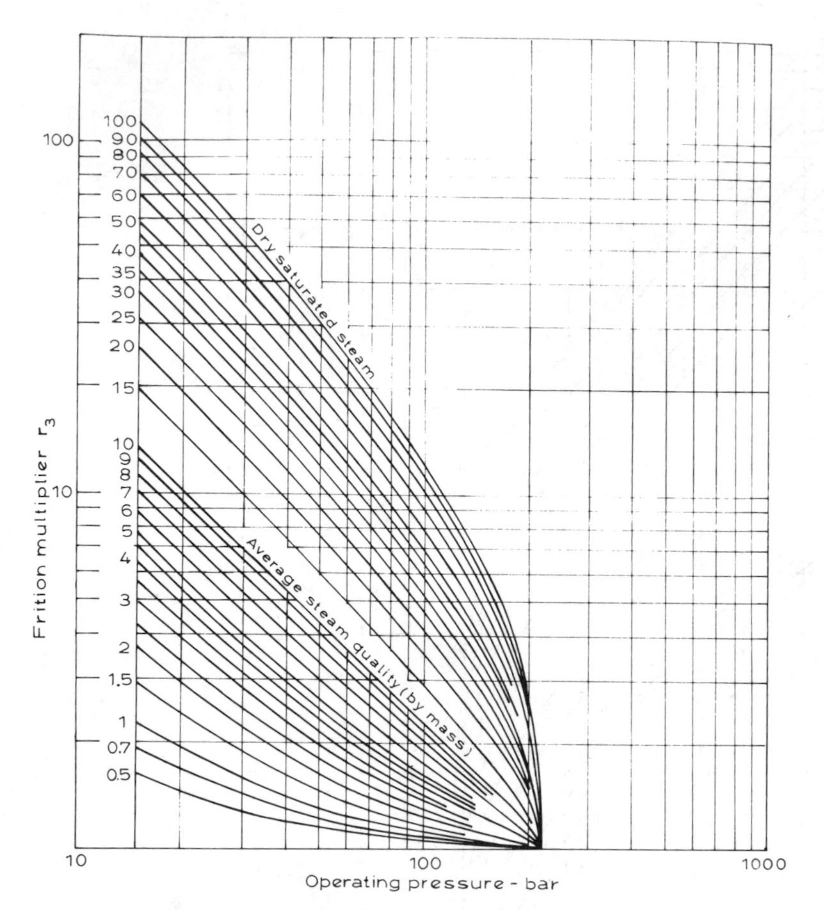


**Figure 8** Friction multiplier related to pressure (Thom, 1970).

## 3.3 Baroczy Method

The Baroczy (1966) method can be used not only for water, but also for sodium, mercury, and Freon-22. The correlation consists of two sets of curves:

1. A plot of $\Phi_{L0}^2$ as a function of a physical property index (Fig. 10)

$$\left(\frac{\mu_L}{\mu_G}\right)^{0.2}\left(\frac{\rho_G}{\rho_L}\right)$$

(for a reference mass velocity, $G_0 = 1356$ kg/m$^2$ s).

2. Plots of a correcting factor of $\Phi_{L0}^2$ when the mass flux in the channel is not the reference mass flux. This correction is a function of the physical property index, quality, and mass velocity (Fig. 11).

## 3.4 Hughmark-Pressburg Method for Gas–Liquid Flow in a Vertical Pipe

Hughmark and Pressburg (1961) carried out tests on vertical upward cocurrent air–liquid flows under isothermal conditions in a test section of 2.54-cm diameter.

Six liquids were used to determine the effect of density, viscosity, and surface tension on holdup and pressure drop. These liquids were water, $Na_2CO_3$ solution, kerosene, oil blends, and trichloroethylene. The method of Hughmark and Pressburg consists of using two plots:

1. A plot of $1-\alpha$ as a function of

$$\left(\frac{G_L}{G_G}\right)^{0.9} \frac{\mu_L^{0.19}\,\sigma^{0.205}\,\rho_G^{0.70}\,\mu_G^{2.75}}{G^{0.435}\,\rho_L^{0.72}}$$

2. A plot of $(\Delta p/\Delta p_{Lo})/L$ as a function of the difference in velocity between the two phases, for different values of a parameter:

$$\psi \triangleq \frac{1}{G^{0.70}\,\mu_L^{0.147}\,\sigma^{0.194}}$$

Calculated total pressure drops show an average absolute error of about 11% over the 563 experimental data points.

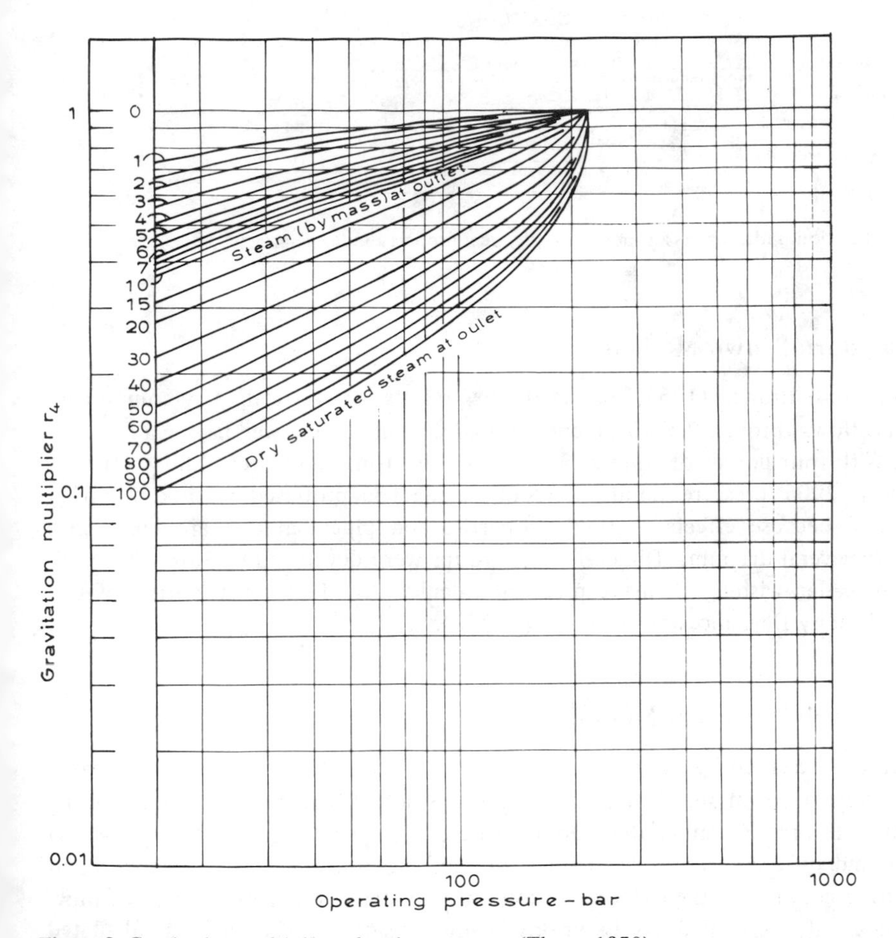


**Figure 9** Gravitation multiplier related to pressure (Thom, 1970).

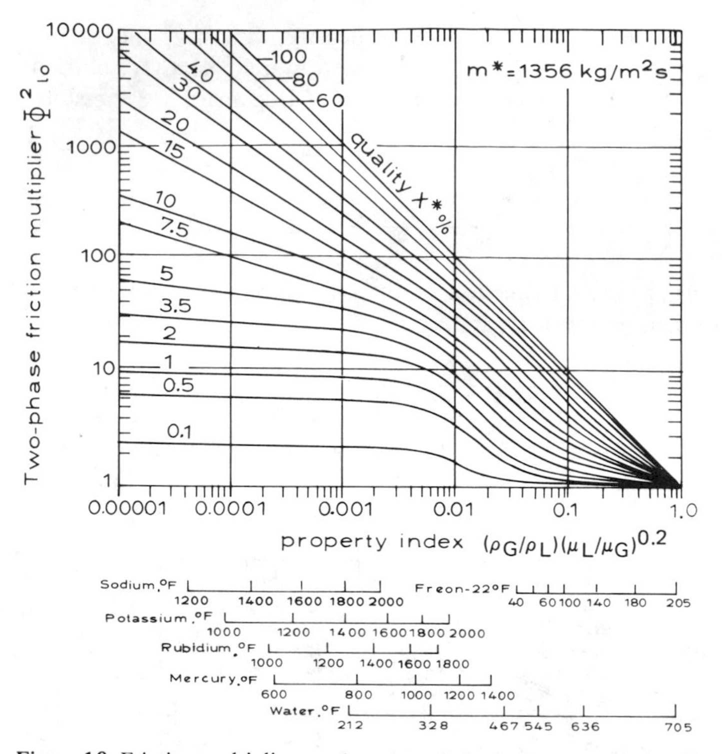


**Figure 10** Friction multiplier as a function of physical property index (Baroczy, 1966).

## 3.5 Hagedorn-Brown Method

Hagedorn and Brown (1965) reported the results of vertical, two-component, two-phase flows through 2.54–3.81 cm nominal-size tubing in a well having a depth of 457 m. With four liquids of widely varying viscosity (oils and water, $0.86 \times 10^{-3}$ to 0.11 Pa s), 2905 pressure points were obtained. The proposed method is rather complicated because effects of fluid properties and pipe diameter are taken into account in several diagrams. Dimensionless groups were developed by Ros (1961). We refer the reader wishing to make practical calculations to the presentation of this method given by De Gance and Atherton (1970b).

## 3.6 Lombardi-Pedrocchi Method

The method presented by Lombardi and Pedrocchi (1972) is the result of a research program carried out at the Centro Informazioni Studi Esferienze (Milan, Italy) using different mixtures (steam-water, argon-water, nitrogen-water, and argon-ethyl alcohol), and different geometries (tubes, annuli, and rod clusters) in adiabatic and heat transfer conditions (the latter condition only for steam-water mixtures); with few exceptions, only vertical ducts in upflow were considered. The authors calculated the gravitation and acceleration pressure drops using the homogeneous model and correlated the friction pressure drop by means of the dimensional expression

$$\left(\frac{dp}{dz}\right)_f = KG^n \frac{\sigma^{0.4}}{\rho_m^{0.86} D^{1.2}} \tag{13}$$

with, for

1. Round tubes: $K = 0.83$, $n = 1.4$
2. Rod clusters and annuli: $K = 0.213$, $n = 1.6$

The validity range of the method (correlation verified on 1400 experimental data obtained from seven different loops) is given below:

| Parameter | Validity range |
|---|---|
| Mass flux | 500–5000 kg/m² s |
| Equivalent diameter | 5–25 mm |
| Channel length | 0.1–4 m |
| $\rho_L/\rho_G$ | 15–100 |
| $\sigma$ | 0.02–0.08 N/m |
| $x$ | 0.01–0.98 |

The authors claim that 87% of all experimental data are ±15% of the predicted values.

## 3.7 Chisholm Method

Chisholm (1972) expanded the Lockhart-Martinelli method and indicated that Eq. (9) for $\Phi_L^2$ can be transformed with sufficient accuracy for engineering purposes to

$$\Phi_{Lo}^2 = 1 + (\Gamma^2 - 1)\{B[x(1-x)]^{(2-n)/2} + x^{2-n}\} \tag{14}$$

where

$$\Gamma^2 \triangleq \frac{(dp/dz)_{Go}}{(dp/dz)_{Lo}} \qquad B \triangleq \frac{C\Gamma - 2^{2-n} + 2}{\Gamma^2 - 1}$$

and $n$ is the exponent of the Blasius law ($n = 0.25$).

# 4 FLOW PATTERN-DEPENDENT CORRELATIONS

A comprehensive review of the pressure drop and holdup correlations developed for each flow pattern may be found in Govier and Aziz (1972). For illustration, we summarize some of these studies below.

Table 1 gives the formulas recommended by Beattie (1973) with reference to the sublayer structure. Friction factor $f$ and dimensionless group Re are related by Colebrook's equation as for single-phase flow:

$$\frac{1}{\sqrt{f}} = 3.48 - 4 \log\left(\frac{2\epsilon}{D} + \frac{9.35}{\mathrm{Re}\sqrt{f}}\right) \tag{15}$$

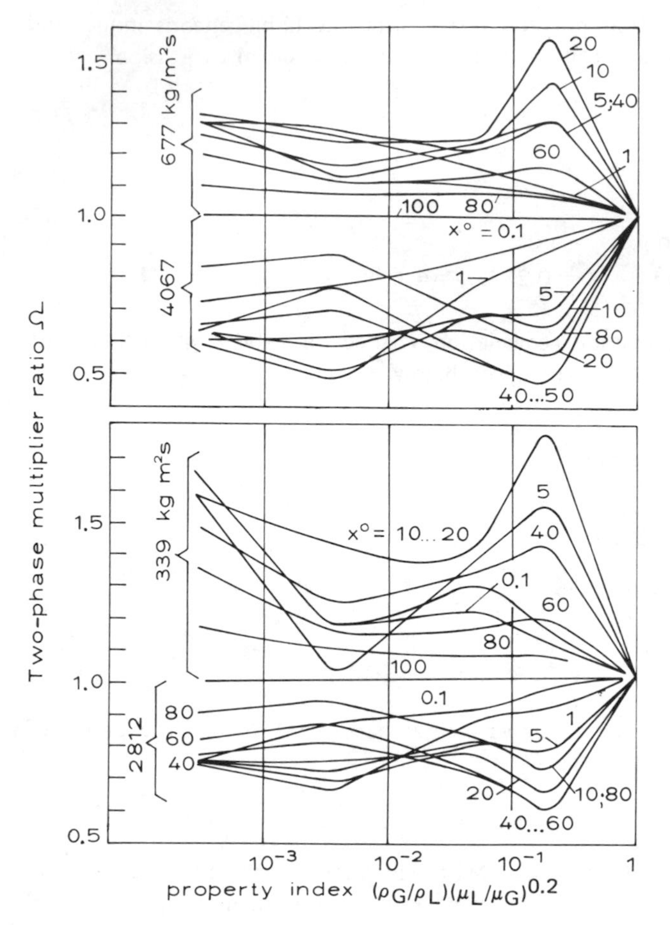


**Figure 11** Correcting factor as a function of property index (Baroczy, 1966).

Re is an appropriate Reynolds number including the quality $X$ of the mixture, except in the attached wall-bubble model, for which it is actually $2.4\sqrt{\mathrm{We}}$, where We is a two-phase Weber number measuring the relative magnitude of inertia and surface tension forces. Table 1 also contains expressions of the two-phase multiplier $\Phi_{Lo}^2$.

## 4.1 Bubbly Flow Regime (Models Based on the Theory of Turbulence)

Bankoff (1960) assumed that there is no local slip in the bubbly flow but rather velocity and void-fraction profiles given by exponential laws, which produce an average slip ratio:

$$k = \frac{\langle w_G \rangle}{\langle w_L \rangle} = \frac{1 - \langle \alpha \rangle}{K - \langle \alpha \rangle}$$

where $K$ is the so-called Armand parameter.

**Table 1 Flow pattern-dependent correlations (Beattie, 1973)[a]**

| Sublayer structure | Example | $f$ | Re | $\Phi^2_{Lo}$ |
|---|---|---|---|---|
| Bubble | Bubble flow | $\dfrac{\rho_l D\, dp/dz}{2G^2[1+X(\rho_l/\rho_g-1)]}$ | $\dfrac{DG}{\mu_l}\dfrac{1+X(\rho_l/\rho_g-1)}{1+X[(3.5\mu_g+2\mu_l)\rho_l/(\mu_g+\mu_l)\rho_g-1]}$ | $\left[1+X\left(\dfrac{\rho_l}{\rho_g}-1\right)\right]^{0.8}\left\{1+X\left[\dfrac{(3.5\mu_g+2\mu_l)\rho_l}{(\mu_g+\mu_l)\rho_g}-1\right]\right\}^{0.2}$ |
| Wavy gas–liquid interface | Annular flow with no entrainment | As above | $\dfrac{DG}{\mu_l}\dfrac{1+X(\rho_l/\rho_g-1)}{1+X(\rho_l\mu_g/\rho_g\mu_l-1)}$ | $\left[1+X\left(\dfrac{\rho_l}{\rho_g}-1\right)\right]^{0.8}\left[1+X\left(\dfrac{\rho_l\mu_g}{\rho_g\mu_l}-1\right)\right]^{0.2}$ |
| Very small bubbles | Flow following obstruction (bubble or annular flow) | As above | $\dfrac{DG}{\mu_l}\dfrac{1+X(\rho_l/\rho_g-1)}{1+X[3.5(\rho_l/\rho_g)-1]}$ | $\left[1+X\left(\dfrac{\rho_l}{\rho_g}-1\right)\right]^{0.8}\left[1+X\left(\dfrac{3.5\rho_l}{\rho_g}-1\right)\right]^{0.2}$ |
| Dry wall | Post "burnout" flow | $\dfrac{\rho_l^2 D\, dp/dz}{2\rho_g G^2[1+X(\rho_l/\rho_g-1)]^2}$ | $\dfrac{DG\rho_g}{\mu_g\rho_l}\left[1+X\left(\dfrac{\rho_l}{\rho_g}-1\right)\right]$ | $\left(\dfrac{\mu_g}{\mu_l}\right)^{0.2}\left(\dfrac{\rho_g}{\rho_l}\right)^{0.8}\left[1+X\left(\dfrac{\rho_l}{\rho_g}-1\right)\right]^{1.8}$ |
| Flow with attached wall bubbles | Boiling flow | $\dfrac{2\rho_l D\, dp/dz}{G^2[1+X(\rho_l/\rho_g-1)]}$ | $2.4\left\{\dfrac{G^2 D[1+X(\rho_l/\rho_g-1)]}{\rho_l\sigma}\right\}^{1/2}$ | Complex |

[a] $X$ = quality; subscript $g$ = gas; subscript $l$ = liquid.

Bankoff obtained the following expression for two-phase wall friction:

$$\tau_w = 0.3164(\mathrm{Re}_L)^{-0.25} \frac{G_L^2}{2\rho_L} \left[1 - \langle\alpha\rangle \left(1 - \frac{\rho_G}{\rho_L}\right)\right]^{0.75} \left[1 - x\left(1 - \frac{\rho_L}{\rho_G}\right)\right] \tag{16}$$

Levy (1963) considered two-phase flow as a compressible and two-dimensional flow and used the concepts of Reynolds stresses and mixing lengths for the velocity and density distribution. He obtained the following expression for wall shear stress:

$$\tau_w = Ay^2 \frac{d\bar{u}}{dy} \frac{d(\bar{\rho}\,\bar{w})}{\mathrm{dy}} \left[1 - \exp\left(-\frac{y}{F}\right)\right]^2 + \mu \frac{d\bar{w}}{dy} \tag{17}$$

where $A$ and $F$ are constants. The reader is referred to the original paper for the application of this method.

Huey and Bryant (1967), dealing with isothermal, homogeneous bubbly flows in horizontal pipes, used a friction length formula that relates the effects of wall friction and pipe length to the state of the fluid and the Mach number, obtaining

$$f = \frac{dp_0/dz}{(2/D)[(1+\delta)/\delta]\gamma p\,\mathrm{Ma}^2\,\{(\gamma\,\mathrm{Ma}^2 - 2)/[2(1-\gamma\,\mathrm{Ma}^2)]\}} \tag{18}$$

where $p_0$ = stagnation pressure
$\gamma$ = ratio of specific heats
$\delta$ = volume flow ratio $Q_G/Q_L$
Ma = Mach number

The Mach number is defined as the ratio of the bulk velocity of the flow to the local speed of sound.

## 4.2 Annular Flow Regime

Silvestri and Varsi (1965) developed a simple approach for predicting film thickness and pressure drop of annular flow assuming Poiseuille flow in the film. An analysis of annular laminar-laminar flow was achieved by Delhaye (1966) with application to liquid-liquid flow.

Assuming that the liquid film consists of a laminar sublayer and a turbulent layer, Ueda (1967) obtained for vertical upflow:

$$\tau_w = \left(\frac{\rho_L \nu_L w_\delta}{\delta} - \frac{\rho_L g \alpha \delta}{2}\right)\left(1 - \frac{\delta}{D}\right)^{-1}$$

where $\nu_L$ = kinematic viscosity of the liquid
$w_\delta$ = velocity of the laminar sublayer at a distance $\delta$ from the wall
$\delta$ = thickness of the laminar sublayer given by $\delta = 50\nu_L/w_\delta$

The term $w_\delta$ is finally determined from semiempirical considerations.

More recently, Friedel et al. (1972) and Friedel (1977a) analyzed wavy annular mist flow and evaluated the following contributions to friction pressure drop:

1. Droplet formation and acceleration
2. Expansion and contraction losses in the gas phase
3. Wave motion
4. Momentum change
5. Wall friction of the liquid phase

They found that most of these terms are negligible compared with the wall frictions and the expansion losses. A comparison of the calculated pressure drops with experimental results obtained in Germany showed a scatter of about ±35%.

## 4.3 Stratified Flow Regime

From simple theoretical considerations, Taitel and Dukler (1976) have proved that holdup and dimensionless pressure drop for stratified flow (horizontal tubes) are functions of the Martinelli parameter $X$ only, under the assumption that $f_G/f_i \cong$ const. Here, $f_G$ denotes the gas wall friction factor and $f_i$ the interfacial friction factor, defined by

$$\tau_i \triangleq f_i \frac{\rho_G(w_G - w_L)^2}{2}$$

where $\tau_i$ is the interfacial shear strength. The further assumption that $f_G = f_i$ gives a result that agrees well with experimental data.

# 5 COMPARATIVE STUDIES AND SIMILARITY ANALYSIS

Dukler et al. (1964a) performed a comparative study of five pressure drop correlations among which are included those of Bankoff and Lockhart-Martinelli. They compiled a data bank and selected 2620 points. An overall comparison (all flow regimes) with all the two-component flow data has shown that the Martinelli correlation is the best. However, deviations increase with the pipe diameter. The expected error for one-component flow is greater than for two-component flow, and the Martinelli correlation remains the best of the five examined correlations.

The same authors (Dukler et al., 1964b) developed a similarity analysis and proposed new correlation parameters (such as Reynolds numbers and friction factors). One of the resulting correlations shows slightly better overall agreement with the data than the Lockhart-Martinelli correlation.

Shiba and Yamazaki (1967) carried out tests on air–water upflows in a 2.54-cm-diameter tube. They noted that their results are in close correlation with those of the Martinelli method. The same conclusion was reached by Pletcher and McManus (1968), who carried out tests on heat transfer and pressure drop in horizontal annular two-phase two-component flow.

De Gance and Atherton (1970a) gave practical information about the computation procedures needed to apply several correlations to horizontal, vertical, and inclined pipes.

Friedel and Mayinger (1973, 1974) carried out measurements with Freon-12 and water in unheated vertical pipes. They noticed substantial differences between experimental results and calculations based on the Lockhart-Martinelli and Baroczy methods (Fig. 12). However, these authors showed that scaling of two-phase friction pressure drop is possible with sufficient accuracy for technical purposes if three dimensionless numbers are identical in the prototype and model.

Simpson et al. (1977) studied the flow of air-water mixtures near the atmospheric pressure in large-diameter (127 and 216 mm), horizontal tubes (±1000 tests). All of the data show a strong flow pattern dependency and require the transition between flow regimes to be carefully defined. None of the friction correlations (homogeneous, Lockhart-Martinelli, Baroczy) adequately reflects the friction pressure drop in the large-diameter tubes.

A data bank with about 6500 average void-fraction and friction pressure drop measurements of vertical upflows in unheated straight pipes was compiled by Friedel (1977b). The author concluded that the Lombardi-Pedrocchi correlation is the best predictor of friction pressure drop in two-phase two-component flows, whereas Baroczy's relationship, corrected by Chisholm, gives the best results in two-phase R12 flow.

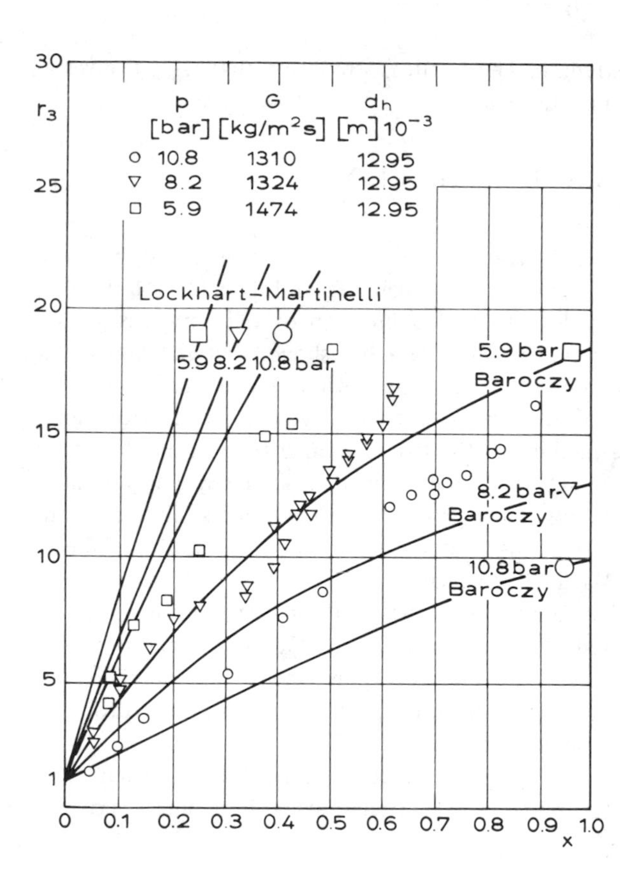


**Figure 12** Friction multiplier related to quality (Friedel and Mayinger, 1974).

## NOMENCLATURE

| | |
|---|---|
| $A$ | constant [Eq. (17)] |
| $\mathfrak{a}$ | cross-sectional area |
| $B$ | parameter [Eq. (14)] |
| $C$ | constant [Eq. (9)] |
| $\mathfrak{C}$ | contour |
| $D$ | diameter |
| $\mathbf{F}$ | gravity force |
| $F$ | constant [Eq. (17)] |
| $f$ | friction factor [Eq. (5)] |
| $G$ | mass velocity |
| $g$ | gravity constant |
| $K$ | constant [Eq. (13)] or Armand parameter (Sec. 4.1) |
| $k$ | average slip ratio (Sec. 4.1) |
| $L$ | channel length |
| $M$ | mass flow rate |
| Ma | Mach number |
| $\mathbf{n}$ | unit normal vector |
| $\mathfrak{P}$ | wetted perimeter |
| $p$ | pressure |
| $Q$ | volumetric flow rate |
| $R$ | volume concentration |
| Re | Reynolds number |
| $r_2, r_3, r_4$ | two-phase multipliers (Sec. 3.2) |
| $t$ | time |
| $w$ | velocity |
| We | Weber number |
| $X$ | Martinelli parameter (Sec. 3.1) |
| $x$ | quality |
| $y$ | distance from the wall |
| $z$ | axial coordinate |
| $\alpha$ | void fraction |
| $\Gamma$ | pressure drop ratio [Eq. (14)] |
| $\gamma$ | ratio of the specific heats |
| $\delta$ | volume flow ratio [Eq. (18)] |
| $\delta$ | thickness of the laminar sublayer |
| $\theta$ | angle of the axial direction with respect to the upward vertical |
| $\lambda$ | liquid fraction |
| $\mu$ | dynamic viscosity |
| $\nu$ | kinematic viscosity |
| $\pi$ | stress tensor component |
| $\rho$ | mass volume |
| $\rho_m$ | mixture mass volume [Eq. (3)] |
| $\sigma$ | surface tension |

$\tau_w$ wall shear stress
$\Phi$ pressure drop ratio (Sec. 3.1)
$\psi$ parameter (Sec. 3.4)

**Subscripts**

$a$ acceleration
$f$ friction
$G$ gas phase
$g$ gravity
$i$ interface
$L$ liquid phase
$l$ laminar
out outlet
$t$ turbulent
$w$ wall

**Operators**

$\langle\ \rangle$ area average
$\bar{\ }$ time average

## REFERENCES


Andeen, G. B. and Griffith, P., Momentum Flux in Two-Phase Flow, *J. Heat Transfer*, vol. 90, no. 2, pp. 211–222, 1968.

Bankoff, S. G., A Variable Density Single-Fluid Model for Two-Phase Flow with Particular Reference to Steam-Water Flow, *J. Heat Transfer*, vol. 82, pp. 265–272, 1960.

Baroczy, C. J., A Systematic Correlation for Two-Phase Pressure Drop, *Chem. Eng. Prog. Symp. Ser.*, vol. 62, no. 64, pp. 232–249, 1966.

Beattie, D. R. H., A Note on the Calculation of Two-Phase Pressure Losses, *Nucl. Eng. Des.*, vol. 25, pp. 395–402, 1973.

Chisholm, D., Friction During the Flow of Two-Phase Mixtures in Smooth Tubes and Channels, NEL rept. 529, 1972.

Cicchitti, A., Lombardi, C., Silvestri, M., Soldaini, G., and Zavattarelli, R., Two-Phase Cooling Experiments–Pressure Drop, Heat Transfer and Burnout Measurements, *Energia Nucleare*, vol. 7, no. 6, pp. 407–525, 1960.

De Gance, A. E. and Atherton, R. W., Horizontal-Flow Correlations, *Chem. Eng.*, vol. 77, July 13, pp. 93–103, 1970a.

De Gance, A. E. and Atherton, R. W., Vertical and Inclined-Flow Correlations, *Chem. Eng.*, vol. 77, Oct. 5, pp. 87–94, 1970b.

Delhaye, J. M., Etude Théorique des Ecoulements Diphasiques du Type Annulaire en Régime Laminaire-laminaire. *C. R. Acad. Sci. Ser. A*, vol. 263, pp. 324–327, 1966.

Dukler, A. E., Wicks, M., and Cleveland, R. G., Frictional Pressure Drop in Two-Phase Flow: A. A Comparison of Existing Correlation for Pressure Loss and Hold up, *AIChE J.*, vol. 10, no. 1, pp. 38–43, 1964a.

Dukler, A. E., Wicks, M., and Cleveland, R. G., Frictional Pressure Drop in Two-Phase Flow: B. An Approach through Similarity Analysis, *AIChE J.*, vol. 10, no. 1, pp. 44–51, 1964b.

Friedel, L., Momentum Exchange and Pressure Drop in Two-Phase Flow, in *Two-Phase Flows and Heat Transfer*, eds. S. Kakaç and F. Mayinger, vol. 1, pp. 239–312, Hemisphere, Washington, D.C., 1977a.

Friedel, L., Mean Void Fraction and Friction Pressure Drop: Comparison of Some Correlations with Experimental Data, European Two-Phase Flow Group Meet., Grenoble, paper A7, 1977b.

Friedel, L., Jebe, P., and Mayinger, F., Correlation for Two-Phase Friction Pressure Drop, European Two-Phase Flow Group Meet., Rome, paper C12, 1972.

Friedel, L. and Mayinger, F., Comparison of Two-Phase Friction Pressure Drop in Water and Freon 12, European Two-Phase Flow Group Meet., Brussels, paper A2, 1973.

Friedel, L. and Mayinger, F., Scaling of Two-Phase Friction Pressure Drop, European Two-Phase Flow Group Meet., Harwell, paper B2, 1974.

Govier, G. W. and Aziz, K., *The Flow of Complex Mixtures in Pipes*, Van Nostrand Reinhold, New York, 1972.

Hagedorn, A. R. and Brown, K. E., Experimental Study of Pressure Gradients Occurring during Continuous Two-Phase Flow in Small-Diameter Vertical Conduits, *J. Pet. Technol.*, vol. 17, pp. 475–484, 1965.

Huey, C. T. and Bryant, R. A. A., Isothermal Homogeneous Two-Phase Flow in Horizontal Pipes, *AIChE J.*, vol. 13, no. 1, pp. 70–77, 1967.

Hughmark, G. A. and Pressburg, B. S., Hold Up and Pressure Drop with Gas-Liquid Flow in a Vertical Pipe, *AIChE J.*, vol. 7, pp. 677–682, 1961.

Isbin, H. S., Moy, J. E., and Da Cruz, A. J. R., Two-Phase Steam-Water Critical Flow, *AIChE J.*, vol. 3, no. 3, pp. 361–365, 1957.

Katsuhara, T., Influence of Wall Roughness on Two-Phase Flow Frictional Pressure Drops (in Japanese), *Trans. JSME*, vol. 24, no. 148, pp. 1050–1056, 1958.

Levy, S., Prediction of Two-Phase Pressure Drop and Density Distribution from Mixing Length Theory, *J. Heat Transfer*, vol. 85, pp. 137–152, 1963.

Lockhart, R. W. and Martinelli, R. C., Proposed Correlation of Data for Isothermal Two-Phase Two-Component Flow in Pipes, AIChE Meet., Buffalo, N.Y., 1947, published in *Chem. Eng. Prog.*, vol. 45, no. 1, pp. 39–48, 1949.

Lombardi, C. and Pedrocchi, E., A Pressure Drop Correlation in Two-Phase Flow, *Energia Nucleare*, vol. 19, no. 2, pp. 91–99, 1972.

Lottes, P. A. and Flinn, W. S., A Method of Analysis of Natural Circulation Boiling Systems, *Nucl. Sci. Eng.*, vol. 1, no. 6, pp. 461–476, 1956.

Martinelli, R. C. and Nelson, D. B., Prediction of Pressure Drop during Forced Circulation Boiling of Water, *Trans. ASME*, vol. 79, pp. 695–702, 1948.

Pletcher, R. H. and McManus, H. N., Heat Transfer and Pressure Drop in Horizontal Annular Two-Phase Two-Component Flow, *Int. J. Heat Mass Transfer*, vol. 11, pp. 1087–1104, 1968.

Richardson, B. L., Some Problems in Horizontal Two-Phase Two-Component Flow, ANL-5949, 1958.

Ros, N. C. J., Simultaneous Flow of Gas and Liquid as Encountered in Well Tubing, *J. Pet. Technol.*, vol. 13, p. 1037, 1961.

Shiba, M. and Yamazaki, Y., A Comparative Study on the Pressure Drop of Air-Water Flow, *Bull. JSME*, vol. 10, no. 38, pp. 290–298, 1967.

Silvestri, M., and Varsi, G., Film Thickness, Pressure Drop and Stability in Annular Two-Phase Flow, *Energia Nucleare*, vol. 12, no. 9, pp. 449–458, 1965.

Simpson, H. C., Rooney, D. H., Grattan, E., and Al-Samarrae, F., Two-Phase Flow in Large Diameter Horizontal Tubes, European Two-Phase Flow Group Meet., Grenoble, paper A6, 1977.

Taitel, Y. and Dukler, A. E., A Theoretical Approach to the Lockhart-Martinelli Correlation for Stratified Flow, *Int. J. Multiphase Flow*, vol. 2, pp. 591–595, 1976.

Thom, J. R. S., Prediction of Pressure Drop during the Circulation of Boiling Water and Steam in Steam Generating Units and Heat Exchange, Conf. on Two-Phase Flow, Ermenonville, Coll. Direction des Etudes et Recherches de E.D.F., Paris, 1970.

Ueda, T., On Upward Flow of Gas-Liquid Mixtures in Vertical Tubes, *Bull. JSME*, vol. 10, no. 42, pp. 989–999, 1967.

Whitcutt, R. D. B. and Chojnowski, B., Two-Phase Pressure Drop in High Pressure Steam Generating Tubes, European Two-Phase Flow Group Meet., Brussels, paper A3, 1973.

CHAPTER

# TWELVE

# PRESSURE DROPS IN ROD BUNDLES

**D. Grand**

In a preceding review on the subject of pressure drops in rod bundles Rouhani (1973) said: "This review gives the author an enviable position regarding the scarcity of the published material on this particular subject." Five years later, the picture has not changed radically. If many studies are available on boiling in rod bundles, on subjects such as mixing or boiling crisis, little has been said about friction pressure drop or more generally the momentum balance in a rod bundle. The main reason for such a lack of information on this subject may well be because in a rod bundle, important heterogeneities in a cross section may exist.

When dealing with a tube, a slug flow model can be assumed; in such a model the evolution of the variables along the channel is described with mean values in a cross section. For a tube model we have to solve a one-dimensional problem, and the momentum equation reduces to its axial component. In a rod bundle, however, great heterogeneities may exist between the interior and exterior regions of the same cross section. In a rod bundle the mean values must be defined on regions smaller than the cross section. The problem is no longer one-dimensional but two- or three-dimensional. Consequently, transverse momentum equations are needed; there is mixing between the regions, and friction pressure drop becomes only one term among others that also need to be modeled. Thus it becomes difficult to isolate the specific role of friction components from all other components that contribute to the momentum balance.

In Sec. 1 we review the important geometric parameters defining a rod bundle and give a qualitative picture of boiling phenomena.

In Sec. 2 we discuss pressure drop for single-phase flows in rod bundles. Indeed, the precise computation of the flow characteristics in single-phase flow is an important step in the knowledge of the boiling that subsequently occurs.

In Sec. 3 we review the so-called mixed flow model, which is quite crude because it assumes homogeneity of the flow variables all over the cross section of the bundle. However, it is the simplest to use and usually works.

In Sec. 4 we derive the momentum balance between two interconnected homogeneous regions. This constitutes the basis of the subchannel analysis used in computer codes such as COBRA, HAMBO, and FLICA.

# 1 FLOW IN A ROD BUNDLE

## 1.1 Definitions

A rod bundle consists of parallel cylinders containing the nuclear fuel pins, arranged on some network held in place by spacers, and contained in a housing (except for pressurized-water reactors). There are mainly two types of networks or arrays:

1. The triangular array, shown in Fig. 1*a*
2. The square array, shown in Fig. 1*b*

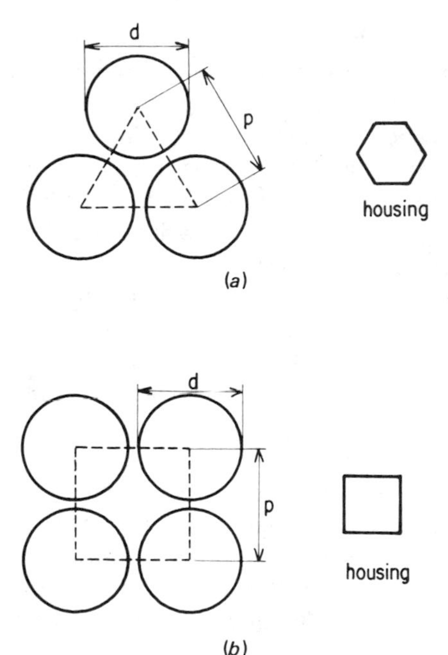


**Figure 1** Types of rod-bundle arrays. (*a*) Triangular array. (*b*) Square array.

The fluid region contained in the elementary triangle or square is called the subchannel. It is the smallest region in which flow variables are homogenized.

The shape of the housing that fits the array is a square for the square array and a hexagon for the triangular array.

The spacing between the outer row of pins and the wall of the duct is the width of the peripheral subchannels. Note the difference in flow area and shape between the inner and outer subchannels. These differences are two of the main causes of heterogeneities in the cross section of the rod bundle.

The pins are held in position by spacers which can be grids or wires wrapped around the pins. Recall that rod bundles are vertical and the coolant flows upward.

### 1.2 Boiling Pattern

The example will be taken from a rod bundle with triangular array and wire-wrap spacers with which the author is familiar. Let us assume a constant heat flux throughout the heated part, a uniform inlet temperature, and a power input such that only liquid is present. Even with these conditions the temperature profile is nonuniform radially, owing to the differences in flow areas of the outer subchannels. Here the heated perimeter is only a fraction of the wetted perimeter. Moreover, friction flow resistance is less important than in the rest of the cross section.

If the power is increased, boiling will start in the hottest subchannels, that is, the central ones. The appearance of boiling locally results in a reduction in flow area, increasing the axial pressure drop. Because of the interconnection between subchannels, flow will be diverted from the boiling subchannels toward those still in single phase. This mechanism, known as internal flow redistribution, is typical of the heterogeneities inside a rod bundle: boiling may be well developed in the central region, whereas subcooled liquid still flows in the outer one. Figure 2 illustrates this point.

Difference in flow areas of subchannels is one cause of heterogeneities, but a nonuniform heat flux may be another cause that leads to the same consequences.

## 2 SINGLE-PHASE FLOW PRESSURE DROP

The use of the friction factor for single-phase flow in circular pipes is a quite good approximation in bare round bundles. However Rehme (1971, 1973) has shown that better correlation of experimental data for single-phase flows could be achieved by using a modified procedure (see Sec. 2.2).

In round bundles with wire-wrap spacers, the presence of the helical wire increases the pressure drop. The friction factor in a tube must be multiplied by a correction factor to give predictions of the pressure drop along the bundle (Sec. 2.3).

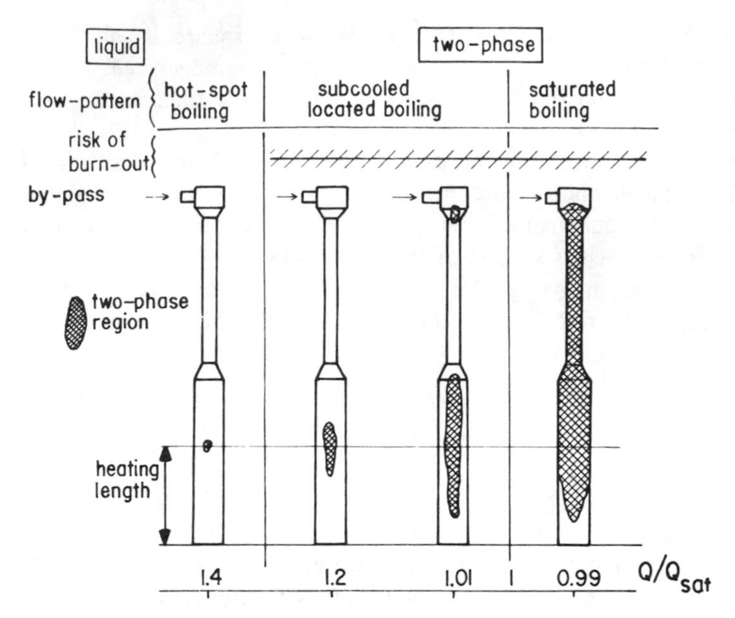


**Figure 2** First stages of boiling during slow flow transients, where $Q_{\mathrm{sat}}$ is the flow rate corresponding to saturation conditions at the exit of the heating length, and $Q$ is the actual flow rate (Costa and Menant, 1976).

## 2.1 Friction Factors in a Pipe

The friction factor $\lambda_0$ for a pipe of diameter $D$ is defined by

$$-\left(\frac{dp}{dz}\right)_f \triangleq \frac{\lambda_0}{D}\frac{\rho}{2}\langle w\rangle^2 \tag{1}$$

where $\langle w\rangle$ is the mean velocity in the cross section and $(dp/dz)_f$ is the friction pressure drop.

### 2.1.1 Laminar flow

$$\lambda_0 = \frac{64}{\mathrm{Re}} \qquad \mathrm{Re} \triangleq \frac{\langle w\rangle D}{\nu} \tag{2}$$

where $\nu$ is the kinematic viscosity.

**2.1.2 Turbulent flow in smooth pipes** The universal distribution law is given by (Schlichting, 1968)

$$\frac{w_{\max}}{v^*} = 2.5 \ln \frac{Dv^*}{2\nu} + 5.5$$

where $v^* \triangleq \sqrt{\tau_w/\rho}$ denotes the friction velocity and $\tau_w$ is the wall shear stress. By integration of the velocity profile, one obtains

$$\sqrt{\frac{8}{\lambda_0}} = 2.5 \ln \frac{\langle w \rangle D}{2\nu} \sqrt{\frac{\lambda_0}{8}} + 5.5 - G_0 \tag{3}$$

where $G_0$ is an integration constant related to the velocities by

$$G_0 \triangleq \frac{w_{max} - \langle w \rangle}{v^*}$$

Its value is $G_0 \simeq 4$ as deduced from Nikuradse's experiments. The Blasius approximation of Eq. (3) is often used:

$$\lambda_0 = 0.316 \text{ Re}^{-0.25} \tag{4}$$

## 2.2 Friction Factors in a Bare Rod Bundle

A generalization of Eq. (1) for rod-bundle geometry is

$$-\left(\frac{dp}{dz}\right)_f \triangleq \frac{\lambda}{D_e} \frac{\rho}{2} \langle w \rangle_T^2 \tag{5}$$

The subscript $T$ refers to values relative to the whole cross section of the channel. Thus $\langle w \rangle_T$ is the axial velocity averaged over the whole cross section and $D_e$ is the corresponding equivalent diameter defined by

$$D_e \triangleq 4 \frac{a_T}{\wp_T}$$

where $a_T$ and $\wp_T$ are the flow area and wetted perimeter of the whole cross section.

The method developed by Rehme (1971, 1973) for calculating $\lambda$ in both laminar and turbulent flow situations in a bare rod bundle is presented below.

**2.2.1 Laminar friction factors (Rehme, 1971)** A generalized form for the friction factor in laminar flow is

$$\lambda = \frac{K_T}{\text{Re}_T} \qquad \text{Re}_T \triangleq \frac{\langle w \rangle_T D_e}{\nu} \tag{6}$$

where $K_T$ is a geometric parameter.

Rehme provided a method for calculating this parameter for both a hexagonal subassembly and a square subassembly. By assuming a uniform pressure distribution in a cross section, he obtained

$$\frac{1}{K_T} = \sum_{i=1}^{N} \frac{1}{K_i} \left(\frac{a_i}{a_T}\right)^3 \left(\frac{\wp_T}{\wp_i}\right)^2 \tag{7}$$

where $a_i$ and $\wp_i$ are the flow area and the wetted perimeter of individual subchannels, respectively.

The summation is extended to all the subchannels. The values $K_i$ for individual

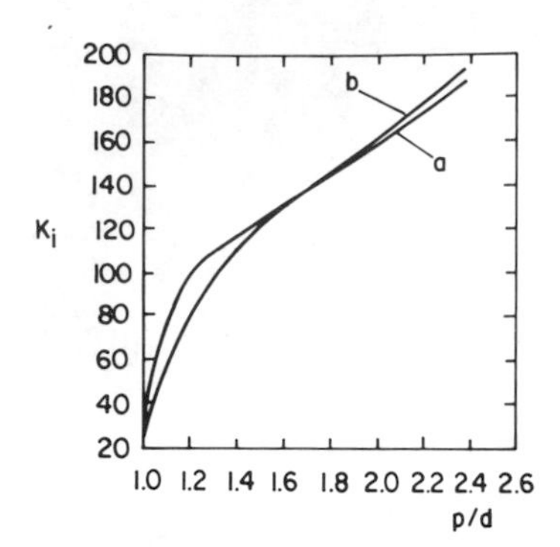


**Figure 3** Dependence of the coefficient $K_i$ on the dimensions of an interior subchannel. Curve a: triangular array; curve b: square array (Rehme, 1971).

subchannels can easily be found because Rehme computed them by numerical integration of the Poisson equation for laminar flow.

Figure 3 gives $K_i(p/d)$ for an interior subchannel.

Figure 4 gives the lines $K_i(p/d,\ e/d) = \text{const}$ for an edge subchannel. Figure 5 plots $K_i(e/d)$ for a corner subchannel. Use of these figures together with Eq. (7) permits the calculation of the geometric parameter $K_T$ for any rod bundle.

Previously, Courtaud et al. (1966) made such a study for a seven-pin bundle in a circular duct.

**2.2.2 Turbulent friction factors (Rehme, 1973)** By assuming that the universal velocity profile still holds, we can generalize the friction law as

$$\sqrt{\frac{8}{\lambda}} = A(K_T)\left(2.5 \ln \frac{\langle w\rangle_T D_e}{2\nu}\sqrt{\frac{\lambda}{8}} + 5.5\right) - G(K_T) \tag{8}$$

$A(K_T)$ and $G(K_T)$ are like $K_T$ geometry parameters independent of the flow conditions, that is, the Reynolds number.

Calculations of the constants $A$ and $G$ would require the integration across the subchannel of the velocity distribution, a calculation that would be quite difficult. A simpler approach is used. It requires one further assumption, namely, that $A$ and $G$ are geometric parameters, functions of $K_T$ only. This permits calculation in simpler flow situations, provided they cover the values of $K_T$ encountered in rod bundles:

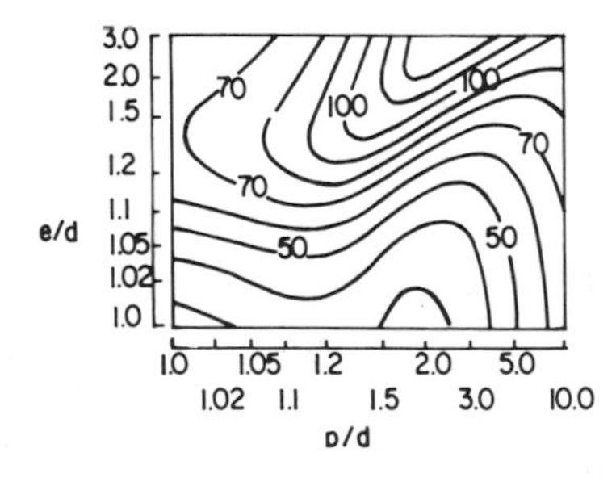


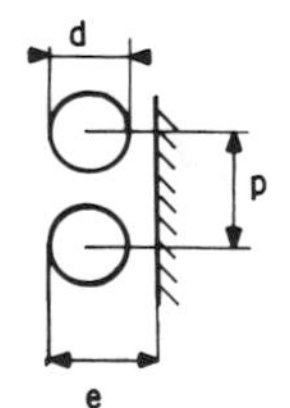


**Figure 4** Contours of $K_i$ for a peripheral subchannel, where $\Delta K_i = 10$ between two consecutive lines (Rehme, 1971).

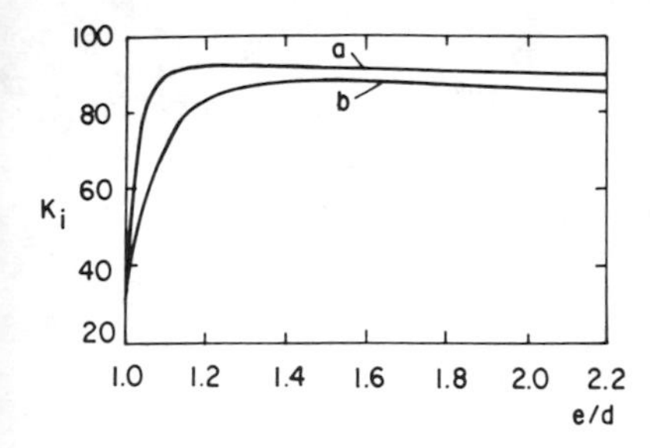


**Figure 5** Dependence of the coefficient $K_i$ on the dimensions of a corner subchannel. Curve a: triangular array; curve b: square array (Rehme, 1971).

1. If $K_T < 64$ the whole rod bundle is modeled by a set of $n$ parallel tubes; then the corresponding values of $A(K_T)$ and $G(K_T)$ are found in Figs. 6 and 7, respectively.
2. If $K_T > 64$ the whole rod bundle is modeled by a single annular channel with zero shear stress on the outer boundary; then, $A(K_T) = 1$ and $G(K_T)$ is found in Fig. 7.

The following procedure must be used to compute the friction factor in a bare rod bundle:

1. Calculate $\mathrm{Re}_T$.
2. Evaluate the $K_i$ using Figs. 3-5.
3. Compute $K_T$ using Eq. (7).
4. If the flow is laminar, compute $\lambda$ using Eq. (6).
5. If the flow is turbulent, evaluate $A$ and $G$ using Figs. 6 and 7 and compute $\lambda$ using Eq. (8).

Figure 8 gives an example of experimental verification of this theory. In Courtaud et al. (1966), an alternative method is presented that uses a multiplicative coefficient of the friction factor in a pipe.

**2.2.3 Correction for surface heating** For a flow in a heated rod bundle, the flow friction factor must be corrected to account for the variation of the viscosity across the thermal boundary layers along the pins.

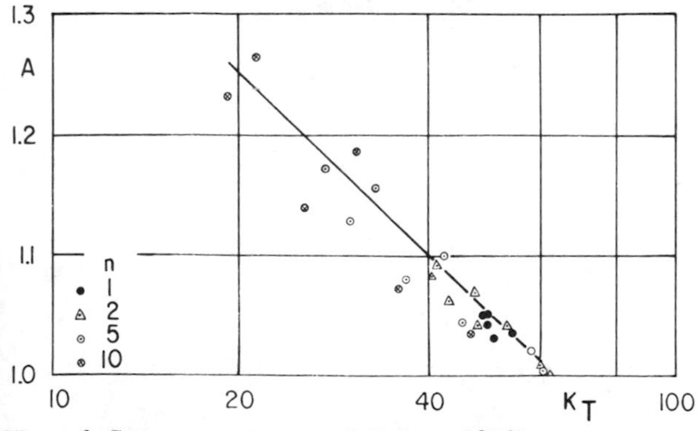


**Figure 6** Geometry parameter $A$ (Rehme, 1973).

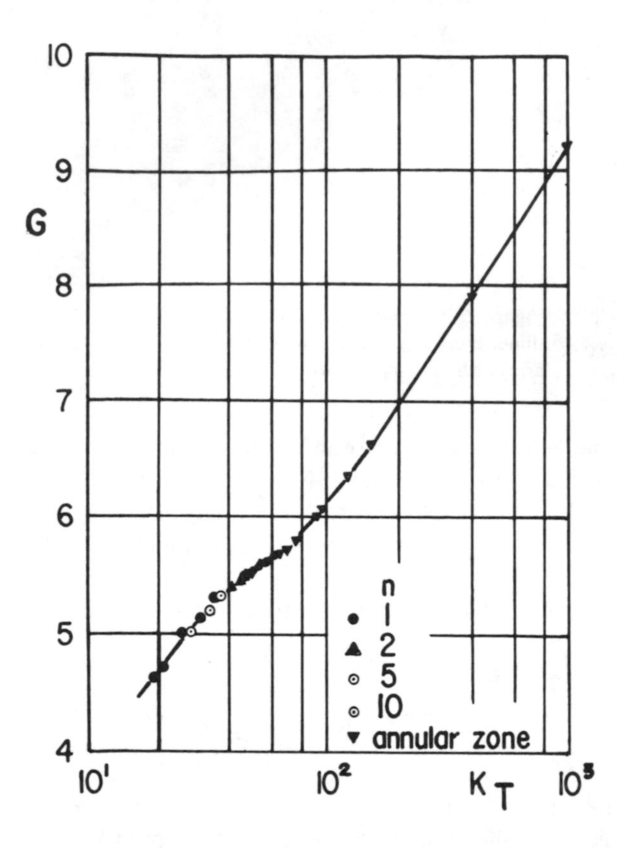


**Figure 7** Geometry parameter $G$ (Rehme, 1973).

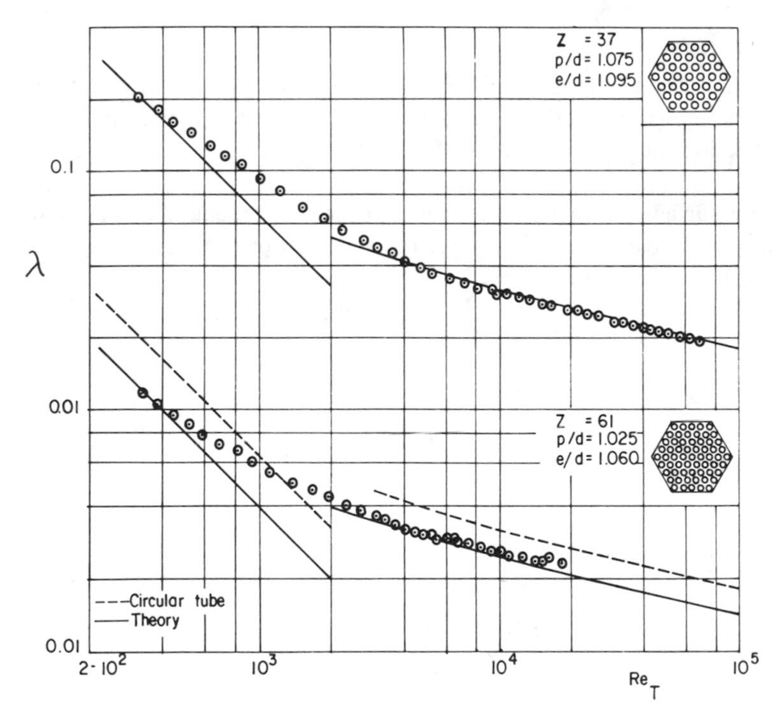


**Figure 8** Rod-bundle friction factor, where $Z$ represents the number of rods (Rehme, 1973).

For water, Tong (1968) suggested the following correction:

$$\frac{\lambda}{\lambda_{\text{iso}}} = \left(\frac{\mu_{\text{wall}}}{\mu_{\text{bulk}}}\right)^{0.6} \tag{9}$$

Equation (9) represents a straight use of Rohsenow and Clark's (1951) result for flow in circular tubes. Keeping this assumption that correction factors developed for a pipe may be applied to rod-bundle geometries, we would suggest the use of the following correlation given by Lafay (1974), which is an important improvement for circular pipes. It is written in terms of the parameter

$$X \triangleq \left(\frac{\mu_{\text{bulk}}}{\mu_{\text{wall}}} - 1\right)\left(\frac{\mu_{\text{bulk}}}{\mu_{\text{wall}}}\right)^{0.17}$$

in the following form:

$$\frac{\lambda}{\lambda_{\text{iso}}} = 1 - 0.5(1 + X)^{F(\text{Re})} \log(1 + X)$$

where $$F(\text{Re}) \triangleq 0.17 - 2 \times 10^{-6}\,\text{Re} + \frac{1800}{\text{Re}}$$

For sodium, because of the high thermal conductivity, the temperature difference is so low that no correction is needed.

In both fluids, this correction is sufficient as long as buoyancy effects do not influence the flow field, which is true in reactors operating under normal conditions.

## 2.3 Friction Factors for Rod Bundles Using Helical Wire-Wrap Spacers

Figure 9 gives the axial variation of the static pressure measured in an edge subchannel of a 19-pin rod bundle with helical wire-wrap spacers.

The wire-wrapped spacer induces significant local pressure perturbations. The pressure variation results from

1. A periodic function of the axial position, its period corresponding to the pitch of helical wire
2. A straight line having a slope that represents the mean pressure gradient

When one is interested in only the global pressure drop between the inlet and outlet, local effects may be ignored. However, the mean pressure gradient will be different from the one in a bare rod bundle.

Novendstern (1972) gave the following procedure for determining the friction pressure drop in that case. First he wrote the friction pressure drop in terms of a friction factor and velocity averaged over an interior subchannel:

$$-\left(\frac{dp}{dz}\right)_f = \frac{\lambda}{D_{e1}} \frac{\rho}{2} \langle w \rangle_1 \tag{10}$$

where subscript 1 denotes an interior subchannel.

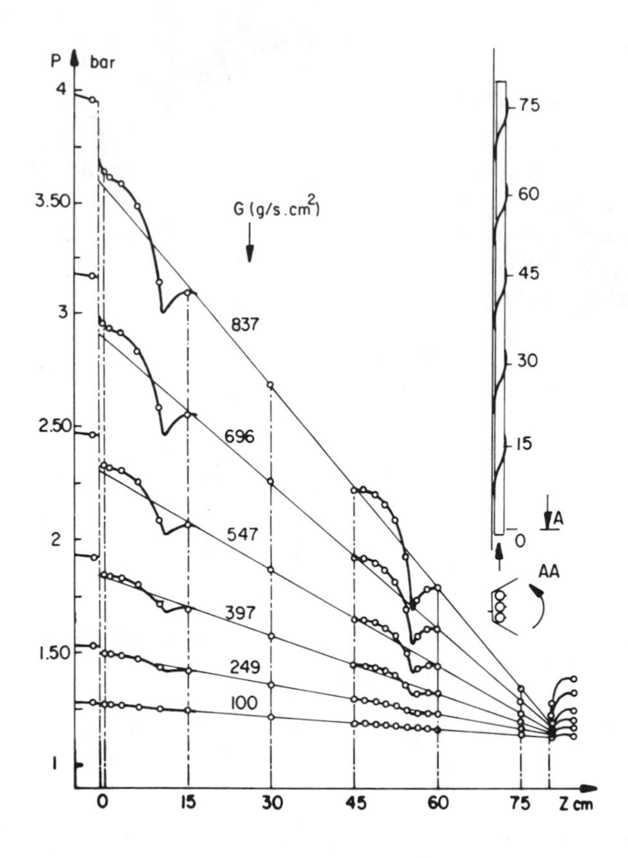


**Figure 9** Axial distribution of the pressure measured in a rod bundle with wire-wrap spacers (Lafay et al., 1975).

This velocity $\langle w \rangle_1$ is related to averaged velocity $\langle w \rangle_T$ by

$$\langle w \rangle_1 = \langle w \rangle_T \frac{a_T}{\sum\limits_{\substack{\text{Interior}\\\text{subchannel}}} a_1 + \sum\limits_{\substack{\text{Edge}\\\text{subchannel}}} a_2 (D_{e2}/D_{e1})^{0.714} + \sum\limits_{\substack{\text{Corner}\\\text{subchannel}}} a_3 (D_{e3}/D_{e1})^{0.714}} \tag{11}$$

where subscripts 2 and 3 refer, respectively, to edge and corner subchannels.

The friction factor is related to the friction factor for smooth pipes by

$$\lambda = \lambda_0 \left[ \frac{1.034}{(p/d)^{0.124}} + \frac{29.7(p/d)^{6.94}\ \mathrm{Re}^{0.086}}{(h/d)^{2.239}} \right] \tag{12}$$

Equation (11), which gives a flow distribution factor, is obtained by assuming uniform pressure across a section of the bundle. Equation (12) is obtained by regression analysis of data. The multiplier of $\lambda_0$ in Eq. (12) depends mainly on geometric factors, that is, the pitch-to-diameter ratio $p/d$ and the ratio of the pitch of the helical wire and diameter $h/d$. The multiplier also depends on flow conditions through the Reynolds number. Its variations with geometric parameters are plotted

in Fig. 10 for a given Reynolds number of 50,000. This Reyno
Eq. (12) for evaluating the multiplier and the friction factor for sm
defined by

$$\mathrm{Re} \triangleq \frac{\langle w \rangle_1 D_{e1}}{\nu}$$

For this correlation to be valid, the bundle should not be loose, a condition th
may be idealistic in a reactor assembly operating with thermal gradients.

## 3 MIXED FLOW MODEL FOR TWO-PHASE FLOW

### 3.1 Procedure

One assumes that the rod bundle behaves in two-phase flow as if it were a single channel with the equivalent diameter $D_e$. The following procedure is then used.

1. Use the friction factor for the rod bundle in single-phase flow as defined in Secs. 2.2 and 2.3.
2. Multiply by the two-phase friction factor multiplier $\phi_l$ obtained for two-phase flows in pipes. The choice of the appropriate correlation should then be made in relation to the range of flow parameters of the rod bundle.

It must be kept in mind that these correlations in pipes were deduced from experiments in which only the total pressure drop is directly available. To get the friction component, both acceleration and elevation components were computed with a flow model and subtracted from the total pressure drop. One must be careful to use the same "package" when making predictions.

### 3.2 Boiling Water

Castellana and Bonilla (1969) reported several studies of water boiling at high pressure in a rod bundle where the authors used such an approach to correlate their

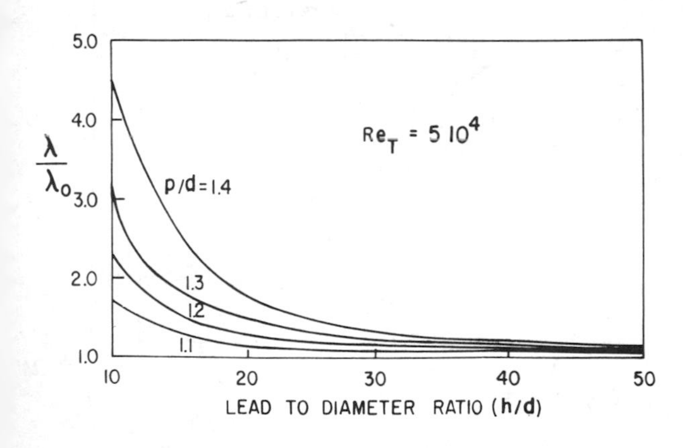


**Figure 10** Friction factor multiplier (Novendstern, 1972).

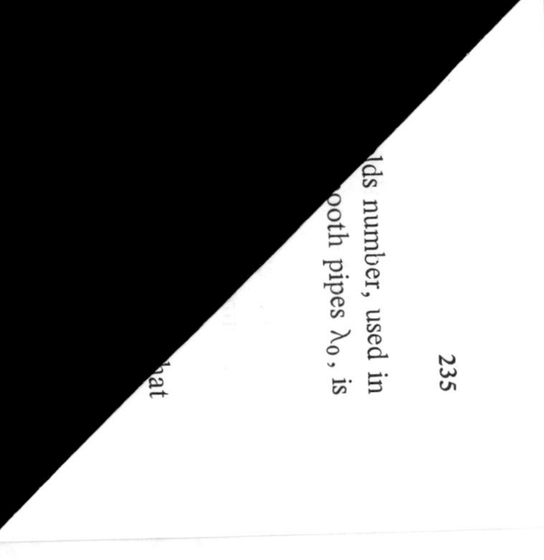


appears that most of these authors used the
ted for the mass-velocity effect. For example,

ıdle in a square array at 40 bar
od bundle in a triangular array at 30 bar

ower pressures (between 6 and 10 bar), Idsinga et

correlation for qualities less than 0.3
n for higher qualities

## 3.3 Boiling Sodium

Kaiser (1977) reported experiments in sodium boiling at atmospheric pressure in a geometry such that a mixed flow model can be successfully applied. It is a seven-rod triangular array. The Lockhart-Martinelli friction multiplier $\phi_l$ is generally used for correlation of sodium boiling pressure drops in pipes. In Fig. 11, a plot of the experimental results in the Lockhart-Martinelli diagram shows good agreement with the theoretical curve. It may seem surprising to use a correlation for air–water flows. The high liquid-vapor density ratio encountered in sodium boiling (as in air–water flows) is apparently the reason for this use.

Experiments at the CEA-Grenoble with a seven-rod triangular array (Menant and Costa, 1974) did not show such an agreement. In this seven-rod array, heterogeneity effects between the subchannels are the origin of the departure from the mixed flow model. This discrepancy results from a different distribution of axial

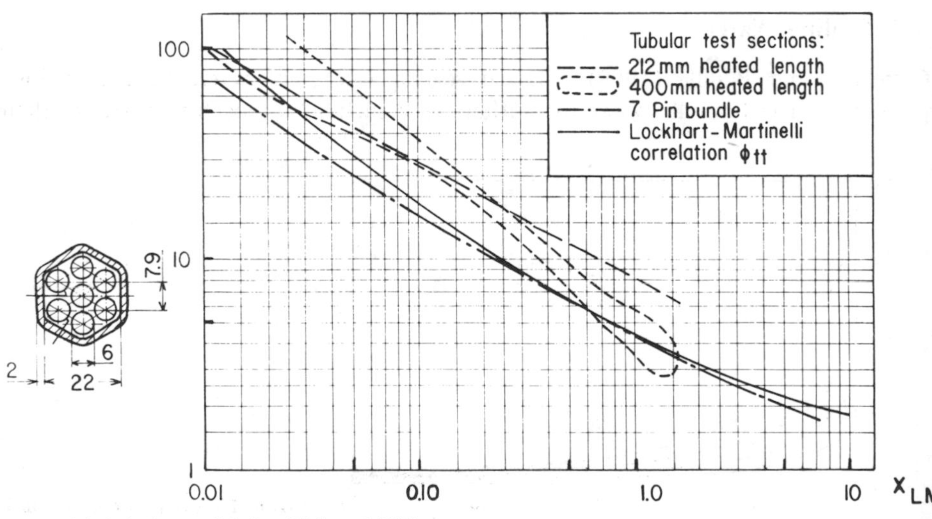


Figure 11 Friction multiplier (Kaiser, 1977).

temperature gradients between the two experiments. A simple calculation demonstrates this fact.

We assume equal friction pressure drop in interior and peripheral subchannels:

$$-\left(\frac{dp}{dz}\right)_f = \lambda_i \rho \frac{w_i^2}{2D_{ei}} = \lambda_p \rho \frac{w_p^2}{2D_{ep}}$$

Inasmuch as the purpose of this calculation is to give orders of magnitude, we can assume the same friction factors, $\lambda_i = \lambda_p$. Thus the velocities are related by

$$\frac{w_i}{w_p} = \sqrt{\frac{D_{ei}}{D_{ep}}} \tag{13}$$

where the subscript $i$ denotes the interior subchannel and $p$ denotes the peripheral subchannel.

Neglecting the exchange between the two regions, we write the energy balance

For interior subchannels

$$\left(\frac{dT}{dz}\right)_i = \frac{\phi \mathcal{P}_{ch,i}}{\rho c w_i a_i}$$

where $\mathcal{P}_{ch}$ denotes the heated perimeter. Here the heated perimeter is equal to the wetted perimeter. Hence

$$\mathcal{P}_{ch,i} = \mathcal{P}_i$$

Thus,

$$\left(\frac{dT}{dz}\right)_i = 4\,\frac{\phi}{\rho c}\,\frac{1}{w_i D_{ei}} \tag{14}$$

For peripheral subchannels the heated perimeter is a fraction $\alpha_p$ of the wetted perimeter, which includes the housing. Hence

$$\mathcal{P}_{ch,p} = \alpha_p \mathcal{P}_p$$

Thus,

$$\left(\frac{dT}{dz}\right)_p = 4\,\frac{\phi}{\rho c}\,\frac{\alpha_p}{w_p D_{ep}} \tag{15}$$

Combining Eqs. (13)–(15) gives

$$\frac{(dT/dz)_i}{(dT/dz)_p} = \frac{1}{\alpha_p}\left(\frac{D_{ep}}{D_{ei}}\right)^{3/2} \tag{16}$$

Table 1 summarizes the results obtained for two types of rod bundle.

**Table 1 Examples of temperature heterogeneities in rod bundles**

| Type of bundle | $D_{ei}$ (cm) | $D_{ep}$ (cm) | $D_{ep}/D_{ei}$ | $\alpha_p$ | $\frac{(dT/dz)_i}{(dT/dz)_p}$ |
|---|---|---|---|---|---|
| GfK bundle | 0.55 | 0.44 | 0.80 | 0.54 | 1.33 |
| CEA-Grenoble bundle | 0.35 | 0.29 | 0.83 | 0.40 | 1.89 |

Because mixing effects have been neglected, these calculations show a trend toward a flatter temperature profile in the GfK bundle than in the CEA-Grenoble bundle. In the GfK bundle the boiling may start at the same axial position in all subchannels, whereas in the CEA-Grenoble bundle the flow diversion, as discussed in Sec. 1.2, is more likely to occur in agreement with the experiments. The resulting heterogeneities in a cross section make the mixed flow model fail.

# 4 SUBCHANNEL ANALYSIS OF TWO-PHASE FLOW

In a pipe, the friction pressure drop is one component of the total pressure drop, the other components being the elevation and the acceleration terms.

In subchannel analysis characterized by interconnected flow regions, the picture becomes more difficult. New components of the pressure drop appear in the axial component of the momentum equation. Furthermore, if each subchannel is characterized by a different pressure, transverse components of the momentum equation, that is on directions lying in the cross section, must be solved together with the axial component.

In Sec. 4.1 the axial and transverse momentum equations for two interconnected subchannels are derived; in Sec. 4.2 the axial approximations for various terms are given.

## 4.1 Momentum Equation

For simplicity of demonstration, we consider two subchannels in a triangular array, which do not have any exchange with the outer flow domain. Thus we will assume that neither transport nor flux occurs along the straight boundaries marked by double lines (Fig. 12).

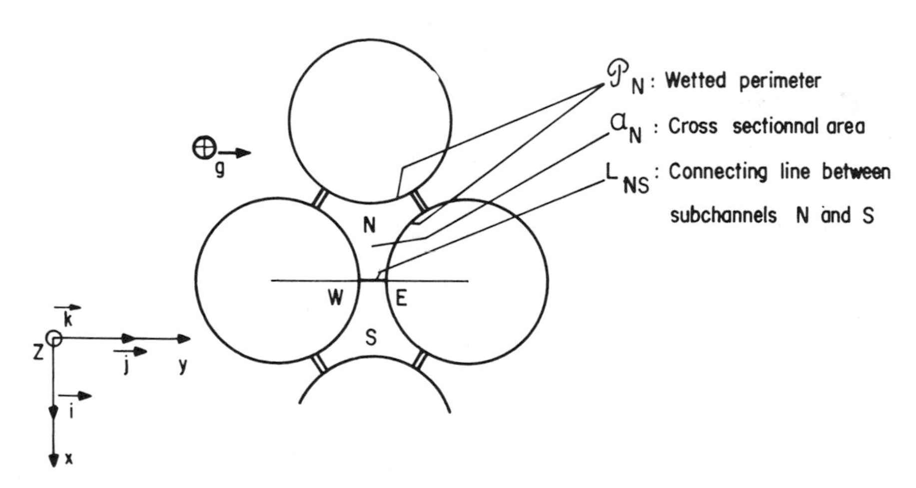


**Figure 12** Coordinate system.

**4.1.1 Axial momentum equation** For steady-state flow of a homogeneous fluid, the axial component of the local statistical-averaged momentum equation is

$$\nabla \cdot (\rho w \mathbf{v}) = -\frac{\partial p}{\partial z} + \nabla \cdot (\tau \cdot \mathbf{k}) - \rho g \tag{17}$$

Owing to the statistical averaging, the stress tensor is the sum of molecular and Reynolds stresses. Integrated over the cross-sectional area $a_N$, it becomes

$$a_N \left( \frac{\partial}{\partial z} \langle \rho w^2 \rangle_N - \langle \rho \rangle_N g + \frac{\partial}{\partial z} \langle p \rangle_N - \frac{\partial}{\partial z} \langle \tau_{zz} \rangle \right)$$

$$= \underbrace{\int_{\mathcal{P}_N} \mathbf{k} \cdot (\mathbf{n}_N \cdot \tau) \, d\mathcal{P}}_{\text{A1}} - \underbrace{\int_{L_{NS}} \rho w \mathbf{v} \cdot \mathbf{n}_N \, dy}_{\text{A2}} + \underbrace{\int_{L_{NS}} (\tau \cdot \mathbf{k}) \cdot \mathbf{n}_N \, dy}_{\text{A3}} \tag{18}$$

The derivation of Eq. (18) from Eq. (17) is obtained with the use of mathematical tools recalled in Chap. 7, and it can easily be verified by the reader.

The area-averaged equation for subchannel $S$ is identical to Eq. (18) by replacing subscript $N$ by $S$. Moreover we have the following identity on the connection line $L_{NS}$:

$$\mathbf{n}_N + \mathbf{n}_S = 0 \tag{19}$$

The terms appearing in the right side of Eq. (18) have the following meaning:

A1 is the friction along the rods.
A2 is the rate of change of momentum due to cross flows between subchannels. Since $\mathbf{n}_N = \mathbf{i}$

$$-\int_{L_{NS}} \rho w \mathbf{v} \cdot \mathbf{n} \, dx = -L_{NS} \langle \rho w u \rangle_{1,L} \tag{20}$$

where the subscript 1 stands for line-averaged quantity.
A3 is the rate of change due to turbulent mixing between subchannels.

A2 and A3 are terms specific to the subchannel geometry. As a consequence of Eq. (19), terms A2 and A3 cancel when we add Eq. (18) for subchannels $N$ and $S$.

If we use the generally accepted hypothesis of space correlation coefficients equal to 1, that is,

$$\langle fg \rangle = \langle f \rangle \langle g \rangle$$

Eq. (18) becomes

$$a_N\left(\frac{\partial}{\partial z}\frac{\langle\rho w\rangle_N^2}{\langle\rho\rangle_N}-\langle\rho\rangle_N g+\frac{\partial}{\partial z}\langle p\rangle_N-\frac{\partial}{\partial z}\langle\tau_{zz}\rangle_N\right)$$

$$=\underbrace{\int_{\wp_N}\mathbf{k}\cdot(\mathbf{n}_N\cdot\tau)\,d\wp}_{A1}\;\underbrace{-L_{NS}\frac{\langle\rho w\rangle_{1,L}\langle\rho u\rangle_{1,L}}{\langle\rho\rangle_{1,L}}}_{A2}+\underbrace{\int_{L_{NS}}(\tau\cdot\mathbf{k})\cdot\mathbf{n}_N\,dy}_{A3} \tag{21}$$

**4.1.2 Transverse momentum equation** The local equation projected over $0x$ is, where $0x$ is the direction normal to the interconnecting line,

$$\nabla\cdot(\rho u\mathbf{v})=-\frac{\partial p}{\partial x}+\nabla\cdot(\tau\cdot\mathbf{i}) \tag{22}$$

Integration over the segment $L_{NS}$ gives

$$L_{NS}\left(\underbrace{\frac{\partial}{\partial z}\langle\rho uw\rangle_{1,L}}_{T1}+\underbrace{\frac{\partial}{\partial x}\langle\rho u^2\rangle_{1,L}}_{T2}+\underbrace{\frac{\partial}{\partial x}\langle p\rangle_{1,L}}_{T3}\right)$$

$$\underbrace{-\frac{\partial}{\partial x}\langle\tau_{xx}\rangle_{1,L}-\frac{\partial}{\partial z}\langle\tau_{xz}\rangle_{1,L}}_{T4}=\underbrace{\sum_{E,W}\mathbf{n}_L\cdot\tau}_{T5} \tag{23}$$

T1 and T2 are the inertial terms in $0z$ and the $0x$ directions, respectively.
T3 is the pressure gradient in the transverse direction.
T4 denotes longitudinal shear stresses, which may be neglected when no flow separation occurs.
T5 is the friction resistance at the rods for transverse flow.

By using the hypotheses made in the derivation of Eq. (21), Eq. (23) may be rewritten in the practical form:

$$L_{NS}\left(\underbrace{\frac{\partial}{\partial z}\frac{\langle\rho u\rangle_{1,L}\langle\rho w\rangle_{1,L}}{\langle\rho\rangle_{1,L}}}_{T1}+\underbrace{\frac{\partial}{\partial x}\frac{\langle\rho u\rangle_{1,L}^2}{\langle\rho\rangle_{1,L}}}_{T2}+\underbrace{\frac{\partial}{\partial x}\langle p\rangle_{1,L}}_{T3}\right)=\underbrace{\sum_{E,W}\mathbf{n}_L\cdot\tau}_{T5} \tag{24}$$

where we have neglected T4.

The main dependent variables are listed below:

$\langle\rho u\rangle_{1,L}$ transverse mass velocity averaged on the gap
$\langle\rho w\rangle_N$ and $\langle\rho w\rangle_S$ axial mass velocities averaged on subchannels
$\langle p\rangle_N$ and $\langle p\rangle_S$ averaged pressure on subchannels

One has the following sets of equations:

2 mass conservation equations (not written here)
2 axial momentum equations
1 transverse momentum equation

Consequently, a closed set of equations is obtained if all the unknown quantities are expressed in terms of the main dependent variables listed above.

## 4.2 Approximation of the Different Terms

**4.2.1 Approximation in the axial momentum equation** *Frictional pressure drop* The procedure universally used to calculate the frictional pressure drop is shown in Secs. 2 and 3, using the equivalent hydraulic diameter of the subchannel.

$$A1 = \frac{\lambda}{D_{e_N}} \frac{\langle \rho w \rangle_N^2}{2\langle \rho \rangle_N} \tag{25}$$

and $\lambda$ given in Secs. 2.2 and 2.3 for single-phase flow, and in Sec. 3 for two-phase flow.

*Effect of transverse flow between subchannels*

$$A2 = \frac{\langle \rho w \rangle_{1,L} \langle \rho u \rangle_{1,L}}{\langle \rho \rangle_{1,L}}$$

The transverse mass flow rate $\langle \rho u \rangle_{1,L}$ will be given by Eq. (24).

The question is how to relate $\langle \rho w \rangle_{1,L} / \langle \rho \rangle_{1,L} = \langle w \rangle_{1,L}$ to quantities averaged on subchannels. There are mainly two ways: First (Rowe, 1973; Rouhani, 1973),

$$\langle w \rangle_{1,L} = \frac{\langle w \rangle_{2,N} + \langle w \rangle_{2,S}}{2} \tag{26}$$

Second, one could take account of the flow direction by giving to $\langle w \rangle_{1,L}$ the value of the subchannel from the origin of the cross flow $\langle \rho u \rangle_{1,L}$ (donor subchannel) (Rowe, 1975).

$$\begin{aligned} \langle w \rangle_{1,L} &= \langle w \rangle_{2,N} \quad \text{if } \langle \rho u \rangle_{1,L} > 0 \\ \langle w \rangle_{1,L} &= \langle w \rangle_{2,S} \quad \text{if } \langle \rho u \rangle_{1,L} < 0 \end{aligned} \tag{27}$$

If it is taken into account that the next step would be the numerical solution of these equations, there is an advantage to using this last formulation, which ensures unconditional numerical stability.

*Turbulent diffusion* The general form of the equation for turbulent diffusion is

$$A3 = -\langle \rho u' \rangle_{1,L} (\langle w \rangle_{2,N} - \langle w \rangle_{2,S}) \tag{28}$$

where $u'$ is a characteristic velocity of turbulent fluctuations, which can be related to an eddy diffusivity by

$$\langle \rho u' \rangle_{1,L} = \frac{\langle \rho \epsilon \rangle_{1,L}}{\Delta_{NS}}$$

where $\Delta_{NS}$ is a characteristic transverse distance between subchannels $N$ and $S$. It is very often the distance between the centroids of the cross-sectional areas.

Rogers and Todreas (1968) reviewed different assumptions regarding single-phase flows. It should be pointed out that their study relates to eddy diffusivity of heat and not momentum. Under the assumption of similar transport of heat and momentum, however, the values of the eddy diffusivities are equal. The general form for the eddy diffusivity $\epsilon$ is

$$\frac{\epsilon}{\nu} = C_\epsilon \text{ Re } \sqrt{\frac{\lambda}{2}} \tag{29}$$

where $\lambda$ is the friction factor and $C_\epsilon$ is an empirical constant.

Lahey and Schraub (1969) pointed out that in two-phase flows the usual procedure is to use the two-phase friction-factor multiplier in Eq. (29) together with mixture variables in Eq. (28). However, they revealed discrepancies with the tendencies of experimental results and suggested an orientation toward correlations based on flow regimes.

Rogers and Todreas (1968) discussed additional effects on turbulent mixing in detail. These effects are of two types:

1. Flow sweeping, which can be induced by helical wire-wrap spacers
2. Flow scattering downstream of grid spacers, or of any device inducing turbulence

**4.2.2 Approximation in the transverse momentum equation** When the characteristic transverse distance $\Delta_{NS}$ is introduced, the transverse pressure gradient T3 becomes:

$$\frac{\partial}{\partial x}\langle p\rangle_{1,L} = \frac{\langle p\rangle_S - \langle p\rangle_N}{\Delta_{NS}}$$

Then Eq. (24) may be rewritten

$$\frac{\langle p\rangle_S - \langle p\rangle_N}{\Delta_{NS}} = \text{T5} - \text{T1} - \text{T2} \tag{30}$$

The term T5 is a frictional resistance at the rods. Its general form is

$$\text{T5} = e_v \frac{\langle\rho u\rangle_{1,L}|\langle\rho u\rangle_{1,L}|}{2\langle\rho\rangle_{1,L}} \tag{31}$$

where $e_v$ is the friction loss factor of the gap.

It was the only term taken into account in the first generation of codes, COBRA I, II, III, FLICA, and HAMBO.

For bare rod bundles $e_v \simeq 0.04$ (Rogers and Todreas, 1968). For bundles with wire-wrap spacers $e_v$ varies axially in sign and magnitude so that it takes into account the presence or absence of the wire in the gap between the subchannels. In two-phase flows the same expression is used with the mixture density instead of the liquid density.

The acceleration term T2 is generally neglected. Only Rouhani (1973) mentioned it as being treated in the Scandinavian SDS code, but no detail was given.

In the acceleration term T1 one replaces the unknown quantity $\langle\rho w\rangle_{1,L}$ by

$$\langle\rho w\rangle_{1,L} = \frac{\langle\rho w\rangle_N + \langle\rho w\rangle_S}{2}$$

and T1 is rewritten as

$$T1 = \frac{\partial}{\partial z}\left(\langle \rho u \rangle_{1,L} \frac{\langle \rho w \rangle_N + \langle \rho w \rangle_S}{2\langle \rho \rangle_{1,L}}\right) \tag{32}$$

This is what is done in COBRA IIIC.

To conclude, if the heterogencities in a cross section of a rod bundle are important, the cross section must be divided into smaller areas, such as subchannels. Compared with the computation of the flow in a straight pipe where the axial momentum equation is sufficient, the computation of the flow in interconnected subchannels requires using the three components of the momentum equation (one axial component and two transverse). Moreover, the axial momentum equation has additional terms representing the rate of change of momentum resulting from cross flows and turbulent mixing between subchannels. The accurate modeling of the whole set of components is a necessity for obtaining good prediction of the flow and pressure fields in the bundle.

## NOMENCLATURE

| | |
|---|---|
| $A$ | constant [Eq. (8)] |
| $a$ | cross-sectional area |
| $c$ | specific heat capacity |
| $D$ | diameter of a tube |
| $D_e$ | equivalent diameter |
| $d$ | diameter of a pin |
| $e$ | width of a peripheral channel (Fig. 4) |
| $e_v$ | friction loss factor of a gap [Eq. (31)] |
| $G$ | constant in the universal law of friction [Eq. (3)]; mass velocity in Fig. 9 |
| $h$ | pitch of the helical wire-wrap spacer |
| $\mathbf{i}$ | unit vector along $0x$ |
| $\mathbf{j}$ | unit vector along $0y$ |
| $K$ | constant of the laminar friction law [Eq. (6)] |
| $\mathbf{k}$ | unit vector along $0z$ |
| $L_{NS}$ | boundary between subchannels $N$ and $S$ |
| $\mathbf{n}$ | normal unit vector |
| $n$ | number of parallel tubes in Rehme's model (Figs. 6 and 7) |
| $\mathcal{P}_i$ | wetted perimeter of the $i$th subchannel |
| $\mathcal{P}_{ch,i}$ | heated perimeter of the $i$th subchannel |
| $p$ | pitch of the array; pressure |
| $Q$ | flow rate in Fig. 2 |
| $Q_{sat}$ | flow rate corresponding to saturation conditions at the exit of the heating length (Fig. 2) |
| Re | Reynolds number [Eq. (2)] |
| $u$ | transverse component of the velocity |
| $u'$ | characteristic velocity of turbulent fluctuations |

| | |
|---|---|
| $\mathbf{v}$ | velocity vector |
| $v^*$ | friction velocity |
| $w$ | axial component of the velocity |
| $x$ | Cartesian coordinate |
| $X$ | parameter for correction for surface heating |
| $X_{LM}$ | Lockhart-Martinelli parameter |
| $y$ | Cartesian coordinate |
| $z$ | axial coordinate |
| $\alpha_p$ | ratio of heated to wetted perimeters |
| $\Delta_{NS}$ | characteristic transverse distance between subchannels $N$ and $S$ |
| $\epsilon$ | eddy diffusivity |
| $\lambda$ | resistance coefficient [Eq. (5)] |
| $\lambda_0$ | resistance coefficient of pipe flow [Eq. (1)] |
| $\mu$ | dynamic viscosity |
| $\nu$ | kinematic viscosity |
| $\rho$ | density |
| $\tau$ | viscous stress tensor |
| $\tau_w$ | wall shear stress |
| $\phi_l$ | two-phase friction-factor multiplier |

**Subscripts**

| | |
|---|---|
| $E$ | extremity of the gap (Fig. 12) |
| $i$ | $i$th subchannel |
| $o$ | circular pipe |
| $N$ | northern subchannel (Fig. 12) |
| $p$ | peripheral subchannel |
| $S$ | southern subchannel (Fig. 12) |
| $T$ | whole cross section of the rod bundle |
| $W$ | extremity of the gap (Fig. 12) |
| 1 | interior subchannel |
| 2 | edge subchannel |
| 3 | corner subchannel |

**Symbols**

| | |
|---|---|
| $\langle \ \rangle$ | area average |
| $\langle \ \rangle_{1,L}$ | line average along $L_{NS}$ |
| $\| \ \|$ | absolute value |

## REFERENCES

Armand, A. A. and Treschev, G. G., Investigation of the Resistance during the Movement of Steam-water Mixtures in a Heater Boiler Pipe at High Pressures, AERE-Lib/Trans. 816, 1959.

Baroczy, C. J., A Systematic Correlation for Two-Phase Pressure Drop, *Chem. Eng. Prog. Symp. Ser.*, vol. 62, pp. 232–249, 1966.

Castellana, F. S. and Bonilla, C. F., Two-Phase Pressure Drop and Heat Transfer in Rod Bundles, in *Two-Phase Flow and Heat Transfer in Rod Bundles*, ed. V. E. Schrock, pp. 15–30, ASME, New York, 1969.

Costa, J. and Menant, B., Sodium Boiling Experiments in a 19-Pin Bundle: Two-Phase Coolant Dynamics, *Proc. Intl. Meet. on Fast Reactor Safety and Related Physics, Chicago, Oct. 5–8*, pp. 1523–1531, 1976.

Courtaud, M., Ricque, R., and Martinet, B., Etude des Pertes de Charge dans des Conduites Circulaires Contenant un Faisceau de Barreaux, *Chem. Eng. Sci.*, vol. 21, pp. 881–893, 1966.

Idsinga, W., Todreas, N., and Bowring, R., An Assessment of Two-Phase Pressure Drop Correlations for Steam-Water Systems, *Int. J. Multiphase Flow*, vol. 3, pp. 401–413, 1977.

Kaiser, A., Two-Phase Pressure Drop Correlations for Sodium: Comparison of Results in Single-Channel and Multi-Channel Geometry, 7th Meet. of the Liquid Metal Boiling Working Group, Petten, Netherlands, June 1–3, 1977.

Lafay, J., Influence de la Variation de la Viscosité avec la Température sur le Frottement avec Transfert de Chaleur en Régime Turbulent Etabli, *Int. J. Heat Mass Transfer*, vol. 17, pp. 815–834, 1974.

Lafay, J., Menant, B., and Barroil, J., Influence of Helical Wire Wrap Spacer System in a Water 19-Rod Bundle, ASME 75-HT-22, 1975.

Lahey, R. T. and Schraub, F. A., Mixing, Flow Regimes and Void Fraction for Two-Phase Flows in Rod Bundles, in *Two-Phase Flow and Heat Transfer in Rod Bundles*, ed. V. E. Schrock, pp. 1–14, ASME, New York, 1969.

Menant, B. and Costa, J., Preliminary Results of Sodium Boiling in a 7-Pin Rod Bundle, 5th Meet. of the Liquid Metal Boiling Working Group, Grenoble, 1974.

Miropol'skiy, Z. L., Shneyerova, R. I., Karamysheva, A. I., Semin, E. T., and Virogradova, M. N., Effect of Heat Flux and Geometry on the Friction Factor in Flow of Water-Steam Mixtures through Channels, *Heat Transfer Sov. Res.*, vol. 1, no. 2, pp. 13–16, 1969.

Novendstern, E. H., Turbulent Flow Pressure Drop Model for Fuel Rod Assemblies Utilizing a Helical Wire-Wrap Spacer System, *Nucl. Eng. Des.*, vol. 22, pp. 19–27, 1972.

Rehme, K., Laminarströmung in Stabbündeln, *Chem. Eng. Tech.*, vol. 43, pp. 962–966, 1971.

Rehme, K., Simple Method of Predicting Friction Factors of Turbulent Flow in Non-circular Channels, *Int. J. Heat Mass Transfer*, vol. 16, pp. 933–950, 1973.

Rogers, J. T. and Todreas, N. E., Coolant Interchannel Mixing in Reactor Fuel Rod Bundles. Single Phase Coolants, in *Single-Phase Coolant Heat Transfer in Rod Bundles*, ed. V. E. Schrock, pp. 1–56, ASME, New York, 1968.

Rohsenow, W. M. and Clark, J. A., Heat transfer and pressure drop data at high heat flux densities to water at high subcritical pressure, *Proc. Heat Transfer and Fluid Mechanics Institute*, Stanford, Calif., pp. 193–208, 1951.

Rouhani, Z., A Review of Momentum Balance in Subchannel Geometry, European Two Phase Flow Group Meet., Brussels, 1973.

Rowe, D. S., A Mathematical Model for Transient Subchannel Analysis of Rod Bundle Nuclear Fuel Elements, *J. Heat Transfer*, vol. 95, pp. 211–217, 1973.

Rowe, D. S., Core Thermal Model Development and Experiments, BNWL-1924 1, 1975.

Schlichting, H., *Boundary Layer Theory*, 6th ed., McGraw-Hill, New York, 1968.

Sher, N. C., Kangas, G. U., and Neusen, K. F., On the Phenomenon of Boiling Flow in a Parallel Rod Array, *Chem. Eng. Prog. Symp. Ser.*, vol. 61, pp. 127–156, 1965.

Tong, L. S., Pressure Drop Performance of a Rod Bundle, in *Single-Phase Coolant Heat Transfer in Rod Bundles*, ed. V. E. Schrock, pp. 57–69, ASME, New York, 1968.

CHAPTER
# THIRTEEN

# SINGULAR PRESSURE DROPS

**M. Giot**

## 1 SUDDEN ENLARGEMENT

In their book on annular two-phase flow, Hewitt and Hall-Taylor (1970) presented a simple correlation that extends the single-phase flow theory. With subscripts 1 and 2 referring to sections 1 and 2 defined in Fig. 1, the momentum balance equation can be written:

$$p_2 A_2 - p_1 A_2 = M_L(w_{L1} - w_{L2}) + M_G(w_{G1} - w_{G2}) \tag{1}$$

or, setting

$$s \triangleq \frac{A_1}{A_2}$$

and using the continuity equation yield for a flow without phase change ($x = \text{const}$):

$$p_2 - p_1 = \frac{G_1^2 s}{\rho_L} \left\{ \left[ \frac{(1-x)^2}{1-\alpha_1} + \frac{\rho_L}{\rho_G} \frac{x^2}{\alpha_1} \right] - s \left[ \frac{(1-x)^2}{1-\alpha_2} + \frac{\rho_L}{\rho_G} \frac{x^2}{\alpha_2} \right] \right\} \tag{2}$$

If the void fractions upstream and downstream are equal [a good approximation in some cases according to Velasco (1975)], then $\alpha_1 = \alpha_2 = \alpha$, and Eq. (2) becomes

$$p_2 - p_1 = \frac{G_1^2 s(1-s)}{\rho_L} \left[ \frac{(1-x)^2}{1-\alpha} + \frac{\rho_L}{\rho_G} \frac{x^2}{\alpha} \right] \tag{3}$$

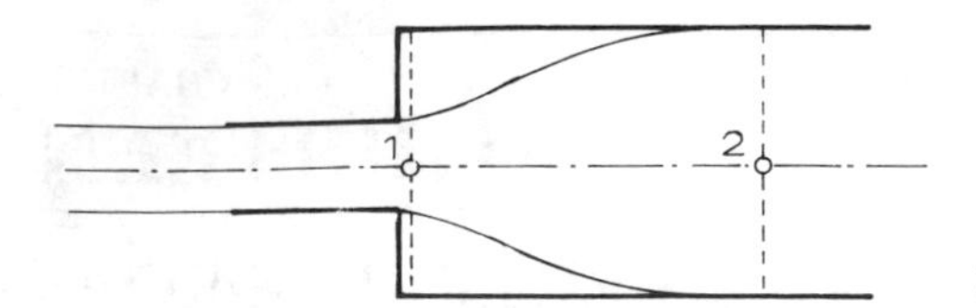


**Figure 1** Flow through a sudden enlargement.

Comparing Eq. (3) with the corresponding equation in single-phase flow reveals that the term in the brackets can be regarded as a two-phase multiplier. Moreover, if one considers the simplified mechanical-energy balance

$$\frac{1}{2}\,[x(w_{G1}^2 - w_{G2}^2) + (1-x)(w_{L1}^2 - w_{L2}^2)] = \tau_f + \int_{p_1}^{p_2} \left(\frac{x}{\rho_G} + \frac{1-x}{\rho_L}\right) dp \qquad (4)$$

where $\tau_f$ denotes the frictional dissipation of energy, one sees that for incompressible two-phase flow with $\alpha_1 = \alpha_2 = \alpha$, Eq. (4) yields

$$\frac{G_1^2}{2}(1-s^2)\left[\frac{x^3}{\alpha^2\rho_G^2} + \frac{(1-x)^3}{(1-\alpha)^2\rho_L^2}\right] = \tau_f + \left(\frac{x}{\rho_G} + \frac{1-x}{\rho_L}\right)(p_2 - p_1) \qquad (5)$$

If one distinguishes a reversible and an irreversible contribution to the pressure difference,

$$\Delta p = \Delta p_R + \Delta p_I$$

eliminating $\Delta p$ from Eq. (5) by means of Eq. (3) leads to

$$\Delta p_R = \frac{G_1^2(1-s^2)}{2}\left[\frac{x^3}{\alpha^2\rho_G^2} + \frac{(1-x)^3}{(1-\alpha)^2\rho_L^2}\right]\left(\frac{1-x}{\rho_L} + \frac{x}{\rho_G}\right)^{-1} \qquad (6)$$

and

$$\Delta p_I = G_1^2(1-s)\left\{ s\left[\frac{(1-x)^2}{(1-\alpha)\rho_L} + \frac{x^2}{\alpha\rho_G}\right] - \frac{1+s}{2}\left[\frac{x^3}{\alpha^2\rho_G^2} + \frac{(1-x)^3}{(1-\alpha)^2\rho_L^2}\right] \times \left(\frac{x}{\rho_G} + \frac{1-x}{\rho_L}\right)^{-1}\right\} \qquad (7)$$

For homogeneous flow, this frictional pressure loss is simply

$$\Delta p_I = -\frac{G_1^2}{2\rho_L}(1-s)^2\left[1 + x\left(\frac{\rho_L}{\rho_G} - 1\right)\right] \qquad (8)$$

One of the assumptions in Eq. (8) is that $\alpha_1 = \alpha_2$. Petrick and Swanson (1959) observed significant void-fraction increases behind the expansion. However, this effect disappears at some distance downstream: for example, $L/D =$ 10–30 in terms of the largest diameter. McGee (1966) considered that values as high as $L/D = 40$–$70$ are necessary to stabilize the void fraction.

According to Chisholm (1968, 1969),

$$\Delta p = \frac{G_1^2}{\rho_L} s(1-s)(1-x)^2 \left(1 + \frac{C}{X} + \frac{1}{X^2}\right) \tag{9}$$

where

$$X^2 \triangleq \left(\frac{1-x}{x}\right)^2 \frac{\rho_G}{\rho_L} \qquad C \triangleq \left[1 + 0.5\left(\frac{\rho_L - \rho_G}{\rho_L}\right)^{0.5}\right]\left[\left(\frac{\rho_L}{\rho_G}\right)^{0.5} + \left(\frac{\rho_G}{\rho_L}\right)^{0.5}\right] \tag{10}$$

According to Velasco (1975), who performed air-water tests in 19-25-mm and 34-40-mm enlargements ($x = 7 \times 10^{-4} \ldots 3 \times 10^{-3}$), the Chisholm model seems to overestimate $\Delta p$, whereas Eq. (2) as well as the McGee formula underestimates $\Delta p$. However Eq. (2) is the one that correlates the results the most precisely.

## 2 SUDDEN CONTRACTION

Consider Fig. 2 and the control volume bounded by sections 1 and 2. As for single-phase flow, the pressure difference between 1 and $c$ is assumed to correspond to the reversible pressure difference, that is, frictional losses occur only during the enlargement of the flow from $c$ to 2 and can be calculated by using Eq. (7) or (8), where 1 is replaced by $c$, and $s$ by the contraction coefficient

$$C_c \triangleq \frac{A_c}{A_2}$$

The following expression is then obtained for irreversible losses:

$$\Delta p_I = \frac{G_2^2}{C_c^2}(1 - C_c)\left\{C_c\left[\frac{(1-x)^2}{(1-\alpha)\rho_L} + \frac{x^2}{\alpha\rho_G}\right] - \frac{1+C_c}{2}\rho_m\left[\frac{x^3}{\alpha^2\rho_G} + \frac{(1-x)^3}{(1-\alpha)^2\rho_L}\right]\right\} \tag{11}$$

where $\rho_m$ is the homogeneous specific volume,

$$\frac{1}{\rho_m} \triangleq \frac{x}{\rho_G} + \frac{1-x}{\rho_L}$$

or, if the flow is assumed to be homogeneous,

$$\Delta p_I = -\frac{G_2^2}{2\rho_L}\left(\frac{1}{C_c} - 1\right)^2\left[1 + x\left(\frac{\rho_L}{\rho_G} - 1\right)\right] \tag{12}$$

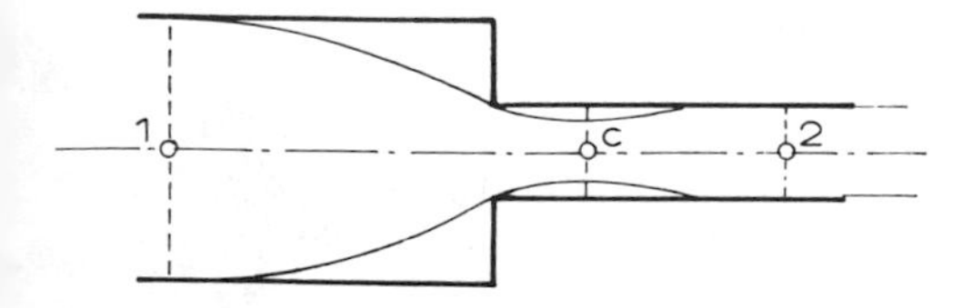


**Figure 2** Flow through a sudden contraction.

The pressure difference is then, according to the homogeneous flow model,

$$p_2 - p_1 = \frac{G_2^2}{2\rho_L}\left[\left(\frac{1}{C_c}-1\right)^2 + 1 - \frac{1}{s^2}\right]\left[1 + x\left(\frac{\rho_L}{\rho_G}-1\right)\right] \tag{13}$$

The contraction coefficient $C_c$ is taken equal to that of single-phase flow, and given, for example, by the Weisbach values as a function of the area ratio $s = A_1/A_2$.

## 3 CONTRACTION–EXPANSION COMBINATION

Harshe et al. (1976) performed tests in a boiling Freon loop using six inserts in order to have several flow area ratios and several restriction lengths.

They concluded that the two-phase pressure drop across well-separated abrupt contraction–expansion combinations may be predicted by a simple summation of the pressure changes through the contraction and expansion (Fig. 3):

$$\begin{aligned}\Delta p = \frac{G_1^2}{2\rho_L}\Bigg(\frac{1}{s^2}\Bigg\{&\frac{\rho_L}{\rho_G}x^2\alpha_1\left(\frac{1}{C_c^2\alpha_3^2}-\frac{1}{\alpha_4^2}\right) + (1-x)^2(1-\bar{\alpha}_1)\left[\frac{1}{C_c^2(1-\alpha_3)^2}\right.\\ &\left.-\frac{1}{(1-\alpha_4)^2}\right]\Bigg\} - \frac{2}{s^2}\Bigg\{\frac{\rho_L}{\rho_G}x^2\left(\frac{1}{C_c\alpha_3}-\frac{1}{\alpha_4}+\frac{s}{\alpha_4}-\frac{s^2}{\alpha_5}\right) + (1-x)^2\left[\frac{1}{C_c(1-\alpha_3)}\right.\\ &\left.-\frac{1}{1-\alpha_4}+\frac{s}{1-\alpha_4}-\frac{s^2}{1-\alpha_5}\right]\Bigg\} + \frac{\rho_L}{\rho_G}x^2\bar{\alpha}_2\left(\frac{1}{s^2\alpha_4^2}-\frac{1}{\alpha_1^2}\right)\\ &+(1-x)^2(1-\bar{\alpha}_2)\left[\frac{1}{s^2(1-\alpha_4)^2}-\frac{1}{(1-\alpha_1)^2}\right]\Bigg)\end{aligned} \tag{14}$$

where $\bar{\alpha}_1 \triangleq (\alpha_3 + \alpha_4)/2$

$\bar{\alpha}_2 \triangleq (\alpha_1 + \alpha_4)/2$

Slip flow was assumed everywhere except at the vena contracta. The values of $\alpha_1$, $\alpha_2$, $\alpha_4$, and $\alpha_5$ were estimated by use of Hughmark's relationship between $x$ and $\alpha$. At the vena contracta, the flow was taken as homogeneous up to a void fraction of

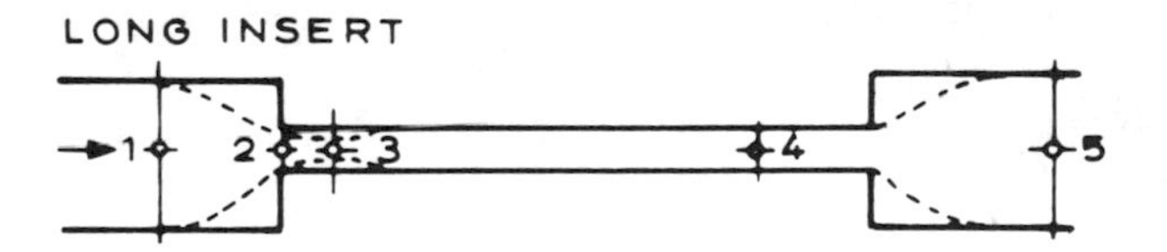


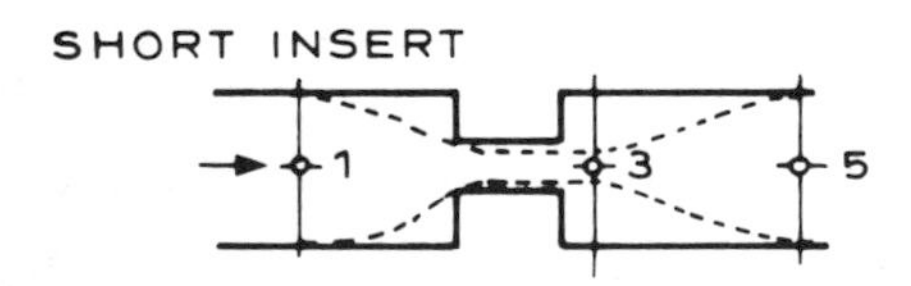


**Figure 3** Flow through contraction–expansion combinations.

0.5. At higher void fractions mixing is incomplete, and the void fraction is obtained from

$$\alpha_3 = \alpha_{\text{slip}} + B(\alpha_{\text{homog}} - \alpha_{\text{slip}}) \tag{15}$$

where $B \triangleq 1.5 - \alpha_4$.

Very short inserts behave differently from long inserts. When the vena contracta occurs outside the restriction, the pressure loss is given by

$$\Delta p = \frac{G_1^2}{2\rho_L s^2} \frac{1}{C_c^2} \left\{ \frac{\rho_L}{\rho_G} x^2 \bar{\alpha}_3 \left( \frac{1}{\alpha_3^2} - \frac{s^2 C_c^2}{\alpha_5^2} \right) + (1-x)^2 (1-\bar{\alpha}_3) \left[ \frac{1}{(1-\alpha_3)^2} - \frac{s^2 C_c^2}{(1-\alpha_5)^2} \right] \right.$$
$$\left. - 2sC_c \left[ \frac{\rho_L}{\rho_G} x^2 \left( \frac{1}{\alpha_3} - \frac{sC_c}{\alpha_5} \right) + (1-x)^2 \left( \frac{1}{1-\alpha_3} - \frac{sC_c}{1-\alpha_5} \right) \right] \right\} \tag{16}$$

where $\bar{\alpha}_3 \triangleq (\alpha_3 + \alpha_5)/2$

If $L/d \leqslant 0.5$, Eq. (16) may be used with the assumption of slip flow at all locations. The Hughmark correlation is recommended to obtain the relationship between $\alpha$ and $x$.

When $L/d > 0.5$, the pressure loss can be predicted by Eq. (16) for short inserts, providing the void fraction at the vena contracta is based on partial mixing.

## 4 ORIFICE

Watson et al. (1967) developed semiempirical correlations between quality, flow rate, and pressure drop across a sharp-edged orifice. These correlations are based on measurements on steam-water and air-water flows whose main parameters are given below.

| Parameter | Value |
|---|---|
| | Steam-water flows |
| Orifice diameter $d$ (mm) | 9.52, 12.70, 15.87, 25.4 |
| Pipe bore $D$ (mm) | 25.91, 50.80 |
| $d/D$ | 0.035, 0.063, 0.098, 0.24, 0.25, 0.378 |
| Orientation of pipe axis | Vertical |
| Store pressure (bar) | 7–13.7 |
| $x$ | 0.001–0.108 |
| | Air-water mixtures |
| Orifice diameter $d$ (mm) | 15.87, 25.4 |
| Pipe bore $D$ (mm) | 50.80 |
| Orientation of pipe axis | Vertical |
| Store pressure (bar) | 1.14–5 |
| $x$ | 0.002–0.540 |

The authors propose the following equation

$$\frac{M_{LS}}{M_L} - 1 = a(YF)^b \tag{17}$$

where $M_L$ is the mass flow rate component in two-phase flow,

$$M_{LS} \triangleq C_L \frac{\pi d^2}{4} \sqrt{2g\rho_L \, \Delta p} \tag{18}$$

is the mass flow rate in single-phase liquid flow with the two-phase pressure drop and normal liquid coefficient of discharge, and

$$Y \triangleq \frac{xF}{1-x} \sqrt{\frac{\rho_L}{\rho_G}}$$

where $F$ provides for expansion of the vapor phase. For steam, it is approximated by

$$F = \begin{cases} 1.70 - 0.70r & \text{if } 1 > r > 0.8 \\ 2.73 - 1.23r & \text{if } 0.8 > r > 0 \end{cases}$$

where $r$ is the ratio of the downstream and upstream pressures (absolute). Finally (in SI units)

$$a \triangleq 6.9d^{0.4} \qquad b = 0.91 - \log a$$

Beattie (1973) proposed a two-phase multiplier:

$$\Phi_{L0}^2 = \left[1 + x\left(\frac{\rho_L}{\rho_G} - 1\right)\right]^{0.8} \left[1 + x\left(\frac{\rho_L}{\rho_G}\frac{\mu_G}{\mu_L} - 1\right)\right]^{0.2} \tag{19}$$

## 5 BENDS, TEES, AND *Y* JUNCTIONS

Chisholm (1967) gave the following expression:

$$\frac{\Delta p}{\Delta p_{L0}} = (1-x)^2 \left(1 + \frac{C}{X} + \frac{1}{X^2}\right) \tag{20}$$

with $$C \triangleq \left[1 + (C_3 - 1)\left(\frac{\rho_L - \rho_V}{\rho_L}\right)^{0.5}\right] \left(\sqrt{\frac{\rho_L}{\rho_V}} + \sqrt{\frac{\rho_V}{\rho_L}}\right)$$

From Fitzsimmons' data on 90° bends, Chisholm deduced the values of $C_3$. They correspond to the curve in Fig. 4.

Formulas and data concerning pressure losses in branch pipes can be found in Kubo and Ueda (1973) and Collier (1977).

A recent paper by Kubie and Gardner (1978) presented an investigation of air–water flow through symmetrical *Y* junctions. The investigation was limited to equal flow in each limb approaching the junction with the air–water flow in the

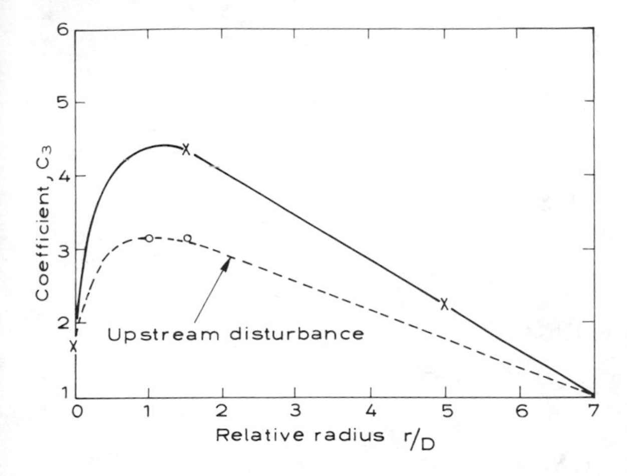


**Figure 4** Coefficient in Eq. (20) (Chisholm, 1967).

inclined limbs being stratified. It was shown that there are basically three flow regimes in the vertical downward flow following the junction:

1. At low water flow rates water falls freely down the center of the vertical outflow limb through air, entraining a small amount of air.
2. As the water flow rate increases, a critical flow rate is reached at which the two streams entering the junction mix so vigorously that air below the junction cannot directly connect with air above it. A two-phase homogeneous mixture of air and water forms in the vertical limb and flows down, occupying the whole cross-sectional area of the limb, and the amount of entrained air rises dramatically.
3. Finally, at some water flow rate the entrained air flow rate falls to a residual level and the water floods the junction and the sloping limbs.

Prediction of transitions and condition of flooding are given in the paper by Kubie and Gardner (1978).

## NOMENCLATURE

| | |
|---|---|
| $A$ | cross-sectional area |
| $a$ | coefficient [Eq. (17)] |
| $B$ | function [Eq. (15)] |
| $b$ | exponent [Eq. (17)] |
| $C$ | coefficient [Eq. (10)] |
| $C_c$ | contraction coefficient |
| $D$ | diameter |
| $d$ | restriction diameter |
| $F$ | function [Eq. (17)] |
| $G$ | mass velocity |
| $g$ | gravity constant |
| $L$ | length of channel |

$M$ mass flow rate
$p$ pressure
$r$ ratio of downstream and upstream pressures
$s$ area ratio
$w$ velocity
$X$ Martinelli parameter (Chap. 11, Sec. 3.1)
$x$ quality
$Y$ function [Eq. (17)]
$\alpha$ void fraction
$\bar{\alpha}$ average void fraction [Eq. (14)]
$\rho$ density
$\rho_m$ homogeneous density
$\tau_f$ energy dissipated by friction
$\Phi_{Lo}$ pressure drop ratio (Chap. 11, Sec. 3.2)

**Subscripts**

$I$ irreversible
$G$ gas phase
$L$ liquid phase
LS single-phase liquid flow
$R$ reversible
$V$ vapor phase

## REFERENCES

Beattie, D. R. H., A Note on the Calculation of Two-Phase Pressure Losses, *Nucl. Eng. Des.*, vol. 25, pp. 395–402, 1973.

Chisholm, D., Pressure Losses in Bends and Tees during Steam-Water Flow, NEL rept. 318, 1967.

Chisholm, D., Prediction of Pressure Losses at Changes of Sections, Bends and Throttling Devices, NEL rept. 388, 1968.

Chisholm, D., Theoretical Aspects of Pressure Changes at Changes of Section during Steam-Water Flow, NEL rept. 418, 1969.

Collier, J. G., Single-Phase and Two-Phase Flow Behaviour in Primary Circuit Components, in *Two-Phase Flows and Heat Transfer*, eds. S. Kakaç and F. Mayinger, vol. 1, pp. 313–355, Hemisphere, Washington, D.C., 1977.

Harshe, B., Husain, A., and Weisman, J., Two-Phase Pressure Drop across Restrictions and Other Abrupt Area Changes, NUREG-0062, 1976.

Hewitt, G. F. and Hall-Taylor, N. S., *Annular Two-phase Flow*, Pergamon Press, New York, 1970.

Kubie, J. and Gardner, G. C., Two-Phase Gas-Liquid Flow through Y-Junctions, *Chem. Eng. Sci.*, vol. 33, pp. 319–329, 1978.

Kubo, T. and Ueda, T., On the Characteristics of Confluent Flow of Gas-Liquid Mixtures in Headers, *Bull. JSME*, vol. 16, no. 99, pp. 1376–1384, 1973.

McGee, J. W., Two-Phase Flow through Abrupt Expansions and Contractions, Ph.D. thesis, Department of Chemical Engineering, University of North Carolina, 1966.

Petrick, M. and Swanson, B. S., Expansion and Contraction of an Air-Water Mixture in Vertical Flow, *AIChE J.*, vol. 5, no. 4, pp. 440–445, 1959.

Velasco, I., L'Ecoulement Diphasique à travers un Elargissement Brusque, Thèse de maitrise, Université Catholique de Louvain, 1975.

Watson, G. G., Vaughan, V. E., and McFarlane, M. W., Two-Phase Pressure Drop with a Sharp-edged Orifice, NEL rept. 290, 1967.

CHAPTER

# FOURTEEN

# BOILING HEAT TRANSFER EQUATIONS

Y. Y. Hsu

## 1 INTRODUCTION

During the postulated loss of coolant accident (LOCA) of light-water reactors, the reactor undergoes the blowdown, refill, and reflood stages, as shown typically in Fig. 1. To assess the safety margin of a reactor during such an accident, it is important to predict the peak clad temperature. Computer codes have been formulated to perform such a task. If the code is written for license evaluation, the assumptions used in the code must conform to licensing criteria and requirements. Such a code is called an evaluation model (EM). If the code is designed to predict, as close as possible, the real thermohydraulic conditions, the best available assumptions are used to estimate the peak clad temperature. Such a code is called the best estimate (BE) model.

Among the input to the thermohydraulic codes, heat transfer equations and switching logics between various heat transfer modes are of vital importance. The purpose of this section is to recommend, based on the current state of knowledge, a set of heat transfer equations and their related switching logics for BE codes.

The blowdown heat transfer will be discussed first, to be followed by the reflood stage. For each part, we will discuss the range of conditions, sensitivities, correlations to be recommended and their data bases, and special features of each region.

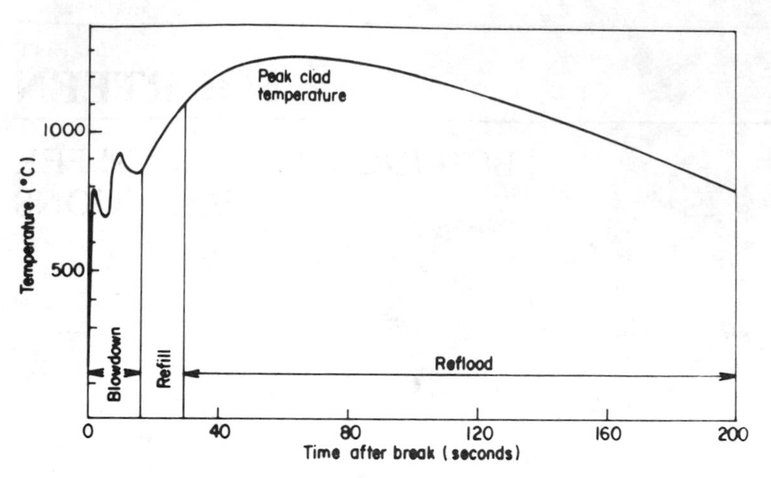


**Figure 1** Conservative estimate of cladding temperature for a double-ended cold-leg break (guillotine) LOCA.

# 2 BLOWDOWN HEAT TRANSFER

## 2.1 Scope

During the blowdown phase, the pressure of a pressurized-water reactor (PWR) vessel first drops precipitously to the vapor pressure corresponding to the normal operation temperature of the coolant. This is the subcooled depressurization period. The subsequent period is the saturated depressurization period, during which the depressurization rate is slower (see Fig. 2). At the same time, coolant rapidly leaves the core, sometimes going through flow reversal or re-reversal, then slows down to a very low rate. The flow rate curve is strongly dependent on the break location and

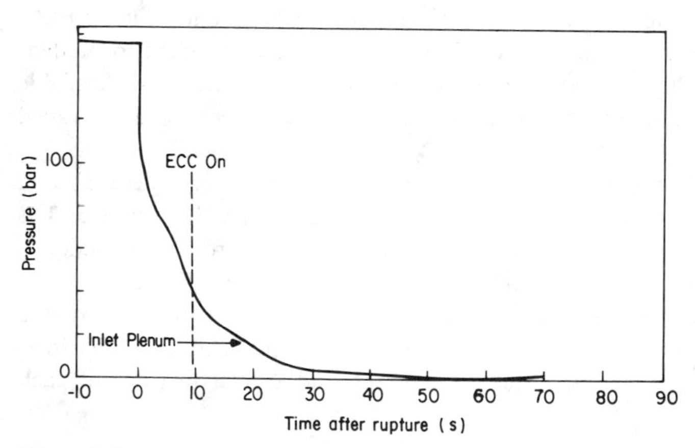


**Figure 2** Pressure history during blowdown of a PWR.

size. The most severe type of break is the double-ended, cold-leg break. A typical flow rate is shown in Fig. 3.

For boiling-water reactors (BWR), there is no subcooled depressurization period. Initial pressure is lower, but there is complication of the uncovering of the jet pump. The pressure and flow behavior are shown in Figs. 4 and 5.

A range of conditions exists during blowdown and the fuel-rod heat transfer undergoes the following stages:

1. Liquid convective cooling
2. Nucleate boiling
3. Onset of critical heat flux (CHF)
4. Post-CHF (transition boiling and film boiling)
5. Steam convective cooling

The prevailing mode is determined by the thermohydraulic conditions, including clad temperature, flow quality/void fraction, and flow rate, which are all functions of the history of the LOCA.

Since many different types of LOCA could be postulated, it is difficult to define precisely the ranges of conditions that could be expected through the whole LOCA history. An attempt was made by Richlen and Nelson (1976) to calculate the expected sequence of events and the attendent condition for double-end, cold-leg guillotine break of a typical PWR. A typical plot of the likely path of the hydraulic condition is shown in Fig. 6. Based on calculations, as illustrated in Fig. 6, and with

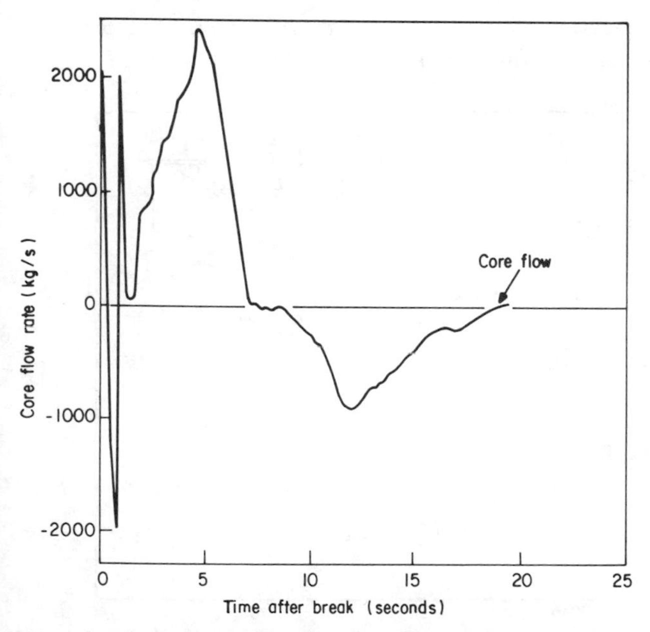


**Figure 3** Calculated core flow rate for a PWR LOCA.

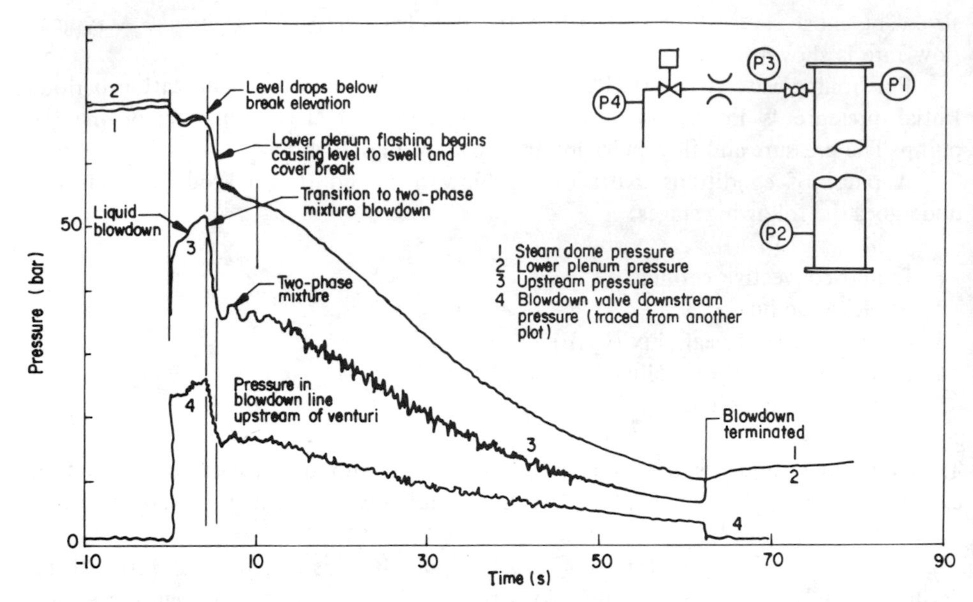


**Figure 4** Pressure histories during blowdown of a BWR bundle.

some engineering judgment, the likely range of conditions for each mode of heat transfer is shown in Table 1.

## 2.2 Sensitivities

The sensitivities of clad temperature to errors in heat transfer coefficient for each mode of cooling can be estimated as follows. The rise of clad superheat $\Delta T$ can be

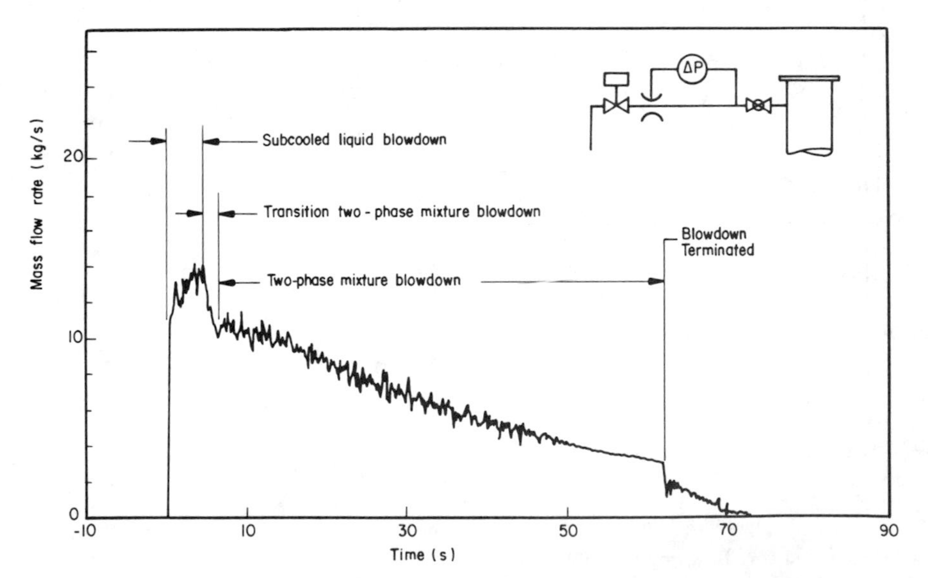


**Figure 5** Bundle flow rate for a BWR blowdown.

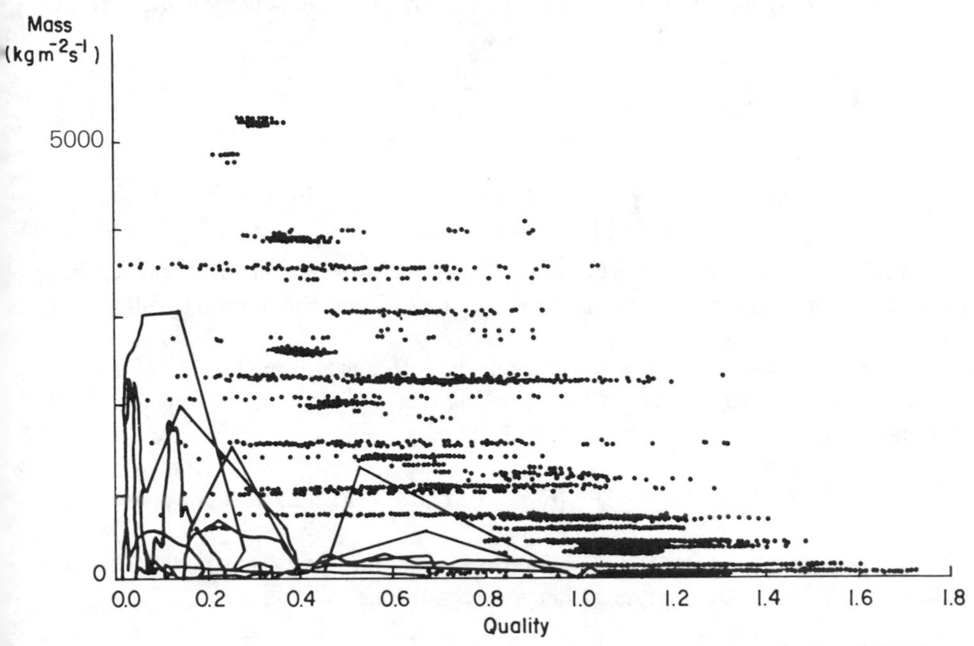


**Figure 6** Mass flux related to quality for an average channel model transient (Richlen and Nelson, 1976). Dots indicate data points for film boiling (tube), and lines indicate paths during blowdown conditions.

related to the heat-generating rate, $q'''$, and the cooling heat transfer coefficients by

$$C_p \pi r^2 \rho \, \Delta z \, \frac{d\,\Delta T}{dt} = q''' \pi r^2 \, \Delta z - h \, \Delta T 2\pi r \, \Delta z \tag{1}$$

or

$$\frac{d\,\Delta T}{dt} = \frac{q'''}{\rho C_p} - \frac{2h}{\rho C_p r} \Delta T \tag{2}$$

**Table 1 Range of conditions of interest to LOCA**

| Mode | $p$ (bar) | $G$ (kg/m$^2$ s) | $\alpha$ | $T_{clad} - T_{sat}$ (K) |
|---|---|---|---|---|
| Forced convection | 138–152 | 0–3400 | $\simeq 0$ | $< 6$ |
| Incipience of boiling | 138–48 | 0–3400 | $\simeq 0$ | 6–11 |
| Nucleate boiling | 138–34 | 0–5400 | 0–0.9 | 6–56 |
| CHF | 117–28 | 0–3400 | 0–1 | 33–56 |
| Transition boiling | 103–14 | 0–4100 | 0–0.5 | 56–280 |
| Film boiling: | | | | |
| Inverted-annular flow | 69–1.4 | 0–6800 | $< 0.4$ | 220–780 |
| Dispersed-droplet flow | 691.4 | 0–4100 | $> 0.4$ | 220–780 |

If $h = h_0 + \delta h$ where $\delta h$ is the error in $h$, the error in clad superheat $\delta\,\Delta T$ can be described as

$$\frac{d\delta\,\Delta T}{dt} = -\frac{2\delta h\,\Delta T}{\rho C_p r} \tag{3}$$

or

$$\Delta(\delta\,\Delta T)_n = -\frac{2}{\rho C_p r}\int_t^{t+\Delta t_n} \delta h\,\Delta T\,dt = -\frac{2}{\rho C_p r}(\delta h)_n\,(\Delta T)_n\,(\Delta t)_n$$

where subscript $n$ refers to the $n$th mode. Thus, the error in clad superheat resulting from the error in $h_n$ is proportional to $(\delta h)_n$, $(\Delta T)_n$, and the duration of the mode $(\Delta t)_n$.

If there is a switching criterion between the $m$th and $n$th modes, and if an error of $(\Delta t)_{mn}$ is expected, the effect is to prolong the $m$th mode at the expense of the $n$th mode. Thus the expected error in clad temperature is

$$\Delta(\delta\,\Delta T)_{mn} = -\frac{2}{\rho C_p r}\,[(\delta h)_m\,(\Delta T)_m - (\delta h)_n\,(\Delta T)_n]\,(\Delta t)_{nm} \tag{4}$$

The relative contribution to clad-temperature error owing to each mode of cooling is shown in Table 2, using a typical range of conditions.

## 2.3 Pre-CHF Heat Transfer Equations

In the following sections, proper equations to be used for each mode of heat transfer will be discussed.

### 2.3.1 Subcooled forced convection *Recommendation of correlation*

1. Dittus-Boelter equation (Dittus and Boelter, 1930)

$$\frac{Dh}{k_l} = 0.023\left(\frac{DG}{\mu_l}\right)^{0.8}\left(\frac{C_{pl}\mu_l}{k_l}\right)^{0.4} \quad \text{for Re} > 2000 \tag{5}$$

where properties are evaluated at bulk temperature.

2. Rohsenow-Choi equation (Rohsenow and Choi, 1961)

$$\frac{Dh}{k_l} = 4 \quad \text{for Re} < 2000 \tag{6}$$

***Data base and limitations in application*** The Dittus-Boelter equation has been widely used for 40 years and has been verified by single-phase, turbulent, forced convection data for many fluids. When the $T_w/T_c$ ratio is high (for example, greater than 3), a temperature ratio multiplier is included. For the blowdown test, the $T_w/T_c$ during the liquid-phase forced convection period is not far from 1. Thus its original form is applicable.

The Dittus-Boelter equation is limited to flows with Re > 2000. The Rohsenow-Choi equation, as shown here, is a compromise between the laminar forced

**Table 2 Relative sensitivity of clad temperature to error in heat transfer coefficient in various modes**

| Mode | $(\Delta T)_n$ (K) | $h_n$ (W/m² K) | $(\delta h)_n$ (W/m² K) | $(\Delta t)_n$ (s) | $(\Delta t)_{mn}$ (s) | $(\Delta T)_n (\delta h)_n (\Delta t)_n$ (J/m²) |
|---|---|---|---|---|---|---|
| Nucleate boiling | 8–33 | 17000–114000 | 3400–23000 | 0.1–0.5 | | 60 |
| CHF | | | | | 0.1–0.3 | 120 |
| Transition boiling | 55–280 | 1100–85000 | 570–28000 | 0.2–0.4 | | 180 |
| Film boiling: | | | | | | |
| Dispersed flow | 220–780 | 850–2800 | 170–850 | 0.5–10 | | 3000 |
| Inverted-annular flow | 220–780 | 57–570 | 11–70 | 30 | | 3000 |

convection equations for uniform heat flux and for uniform temperature, which were analytically derived. The Rohsenow-Choi equation is limited to laminar-flow forced convection with Re < 2000.

*Discussion* The Dittus-Boelter equation is well-supported by a broad data base, with error less than 20%. The laminar-flow convection equation is analytically derived in a region where analysis has been shown to be generally valid, and thus the laminar equation can be used with confidence.

**2.3.2 Nucleate boiling** ***Recommended correlation*** Chen's equation for nucleate boiling (Chen, 1966) reads

$$h = 0.00122 \frac{k_l^{0.79} C_{pl}^{0.45} \rho_l^{0.49}}{\sigma^{0.5} \mu_l^{0.29} H_{fg}^{0.24} \rho_g^{0.24}} (T_w - T_s)^{0.24} \, \Delta p^{0.75} S$$

$$+ 0.023 \left[\frac{DG(1-x)}{\mu_l}\right]^{0.8} \left(\frac{C_{pl}\mu_l}{k_l}\right)^{0.4} \frac{k_l}{D} F \tag{7}$$

Equation (7) contains two parts. The first part is the contribution by microscopic agitation by bubbles and is based on the Forster-Zuber equation (Forster and Zuber, 1955) with a suppression factor of $S$, to account for reduction of bubble nucleation activities when flow is high. The second part is the contribution by macroscopic forced convection and is based upon Dittus-Boelter equation with a modifying factor $F$. The $F$ and $S$ factors are shown in Figs. 7 and 8. One attractive feature of $F$ is its automatic transition to forced convection of steam when quality becomes high and flow is in the form of annular flow with thin evaporating liquid film on the wall. The competing correlations are that by Thom et al. (1965-1966) and Schrock and Grossman (1959): The Thom correlation is

$$q'' = 1.97 \times 10^{-3} (T_w - T_s)^2 e^{0.23p} \tag{8}$$

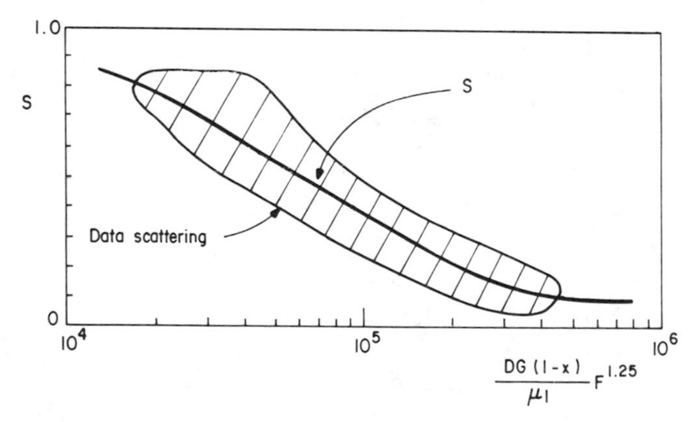


**Figure 7** Dimensionless function for nucleate boiling in Chen's correlation.

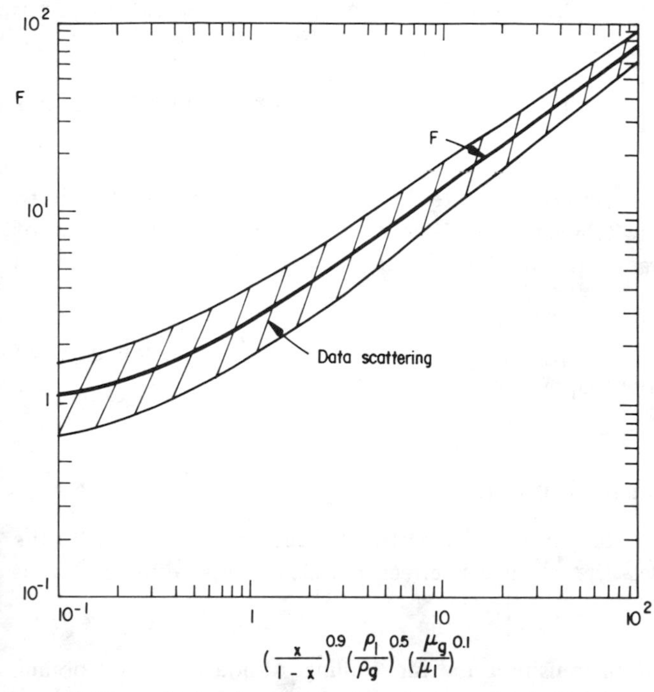


**Figure 8** Dimensionless function for forced convection in Chen's correlation.

where $p$ is in megapascals, $T$ in kelvins, and $q''$ in megawatts per square meter. The Schrock-Grossman correlation is

$$h = (2.5)(0.023)\frac{k_f}{D_e}\mathrm{Pr}_f^{0.4}\ [\mathrm{Re}_f\,(1-x)]^{0.8}\left[\left(\frac{x}{1-x}\right)^{0.9}\left(\frac{\mu_g}{\mu_f}\right)^{0.1}\left(\frac{\rho_f}{\rho_g}\right)^{0.5}\right]^{0.75} \tag{9}$$

***Range of data base and error*** For Chen's equation, the data base is

$$p = 0.09\text{-}3.45 \text{ MPa (original), and } 6.9 \text{ MPa (extended)}$$

$$G = (0.054\text{-}4.07) \times 10^3 \ \text{kg/m}^2\,\text{s}$$

$$x = 0\text{-}0.7$$

The original upper limit of the data base was 3.45 MPa but it has been tested to be valid at 6.90 MPa using General Electric data. The data base for Thom's equation is

$$p = 5.3\text{-}13.8 \text{ MPa}$$

$$G = (1.1\text{-}3.8) \times 10^3 \ \text{kg/m}^2\,\text{s}$$

and the data base for Schrock and Grossman is

$$p = 0.3\text{-}3.5 \text{ MPa}$$

$$G = (0.2\text{–}4.0) \times 10^3 \text{ kg/m}^2 \text{ s}$$

$$x = 0.03\text{–}0.50$$

The comparison of error of various correlations is shown in Table 3. Chen's equation appears to be the best.

***Discussion*** Thom's equation has no flow rate dependence and is not applicable to high quality. Therefore, a switchover from Thom's correlation to the correlation of Schrock and Grossman was required. The justification of recommending Chen's correlation is that

1. It covers a wide range of quality and makes a smooth transition between nucleate boiling and flow evaporation modes.
2. The error is small.

## 2.4 Post-CHF Heat Transfer Equations

As shown in the sensitivity comparison, the post-CHF heat transfer, especially the film boiling regime, causes the strongest effect on clad temperature and thus warrants special attention.

**2.4.1 Equations including both transition and film boiling** Although the two boiling regions can be treated separately, it is more convenient to combine them in the form

$$q'' = q_l''F_l + q_v''F_v \tag{10}$$

where $q_l''$ and $q_v''$ are contributions to heat flux by liquid and vapor, respectively, while $F_l$ and $F_v$ are respective weighing factors. Various authors use various expressions for $q_l''$, $F_l$, $q_v''$. Following are those used by Tong and Young (1974),

**Table 3 Comparison of nucleate boiling correlations (Chen, 1966)**

| | Average percentage deviations for correlations | | | | |
|---|---|---|---|---|---|
| Data | Dengler and Addoms | Guerrieri and Talty | Bennett et al. | Schrock and Grossman | Chen |
| Dengler and Addoms (water) | 30.5 | 62.3 | 20.0 | 20.3 | 14.7 |
| Schrock and Grossman (water) | 89.5 | 16.40 | 24.9 | 20.0 | 15.1 |
| Sani (water) | 26.9 | 70.3 | 26.5 | 48.6 | 8.5 |
| Bennett et al. (water) | 17.9 | 61.8 | 11.9 | 14.6 | 10.8 |
| Guerrieri and Talty (methanol) | 42.5 | 9.5 | 64.8 | 62.5 | 11.3 |
| Guerrieri and Talty (cyclohexane) | 39.8 | 11.1 | 65.9 | 50.7 | 13.6 |
| Guerrieri and Talty (benzene) | 65.1 | 8.6 | 56.4 | 40.1 | 6.3 |
| Guerrieri and Talty (heptane) | 61.2 | 12.3 | 58.0 | 31.8 | 11.0 |
| Guerrieri and Talty (pentane) | 66.6 | 9.4 | 59.2 | 35.8 | 11.9 |
| Combined average for all data | 38.1 | 52.6 | 32.6 | 31.7 | 11.0 |

Chen et al. (1977), and Condie and Bengston (given by Nelson, 1975). The data base available is shown in Table 4 (Groeneveld and Gardiner, 1977).

*Tong-Young equation*

$$q_l'' = q_{\mathrm{NB}}'' \tag{11}$$

$$F_l = \exp\left(-F_{\mathrm{TY}}\right) \tag{12}$$

$$F_{\mathrm{TY}} = 0.039 \frac{x_e^{2/3}}{dx_e/dz} (0.018\ \Delta T_{\mathrm{sat}})^{1 + 0.0029\ \Delta T_{\mathrm{sat}}} \tag{13}$$

where $z$ is in meters and $\Delta T_{\mathrm{sat}}$ is in kelvins.

Later $q_l''$ for the Tong-Young equation was modified by Hsu (1975) as

$$q_l'' F_l = q_{\mathrm{TB}}'' = q_{\mathrm{CHF,HB}}'' \exp\left(-F_{\mathrm{TY}}\right) \tag{14}$$

where $q_{\mathrm{CHF,HB}}''$ is given by Eq. (34).

The Tong-Young data base is from Harwell Atomic Energy Research Establishment (AERE)

$$p = 3.5\text{–}9.7 \text{ MPa}$$

$$G = (0.7\text{–}4) \times 10^3 \text{ kg/m}^2\text{ s}$$

$$x = 0.15\text{-}1.10$$

The comparison of the Tong-Young equation with data is shown in Fig. 9.

***Chen's equation*** Chen's equation is based on a large data base of tube data in the NRC data bank. The heat flux for liquid $q_l''$ is based on the phenomenological argument of intermittent contact and onset of nucleation boiling and departure. For the liquid contribution

$$q_l'' = q_{lc}'' \tag{15}$$

where

$$q_{lc}'' = \frac{\Phi_1 + \Phi_2}{t_1 + t_{12} + t_2} \tag{16}$$

and

$$F_l = \exp\left[-\lambda_l (1.8\ \Delta T_{\mathrm{sat}})^{1/2}\right] \tag{17}$$

where $\Phi_1$ and $\Phi_2$ are the heat quantities transferred during prenucleation and evaporation periods, respectively, and $t_1$, $t_{12}$, and $t_2$ are the time periods of prenucleation, bubble growth, and evaporation, respectively, and $\lambda_l$ is defined as

$$\lambda_l = f_l(G, \alpha)$$

The details can be found in Chen et al. (1977, p. 90 et seq.). However, the fraction of contact, $F_l$, was determined empirically. The comparison of the Chen equation with data is shown in Fig. 10.

The data base is

$$p = 0.7\text{–}13.8 \text{ MPa}$$

$$G = (0.07\text{-}5.2) \times 10^3 \text{ kg/m}^2\text{ s}$$

## Table 4 Transition boiling data of water in a forced convective system[a]

| Geometry | Reference | Range of data | | | | Comments |
|---|---|---|---|---|---|---|
| | | $P$ (MPa) | $G$ ($10^3$ kg/m² s) | $q''$ ($10^3$ kW/m²) | Subcooling (°C) or quality | |
| Annulus: 0.64 cm ID 6.35 cm OD | Ellion | 0.110–0.413 | 0.33–1.69 | 1.47–1.96 | 28–56°C | Heat flux-controlled system with stabilizing fluid, $L_H = 7.62$ cm |
| Tube: $D_e = 0.386$ cm | McDonough et al. | 5.51 8.27 13.78 | 0.27–2.04 | 0.32–3.78 | Subcooled and low quality | NaK used as heating fluid. $T_w$ inferred from heat transfer correlation for NaK. Data no longer available |
| Annulus: 0.013 cm ID 1.21 cm OD | Peterson | 0.101 | 0.64–1.93 | 0.41–1.99 | Saturated | Heat flux controlled by electronic feed-back, $L_H = 5.08$ cm |
| Tube: $D_e = 1.25$ cm | Plummer | 6.89 | 0.07–0.34 | 0.06–0.27 | $x = 0.30$–$1.00$ | Transient test $L_H = 10.16$ cm |
| Annulus: 1.37 cm ID 2.54 cm OD | Ramu and Weisman | 0.172–0.206 | 0.02–0.05 | 0.03–0.26 | $x = 0$–$0.50$ | Hg used as heating fluid, $x$ not reported. Limited range in $T_w$ |
| Rod bundles: $D_e = 1.27$ cm | Westinghouse FLECHT Cadeck et al. | 0.103–0.620 | 0.05–0.25 | 0.01–0.27 | 0–78°C | Transient test. $\Delta T_{sub}$ or $x$ unknown. |
| Tube: $D_e = 1.27$ cm | Cheng and Ng | 0.103 | 0.19 | 0.016–0.158 | 0–26°C | Transient and steady-state test, high inertia. Copper block, $L_H = 10.16$ cm |
| Tube: $D_e = 1.27$ cm | Fung | 0.101 | 0.068–1.35 | 0.008–1.89 | 0–76°C | Similar tests to Cheng and Ng |
| Tube: $D_e = 1.27$ cm | Newbold et al. | 0.303 | 0.016–1.25 | 0.016–0.948 | 0–80°C | Similar tests to Cheng and Ng, however, guard heaters were employed to reduce axial conduction |

[a]From Groeneveld and Gardiner (1977).

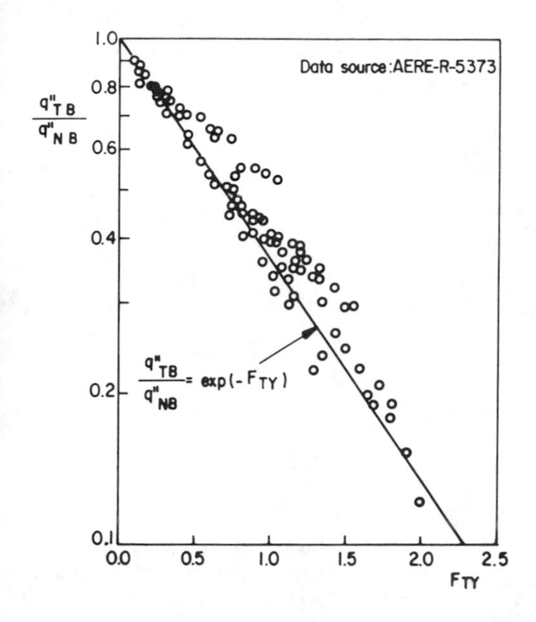


**Figure 9** Correlation for transition boiling heat flux decay as a function of wall superheat and equilibrium quality.

$$\Delta T = 90\text{–}540^\circ\text{C}$$

$$x = 0.2\text{–}1.0$$

***Condie-Bengston's equation (modified) for transition boiling***

$$q''_{TB} = q''_{\text{CHF,HB}} \exp(-0.9\ \Delta T_{\text{sat}}) \tag{18}$$

where $\Delta T_{\text{sat}}$ in K.

The Condie-Bengston equation (modified) for transition boiling is shown in Fig. 11 (Nelson, 1975). The data base of the Condie-Bengston equation is the same as that of Chen.

***Hsu's equation (Hsu, 1975)***

$$h_{\text{H}} = 1456p^{0.558} \exp(-0.003758p^{0.1733}\ \Delta T_{\text{sat}}) \tag{19}$$

where $p$ is in megapascals, $\Delta T_{\text{sat}}$ is in kelvins, and $h$ is in watts per square meter per kelvin. Equation (19) is based on FLECHT and Semiscale heat transfer data (13 runs, 200-300 points), where $p = 0.103\text{–}0.62$ MPa, the inlet velocity $< 25$ cm/s, and $\alpha < 0.5$.

A typical comparison of the Hsu equation with data is shown in Fig. 12.

***Log-linear interpolation method*** Finally, a simplified method of log-linear interpolation between the $q''_{\text{max}}$ and $q''_{\text{min}}$ points (at Leidenfrost temperature) was proposed by an MIT group (Bjornard, 1977). A comparison of various correlations with data is shown in Fig. 13 (Groeneveld and Gardiner, 1977).

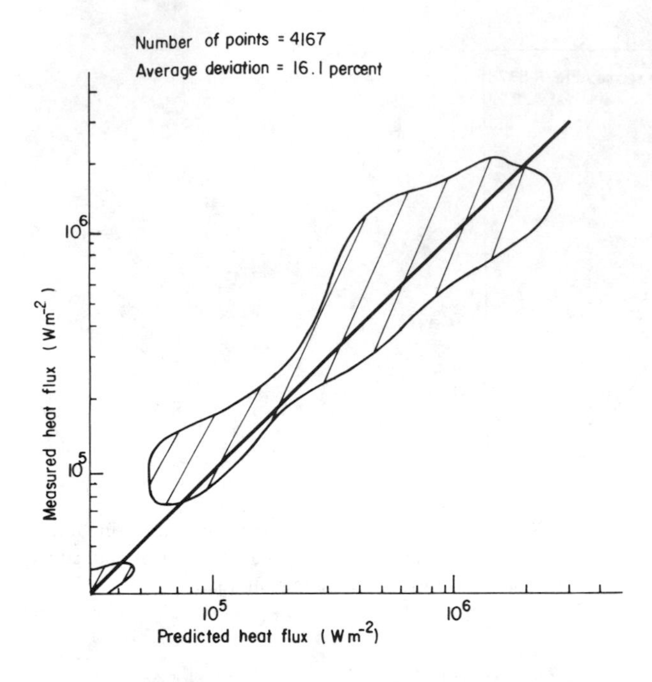


**Figure 10** Comparison of correlation with post-CHF data to pressure of 19.5 MPa (Chen et al., 1977).

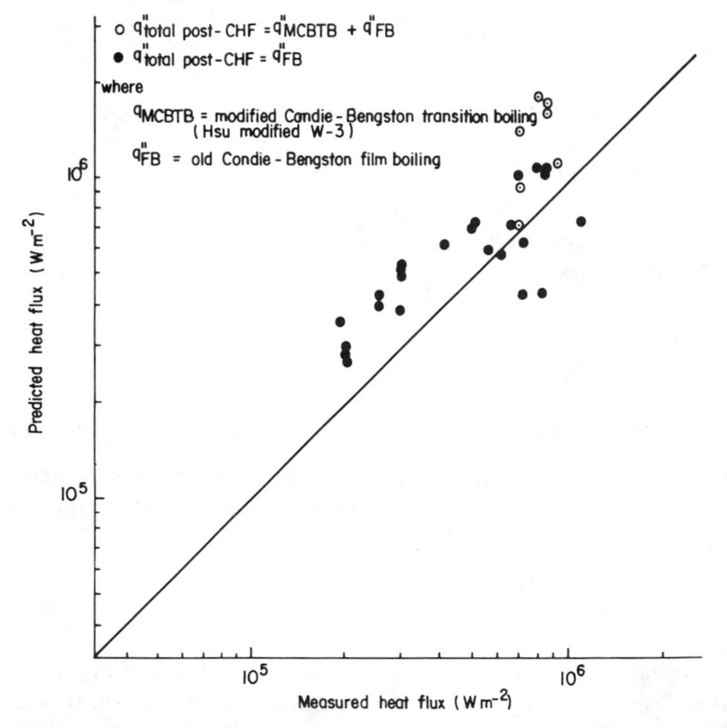


**Figure 11** Comparison of modified Condie-Bengston transition term and original Condie-Bengston film boiling to data in transition region.

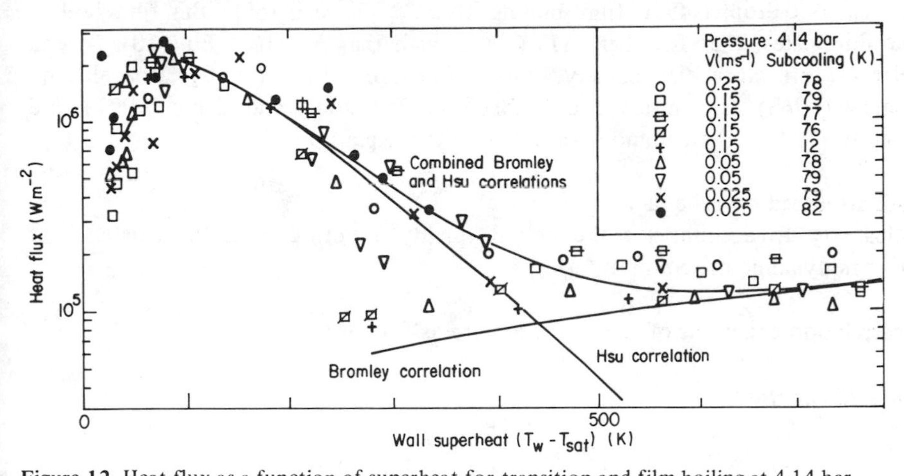


**Figure 12** Heat flux as a function of superheat for transition and film boiling at 4.14 bar.

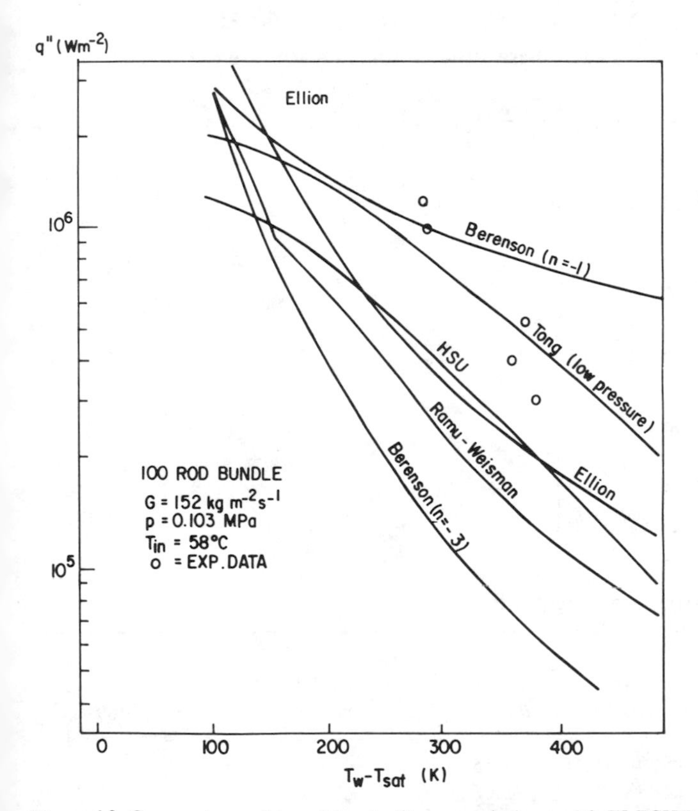


**Figure 13** Comparison of transition boiling correlations with FLECHT data (Groeneveld and Gardiner, 1977).

**2.4.2 Dispersed-droplets-flow film boiling** If $\Delta T_{sat}$ is high, only the film boiling regime dominates, that is, when $\Delta T_{sat}$ is so high that $F_l \to 0$ in Eq. (10). Several correlations are available, namely, those by Groeneveld (1969), Dougall and Rohsenow (1963), Groeneveld and Delorme (1976), Jones and Zuber (1977), and Chen et al. (1977). The common features of these equations are

1. All are based on tube data.
2. Most try to account for the actual quality resulting from the existence of thermodynamic nonequilibrium.

The correlation equations of these authors are as follows:

*Groeneveld equation*

$$h = a \frac{k_g}{D_e} \left\{ \mathrm{Re}_g \left[ x + \frac{\rho_g}{\rho_f}(1-x) \right] \right\}^b \mathrm{Pr}_{\mathrm{wall}}^c Y^d (q'')^e \left( \frac{D_h}{D_e} \right)^f \times \left( \frac{\mu_g}{\mu_{\mathrm{wall}}} \right)^g \left( \frac{\rho_f}{\rho_g} \right)^i \left[ \alpha + \frac{\rho_f}{\rho_g}(1-\alpha) \right]^j \tag{20}$$

where $Y$ (known as Miropolskiy's two-phase flow factor) and $\mathrm{Re}_g$ are defined as

$$Y \triangleq 1 - 0.1 \left( \frac{\rho_f}{\rho_g} - 1 \right)^{0.4} (1-x)^{0.4} \tag{21}$$

and the coefficients are shown in Table 5. $\mathrm{Re}_g \triangleq \dfrac{GD_h}{\mu_g}$

*Dougall-Rohsenow equation*

$$h = 0.23 \frac{k_g}{D} \left\{ \mathrm{Re}_g \left[ x + (1-x) \frac{\rho_g}{\rho_f} \right] \right\}^{0.8} \mathrm{Pr}_g^{0.4} \qquad \text{if } x > 1.0, \text{ set } x = 1.0 \tag{22}$$

*Groeneveld-Delorme equation*

$$\mathrm{Nu}_{gf} \triangleq \frac{hD}{k_{gf}} = 0.008348 \left\{ \frac{GD}{\mu_{gf}} \left[ x_{ac} + \frac{\rho_v}{\rho_l}(1 - x_{ac}) \right] \right\}^{0.8774} \mathrm{Pr}_{gf}^{0.6112} \tag{23}$$

**Table 5 Coefficients in Groeneveld's equation**

| Geometry | $a$ | $b$ | $c$ | $d$ | $e, f, g$ $i, j$ | No. of points | Rms error, % | Eq. no. in Groeneveld (1969) |
|---|---|---|---|---|---|---|---|---|
| Tubes | $1.09 \times 10^{-3}$ | 0.989 | 1.41 | −1.15 | 0 | 438 | 11.5 | 5.5 |
| Annuli | $5.20 \times 10^{-2}$ | 0.688 | 1.26 | −1.06 | 0 | 266 | 6.9 | 5.7 |
| Tubes and annuli | $3.27 \times 10^{-3}$ | 0.901 | 1.32 | −1.50 | 0 | 704 | 12.4 | 5.9 |

where subscript $l$ = saturated liquid
subscript $g$ = saturated vapor
subscripts $gf$ = vapor film temperature (average of wall and bulk vapor temperature)

$$x_{ac} = H_{fg} x_e / (H_{vac} - H_l)$$

$$\frac{H_{vac} - H_{ve}}{H_{fg}} = \exp(-\tan \Psi) \exp[-(3\alpha_{hom})^{-4}]$$

$$\Psi = 0.13864 \, \mathrm{Pr}^{0.2031} \, \mathrm{Re}_{hom}^{0.20006} \left( \frac{q'' D C_{pve}}{k_{ve} H_{fg}} \right)^{-0.09232} (1.3072 - 1.0833 x_e + 0.8455 x_e^2)$$

if $\Psi < 0$, $\Psi = 0$

if $\Psi > \frac{\pi}{2}$, $\Psi = \frac{\pi}{2}$

$$\mathrm{Re}_{hom} = \frac{GD}{\mu_{ve}} \left[ x + \left( \frac{\rho_v}{\rho_l} \right) (1 - x) \right]$$

***Jones-Zuber equation***

$$\frac{d(x_e - x)}{dx_e} + \mathrm{N}_{SR}(x_e - x) = 1$$

$$\mathrm{N}_{SR} = \begin{cases} 14.3 \mathrm{Bo} \left( \frac{S}{\sqrt{p_r}} \right)^2 & \text{if } \frac{S}{\sqrt{p_r}} < 0.22 \\ 1.23 \mathrm{Bo} \left( \frac{S}{\sqrt{p_r}} \right)^{3/8} & \text{if } \frac{S}{\sqrt{p_r}} > 0.22 \end{cases} \tag{24}$$

where $S$ is a superheat relaxation number based on computed local predictions. The actual quality enables the vapor temperature to be determined. Finally, the wall temperature is computed by the following heat transfer correlation:

$$\frac{q''}{(T_w - T_v)} \frac{D}{k_v} = 0.023 \left( \frac{GDx}{\mu_v \alpha} \right)^{0.8} \mathrm{Pr}_v^{0.4}$$

***Chen's film boiling equation***

$$q''_{vc} = h_{vc}(T_w - T_v)$$

$$h_{vc} = \frac{f}{2} G_T x_{ac} C_{pv} \, \mathrm{Pr}_v^{-2/3} \tag{25}$$

with the physical properties evaluated at the film temperature $(T_w + T_v)/2$.

$$f = 0.037 \left( \frac{D \rho_v \langle j \rangle}{\mu_v} \right)^{-0.17}$$

$$\langle j \rangle = G_T \left( \frac{x_{ac}}{\rho_v} + \frac{1 - x_{ac}}{\rho_l} \right) \tag{26}$$

The nonequilibrium quality and vapor temperature are obtained from

$$\frac{x_{ac}}{x_e} \equiv \frac{H_{fg}(p, T_s)}{H_v(p, T_v) - H_l(p, T_s)} = 1 - B(p)T_d \tag{27}$$

where

$$T_d \triangleq \frac{T_v - T_s}{T_w - T} \tag{28}$$

$$B(p) = \frac{0.26}{1.15 - p_r} \quad \text{for water from 4 to 200 bar}$$

The data base is same for all the above equations, that is, the NRC Data Bank with the following range of conditions (see Table 6) for:

| | |
|---|---|
| Geometry | Vertical tube |
| Flow | Upward |
| Experimental method | Heat flux controlled, uniform heat flux at the wall |

Equations (20) and (22)-(25) are all for dispersed droplet flow. The first two correlations used equilibrium quality while the last three use an empirically determined nonequilibrium quality. These three equations [Eqs. (23)-(25)] recognize the existence of nonequilibrium enthalpy distribution and thus conceptually give a closer description of true physical phenomena.

The comparison of each correlation with data is shown in Figs. 14-18. The error bands are shown in Table 7.

**2.4.3 Inverted annulus-flow film boiling** When the void fraction is low, the film boiling is of the inverted annulus type, that is, the low void core is separated from the heating surface by a thin layer of vapor film. For this flow region, several equations were proposed, including the classic Bromley equation for the vertical surface, the Ellion equation, and the Hsu-Westwater equation. It was shown by

**Table 6 Range of conditions for data in NRC data bank**

| System pressure (MPa) | Tube diameter (cm) | Mass flux (kg/m² s) | Equilibrium quality | Heat flux (MW/m²) |
|---|---|---|---|---|
| 0.42–10.44 | 1.27 | 40–680 | 0.675–1.728 | 0.1–0.66 |
| 6.89 | 1.26 | 380–5230 | 0.30–0.9 | 0.35–2.0 |
| 6.77–7.03 | 1.26 | 1110–1870 | 0.516–1.083 | 0.13–1.5 |
| 0.689 | 0.49 | 1080–3940 | 0.383–0.90 | 0.14–1.6 |
| 16.60–19.50 | 0.91–0.25 | 2030–3370 | 0.16–0.96 | 0.89–1.7 |
| 6.88–7.28 | 0.59 | 1100–3020 | 0.456–1.238 | 0.20–1.7 |
| 0.64–7.06 | 1.27 | 16.3–1020 | 0.392–1.634 | 0.03–1.0 |
| 13.99–19.48 | 1.00–2.00 | 690–3520 | 0.151–1.270 | 0.26–1.7 |

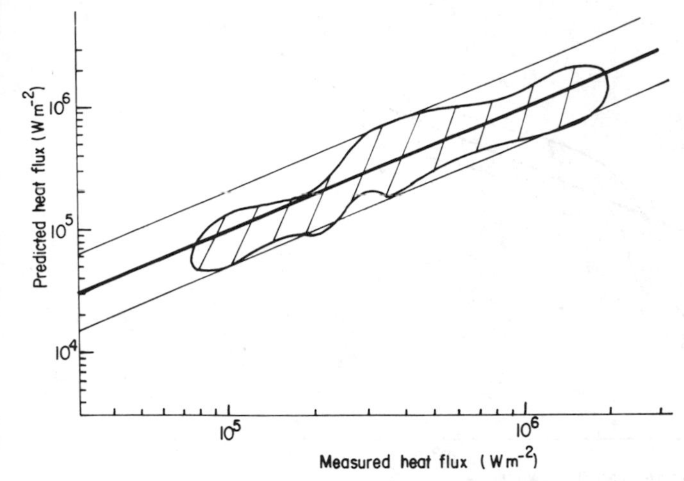


**Figure 14** Predicted heat flux compared with measured heat flux for Groeneveld (1969).

several tests (Hsu, 1975; Leonard et al., 1978) that the data can best be correlated by a modified Bromley equation for film boiling.

$$h_{\text{mod Bromley}} = 0.62 \left[ \frac{k_{vf}^3 \rho_{vf}(\rho_l - \rho_{vf}) H_{fg} g}{\lambda_c \mu_{vf}(T_w - T_s)} \right]^{1/4}$$

$$\lambda_c = 2\pi \left[ \frac{\sigma}{g(\rho_l - \rho_{vf})} \right]^{1/2} \tag{29}$$

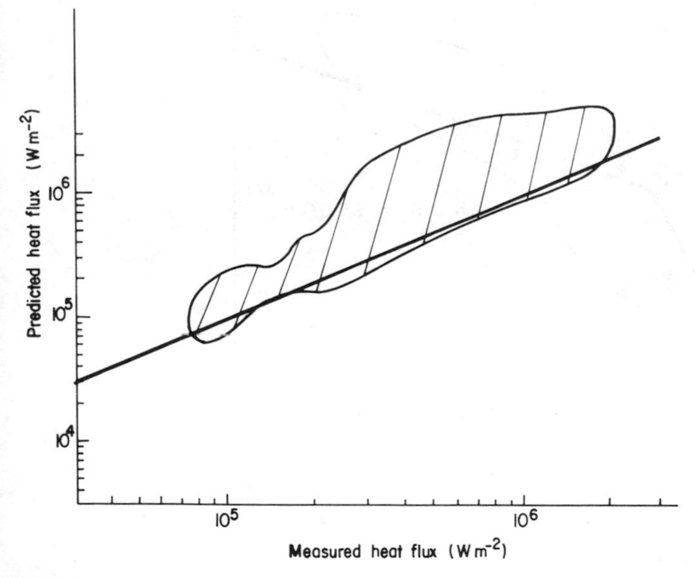


**Figure 15** Predicted heat flux compared with measured heat flux for Dougall-Rohsenow (1963).

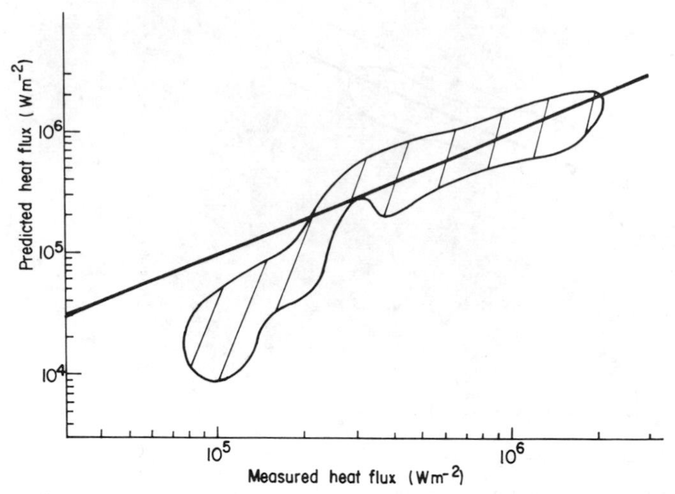


**Figure 16** Predicted heat flux compared with measured heat flux for Groeneveld and Delorme (1976) nonequilibrium correlation.

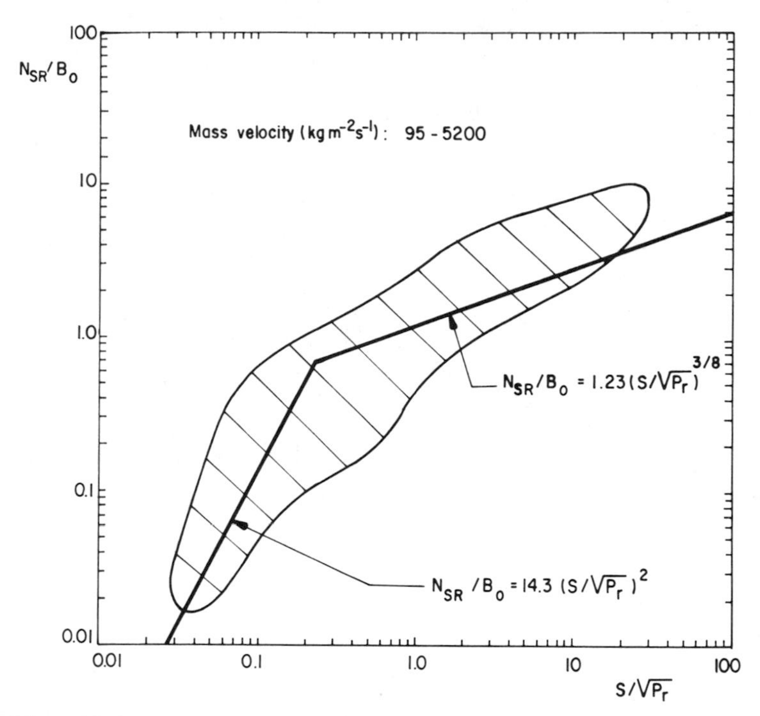


**Figure 17** Jones and Zuber (1977) correlation.

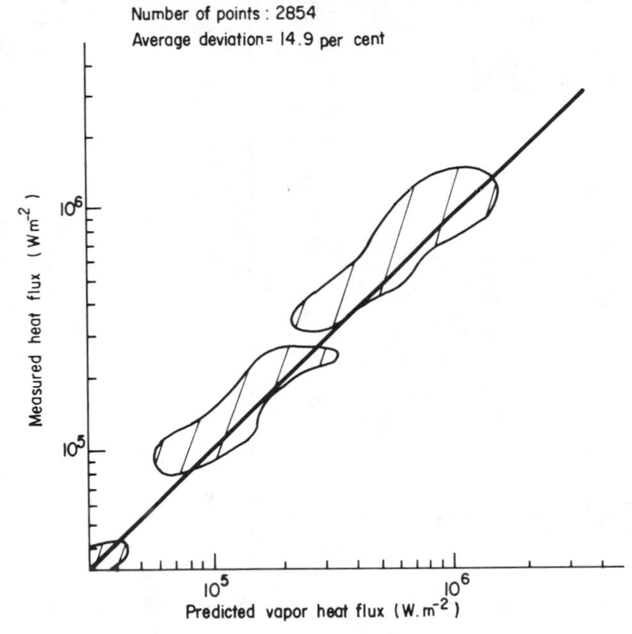


**Figure 18** Comparison of calculated vapor heat flux with measured total heat flux for flow film data (Chen et al., 1977).

where $\lambda_c$ is Taylor's wavelength. The comparison with data is shown in Figs. 12 and 19. The success of the modified Bromley equation to predict heat transfer of film boiling on a vertical rod is consistent with the findings of film boiling on large horizontal cylinders and large spheres (Breen and Westwater, 1962; Baumeister and Hamill, 1967), where the characteristic length is again represented by $\lambda_c$ when the vertical dimension is large.

It was recently shown by Andersen (given by Leonard et al., 1978) that an analysis based on the Helmholtz instability concept leads into a $\lambda_c$ equation as follows:

$$\lambda_c = 16.24 \left[ \frac{\sigma^4 H_{fg}^3 \mu_g^5}{\rho_g(\rho_f - \rho_g)^5 g^5 k_g^3 (T_w - T_s)^3} \right]^{1/11} \tag{30}$$

**Table 7 Error bands for film boiling correlations as compared with table data**

| Source | Upper limit, % | Lower limit, % |
|---|---|---|
| Groeneveld Eq. (5.7) (1969) | 100 | –50 |
| Dougall-Rohsenow (1963) | 200 | –10 |
| Groeneveld-Delorme (1976) | 50 | –60 |
| Chen et al. (1977) | 50 | –30 |
| Jones-Zuber (1977) | 150 | –50 |

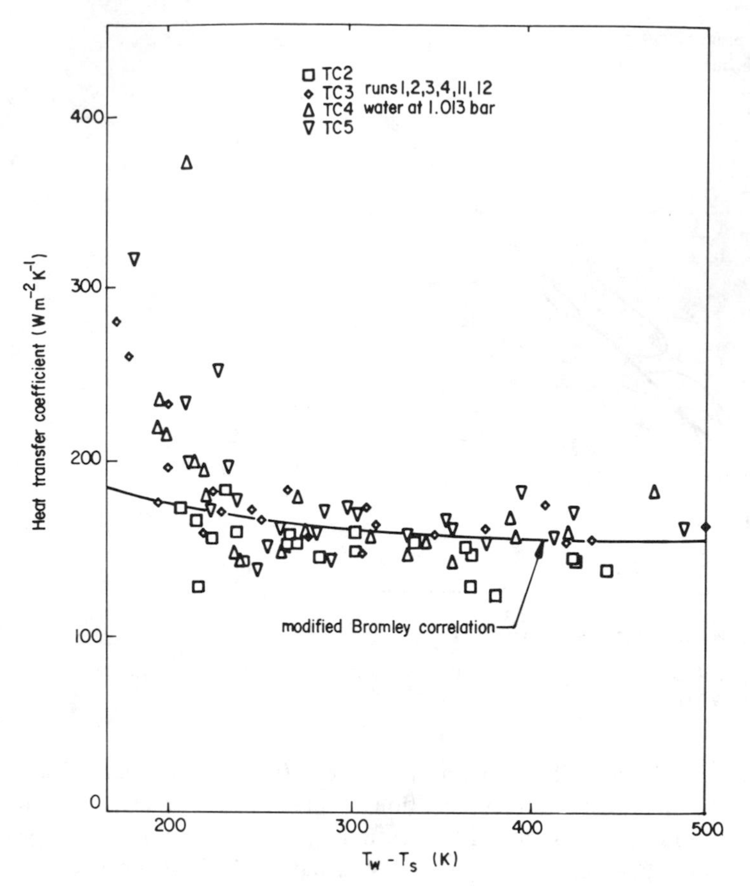


**Figure 19** Comparison of saturated film boiling data with modified Bromley correlation (NEDO-20566-1/REV. 1).

It is interesting to note that the Andersen equation and modified Bromley equation predict virtually the same heat transfer coefficient as shown in Fig. 20.

However, the modified Bromley equation is easier to use. The data base for modified Bromley equation is

$p = 0.1\text{-}0.7$ MPa
$q'' = 30\text{-}130$ kW/m$^2$
$\Delta T = 278\text{-}778$ K
Subcooling $< 77.9$ K
Velocity $< 0.3$ m/s
Void $< 0.4$

**2.4.4 Discussion** For inverted annular-flow film boiling, it is obvious that the modified Bromley equation should be recommended, although the Andersen equation can also be used.

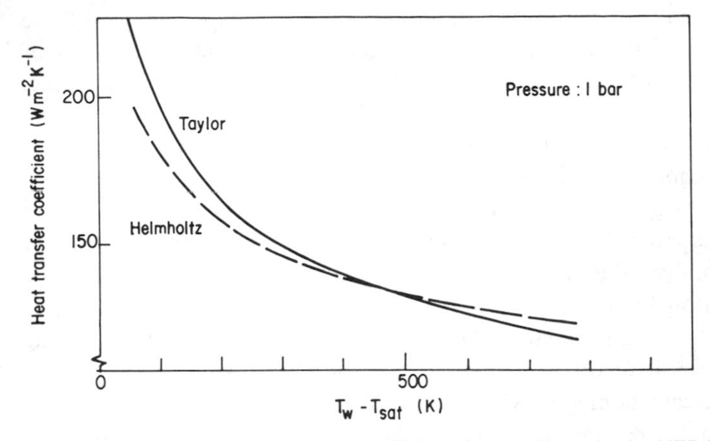


**Figure 20** A comparison of Taylor and Helmholtz stability criteria (NEDO-20566-1/REV. 1).

For dispersed-droplet-flow film boiling, the choice is not very simple. Most of the data in the film boiling base were obtained from the flow inside tubes, while reactor geometry is of the bundle type. Thus, the equations that give the best fit to large tube data does not necessarily mean the best fit for bundle geometry. The Groeneveld equation [Eq. (20)], the Chen nonequilibrium equation [Eq. (25)], and the Jones-Zuber equation [Eq. (24)] all seem to give satisfactory prediction for tube data and can all be used. The error bands of each equation are shown in Table 7.

The Dougall-Rohsenow equation, in general, overpredicts tube data. However, it is also found, from limited bundle data, that film boiling heat transfer from the bundle is usually higher than that from the tube. Thus, the Dougall-Rohsenow equation seems to predict bundle data better. This is shown in Fig. 21. The success of the Dougall-Rohsenow equation to predict bundle data may indicate that the

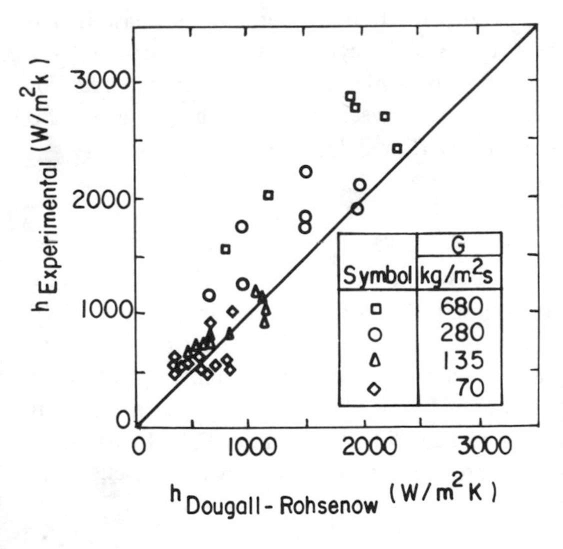


**Figure 21** Comparison of measured heat transfer coefficients during stable film boiling in rod bundles with the predictions of the Dougall-Rohsenow correlation.

quality for flow in bundle is closer to equilibrium value, perhaps because of the effect of the grid spacer. More bundle data are needed to verify the applicability of the Dougall-Rohsenow equation to film boiling from bundle. For the time being, the Dougall-Rohsenow is tentatively recommended for droplet-flow-dispersed film boiling in bundle. Such a choice is supported by recent blowdown heat transfer data of Oak Ridge National Laboratory (Craddick et al., 1978).

For transition boiling, the choice is less obvious. Although each transition boiling correlation is supported by its own data base, it was shown by Groeneveld that none is really universally satisfactory. Fortunately, during blowdown, the duration of the transition boiling mode is short; thus clad temperature is not highly sensitive to errors in transition boiling. On the other hand, a transition boiling correlation should be sensitive to the effect of mass flux and quality to allow the onset of return to nucleate boiling (RNB) when sufficient coolant is present. Under such a premise, the choice of transition boiling should be primarily based upon the capability of each correlation to account for the effect of flow rate, $\Delta T$, quality, and pressure. From such criteria, the log-linear approach should be ruled out first since the minimum point is always at Leidenfrost temperature, independent of flow rate and quality. Chen's correlation is not very sensitive to mass flux either.

Only Tong-Young's equation is sensitive enough to change of mass flux and quality that it will sponsor return to nucleate boiling when coolant flow is increased. Furthermore, the Tong-Young equation is based on data covering the blowdown range. With such considerations, the Tong-Young equation is tentatively recommended until more transition boiling bundle data become available and a better correlation can be formulated. It should be noted that $q_v''$ in the Tong-Young equation has not been shown to be as generally satisfactory as it is in the Groeneveld or Chen (for tube data) or Dougall-Rohsenow (for bundle data) equations. Thus, these equations [Eqs. (20), (22), and (25)] should still be used as $q_v''$. If quality is low so that flow pattern is of inverted annular type, modified Bromley's equation should be used.

The lower limit for the Tong-Young transition boiling-equation data base is 680 kg/m$^2$ s. In fact, if the equation were extrapolated beyond its data base to $G = 0$, the decay factor $F$ becomes 1, which is not reasonable. Thus, it is proposed that for $G$ less than 680 kg/m$^2$ s, $q_l$ be interpolated between $q_l$ of Tong-Young at 680 kg/m$^2$ s and that of Hsu, which is based on low flow data.

## 2.5 Criterion for Transient CHF

It has been known that there are two basic kinds of CHF during steady-state operation: the departure from nucleate boiling (DNB) type and the dryout type. DNB prevails in the low-quality flow and is determined by heat flux, while dryout prevails in high quality and is determined by critical quality (or void). The typical DNB type of CHF correlation is the *W*-3 equation of Westinghouse (Tong, 1967) while the dryout type is typified by the Harwell model (Fig. 22) (Whalley et al., 1973). In transient CHF during blowdown, both types can be present. This condition will be discussed in more detail in the following sections.

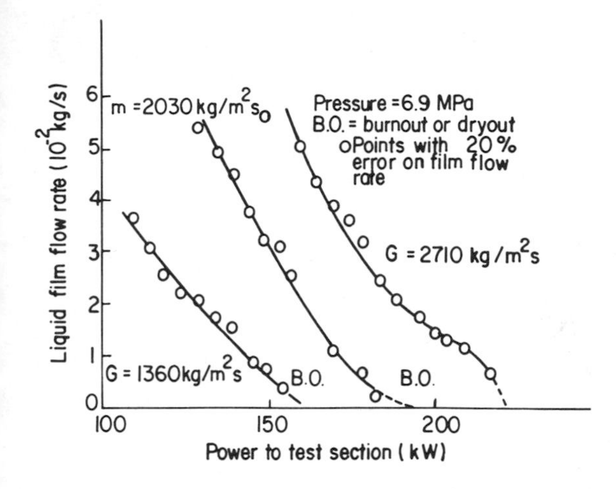


**Figure 22** Film flow rate data for evaporation of water in a vertical tube at high pressure.

**2.5.1 Transient CHF phenomenon** As stated previously, during blowdown flashing occurs, resulting in a large void. Thus although steady-state CHF correlations could still be used if the transient is not very fast and if local instantaneous flow parameters can be calculated, it is important to look at transient CHF in the light of the peculiar thermohydraulic behavior of blowdown. The mass flux and quality at various levels of a Semiscale test have been calculated by interpolating from measured quantities at the inlet and exit through code computation (Figs. 23 and 24) (Snider, 1977).

Note the high-quality region throughout the vessel after 0.7 s. Because of the predominance of the high-quality region in fast blowdown, it is unlikely that the DNB type of CHF can be applied. However, if the break is small so that flashing does not occur very fast and if the original subcooling is high, the DNB-type CHF should still be considered as a possibility, especially in peak power zone. But for large breaks with strong flashing, it is more likely to have dryout type of CHF.

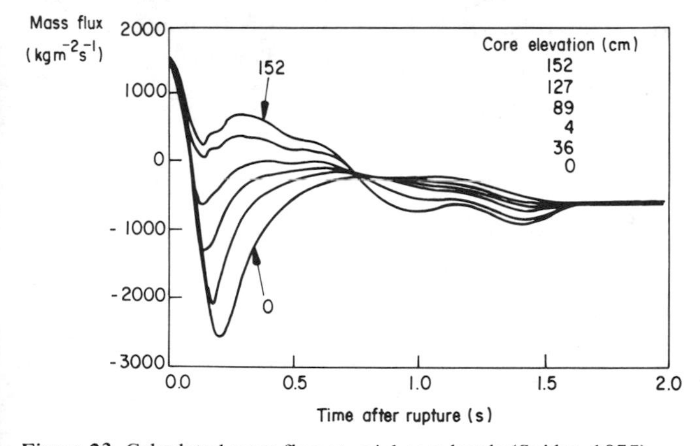


**Figure 23** Calculated mass flux at axial core levels (Snider, 1977).

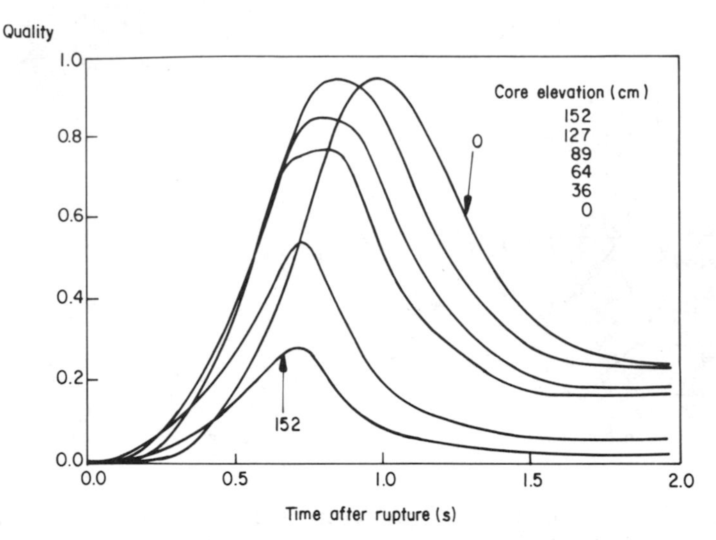


**Figure 24** Calculated quality at axial core levels (Snider, 1977).

As shown in Fig. 25, quality rises sharply within a 0.2 s span, which is within the uncertainty band of heat flux spread. Thus, if one uses critical quality or critical void as the criterion, the time to CHF is bound to be predicted within 0.2 s.

A third possible mechanism has been advanced (Henry and Leung, 1977), which postulated the formation of a vapor blanket in the region where there was no boiling during the steady-state operation prior to blowdown. In other words, it was postulated that instantaneous flashing occurs in the region where there were no preexisting nucleation sites, resulting in early CHF, while in the region where ebullition already is ongoing, flashing would not cause vapor blanketing. The theory is intriguing not so much in predicting the onset of early CHF but in precluding early CHF in the previous bubbling region. Experimental evidence of such a vapor-blanketing mechanism was proposed by Henry and Leung in Fig. 26. The general applicability of such a model is currently under thorough examination.

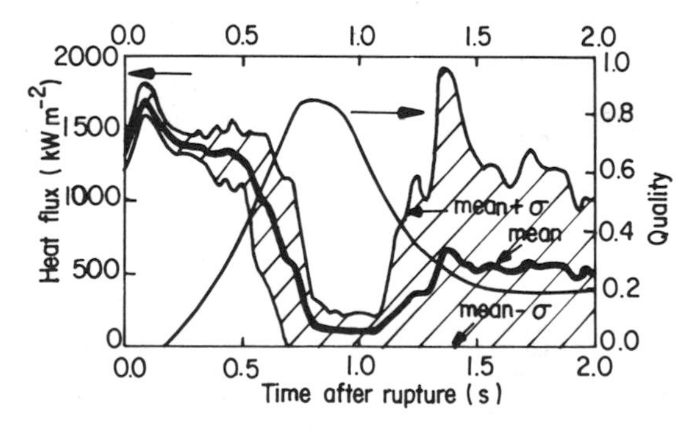


**Figure 25** Heat flux during blowdown and the quality calculated by COBRA-IV (Snider, 1977).

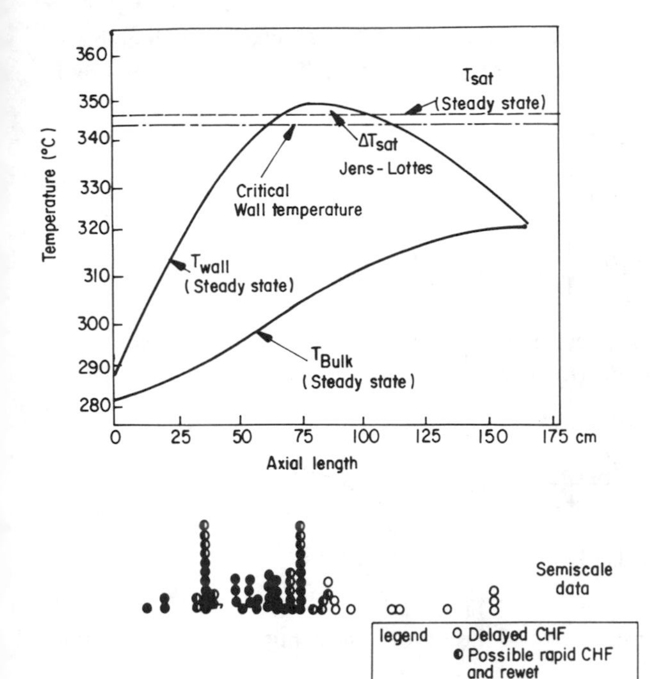


**Figure 26** Semiscale test S-02-9 at initial pressure 15.6 MPa (Henry and Leung, 1977).

**2.5.2 Correlations** ***DNB-type*** For the DNB type, for lack of experimental data, an existing steady-state DNB CHF equation such as *W*-3 (Tong, 1967) can be used with

$$q''_{W\text{-}3,\,x=0} = 3.013(2.1942 - 0.0766p)[1.037 + (1.095 \times 10^{-4}G)] \times [0.2664 + 0.8357 \exp(-124D_e)] \tag{31}$$

with $D_e$ being the equivalent hydraulic diameter in meters, $G$ in kilograms per square meter per second, $p$ in megapascals, and $q''$ in megawatts per square meter, with the understanding that steady-state correlation in general will underpredict experimental data.

***Dryout type*** For dryout transient CHF, the simplest is that proposed by Slifer and Hench (1971):

$$q'' = 3.15(0.84 - x) \tag{32}$$

with $q''$ in megawatts per square meter. As mentioned previously, any equation involving critical quality is bound to hit the experimental time of CHF closely if the transient is rapid. For a slow transient, more data of bundle CHF are needed before a detailed examination can be made. Henry's other model can also be considered as a special dryout type of CHF and warrants closer examination.

***DNB-dryout combination*** Since CHF can be either DNB or dryout, it is desirable to have some correlation to account for both mechanisms. Two correlations are in existence:

1. Zuber-Griffith equation (Fig. 27) (given in Smith and Griffith, 1976):

$$q''_{\mathrm{CHF}} = q''_{\mathrm{Zuber}}(1-\alpha) \tag{33}$$

where
$$q''_{\mathrm{Zuber}} = 0.13\rho_g H_{fg}\left[\frac{\sigma g(\rho_l-\rho_g)}{\rho_g^2}\right]^{1/4}\left(\frac{\rho_l}{\rho_l+\rho_g}\right)^{1/2}$$

which is applicable for low flow rate when pool boiling can be approximated.

2. Another correlation is the Hsu-Beckner equation (Hsu and Beckner, 1977). It is expressed in the form

$$\frac{q''_{\mathrm{CHF}} - q''_{\mathrm{steam}}}{q''_{W\text{-}3,x=0} - q''_{\mathrm{steam}}} = [1.76(0.96-\alpha)]^{0.5} \tag{34}$$

The interesting features of the Hsu-Beckner correlation are

a. It provides continuous transition from the dryout mechanism to the DNB-type mechanism.
b. It automatically reduces to steam cooling when $\alpha$ is high.
c. It indicates that at low void region, transient CHF is 30% higher than that predicted by the steady-state $W$-3 equation.

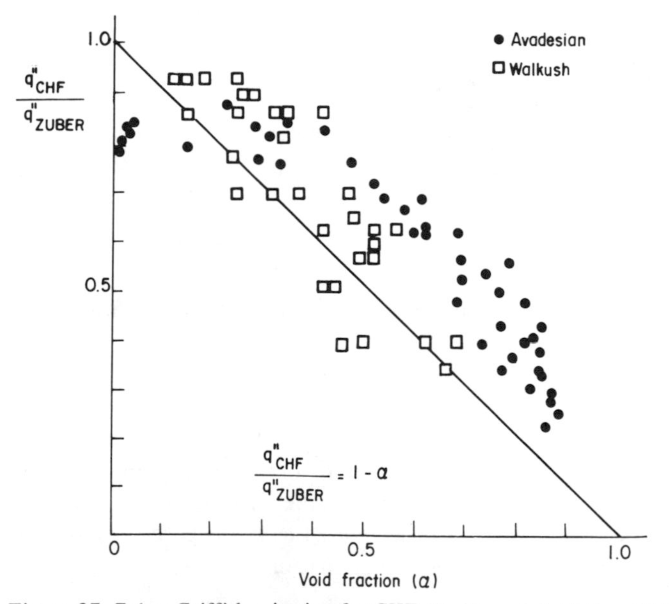


**Figure 27** Zuber-Griffith criterion for CHF (Smith and Griffith, 1976).

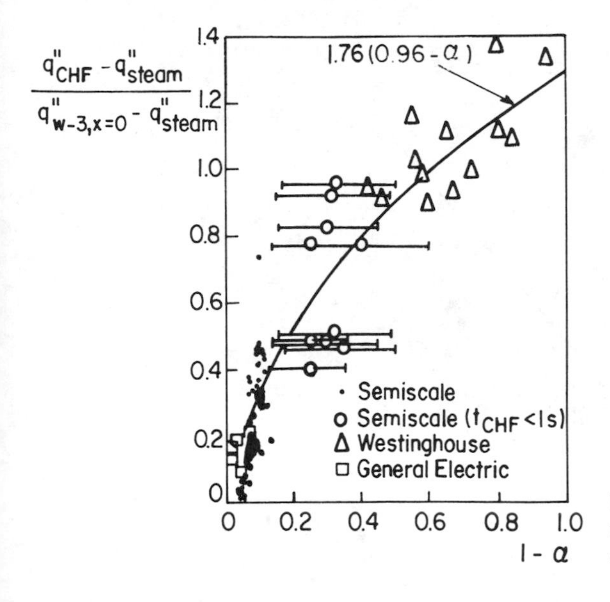


**Figure 28** Critical heat flux correlation of Hsu and Beckner for blowdown.

Comparison of the above equation with experimental data is shown in Figs. 28 and 29. As mentioned, due to the uncertainty of the heat flux and quality calculation, a spread of at least 0.2 s is expected.

**2.5.3 Discussion** The CHF mechanisms can be summarized in the following table.

| Type | Quality | Flow | Temperature | Heat flux |
|---|---|---|---|---|
| DNB | Low | High | $T > T_{sat}$ | $q > q_{DNB}$ |
| Dryout | $\alpha \to 1$ high | Low | Any $T$ | Any $q$ |
| Transient rapid CHF | Zero first then high | High then low | $T_w > 316°C$ | Any $q$ |

It should be noted that both the Zuber-Griffith and Hsu-Beckner equations are using void fraction, while the Slifer-Hence transient CHF equation uses quality as the criterion. Conceptually, the void fraction is a more logical choice since the quality definition breaks down when countercurrent flow occurs. At present, most operating codes only assume flow with no slip, thus quality does not pose a problem. In future advanced codes, when drift flux is used, use of quality in low flow or countercurrent flow will pose problems for the computer.

It should be further stressed that since most tests are of the rapid-pressure transient type, not enough data are available to correlate the DNB type of transient CHF.

A third problem in transients is the uncertainty of codes in calculating hydraulic parameters. Only one-dimensional, equilibrium, no-slip models are used so far. Thus, CHF correlation cannot be expected to be more accurate than the

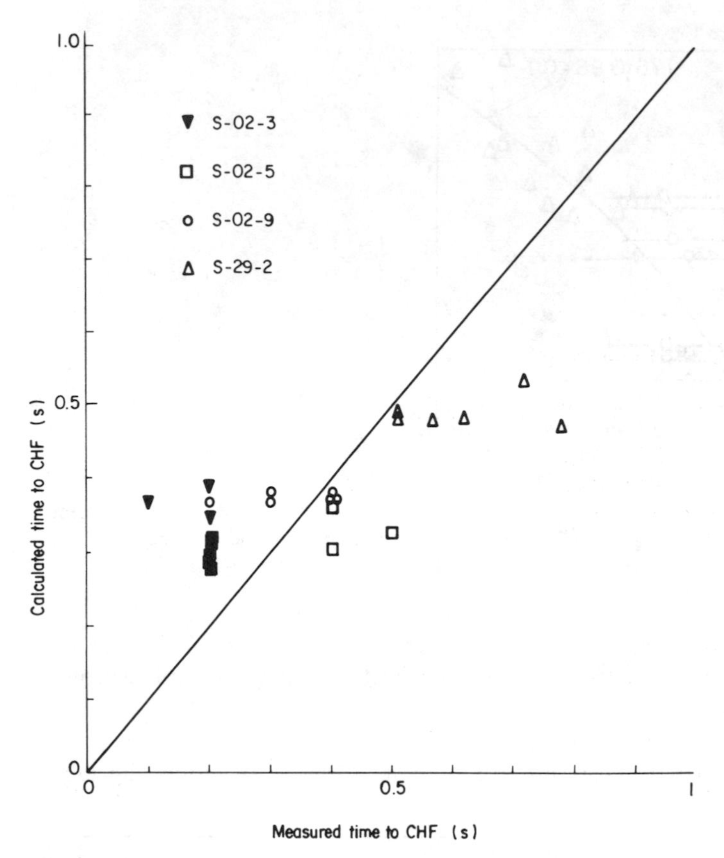


**Figure 29** Comparison of measured-to-calculated time to CHF for TCHF1 correlation for local conditions calculated by SCORE-EVET using compressed data. Shaded symbols represent the hot pins for those cases that had the four center pins at a higher power.

hydraulic data; it is hoped that with future two-dimensional, nonequilibrium, drift-flux models and new in-core measurements, hydraulic parameters can be calculated with more certainty, and thus improve CHF correlations.

# 3 REFLOOD HEAT TRANSFER

## 3.1 Introduction

**3.1.1 General description** The heat transfer modes for reflood are shown in Fig. 30 and relate to elevation $z$ as follows:

$z < Z_q$ = subcooled convection and nucleate boiling
$Z_q$ = quench front
$Z_q < z < Z_f$ = transition boiling and inverted-annulus film boiling

$Z_f$ = froth front
$Z_f < z$ = dispersed-flow film boiling

Note that $Z_f - Z_q$ decreases with flooding rate.

The advancing of the quench front can be determined from the rewetting package if coolant heat transfer coefficients and solid thermal properties are known. The advancing of the froth front can be determined from flood and entrainment rates. If $Z_q < Z_f$ a transition and an inverted-annulus film region exist. If $Z_q \geqslant Z_f$, the inverted-annulus film boiling regime does not exist, and the transition boiling zone is very short. If the entrainment is satisfied, the dispersed-flow film boiling region exists, and there are droplets in flow.

If significant entrainment is not achieved, the steam flow downstream of the froth front carries only a limited quantity of droplets.

Because of the integral nature of the reflood process, the thermohydraulic process in each zone is related to the upstream conditions and previous history. In the subsequent discussions the rewetting, the entrainment criteria, the heat transfer equations, and the void-fraction equations will be introduced for each zone, proceeding upward from the bottom of the core. The steam flow is integrated along the evaporating length.

**3.1.2 Range of parameters and sensitivity of clad temperatures to parameters** The ranges of parameters for each regime of heat transfer are shown in Table 8.

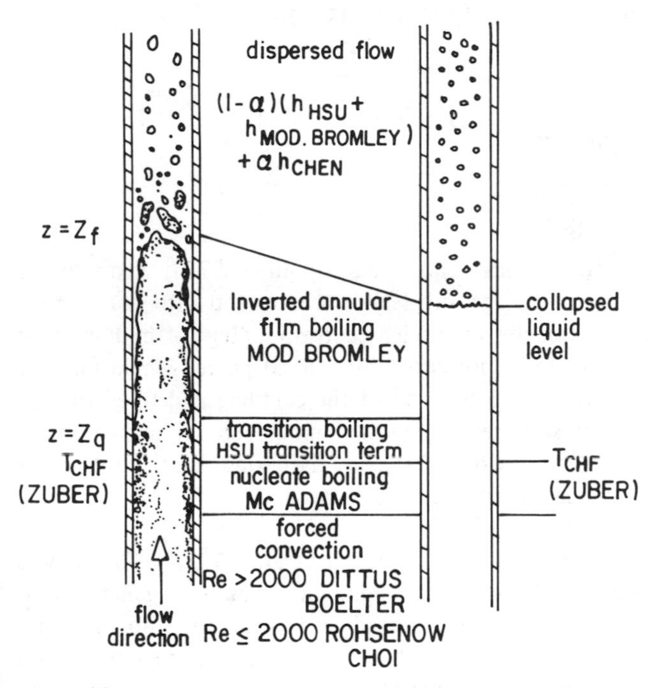


**Figure 30** Heat transfer regimes in emergency core cooling by bottom reflooding and predictive correlations.

**Table 8 Range of conditions for reflood heat transfer**

| | $p$ (bar) | $V$ (m/s) | $\alpha$ | $T_w - T_{sat}$ (°C) |
|---|---|---|---|---|
| Nucleate boiling | 1–6 | 0.02–0.3 | $\sim 0$ | $< 30/50$ |
| Transition boiling | 1–6 | 0.02–0.3 | $\sim 0$ | 30/50–140/200 |
| Inverted annulus-flow film boiling | 1–6 | $< 1$ | $< 0.3/0.4$ | 150/200–400/500 |
| Dispersed-flow film boiling | 1–6 | 7–25 | $> 0.3/0.4$ | $> 400/500$ |

Since each regime of reflood is closely related to the upstream and downstream zones, it is difficult to assign a particular sensitivity to clad temperature to a given regime. Nevertheless, the sensitivity of clad temperatures to each parameter can be roughly classified as follows:

1. Most sensitive parameters
   a. Input flood rate
   b. Distribution of enthalpy between sensible heat and latent heat
   c. Leidenfrost temperature (or slopes of the transition boiling regime)
2. Moderately sensitive parameters
   a. Criterion for froth front
   b. Quality and void in inverted annular film boiling zone
3. Less sensitive parameters
   a. Subcooling
   b. CHF point
   c. Nucleate boiling heat transfer

## 3.2 Recommended BE Package

**3.2.1 Rewetting at quench front** Elias and Yadigaroglu (1977), proposed a rewetting package (Fig. 31). This package divides the heater rod into many small segments, each subject to a heat transfer coefficient corresponding to the local heat transfer mode with each segment also represented by a local power-generation rate and a local gap resistance. The thermal properties of the cladding and filler are also included. This package is the most general rewetting model so far and properly accounts for the complicated interaction between the solid and coolant. A quench-front velocity can be calculated from this package.

**3.2.2 Entrainment criteria** From FLECHT data (Rosal et al., 1975, 1977), it was found that in the early stage of reflood, before enough steam was generated to entrain water, the steam became highly superheated as it proceeded downstream. However, at a certain point, a significant amount of water was entrained and steam superheat was reduced. Typical results are shown in Fig. 32.

Steam velocity near the froth front can be approximately calculated from steam

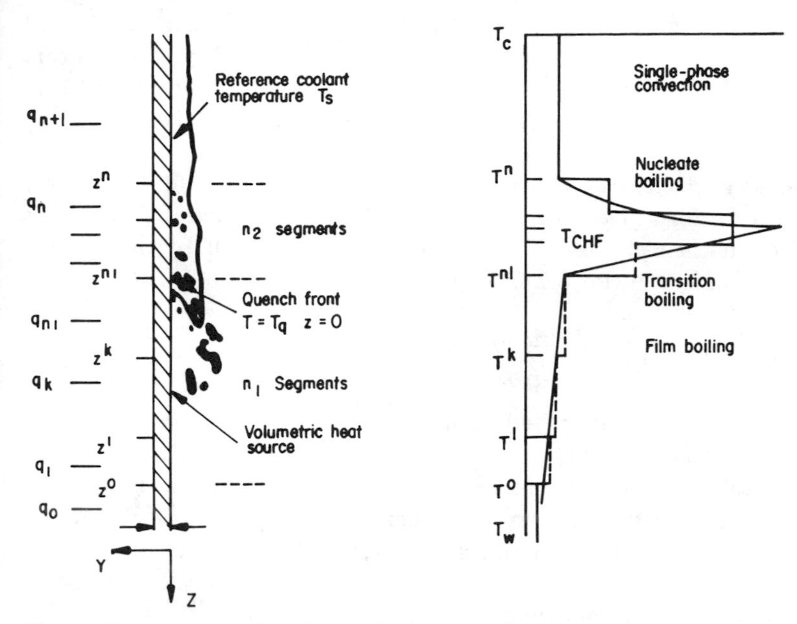


**Figure 31** Rewetting of a hot surface by a falling liquid film and the model of Elias and Yadigaroglu (1977).

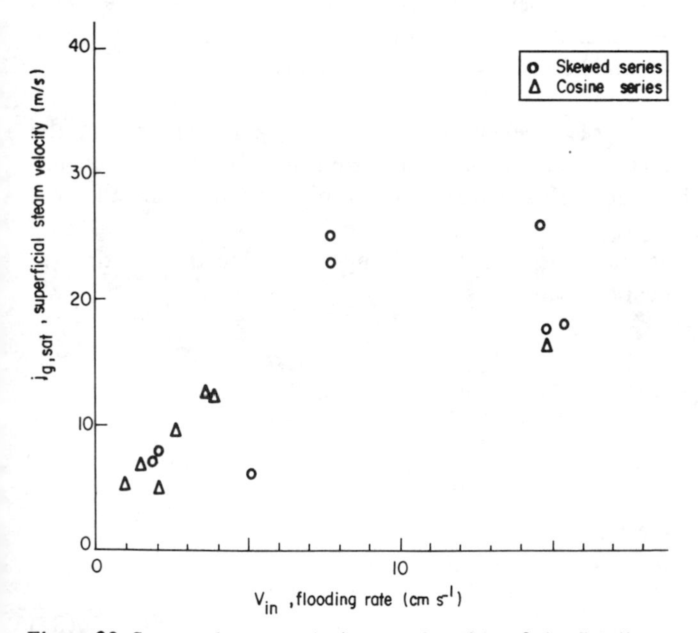


**Figure 32** Saturated steam velocity as a function of the flooding rate for FLECHT low-flooding-rate tests.

effluent-rate data and the saturated steam density (since steam is not superheated near froth front). The velocity of saturated steam at the time of maximum steam superheat downstream is given in Fig. 32. It is observed that the froth front steam velocity was in the range of 6–8 m/s when the downstream superheat began to drop.

The entrainment velocity can be determined by balancing the gravitational force against the drag force for a given drop size. FLECHT movies showed that the drops flying up in the subchannels have a diameter of about one-half of the channel width, or about 1.5 mm. Thus for 0.277 MPa, the velocity should be

$$V_g - V_f = \left(\frac{4gd}{3C_D}\frac{\rho_f - \rho_g}{\rho_g}\right)^{1/2} = 5.2 \text{ m/s} \tag{35}$$

with a drag coefficient $C_D = 0.45$ and a drop diameter $d = 1.5$ mm. This value compares well with the lower bound of the entrainment velocity.

Although in reality the droplet entrainment is a gradual process with drop size and drop population increasing with steam velocity, for simplicity, we will divide the downstream flow pattern into two kinds:

| | |
|---|---|
| If $V_{g,\text{sat}} < V_{\text{entrainment}}$ | No major entrainment occurs. |
| If $V_{g,\text{sat}} > V_{\text{entrainment}}$ | All the water above that elevation will be entrained. |

where $V_{g,\text{sat}}$ is determined from the enthalpy balance. This entrainment criterion differs from the critical Weber number criterion in that a drop size is given instead of being determined by a Weber number. The justification of using a definite size is provided by experimental evidence and is based on the argument that the presence of grid spacers will tend to filter the drops into a certain range of sizes.

**3.2.3 Thermohydraulics below the quench line** *Heat transfer* In this region, the wall temperature is already cooled to near saturation, and there is not much change in the sensible heat of the heater. Heat going to the coolant is practically the heat generated, i.e., $q'' = q''_{\text{generated}}$.

There is no need to be too concerned about proper selection of heat transfer equations. The following equations can be used:

1. Forced convection
   a. Dittus-Boelter equation (Dittus and Boelter, 1930):

$$\frac{Dh}{k_l} = 0.023\left(\frac{DG}{\mu_l}\right)^{0.8}\left(\frac{C_{pl}\mu_l}{k_l}\right)^{0.4} \quad \text{for Re} > 2000 \tag{36}$$

   where properties are evaluated at bulk temperature.
   b. Rohsenow-Choi equation (Rohsenow and Choi, 1961):

$$\frac{Dh}{k_l} = 4 \quad \text{for Re} < 2000 \tag{37}$$

   where the Reynolds number $DG/\mu_l$ is evaluated at bulk temperature.

2. Nucleate boiling is initiated when $T_w = T_{sat}$. The heat transfer is given by McAdams equation for nucleate boiling (McAdams, 1954)

$$q'' = 2.26(T_w - T_{sat})^{3.86} \qquad \text{or} \qquad h_{NB} = 2.26(T_w - T_{sat})^{2.86} \tag{38}$$

   where $T$ is in kelvins, $q''$ is in watts per square meter, and $h$ is in watts per square meter per kelvin.

***Void fraction*** Yeh's equation for void fraction has been shown to be valid for this region (Fig. 33) (Cunningham and Yeh, 1973):

$$\alpha(z) = 0.925 \left(\frac{\rho_g}{\rho_f}\right)^{0.239} \left[\frac{u_g(z)}{u_{bcr}}\right]^a \left[\frac{u_g(z)}{u_g(z) + u_l(z)}\right]^{0.6} \tag{39}$$

where $a = 0.67$ if $u_g(z)/u_{bcr} < 1$
$a = 0.47$ if $u_g(z)/u_{bcr} \geqslant 1$
$u_{bcr} = \frac{2}{3}\sqrt{gR_{bcr}}$
$R_{bcr} = (1.53/\frac{2}{3})^2 \sqrt{g\rho_l}$

**3.2.4 Quench front** The quench front is really a zone where the solid-liquid contact transits from 0% to some appreciable value, say, 70% (because of the presence of bubbles, solid-liquid contact is less than 100% even in nucleate boiling). This is the zone where transition boiling takes place. However, since the change from film boiling to transition is very gradual, it is difficult to give a clear demarcation line, and the Leidenfrost temperature is not well defined. Thus, for the sake of clarity, the quench front is defined as the regime where CHF is attained (that is, at $Z = Z_q$ with $q'' = q''_{\text{Zuber}}$)

$$q''_{\text{Zuber}} = 0.13\rho_g H_{fg} \left[\frac{\sigma g(\rho_l - \rho_g)}{\rho_g^2}\right]^{1/4} \left[\frac{\rho_l}{\rho_l + \rho_g}\right]^{1/2} \tag{40}$$

The $(\Delta T)_{\text{CHF}}$ is defined as the $\Delta T$ at which McAdams equation intersects $q''_{\text{Zuber}}$ (Zuber, 1958).

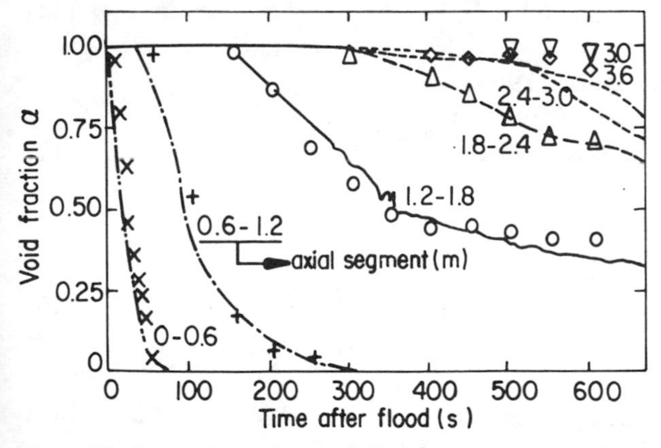


**Figure 33** Comparison of the calculated and the measured average void fraction for FLECHT low-flooding-rate test 02833. Dots indicate calculated values and lines indicate measured values.

**3.2.5 Inverted-annulus film boiling and transition boiling** This section deals with boiling only for a high flooding rate, where $Z_f > Z_q$. For a low flooding rate, see Sec. 3.2.7.

***Heat transfer*** The combination of Hsu's transition boiling and modified Bromley equations for film boiling can be used (Hsu, 1975):

$$h = h_{\mathrm{H}} + h_{\mathrm{mod\ Bromley}}$$

$$h = 1456p^{0.558} \exp(-0.003758p^{0.1733}\, \Delta T_{\mathrm{sat}}) + 0.62 \left[ \frac{k_{vf}^3 \rho_{vf}(\rho_l - \rho_{vf}) H_{fg} g}{\Delta T_{\mathrm{sat}} \mu_{vf}} \frac{1}{2\pi} \sqrt{\frac{g(\rho_l - \rho_g)}{\sigma}} \right]^{1/4} \tag{41}$$

In Eq. (41) all the variables are in SI units except $p$, which is in megapascals.

Comparison of Eq. (41) with experimental data of FLECHT and Semiscale is shown in Fig. 12. The quantity $h_{\mathrm{H}}$ was based on low void-fraction data. For high void-fraction cases, $h_{\mathrm{H}}$ tends to overpredict the data. A correction may have to be included if higher accuracy is sought. One possible form to use for the whole regime above quench front is

$$h = (1 - \alpha) h_{\mathrm{H}} + h_{\mathrm{mod\ Bromley}} + \alpha h_{\mathrm{steam}} \tag{42}$$

where $h_{\mathrm{steam}}$ is given by the Dittus-Boelter or Rohsenow-Choi equation.

However, Eq. (42) is still to be verified experimentally, and until it is, Eq. (41) should be used.

***Void fraction*** In this regime, the flow can be considered to consist of a bubbly core and a film boiling vapor annulus. The flow equations are:

1. Vapor annulus:

$$\delta = \frac{k_g}{h} \tag{43}$$

$$W_{va} = 2.1 r\, \delta^3 \frac{g \rho_g (\rho_l - \rho_g)}{\mu_g} \tag{44}$$

2. Bubble core (Zuber and Findlay, 1964):

$$\frac{u_{vs}}{\alpha_{\mathrm{core}}} = 1.2 u_{\mathrm{core}} + 1.53 \left( \frac{\sigma g\, \Delta\rho}{\rho_l^2} \right)^{1/4} \tag{45}$$

3. The total quality and void at $z$ are

$$x = \frac{1}{W_T} (u_{vs} \rho_v A_{\mathrm{core}} + W_{va}) = x(Z_q) + \frac{1}{W_T} \int_{Z_q}^{z} \frac{2\pi r q''}{H_{fg}}\, dz \tag{46}$$

$$\alpha = (\alpha_{\mathrm{core}} A_{\mathrm{core}} + 2\pi r\, \delta) \frac{1}{A_{\mathrm{cross\ section}}} \tag{47}$$

**3.2.6 Mass balance at transition between inverted-annulus region and dispersed-flow region ($z = Z_f$)** The froth front is the elevation where steam velocity attains entrainment velocity. The water is entrained in the steam at the rate

$$\dot{m}_{\text{entrained}} = \dot{m}_{\text{in}} - (V_{\text{steam}} \alpha \rho_{v,\text{sat}}) - \frac{d}{dt} \int_0^{Z_f} (1-\alpha)\rho_l \, dz \tag{48}$$

with $\alpha = 0.4$ (which is the void fraction of a cubic array of spheres),

$$V_{\text{steam}} = V + V_{\text{entrainment}} \tag{49}$$

$$V_{\text{entrainment}} = \left[ \frac{4d(\rho_l - \rho_v)g}{3C_d \rho_v} \right]^{1/2} \tag{50}$$

The entrained water is assumed to form droplets of diameter 1.5 mm traveling at a velocity of

$$V_d = \int_{Z_f}^{z} \frac{dV_d}{dz} \, dz \tag{51}$$

where

$$\frac{dV_d}{dz} = \frac{1}{V_d} \left[ \frac{3\rho_g (V_g - V_d)^2 C_D}{4\rho_l d} - \frac{g(\rho_l - \rho_g)}{\rho_l} \right]$$

$$\frac{dV_d}{dz} \approx \frac{1}{V_d} \frac{3\rho_g V_g^2 C_D}{4\rho_l d} \tag{52}$$

with $d = 1.5$ mm and $C_D = 0.45$. The drop population per unit volume is

$$n = \frac{6(1-\alpha)}{\pi d^2} \tag{53}$$

***Grid-spacer correction*** It is assumed that the drops do not break up into smaller drops. Instead the size just becomes smaller as a result of evaporation as the drops travel downstream. However, when the drops reach a grid spacer, they impinge on the grid, form streamers, and reemerge downstream of the grids as new drops of 1.5 mm diameter, and in fewer numbers. The drop velocity then starts from zero again:

$$V_d = \int_{z_{\text{grid}}}^{z} \frac{dV_d}{dz} \, dz \tag{54}$$

The result is more efficient cooling of the rod downstream since slip is higher when drops are moving slowly.

**3.2.7 Dispersed-flow film boiling** In this flow regime, the vapor phase is the continuous phase, with droplets dispersed in the flow. When the steam flow is low, the drops are small, the population density is low, and the contribution of drops to cooling is small. When the steam flow is high, the drops are entrained, and the role of cooling by droplets cannot be neglected.

***Heat transfer*** For low steam rates, heat transfer will be activated by forced convective cooling of steam. However, for a high steam rate, significant entrainment takes place, and Kirchner's REFLUX program should be used (Kirchner, 1976). Thus, the recommended heat transfer equations are

For $V_{\text{steam}} < V_{\text{entrainment}}$

$$q'' = h_{\text{steam}}(T_w - T_g) \tag{55}$$

where $T_g$ is the bulk vapor temperature.

$$h_{\text{steam}} = \max \begin{pmatrix} h_{\text{DB}} \\ h_{\text{RC}} \end{pmatrix} \tag{56}$$

where both $h_{\text{DB}}$ and $h_{\text{RC}}$ are evaluated with steam properties at saturation temperature.

For $V_{\text{steam}} > V_{\text{entrainment}}$

$$q''_{\text{wall}} = h_{\text{steam}}(T_w - T_g) + q''_{\text{radiation, wall-to-drop}} + q''_{\text{radiation, wall-to-vapor}} \tag{57}$$

with

$$q''_{\text{radiation, wall-to-drop}} = F_{w\delta}\,\sigma(T_w^4 - T_{\text{sat}}^4) \tag{58}$$

$$q''_{\text{radiation, wall-to-vapor}} = F_{wg}\,\sigma(T_w^4 - T_g^4) \tag{59}$$

where $\sigma$ is the Stefan-Boltzmann constant. The heat transfer to drop is given by

$$q''_{\text{vapor-to-drop}} = \frac{k_g}{d}\,(2 + 0.74\ \text{Re}_\delta^{0.5}\ \text{Pr}^{0.33})(T_g - T_{\text{sat}}) \tag{60}$$

where

$$\text{Re}_\delta \triangleq \frac{\rho_g V_g d}{\mu_g} \tag{61}$$

$$q''_{\text{radiation, vapor-to-drop}} = F_{g\delta}\,\sigma(T_w^4 - T_{\text{sat}}^4) \tag{62}$$

where $F_{w\delta}$, $F_{wg}$ and $F_{g\delta}$ are the gray-body factors (Sun et al., 1975).

***Void fraction and steam temperature*** For the change of drop size a modification of the drop diameter gradient for dispersed flow suggested by Groeneveld yields

$$\frac{d}{dz}d = -\left[\frac{2q''_{\text{vapor-to-drop}}}{(H_g - H_{f\,\text{sat}})\rho_f V_l} + \frac{4q''_{\text{wall-to-all-drops}}}{3(H_g - H_{f\,\text{sat}})\rho_f(1-\alpha)V_l}\,\frac{d}{D_h}\right] \tag{63}$$

The corresponding change of void fraction is

$$\frac{d\alpha}{dz} = -\frac{3(1-\alpha)}{d}\,\frac{d}{dz}d \tag{64}$$

The change of steam temperature is

$$C_{pg} \frac{dT_g}{dz} = \frac{4}{D_h V_{\text{steam}} \rho_g} q''_{\text{wall}} + \frac{3(1-\alpha)}{d} \frac{\rho_l}{\rho_g} H_{fg} \frac{d}{dz} d \tag{65}$$

## 3.3 Concluding Remarks

There are three areas in reflood for which definitive answers are not yet available, namely,

1. The criterion for transition from the inverted-annular region to the dispersed region
2. The proper partition of enthalpy between latent heat and sensible heat in dispersed droplet flow
3. The determination of Leidenfrost temperature where rewetting initiates

The best estimate package proposed in this section recommends some methods of calculating the important thermohydraulic parameters for reflood heat transfer, but they will have to be verified later by bench experimental results.

## NOMENCLATURE

| | |
|---|---|
| $a$ | constant |
| $B$ | parameter used by Chen et al. (1977) |
| Bo | Boussinesq number |
| $C_D$ | drag coefficient |
| $C_p$ | specific heat at constant pressure |
| $D$ | diameter of pipe or rod |
| $D_e$ | equivalent hydraulic diameter |
| $D_h$ | equivalent heated diameter |
| $d$ | drop diameter |
| $F$ | factor used by Chen (1966) |
| $F_L, F_V$ | weighing factors for liquid and vapor contributions |
| $F_W$ | radiation viewing factor |
| $f$ | function used by Chen et al. (1977); friction factor |
| $G$ | mass flux |
| $g$ | gravitational acceleration |
| $h$ | heat transfer coefficient |
| $H$ | enthalpy |
| $H_{fg}$ | heat of evaporation |
| $\langle j \rangle$ | mixture volumetric flux |
| $k$ | thermal conductivity |
| $\dot{m}$ | mass flux |
| Nu | Nusselt number |
| $N_{SR}$ | superheat relaxation number |
| Pr | Prandtl number |

| | |
|---|---|
| $\mathrm{Pr}_g$ | vapor Prandtl number calculated at the saturation temperature |
| $\mathrm{Pr}_{\mathrm{wall}}$ | vapor Prandtl number calculated at the wall temperature |
| $p$ | pressure |
| $p_r$ | reduced pressure |
| $q$ | heat flux |
| $q''$ | heat flux density |
| $q'''$ | heat-generating rate |
| $R$ | radius |
| Re | Reynolds number |
| $r$ | radius |
| $S$ | factor used by Chen (1966); parameter used by Jones and Zuber (1977) |
| $T$ | temperature |
| $T_d$ | temperature ratio indicating departure from equilibrium |
| $t$ | time |
| $u$ | superficial velocity |
| $V$ | velocity |
| $W$ | mass flow rate |
| $x$ | quality |
| $Y$ | Miropolskiy's factor [Eq. (21)] |
| $z$ | axial coordinate |
| $Z_q$ | quench front |
| $Z_f$ | froth front |
| $\alpha$ | void fraction |
| $\Delta p$ | difference in vapor pressure corresponding to $T_w - T_s$ |
| $\Delta T_{\mathrm{sat}}$ | $T_w - T_s$ |
| $\delta$ | film thickness |
| $\lambda$ | wavelength |
| $\lambda_L$ | function used by Chen et al. (1977) |
| $\mu$ | viscosity |
| $\mu_{\mathrm{wall}}$ | vapor viscosity evaluated at the wall temperature |
| $\rho$ | density |
| $\sigma$ | surface tension; Stefan-Boltzmann constant |
| $\Phi$ | heat transfer per unit area |
| $\Psi$ | parameter used by Groeneveld and Delorme (1976) |

**Subscripts**

| | |
|---|---|
| ac | actual |
| $bcr$ | bubble critical radius |
| $c$ | critical |
| CHF | critical heat flux |
| DB | Dittus-Boelter |
| $d$ | droplet |
| $e$ | equilibrium |
| $f$ | froth, film, liquid |
| $g$ | gas |
| H | Hsu |

| | |
|---|---|
| HB | Hsu-Beckner |
| hom | homogeneous |
| in | inlet |
| $l$ | saturated liquid |
| $lc$ | liquid contact |
| NB | nucleate boiling |
| $q$ | quench |
| RC | Rohsenow-Choi |
| $r$ | reduced |
| $s$ | saturation |
| sat | saturation |
| $T$ | total |
| $TB$ | transition boiling |
| TY | Tong-Young |
| $v$ | vapor |
| $va$ | vapor annulus |
| $vac$ | actual vapor value |
| $ve$ | equilibrium vapor value |
| $vf$ | film vapor value |
| $vs$ | vapor in the core |
| $w$ | wall |
| wall | physical properties of the liquid evaluated at the wall temperature |
| $\delta$ | drop |

## REFERENCES


Baumeister, K. J. and Hamill, T. T., Laminar Flow Analysis of Film Boiling from a Horizontal Wire, NASA TN D-4035, 1967.

Bjornard, T. A., Blowdown Heat Transfer in a Pressurized Water Reactor, Ph.D. thesis, Department of Mechanical Engineering, Massachusetts Institute of Technology, 1977.

Breen, B. P. and Westwater, J. W., Effect of Diameter of Horizontal Tubes on Film Boiling Heat Transfer, *Chem. Eng. Prog. Symp. Ser.*, vol. 58, pp. 67–72, 1962.

Chen, J., A Correlation for Boiling Heat Transfer to Saturated Fluids in Convective Flow, *I & EC Process Design Dev.*, vol. 5, pp. 322–329, 1966.

Chen, J. C., Sundaram, R. K., and Ozkaynak, F. T., A Phenomenological Correlation for Post-CHF Heat Transfer, NUREG-0237, 1977.

Craddick, W. G., Hyman, C. R., Mullins, C. B., Hedrick, R. A., and Turnage, K. G., PWR Blowdown Heat Transfer Separate-Effects Program Data Evaluation Report–Heat Transfer for THTF Test Series 100, ORNL/NUREG-45, 1978.

Cunningham, J. P. and Yeh, H. C., Experiments and Void Correlation for PWR Small-Break LOCA Conditions, *ANS Trans.*, vol. 17, pp. 369–370, 1973.

Dittus, F. W. and Boelter, L. M. K., Heat Transfer in Automobile Radiators of Tubular Type, Publications in Engineering, University of California, Berkeley, p. 443, 1930.

Dougall, R. S. and Rohsenow, W. M., Flow Boiling on the Inside of Vertical Tubes with Upward Flow of the Fluid at Low Qualities, MIT rept. 9079-26, 1963.

Elias, E. and Yadigaroglu, G., A General One Dimensional Model for Conduction-controlled Rewetting of a Surface, *Nuclear Eng. Des.*, vol. 42, pp. 185–194, 1977.

Forster, H. K. and Zuber, N., Bubble Dynamics and Boiling Heat Transfer, *AIChE J.*, vol. 1, pp. 532–535, 1955.

Groeneveld, D. C. (revised by E. O. Moeck), An Investigation and Heat Transfer in the Liquid Deficient Regime, AECL 3281, 1969.

Groeneveld, D. C. and Delorme, G. G. J., Prediction of Thermal Non-Equilibrium in the Post-dryout Regime, *Nucl. Eng. Des.*, vol. 36, pp. 17–26, 1976.

Groeneveld, D. C. and Gardiner, S. R. M., Post-CHF Heat Transfer under Forced Convective Conditions, in *Thermal and Hydraulic Aspects of Nuclear Reactor Safety,* vol. 1, *Light Water Reactors*, eds. O. C. Jones, and S. G. Bankoff, pp. 43–73, ASME, New York, 1977.

Henry, R. E. and Leung, J. C. M., A Mechanism for Transient Critical Heat Flux, *Proc. ANS Special Topic Meet. on Thermal Reactor Safety, Sun Valley, Idaho*, vol. 2, p. 692, 1977.

Hsu, Y. Y., A Tentative Correlation for the Regime of Transition Boiling and Film Boiling During Reflood, paper presented at 3d WRSR Information Meet., USNRC, 1975.

Hsu, Y. Y. and Beckner, W. D., A Correlation for the Onset of Transient CHF, paper presented at 5th WRSR Information Meet., USNRC, 1977.

Jones, O. C., Jr. and Zuber, N., Post-CHF Heat Transfer: A Non-Equilibrium, Relaxation Model, ASME paper 77-HT-79, 1977.

Kirchner, W. L., Reflood Heat Transfer in a Light Water Reactor, MIT, NUREG-0106, NRC-24, 1976.

Leonard, J. E., Sun, K. H., Andersen, J. G. M., Dix, G. E., and Yuoh, T., Calculation of Low Flow Film Boiling Heat Transfer for BWR LOCA Analysis, G.E. rept. NEDO-20566-1, Rev. 1, 1978.

McAdams, W. H., *Heat Transfer*, p. 378, McGraw-Hill, New York, 1954.

Nelson, R., Idaho National Engineering Laboratory, Private communication, 1975.

Richlen, S. L. and Nelson, R. A., Comparisons of Non-Equilibrium Correlations to Post-CHF Tube Data, RES-76-168, Idaho National Engineering Laboratory, 1976.

Rohsenow, W. M. and Choi, H., *Heat, Mass and Momentum Transfer,* pp. 141–142, Prentice Hall, Englewood Cliffs, N.J., 1961.

Rosal, E. R., Hochreiter, L. E., McGuire, M. F., and Krepinerich, M. C., FLECHT Low Flooding Rate Cosine Test Series Data Report, Westinghouse Electric rept. WCAP-8651, 1975.

Rosal, E. R., Conway, C. E., and Krepinerich, M. C., FLECHT Low Flooding Rate Skewed Test Series Data Report, Westinghouse Electric rept. WCAP-9108, 1977.

Schrock, V. E. and Grossman, L. M., Forced Convection Boiling Studies. Forced Convection Vaporization Project, USAEC rept. TID-14632, 1959.

Slifer, B. C. and Hench, J. E., Loss-of-Coolant Accident and Emergency Core Cooling Models for G.E. BWR, G.E. rept. NEDO-10329, 1971.

Smith, R. A. and Griffith, P., A Simple Model for Estimating Time to CHF in a PWR LOCA, Natl. Heat Transfer Conf. ASME paper 76-HT-9, 1976.

Snider, D. M., Analysis of Thermal-Hydraulic Phenomena Resulting in Early Critical Heat Flux and Rewet in the Semiscale Core, TREE-NUREG-1073, 1977.

Sun, K. H., Gonzalez-Santalo, J. M., and Tien, C. L., Calculations of Combined Radiation and Convection Heat Transfer and Rod Bundle under Emergency Cooling Conditions, National Heat Transfer Conf., ASME paper 75-HT-64, 1975.

Thom, J. R. S., Walker, W. M., Fallon, T. A., and Reisting, G. F. S., Boiling in Subcooled Water during Flow up Heated Tubes or Annuli, *Proc. IME* (*London*), vol. 180, pt. 3C, pp. 226–246, 1965–1966.

Tong, L. S., Prediction of Departure from Nucleate Boiling for and Axially Non-uniform Heat Flux Distribution, *J. Nucl. Eng.*, vol. 21, pp. 241–248, 1967.

Tong, L. S. and Young, J. D., A Phenomenological Transition and Film Boiling Heat Transfer Correlations, *Heat Transfer 1974, Proc. 5th Int. Heat Transfer Conf., Tokyo, Japan*, vol. 4, pp. 120–124, 1974.

Whalley, P. B., Hutchinson, P., and Hewitt, G. F., The Calculation of Critical Heat Flux in Forced Convection Boiling, AERE-R-7420, 1973.

Zuber, N., Stability of Boiling Heat Transfer, *Trans. ASME*, vol. 80, pp. 711–720, 1958.

Zuber, N. and Findlay, J. A., The Effects of Non-uniform Flow and Concentration Distribution and the Effects of the Local Relative Velocity on the Average Volumetric Concentration in Two-Phase Flow, G.E. rept. GEAP-4542, 1964.

CHAPTER

# FIFTEEN

# CONDENSATION HEAT TRANSFER

Y. Y. Hsu

The classic condensation problems usually are concerned with film condensation and dropwise condensation. In reactor safety-related issues, these are of less interest. The configurations encountered in reactor safety are mostly of the direct contact type, including

1. Condensation of vapor bubbles in the liquid [steam-water mixing during emergency core cooling (ECC)]
2. Condensation of vapor jets in the liquid (steam-water mixing in the containment)
3. Condensation of vapor on subcooled droplets (ECC top-spray)
4. Condensation between two streams (more or less parallel, in the downcomer)

The range of conditions is

Pressure: 0.1-5 MPa
Water temperature: 20°C–saturation
Vapor temperature: 200°C–saturation

In the past, condensation heat transfer has not been studied as much in reactor safety research as has boiling heat transfer. We have started a few research programs recently, but there is still not enough information. Thus, the listed equations for condensation are only tentative recommendations subject to future verification.

# 1 CONDENSATION OF BUBBLES AND JETS

## 1.1 Condensation of Bubbles

Condensation of vapor bubbles is usually treated in the same theoretical manner as that of bubble growth. Many equations have been developed for bubble growth and collapse, notably by Forster and Zuber (1954), Plesset and Zwick (1954), Florschuetz and Chao (1965), Prisnyakov (1971), Theofanous et al. (1970), Hewitt and Parker (1968), Cho and Seban (1969), Zuber (1961), and Akiyama (1973). In general, there are two stages of bubble collapse: For large bubble size, the rate is controlled by heat transfer, and when the bubble is nearly completely collapsed, the rate is inertia-controlled. Since the heat transfer-controlling stage covers most of the bubble life, we should be more interested in this stage. The equations commonly used are

1. Florschuetz and Chao (1965) (using the Plesset and Zwick integral):

$$3\bar{\tau} = 2\frac{R_0}{R} + \left(\frac{R}{R_0}\right)^2 - 3$$

where

$$\bar{\tau} \triangleq \frac{4}{\pi}\left[\frac{C_{pl}\rho_l(T_{\text{sat}} - T_f)}{\rho_g H_{fg}}\right]^2 \frac{\kappa_l t}{R_0^2} \tag{1}$$

2. Florschuetz and Chao (using plane interface):

$$\frac{R}{R_0} = 1 - \bar{\tau}^{1/2} \tag{2}$$

3. Prisnyakov (1971):

$$\frac{R}{R_0} = 1 - 2\epsilon\bar{\tau}^{1/2} \tag{3}$$

where $\epsilon$ is a correcting factor given by

$$\epsilon^{-1} \triangleq 1 - \frac{\rho_g}{\rho_l} + 2\frac{H_l}{H_{fg}} \tag{4}$$

4. Zuber (1961):

$$\frac{R}{R_m} = \left(\frac{t}{t_m}\right)^{1/2}\left[2 - \left(\frac{t}{t_m}\right)^{1/2}\right] \tag{5}$$

Comparison of Eqs. (1)–(5) with the bubble rate (for bubbles formed on the wall) has been made by Bucher and Nordmann (1978), as shown in Fig. 1.

## 1.2 Bubbles from Nozzle

The bubble collapse rate can be computed from Zuber's equation or from equations derived from experimental data to take into account velocity. Akiyama's equation is found to be good for 10°C subcooling (Fig. 2):

$$\frac{R}{R_0} = \left(1 - 1.06 \frac{Gt}{R_0^{1.4}}\right)^{1/1.4}$$

$$G \triangleq 0.37 \frac{k_l v^{0.6}\ \Delta T_{sub}\ \mathrm{Pr}_l^{1/3}}{H_{fg}\rho_v \nu_l^{0.6}} \tag{6}$$

$$v = 1.18 \left[\frac{\sigma g(\rho_f - \rho_v)}{\rho_f^2}\right]^{1/4}$$

A quasi-steady-state equation has also been derived by Moalem and Sideman (1973):

$$\mathrm{Nu} = \frac{2Rh}{k_f} = \frac{2}{\sqrt{\pi}} \left(\frac{2Rv}{\kappa_f}\right)^{1/2} \tag{7}$$

However, it was found by Brucker and Sparrow (1977) that McAdam's equation for solid spheres can be used for condensation:

$$\mathrm{Nu} = \frac{2R_0 h}{k_f} = 0.37 \left(\frac{\rho_f v 2R_0}{\mu_f}\right)^{0.6} \tag{8}$$

within 50%. The data base is with subcooling of 15–100°C, pressure 10.3-62.1 bar, and Peclet number 2000–3000.

## 1.3 Jets from Nozzle

A study specifically aimed at condensation of vapor jets into flowing subcooled liquids was done by Young et al. (1974) (Fig. 3). They correlated heat transfer in the liquid side by the form

$$\mathrm{St} = 6.5\ \mathrm{Re}^{-0.40} \tag{9}$$

where

$$\mathrm{St} \triangleq \frac{h_f}{\rho_f C_{pf} V^*} \qquad \mathrm{Re} \triangleq \frac{V^* D}{\nu_f}$$

$$V^* \triangleq v_i - v_f$$

$D$ being the nozzle diameter and $v_i$, the interfacial velocity, given by

$$v_i = \frac{(\epsilon_{0g} + m/2)v_g + (\epsilon_{0f} - m/2)v_f}{\epsilon_{0g} + \epsilon_{0f}}$$

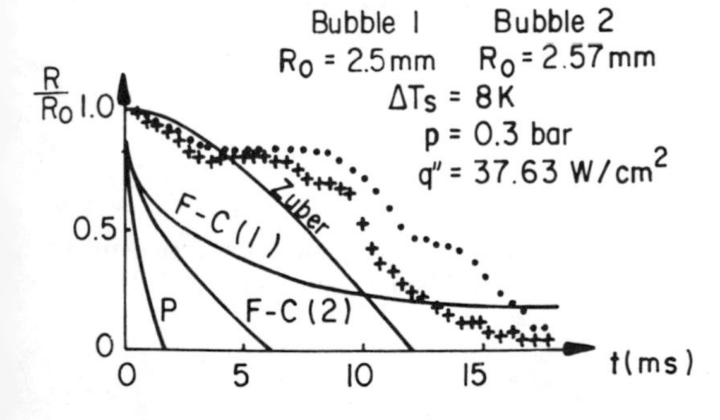


**Figure 1** Bubble condensation in subcooled pool boiling; theoretical and experimental results (Bucher and Nordmann, 1978).

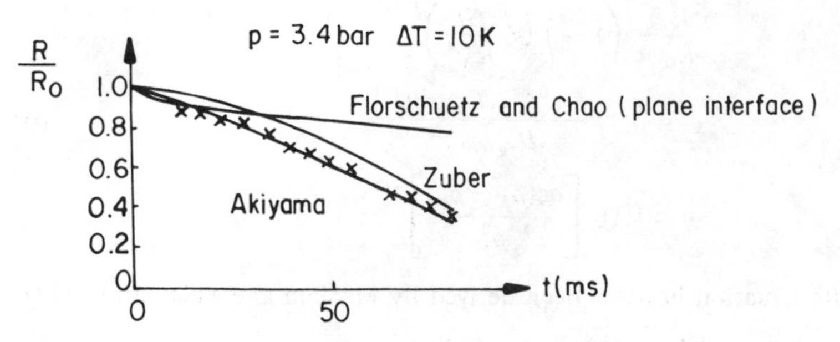


**Figure 2** Bubble growth and collapse at nozzles (Bucher and Nordmann, 1978).

where $m$ is the mass flux due to condensation and $\epsilon_0$ the Reynolds flux obtained from

$$\epsilon_{0g} = \tfrac{1}{8} f_g \rho_g (v_g - v_i) \qquad \epsilon_{0f} = \tfrac{1}{8} f_f \rho_f (v_f - v_i)$$

with $f$ the usual pipe friction factor.

The comparison with data is shown in Fig. 4. The data base is in the range

Steam flow: $1.1 \times 10^{-2}$ kg/s-$2.4 \times 10^{-2}$ kg/s
Stagnant pressure of steam: 230–530 kN/m$^2$
Nozzle pressure ratio: 0.2–0.6
Nozzle diameter: 6.4 mm
Water velocity: 3 m/s
Water temperature: 310 K
Scatter: $\sim \pm 5\%$

At the present time, Eq. (9) is recommended for condensation of vapor jets from nozzles.

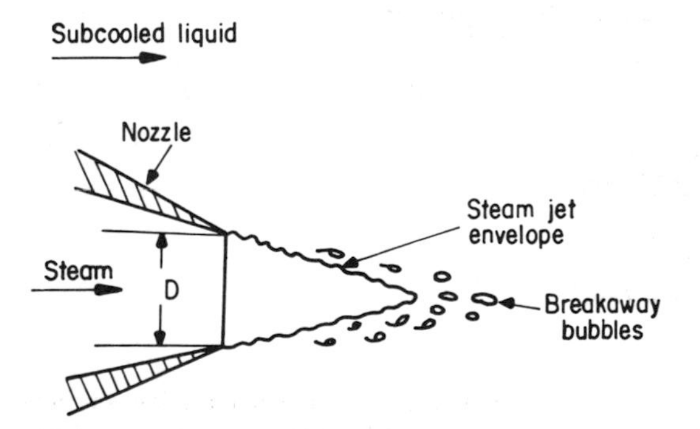


**Figure 3** Steam jet condensing in subcooled liquid (Young et al., 1974).

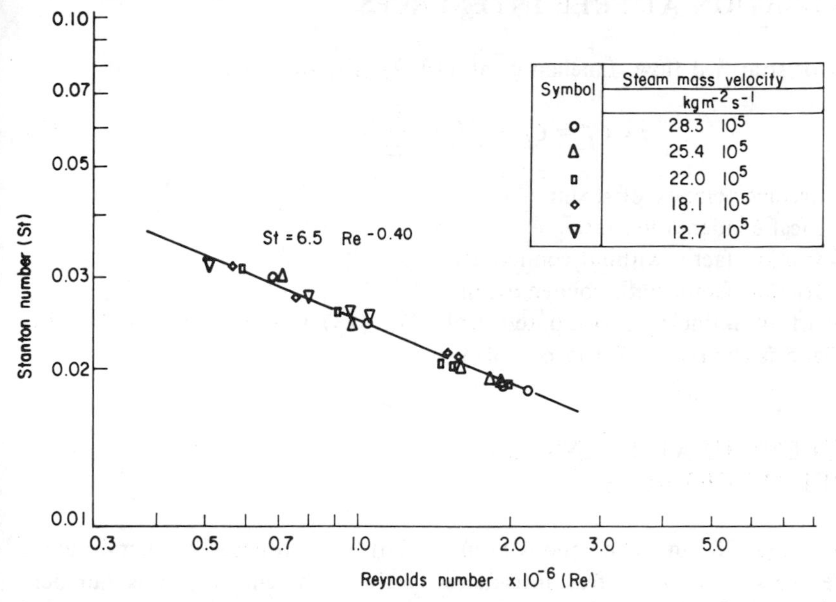


**Figure 4** Steam-jet Stanton number compared with the Reynolds number (Young et al., 1974).

# 2 CONDENSATION ON DROPS

## 2.1 Saturated Vapor on Subcooled Water

The classic heat conduction solution inside a solid sphere can be used to determine heat transfer to a drop (Andersen, 1976):

$$h_{i-f} = \frac{q''_{i-f}}{T_{\text{sat}} - T_{\text{drop av}}} = \frac{\pi^2}{3}\frac{k_f}{R} \tag{10}$$

## 2.2 Superheated Vapor on Drops

Lee and Ryley (1968) have obtained a correlation in the form

$$\frac{q''_{v-i}2R}{(T_v - T_{\text{sat}})k_v} = 2 + 0.74\left(\frac{2Rv\rho_v}{\mu_v}\right)^{0.5}\text{Pr}_v^{0.33} \tag{11}$$

Thus the condensation rate is

$$\dot{m}_{\text{drop}} = \frac{(q''_{i-f} - q''_{v-i})4\pi R^2}{H_{fg}} \tag{12}$$

## 3 CONDENSATION AT FREE INTERFACES

For annular or stratified flow, Linehan et al. (1969) proposed that

$$C_f^* = C_f - \frac{2}{\rho_v V_v} \frac{dW_v}{dz} \tag{13}$$

where $V_v$ = average velocity of steam
$dW_v/dz$ = local condensation rate
$C_f$ = friction factor without condensation
$C_f^*$ = friction factor with condensation

The change of film thickness computed from Eq. (13) is shown in Fig. 5. Also shown in Fig. 5 is the range of data conditions.

## 4 TURBULENT HEAT TRANSFER AT FREE INTERFACES

Brumfield et al. (1975) and Theofanous et al. (1976) have proposed a general model to correlate mass transfer at free interfaces by the turbulent Reynolds number.

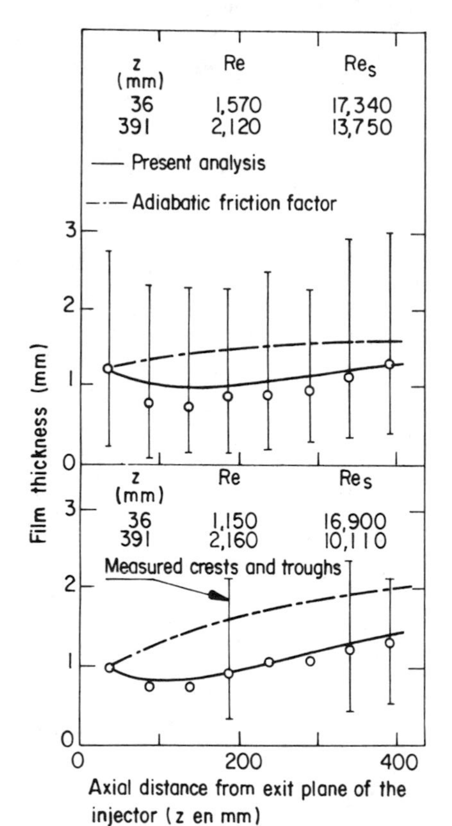


**Figure 5** Effect of model of interface shear stress on predictions of film thickness (Linehan et al., 1969). Re: film Reynolds number; $Re_S$: steam Reynolds number.

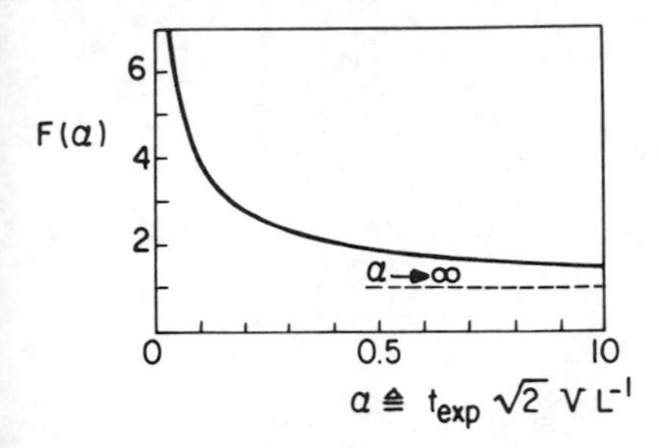


**Figure 6** The function $F(\alpha)$ (Brumfield et al., 1975).

Bankoff (1977) proposed that Theofanous' model be converted to determine condensation heat transfer.

$$\begin{aligned} \mathrm{Nu} &= 0.25\,\mathrm{Re}_t^{3/4}\,\mathrm{Pr}^{1/2} \qquad \text{for } \mathrm{Re}_t > 500 \\ \mathrm{Nu} &= 0.7F(\alpha)\,\mathrm{Re}_t^{1/2}\,\mathrm{Pr}^{1/2} \qquad \text{for } \mathrm{Re}_t < 500 \end{aligned} \tag{14}$$

where $\mathrm{Nu} \triangleq h_f L/k_f$, with $L$ being the macroscale

$\mathrm{Re}_t \triangleq LV/\nu_f$, with $V$ being the turbulent intensity

$F(\alpha)$ is shown in Fig. 6 where $\alpha \triangleq t_{\text{exposure}} \sqrt{2}VL^{-1}$. The validity of such a model for condensation application is being checked at Northwestern University.

## 5 CONCLUDING REMARKS

In comparison with boiling heat transfer, the condensation data base is much smaller for reactor safety application. Although many equations are available, their applicability remains to be verified by more data. Another complication is the very chaotic feature of the interface during condensation. For most of the time, both the heat transfer coefficient and the interfacial area per unit volume are difficult to measure, let alone predict. A great deal of fundamental modeling is needed before we can reach a high level of confidence.

## NOMENCLATURE

| | |
|---|---|
| $C_f$ | friction factor |
| $C_p$ | specific heat at constant pressure |
| $D$ | diameter of pipe, rod, or nozzle |
| $F$ | factor used by Brumfield et al. (1975) |
| $f$ | friction factor |
| $G$ | parameter used by Akiyama (1973) |
| $g$ | gravitational acceleration |
| $H$ | specific enthalpy |
| $H_{fg}$ | heat of evaporation |
| $h$ | heat transfer coefficient |

$k$ thermal conductivity
$L$ macroscale
$m$ mass flux
Nu Nusselt number
Pr Prandtl number
$q''$ heat flux density
$R$ radius
Re Reynolds number
$R_m$ maximum bubble radius
St Stanton number
$T$ temperature
$t$ time
$t_m$ time needed for a bubble to reach its maximum radius
$V$ average velocity; turbulent intensity
$V^*$ characteristic velocity
$v$ bubble or droplet velocity relative to the liquid phase
$W$ mass flow rate per unit width of channel
$z$ axial coordinate
$\alpha$ Brumfield parameter
$\epsilon$ correction factor used by Prisnyakov (1971)
$\epsilon_0$ Reynolds flux
$\kappa$ thermal diffusivity
$\mu$ viscosity
$\nu$ kinematic viscosity
$\rho$ density
$\sigma$ surface tension
$\bar{\tau}$ dimensionless time [Eq. (1)]

**Subscripts**

av average
$f$ liquid
$g$ gas
$i$ interface
$l$ liquid
sat saturation
sub subcooled
$t$ turbulent
$v$ vapor
0 initial

## REFERENCES

Akiyama, M., Bubble Collapse in Subcooled Boiling, *Bull. JSME*, vol. 16, no. 93, pp. 570-575, 1973.

Andersen, J. G. M., Corecool: A Model for Temperature Distribution and Two Phase Flow in a Fuel Element Under LOCA Conditions, General Electric rept. NEDO-21325, 1976.

Bankoff, S. G., Northwestern University, Private communication, 1977.

Brucker, G. G. and Sparrow, E. M., Direct Contact Condensation of Steam Bubbles in Water at High Pressure, *Int. J. Heat Mass Transfer*, vol. 20, pp. 371-381, 1977.

Brumfield, L. K., Houze, R. N., and Theofanous, T. G., Turbulent Mass Transfer at Free, Gas Liquid Interfaces, with Applications to Film Flow, *Int. J. Heat Mass Transfer*, vol. 18, pp. 1077-1091, 1975.

Bucher, B. and Nordmann, D., Investigations of Subcooled Boiling Problems, in *Two-Phase Transport and Reactor Safety*, eds. R. N. Veziroǧlu and S. Kakaç, vol. 1, pp. 31-49, Hemisphere, Washington, D.C., 1978.

Cho, S. M. and Seban, R. A., On Some Aspects of Steam Bubble Collapse, *J. Heat Transfer, Trans. ASME, ser. C*, vol. 91, pp. 537-542, 1969.

Florschuetz, L. W. and Chao, B. T., On the Mechanism of Vapor Bubble Collapse, *J. Heat Transfer, Trans. ASME, ser. C*, vol. 87, pp. 209-220, 1965.

Forster, H. K. and Zuber, N., Growth of a Vapor Bubble in a Superheated Liquid, *J. Appl. Phys.*, vol. 25, pp. 474-478, 1954.

Hewitt, H. D. and Parker, J. D., Bubble Growth and Collapse in Liquid Nitrogen, *J. Heat Transfer, Trans. ASME, ser. C*, vol. 90, pp. 22-26, 1968.

Lee, K. and Ryley, D. J., The Evaporation of Water Droplets in Superheated Steam, *J. Heat Transfer, Trans. ASME, ser. C*, vol. 90, pp. 445-451, 1968.

Linehan, J. H., Petrick, M., and El-Wakil, M. M., On the Interface Shear Stress in Annular Flow Condensation, *J. Heat Transfer, Trans. ASME, ser. C*, vol. 91, pp. 450-452, 1969.

Moalem, D. and Sideman, S., The Effect of Motion on Bubble Collapse, *Int. J. Heat Mass Transfer*, vol. 16, pp. 2321-2329, 1973.

Plesset, M. S. and Zwick, S. A., The Growth of Vapor Bubbles in Superheated Liquids, *J. Appl. Phys.*, vol. 25, pp. 493-500, 1954.

Prisnyakov, V. F., Condensation of Vapor Bubbles in Liquid, *Int. J. Heat Mass Transfer*, vol. 14, pp. 353-356, 1971.

Theofanous, T. G., Biasi, L., Isbin, H. S., and Fauske, H. K., Nonequilibrium bubble collapse: A Theoretical Study, *Chem. Eng. Prog. Symp. Ser.*, vol. 66, no. 102, pp. 37-47, 1970.

Theofanous, T. G., Houze, R. N., and Brumfield, L. K., Turbulent Mass Transfer at Free, Gas-Liquid Interfaces, with Application to Open Channel, Bubble and Jet Flows, *Int. J. Heat Mass Transfer*, vol. 19, pp. 613-624, 1976.

Young, R. J., Yang, K. T., and Novotny, J. L., Vapor-Liquid Interaction in a High Velocity Vapor Jet Condensing in a Coaxial Water Flow, *Heat Transfer 1974, Proc. 5th Int. Heat Transfer Conf., Tokyo, JSME*, vol. 3, pp. 226-230, 1974.

Zuber, N., The Dynamics of Vapor Bubbles in Nonuniform Temperature Fields, *Int. J. Heat Mass Transfer*, vol. 2, pp. 83-105, 1961.

CHAPTER

# SIXTEEN

# REGIME TRANSITIONS IN BOILING HEAT TRANSFER

G. Yadigaroglu

Several boiling heat transfer regimes, each characterized by widely differing heat transfer mechanisms and values of the heat transfer coefficient have been known to exist in nuclear reactors under normal and accident conditions. Regime transitions leading from one regime to another often have dramatic effects on the behavior of the boiling system, both in pool-boiling as well as in flow-boiling situations.

The various boiling heat transfer regimes may coexist in different axial regions along a boiling channel or may succeed one another in time during a transient. Figure 1 shows the different heat transfer and hydrodynamic regimes and the corresponding transition points at steady state in a typical boiling system such as a boiling-water reactor (BWR) channel or steam generator tube. Clearly all the regimes are not present in every system all the time; for example, dryout does not occur in BWRs under normal operating conditions. Figure 2 shows another set of heat transfer and hydrodynamic regimes encountered during the reflooding phase of the loss of coolant accident (LOCA) in light-water reactors (LWR).

The various transition points are defined in this introductory section in brief, general terms, that is, without reference to any particular system. The physical processes at the transition points are then discussed at greater length in Secs. 2-6. The heat transfer mechanisms in the various heat transfer regimes and their analysis are treated in other chapters of this book.

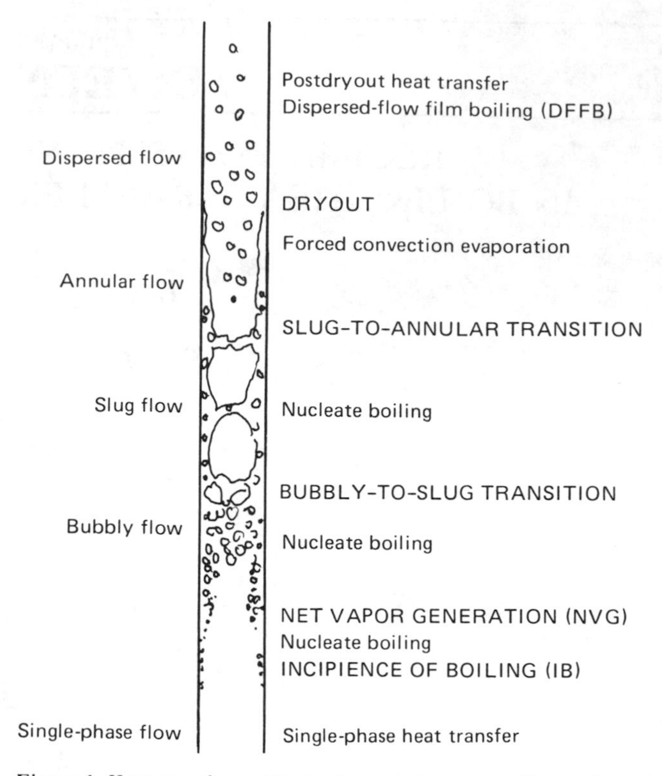


**Figure 1** Heat transfer and hydrodynamic two-phase flow regimes in a boiling channel.

The discussion includes the incipience of boiling (IB) point; the net vapor generation (NVG) point; the burnout point, or critical heat flux transition (CHF); and rewetting.

Hydrodynamic two-phase flow regime transitions, such as the slug-to-annular flow transition or the bubbly-to-slug flow regime change are beyond the scope of this review. The transition from an inverted-annular film boiling (IAFB) to a dispersed-flow film boiling regime (DFFB) was included for completeness in spite of the fact that it is essentially a hydrodynamic flow regime transition since it controls the mode of heat transfer in postdryout heat transfer. Regarding the NVG point, there is no abrupt transition with regard to heat transfer at that point; the transition is from an essentially single-phase flow regime (with a negligible amount of voids on the wall) to a two-phase bubbly flow regime.

## 1 TRANSITION POINTS

The change of the heat transfer regime from purely single-phase natural or forced convection to nucleate boiling occurs at the incipience of boiling. The conditions at this point can be predicted reasonably well on the basis of the theories of nucleation from cavities on the heated surface.

In subcooled flow boiling, the void fraction remains very low until vapor generation at the heated wall overcomes the effects of condensation in the subcooled core of the flow. Thus the void fraction seems to depart from zero at a point, commonly referred to as the net vapor generation point. Several theories have been advanced to predict the conditions at that point; these are discussed in Sec. 3.

The transition from a nucleate boiling regime characterized by a very high value of the heat transfer coefficient to a film boiling regime for which heat transfer is poor marks the point of departure from nucleate boiling (DNB). For high-quality two-phase flows the corresponding transition from forced convection vaporization to dispersed-flow film boiling occurs generally at dryout of the wall. Dryout and departure-from-nucleate-boiling phenomena have also been referred to as the critical heat flux condition, burnout, boiling crisis, etc. Extensive research in this area has not led yet to any comprehensive model of the critical heat flux phenomena in flow boiling although models applicable under a limited range of conditions have had success in predicting the trends of the data. Although reasonably accurate

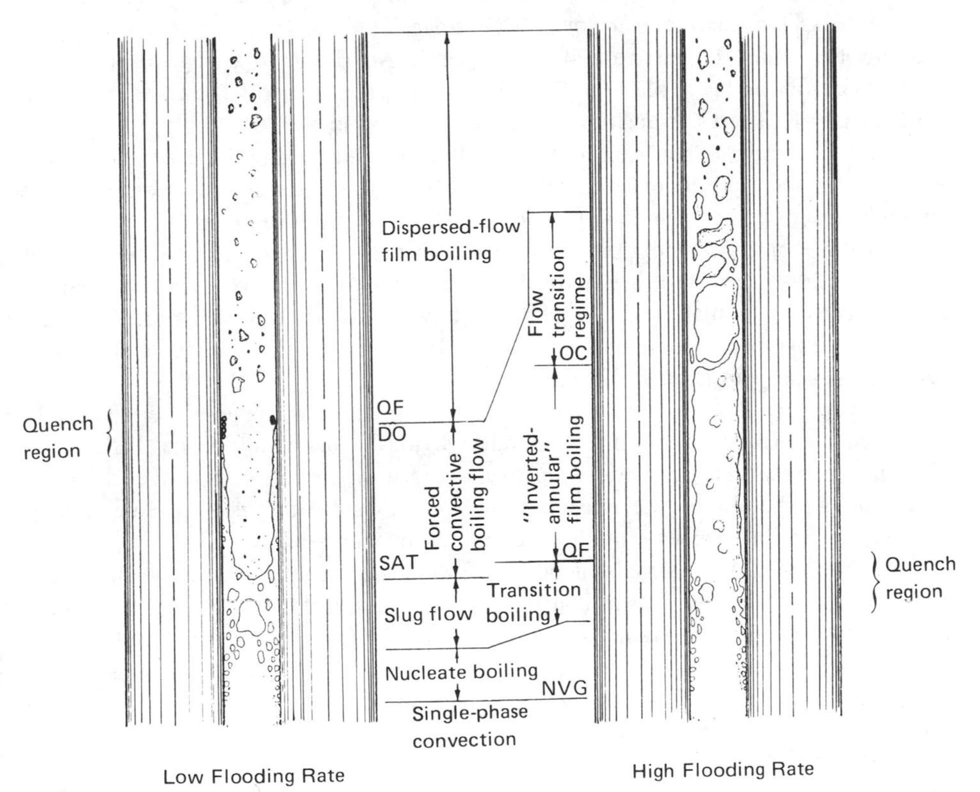


**Figure 2** Two-phase heat transfer and hydrodynamic regimes during reflooding. NVG, net vapor generation point; SAT, slug-to-annular transition point; QF, quench front; DO, dryout point; and OC, onset-of-carryover point.

theoretical explanations and models of departure from nucleate boiling in pool boiling situations exist, for practical purposes we must still rely on empirical correlations of flow-boiling critical heat fluxes. Burnout is discussed in Sec. 4.

The rewetting transition occurs when an initially hot wall is progressively cooled by a liquid film or a flowing two-phase mixture. Rewetting is detected by a dramatic and sudden increase of the heat transfer coefficient, characteristic of the transition from film boiling to nucleate boiling. Rewetting models based on axial-conduction mechanisms in the wall have been successful in explaining the data trends. Several unanswered questions remain, however, regarding the exact nature of the phenomena taking place at the rewetting front, the existence and the dependence of the rewetting temperature on system parameters, and the values of the heat transfer coefficients in the immediate vicinity of the quench front, as discussed in Sec. 5.

An inverted-annular film boiling heat transfer regime may exist during reflooding or in other film boiling situations, for example, in cryogenic systems. When the velocity of the vapor in the vapor film covering the heated surface exceeds a certain value, the interface between the vapor film and the liquid core may become unstable, and the liquid core may be dispersed. This condition is referred to as the onset of carryover (OC) since liquid drops or chunks are carried over by the vapor downstream of that point. No dramatic change of the heat transfer coefficient takes place at OC, but nevertheless it marks the transition from IAFB to DFFB. The OC transition is briefly discussed in Sec. 6.

Often the regime transitions are controlled by mechanisms independent of the heat transfer mechanisms existing before and after the transition. For example the dryout transition in high-quality annular flow is produced essentially by the disappearance of the liquid film on the wall; factors for analyzing the behavior of the liquid film on the wall are quite different from factors governing heat transfer in the preceding annular flow and in the following dispersed-flow regime. Thus the dryout point is not automatically predicted from annular-flow, forced-convection-evaporation heat transfer correlations.

In rewetting, the progression of the quench front is believed to be largely controlled by axial conduction in the wall. Obviously this condition has little to do with the heat transfer mechanisms upstream and downstream of the quench front. Again, the prediction of the quench front cannot be treated as a limiting case of heat transfer in the regimes on either side of the quench front.

Because of the relative independence of the mechanisms controlling the transitions from the mechanisms controlling the heat transfer process, special ad-hoc methods must be used to predict these transitions. Thus the first task in the analysis of a boiling system is to define the transition points, that is, the boundaries of the regions where different heat transfer and hydrodynamic regimes exist. This is also our task in this chapter.

## 2 INCIPIENCE OF BOILING

Well-established results of the theory of nucleation from cavities on a heated wall will be used in this section to derive criteria for the incipience of boiling.

We will start by recalling the fundamental relationships for the superheat of the liquid required to maintain a bubble of radius $r$ in equilibrium in an isothermal liquid at temperature $T$ and pressure $p_L$. The state of the vapor in the bubble is very near to saturation, as for example, shown by Rohsenow (1973). Static equilibrium of the bubble then yields

$$p_G - p_L = p_{sat}(T) - p_L = \frac{2\sigma}{r} \tag{1}$$

where $\sigma$ is the surface tension. This expression allows one to calculate the temperature $T^*$ of the medium required to sustain a bubble of radius $r$ at equilibrium. This temperature can be obtained using Eq. (1) and tables giving the physical properties of the fluids; it can also be approximated by the following analytical expression derived using the Clausius-Clapeyron relation:

$$T^* = T_{sat}(p_L)\left(1 + \frac{v_{LG}}{h_{LG}}\frac{2\sigma}{r}\right) \tag{2}$$

where $v_{LG}$ is the difference of the specific volumes between the vapor and the liquid, and $h_{LG}$ is the latent heat of vaporization.

We will consider now nucleation of bubbles from cavities and their growth in the temperature gradient near a heated wall. For a bubble attached to a cavity on the heated surface, it is clear that a simple geometric relationship exists between the radius of the bubble and the dimension of the mouth of the cavity, although this relationship depends on the contact angle and on the geometry of the cavity. Thus the bubble radius entering in Eq. (2) for the superheat required for nucleation is closely related to the radius of the mouth of the cavity. Hsu (1962) postulated that the temperature of the liquid surrounding the top of the bubble should equal or exceed $T^*$ for nucleation to occur. The following method, based on the ideas of Hsu and Graham (1961), Hsu (1962), Bergles and Rohsenow (1964), and Davis and Anderson (1966), results in the derivation of criteria for the point of incipience of boiling.

Near the heated wall the temperature gradient is practically linear and the temperature distribution (Fig. 3) is given by

$$T_L(y) = T_w - \frac{q''}{k_L}y \tag{3}$$

where $T_w$ is the wall temperature, $q''$ is the heat flux, and $k_L$ is the thermal conductivity of the liquid. Figure 3 also shows Eq. (2) plotted with $r$ replaced by $y/n$, where $n$ is a factor of the order of unity, accounting for bubble geometry and for the fact that the temperature distribution near the wall is perturbed by the presence of the bubble. According to the postulate of Hsu, the incipience of boiling will occur when the two curves touch. At the point of tangency, at $y = y^*$, we will have

$$T_L = T^* \qquad \frac{dT_L}{dy} = \frac{dT^*}{dy}$$

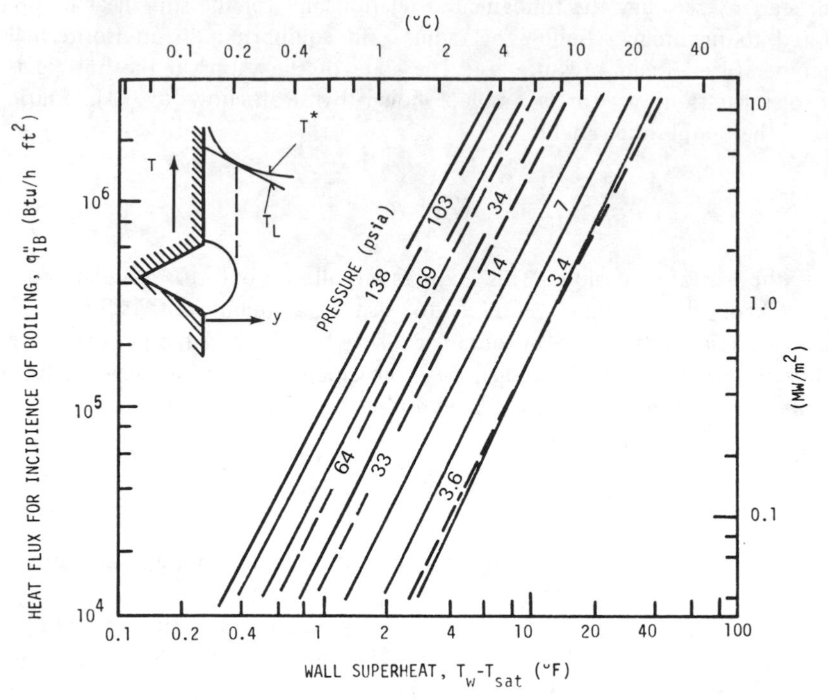


**Figure 3** Relationship between heat flux and wall superheat at the incipience of boiling. Solid lines represent the Bergles and Rohsenow (1964) criterion, Eq. (7); dashed lines represent Eq. (5).

Using the equality of the gradients and Eqs. (2) and (3), we obtain the distance from the wall $y^*$ as

$$y^* = \left(T_{\text{sat}} \frac{v_{LG}}{h_{LG}} 2\sigma n \frac{k_L}{q''}\right)^{1/2} \tag{4}$$

Finally, using the equality of the temperatures at $y = y^*$ yields a relationship between the wall superheat and the heat flux at the incipience of boiling:

$$q''_{\text{IB}} = \frac{k_L h_{LG}}{8 T_{\text{sat}} v_{LG} \sigma n} (T_w - T_{\text{sat}})^2_{\text{IB}} \tag{5}$$

We have assumed so far that the range of cavity sizes available on the heated surface includes the value $r^* = y^*/n$ so that incipience will occur at the point of tangency. If this is not true, and the largest available cavity on the surface has a radius $r_{\max} < r^*$, the incipience of boiling will be delayed until a sufficient superheat becomes available for that particular cavity to nucleate. In this case the heat flux at incipience will be given by the intersection of the two curves described by Eqs. (2) and (3) with $r = r_{\max}$:

$$q''_{\text{IB}} = \frac{k_L (T_w - T_{\text{sat}})}{r_{\max} n} - T_{\text{sat}} \frac{v_{LG}}{h_{LG}} \frac{2\sigma k_L}{r^2_{\max} n} \tag{6}$$

Thus Eq. (5) seems to be only a lower bound for the heat flux at IB. In fact, for a high value of the heat transfer coefficient, high subcooling and heat flux, or low thermal conductivity, the temperature gradient in the boundary layer will be steep, and the point of tangency will be situated near the wall; a small cavity radius will be required to produce incipience of boiling. Since small cavities are generally available on any surface, the IB point will be predicted correctly by Eq. (5). This is the case, for example, of forced convection boiling of water.

If on the contrary, the heat transfer coefficient, subcooling, and heat flux are low or if the thermal conductivity is extremely high as for liquid metals, the tangency criterion will yield a large cavity size that might not be available on a normal, smooth surface. In this case nucleation will be delayed as explained above.

Bergles and Rohsenow (1964) initially solved the tangency problem graphically and correlated the results by the following widely used formula, valid for water between 1 and 138 bar:

$$q''_{\mathrm{IB}} = 1083p^{1.156}\,[1.8(T_w - T_{\mathrm{sat}})]^{2.16/p^{0.0234}} \tag{7}$$

where the pressure, temperatures, and heat flux are expressed in bar, degrees Celsius, and watts per square meter respectively. Equation (7) is plotted in Fig. 3 for various pressures and agrees closely with Eq. (5) with $n = 1$.

Frost and Dzakowic (1967) extended the treatment of IB described above to fluids other than water by putting the value of $n$ in Eq. (5) equal to the square of the Prandtl number of the fluid and by investigating the effect of pressure. Brown (1967) also found that the value of $n$ for different fluids varied between 1 and 3.

If an inert gas is present in the liquid and consequently in the bubbles, Eq. (1) must be modified to account for the partial pressure $p_a$ of the inert gas in the bubbles,

$$p_a + p_G - p_L = \frac{2\sigma}{r} \tag{8}$$

Béhar et al. (1966) noted that for certain fluids such as water, for which solubility of gases decreases with temperature, the first bubbles appearing as the heat flux is increased contain essentially inert gases; they refer to this nucleation mechanism as degassing.

The wall superheat at which degassing starts may be well below the superheat normally required for nucleation and depends on the amount of dissolved gases. Murphy and Bergles (1972) showed that for a liquid containing dissolved gas, at equal pressure, the thermodynamic equilibrium or "gassy" saturation temperature is significantly lower than that for a pure liquid. They showed that the incipience criterion, Eq. (5), becomes

$$q''_{\mathrm{IB}} = \frac{k_L}{8\sigma}(T_w - T_{\mathrm{sat},g})^2 \left.\frac{dp_{\mathrm{sat}}}{dT}\right|_{p_L,\,\mathrm{sat},g} \tag{9}$$

The $dp_{\mathrm{sat}}/dT$ term describing the behavior of the two-component system along the equilibrium line appears now explicitly since it was not approximated through use of the Clausius-Clapeyron relation, as usual [see Eq. (2)].

Several authors, including Béhar et al. (1966), Murphy and Bergles (1972), and Yin and Abdelmessih (1976), described a delay in nucleation and a hysteresis phenomenon observed in relation to the IB of organic liquids, fluorocarbons, and probably other fluids. As the heat flux is increased, the IB appears and is accompanied by a sudden reduction in wall temperature, as shown in Fig. 4. For decreasing heat flux the transition is smooth. The delay of the IB is due to deactivation of the largest available cavities on the surface by flooding of the liquid; this happens when the liquid wets the surface well. The criteria for incipience of boiling given above apply only to decreasing heat flux, as shown in Fig. 4 and by Yin and Abdelmessih (1976).

Nucleation delays leading to large superheats are often observed in liquid-metal systems. Henry et al. (1974) attributed these to deactivation of the cavities and pointed out the importance of the presence of inert gases in the system. Since inert gases diffuse in and out of cavities slowly, the operating history of the system has an important effect on the incipience of boiling.

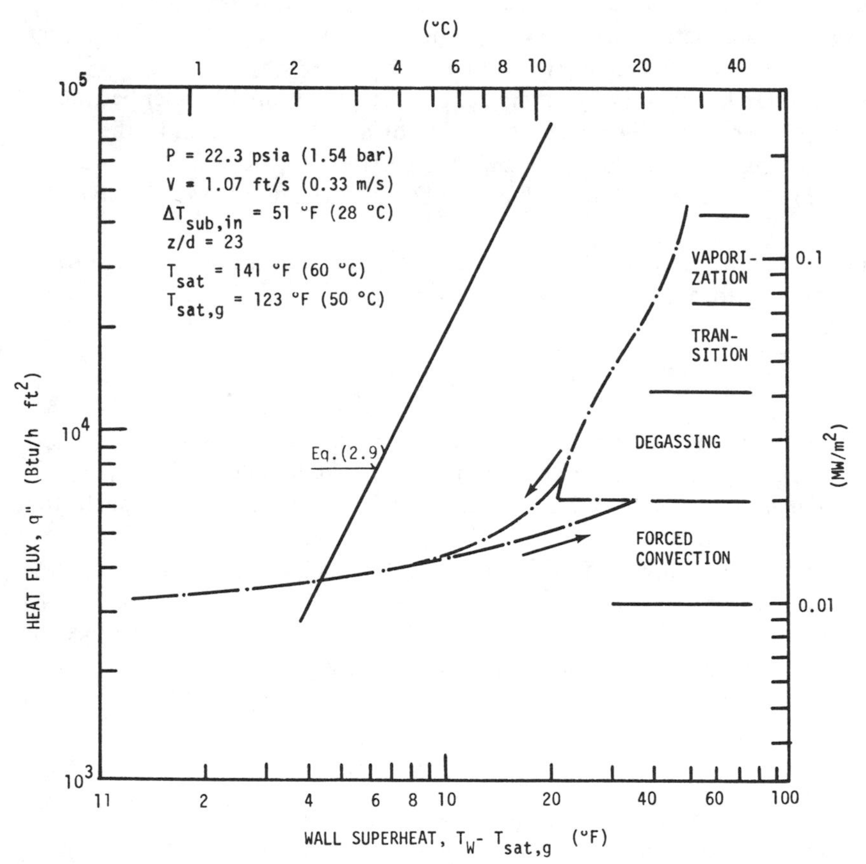


**Figure 4** Boiling curve obtained with gassy coolant by Murphy and Bergles (1972) showing hysteresis around the IB point.

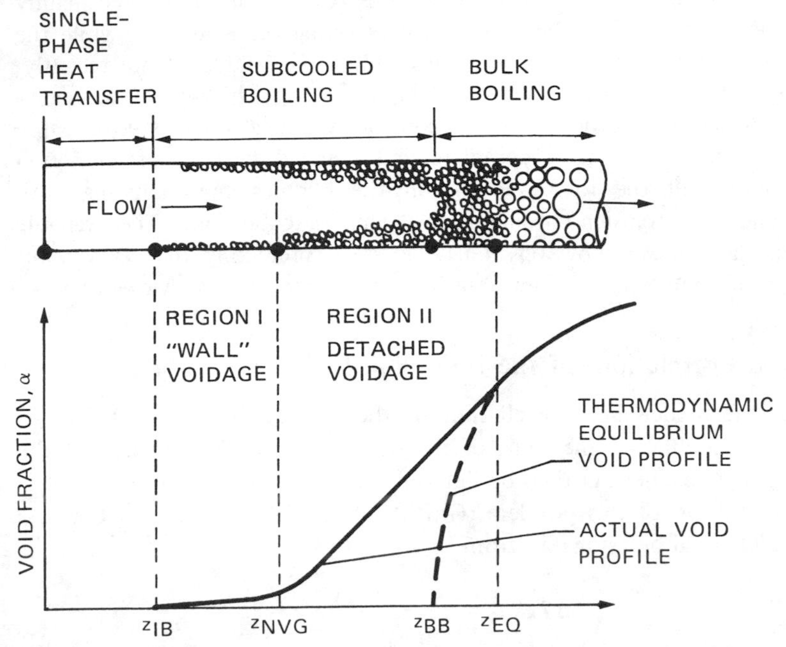


**Figure 5** Variation of the void fraction along a heated channel. IB, incipience of boiling; NVG, net vapor generation; BB, bulk boiling boundary; and EQ, thermodynamic equilibrium.

## 3 THE POINT OF NET VAPOR GENERATION

The point of NVG is defined as the point in the channel where the subcooled-boiling void fraction starts growing appreciably (see Fig. 5). Upstream of this point, in the wall-voidage region, the void fraction remains very small; moreover the bubbles stick to the wall or tend to travel in a narrow layer close to the wall. The growth of the bubble layer is limited by the high subcooling prevailing in the wall voidage region. As the subcooling decreases along the channel, a point is reached where the thermodynamic and hydrodynamic conditions are such that a rapid increase in vapor fraction is initiated.

A number of conditions must be satisfied for the persistence and growth of the void fraction in the bulk of subcooled flows: The bubbles must first detach from the cavities where they were created; they must also be ejected from the wall into the main stream; and finally the liquid core temperature must be sufficiently close to saturation so that vapor generation can overcome the effects of bubble condensation.

The NVG point has been identified as the point of initial bubble detachment from the wall (BD) by Staub (1968) and Levy (1967), while Dix (1971) has shown that at low velocities the NVG point is only slightly downstream of the initial bubble ejection point (BE), that is, the point where bubbles sliding along the wall are ejected into the main stream. In both cases the NVG point is determined mainly by hydrodynamic considerations (balance of forces acting on a bubble), while the heat flux determines the temperature field surrounding the bubbles. More recently, Saha and Zuber (1974) correlated NVG data by considering a high-mass-flux region where the phenomena are hydrodynamically controlled, and a low-mass-flux region where thermal control prevails. A complete and detailed analysis of the NVG problem, considering all the forces acting on the bubbles and including both hydrodynamic and heat transfer considerations, remains to be done. The available models and correlations are, however, fairly good in predicting the NVG point within the range of conditions of their data base. These are briefly reviewed below.

## 3.1 Models and Correlations of the NVG Point

The purpose of the models and correlations of the NVG point is to provide an estimate of the subcooling of the bulk of the fluid at NVG. The location of the NVG point along the channel can then be determined from a heat balance.

In the first analysis of this problem, Griffith et al. (1958) suggested that the subcooling at NVG could be estimated from

$$\Delta T_{\mathrm{NVG}} = \frac{q''}{5h_{fo}}$$

where $q''$ is the heat flux, and $h_{fo}$ is the single-phase heat transfer coefficient, calculated assuming no boiling and no presence of vapor.

Bowring (1962) derived the following empirical correlation for water, valid over the pressure range of 11–138 bar:

$$\Delta T_{\mathrm{NVG}} = \eta \frac{q''}{G v_L}$$

where $G$ is the mass flux and $v_L$ the specific volume of the liquid. The pressure effect is correlated by $\eta$. When the variables are expressed in SI units (°C, W, m, kg, s) and the pressure $p$ in bar, $\eta$ is equal to

$$\eta = (14.0 + 0.1p) \times 10^{-6}$$

Levy (1967) and Staub (1968) presented quite similar methods for the prediction of the NVG point. They both identify the NVG point as the point of bubble detachment and derive criteria for the subcooling at NVG starting from a force balance on a bubble detaching from the wall. They examined and listed the forces acting on a bubble attached to the wall but retained only the important ones to simplify the analysis. Levy considered surface tension and drag but neglected buoyancy forces even at low mass fluxes, while Staub considered all three forces. The authors then made the assumptions that the temperature profile near the wall is given by the universal temperature profile for all-liquid flow and that the fluid temperature at the tip of the bubble is equal to the saturation temperature. The force balance yields the diameter of the bubbles at detachment, and the bulk subcooling can be calculated using the universal temperature distribution in the liquid. The resulting expressions are complex because of the complicated formulas used to describe the temperature profile; the subcooling at NVG is given in terms of the Prandtl number of the fluid and is proportional to the group

$$\frac{q''}{c_L G(f/2)^{1/2}}$$

where $c_L$ is the specific heat of the liquid, and $f$ is the friction factor. Madejski (1970) presented an analysis in the same lines including the effects of all forces acting on the bubble. Betten and Paul (1976) modified the Staub-Levy analysis by assuming that bubble agitation near the wall produces localized turbulence equivalent to that of the turbulent core; they also introduced a correction to account for the effect of high Prandtl numbers.

Data and occasionally empirical correlations of the subcooling at NVG, valid over limited ranges of conditions, are presented for water by Rouhani (1968) and Ahmad (1970) (high $p$, high $G$); Costa (1967) (low $p$, high $G$); Sekoguchi et al. (1974) (low $p$, low $G$); Maitra and Subba Raju (1975) (low $p$, very low $G$); and for Freon-114 by Dix (1971) (low $G$).* Ünal (1975), reviving the formula proposed by Griffith et al. (1958), claimed that the numerical coefficient can be adjusted according to the velocity range to fit a variety of water and Freon data. In all correlations the subcooling at NVG is proportional to the ratio $q''/G$, with the mass flux occasionally raised to a fractional power.

Saha and Zuber (1974) examined a variety of NVG data and concluded that the bubble detachment criterion does not sufficiently describe the physical phenomena and that thermal local conditions at the NVG point must also be considered. The point of NVG is determined by the relative magnitude of the rates of vaporization and condensation near the wall. Vaporization is proportional to the film temperature difference or to $q''/h$, while condensation is proportional to the subcooling $\Delta T_{\mathrm{NVG}}$. At low mass fluxes, forming the ratio of these driving forces, we obtain a Nusselt number as the relevant parameter in the thermally controlled region:

$$\mathrm{Nu} = \frac{q''D}{\Delta T_{\mathrm{NVG}} k_L}$$

At high mass fluxes the phenomena are hydrodynamically controlled by bubble detachment considerations, and the Reynolds and Prandtl numbers also become important. Assuming that the Reynolds analogy between momentum and heat transport holds, the relevant dimensionless number is the Stanton number,

$$\mathrm{St} = \frac{\mathrm{Nu}}{\mathrm{Re\,Pr}} = \frac{q''}{G c_L\, \Delta T_{\mathrm{NVG}}}$$

The Péclet number, scaling the effects of flow velocity to diffusion velocity normal to the surface, is then introduced. When NVG data are plotted in the Stanton versus Péclet number plane, Fig. 6, the thermally and hydrodynamically controlled regions appear. Saha and Zuber correlated available data by

*Low $p$ = atmospheric or near-atmospheric pressure; low $G$ = below ~1000 kg/m² s.

$$\text{Nu} = 45 \quad \text{for Pe} < 70{,}000 \text{ (low-mass-flux, thermally controlled region)}$$

$$\text{St} = \frac{\text{Nu}}{\text{Re Pr}} = 0.0065 \quad \text{for Pe} > 70{,}000 \text{ (high-mass-flux, hydrodynamically controlled region)}$$

Saha and Zuber's correlation confirms the $q''/G$ dependence at high mass fluxes, but shows that the subcooling at NVG becomes independent of mass flux at low mass fluxes.

The correlation is convenient to use and is probably the best available at the present time. The data base for this correlation is fairly broad but includes no data at very low mass fluxes. Indeed the correlation seems to predict unreasonably low subcoolings for low-mass-flux, near-atmospheric water systems.

The cautioning remarks made earlier with respect to the influence of dissolved gases on the point of incipience of boiling apply equally well to NVG considerations (Staub, 1968). Premature NVG as a result of dissolved gases should be expected, especially in low-pressure systems. Since the amount of dissolved gases is difficult to control, the accuracy of any model of NVG phenomena might in practice be limited.

There are no observations or analyses of the bahavior of the NVG point during flow and/or heat flux transients. Roy and Yadigaroglu (1976) have recently shown that the velocity and temperature profiles near the heated wall may be quite different from those of the quasi-steady-state profiles, even for relatively slow transients. Since the shape of these profiles near the wall determines the occurrence

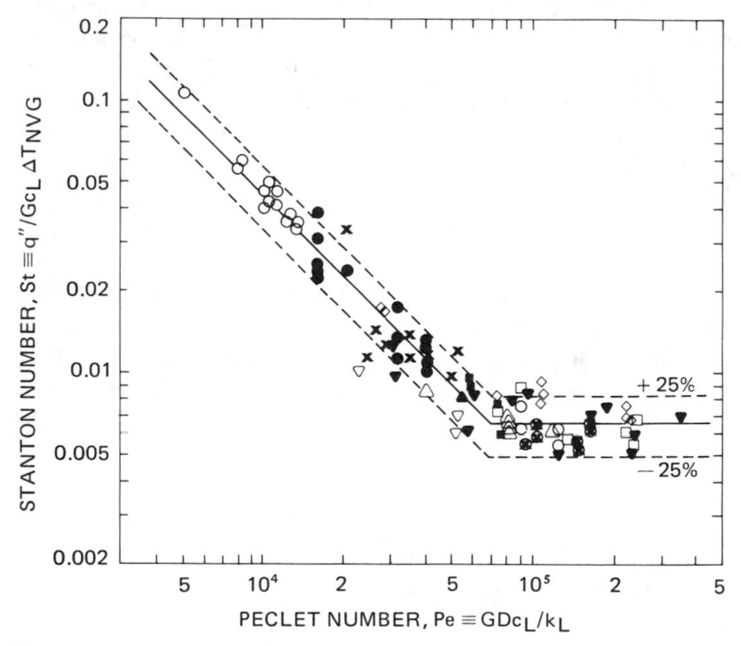


**Figure 6** Comparison of Péclet number and Stanton number at the point of NVG (Saha and Zuber, 1974).

of NVG, it seems that transient NVG prediction based on quasi-steady-state application of available correlations is probably unreliable.

## 4 BURNOUT

In boiling heat transfer, as the wall temperature is increased, the heat flux to the fluid increases with some power (generally greater than 1) of the wall superheat, $T_w - T_{sat}$. At some point, however, a maximum heat flux condition is reached. In temperature-controlled systems such as steam generators, the heat flux beyond this point decreases with subsequent increases of the wall superheat. In heat flux-controlled systems, such as electrically heated test sections and nuclear fuel rods, since the negative-slope part of the heat flux-wall-superheat characteristic is dynamically unstable, a dramatic and sudden large increase of the wall temperature is observed as the heat flux is increased to a value slightly above this critical heat flux. These phenomena have been generically referred to as burnout, boiling crisis, calefaction, or boiling transition.

The physical phenomena occurring at burnout, analytical models, and empirical correlations used for predictive purposes are reviewed below. Burnout in pool-boiling situations is considered in Sec. 4.1, while burnout in flow boiling is treated in Sec. 4.2. Only steady-state situations are examined here since transient burnout is discussed in Chap. 14.

### 4.1 Burnout in Pool Boiling

The burnout mechanism in pool boiling has been studied extensively and is fairly well understood; correlations can be used to predict the critical heat flux for all liquids.

For classic saturated pool boiling, in the nucleate boiling region, as the heat flux is gradually increased, more and more nucleation sites are activated until the entire heater surface becomes blanketed with vapor. In this case burnout can appropriately be referred to also as departure from nucleate boiling (DNB). In heat flux-controlled systems DNB can lead to physical burnout of the test section if film boiling temperatures are incompatible with the heater materials.

**4.1.1 Physical mechanism and analytical modeling** Several detailed mechanisms for DNB in pool boiling have been advanced, but burnout is essentially a result of the inability of the liquid to reach the heated surface. This condition can be analytically modeled and explained in a number of alternative ways.

According to Rohsenow and Griffith's (1956) bubble packing model, the number of nucleation sites increases with the heat flux and at DNB becomes so large that neighboring bubbles or columns of vapor coalesce and the vapor covers completely the surface.

Explanations of the burnout mechanism based on hydrodynamic (Helmholtz) instability considerations have been proposed by Zuber (1958), Chang and Snyder

(1960), Moissis and Berenson (1963), and Lienhard and Dhir (1973). At high heat fluxes "jets" of vapor leave the heated surface; the liquid has to reach the surface by flowing in the areas left between the vapor columns. This situation (parallel flow of two fluids having largely different densities) leads eventually to an instability and breakdown of the vapor-liquid interface as the relative velocity of the vapor with respect to the liquid attains a certain critical value.

Finally, in liquid-suspension and flooding models (Wallis, 1961; Griffith et al., 1972) the boiling crisis is visualized as a flow pattern transition where the liquid filaments bringing the liquid to the heated surface break up and the drops that are created remain suspended or "fluidized" in the vapor stream. This situation is similar to the "flooding" observed in countercurrent two-phase flows. The fluidization and flooding models are not necessarily in disagreement with the hydrodynamic instability models; the condition they describe follows after breakdown of the gas-liquid interface as predicted by the hydrodynamic considerations.

In all these models the controlling parameter is the volumetric flux of the vapor away from the heated surface

$$\frac{q''_{\text{crit}}}{\rho_G h_{LG}}$$

where $\rho_G$ is the density of the vapor and $h_{LG}$ is the latent heat of vaporization.

This volumetric flux or superficial vapor velocity is comparable to the critical relative velocity for the Helmholtz instability $U$, which is given in terms of a Weber number

$$\text{We} = \frac{\rho_G U^2 \lambda}{\sigma} = \text{const}$$

where $\sigma$ is the surface tension and $\lambda$ the unstable wavelength. The unstable wavelength in turn is given by expressions having the form

$$\lambda = \text{const} \sqrt{\frac{\sigma}{g(\rho_L - \rho_G)}} \tag{10}$$

Combining the two, we obtain the term

$$\left[\frac{g\sigma(\rho_L - \rho_G)}{\rho_G^2}\right]^{1/4}$$

Thus the correlations usually take the form

$$\frac{q''_{\text{crit}}}{h_{LG}\rho_G} = \text{const} \left[\frac{g\sigma(\rho_L - \rho_G)}{\rho_G^2}\right]^{1/4} f\left(\frac{\rho_L}{\rho_G}\right) \tag{11}$$

The last factor, which is a function of the ratio of densities, is usually very close to 1, except in the vicinity of the thermodynamic critical region. No data are available, however, to test the differences of the various models in this region.

For liquid metals Eq. (11) is corrected, either by multiplication with another dimensionless quantity or by addition of a corrective factor, as shown in Table 1.

**Table 1 Pool-boiling critical heat flux correlations (horizontal surfaces or horizontal cylinders)**

General form: $$\frac{q''_{crit}}{\rho_G h_{LG}}\left[\frac{g\sigma(\rho_L-\rho_G)}{\rho_G^2}\right]^{-1/4}=Cf\left(\frac{\rho_L}{\rho_G}\right)$$

| Author | $Cf\left(\frac{\rho_L}{\rho_G}\right)$ | Remarks |
|---|---|---|
| Kutateladze (1952) | 0.16 | Obtained through dimensional analysis |
| Zuber (1958) | $\frac{\pi}{24}\left(\frac{\rho_L+\rho_G}{\rho_L}\right)^{1/2}$ | $\frac{\pi}{24}=0.131$ |
| Zuber (1961) | 0.12–0.157 | |
| Chang and Snyder (1960) | $0.145\left(\frac{\rho_L+\rho_G}{\rho_L}\right)^{1/2}$ | |
| Moissis and Berenson (1963) | $0.18\frac{[(\rho_L+\rho_G)/\rho_L]^{1/2}}{1+2(\rho_G/\rho_L)^{1/2}+\rho_G/\rho_L}$ | |
| Borishanskii (1956) | $0.13+4\left\{\frac{\rho_L\sigma^{3/2}}{\mu_L^2[g(\rho_L-\rho_G)]^{1/2}}\right\}^{-0.4}$ | Includes effect of viscosity through dimensional analysis |

General form: $$\frac{q''_{crit}}{\rho_G h_{LG}}=F$$

| Author | $F$ | Remarks |
|---|---|---|
| Rohsenow and Griffith (1956) | $12.1\times10^{-3}\left(\frac{\rho_L-\rho_G}{\rho_G}\right)^{0.6}$ (m/s) | Dimensional, for water only |
| Noyes (1963) | $0.144\left(\frac{\rho_L-\rho_G}{\rho_G}\right)^{1/2}\left(\frac{g\sigma}{\rho_L}\right)^{1/4}(\mathrm{Pr}_L)^{-0.245}$ | Liquid metals only |
| Noyes and Lurie (1966) | $0.16\left[\frac{\sigma g(\rho_L-\rho_G)}{\rho_G^2}\right]^{1/4}+\frac{K_{NL}}{\rho_G h_{LG}}$ | Liquid metals only |

$K_{NL}$ is a constant depending on the fluid:
For sodium (0.014–1.5 bar), $K_{NL}=1.26$ MW/m²
For potassium (0.01–1.5 bar), $K_{NL}=0.95$ MW/m²

Disregarding the effects of nuisance variables, such as surface finish, dissolved-gas content, and fouling, which have a small effect, in saturated pool boiling the left side of Eq. (11) is function of system pressure only. The constants and $f(\rho_L/\rho_G)$ vary according to the analytical model on which the empirical correlations are based.

Lienhard and Schrock (1963), Lienhard and Watanabe (1966), and Cobb and Park (1969) used the law of corresponding states to correlate CHF data for a variety of fluids. The pressure dependence is accounted for by universal correlations in terms of reduced thermodynamic variables.

Several of the available saturated pool-boiling correlations are given in Table 1. Figure 7 shows the predicted variation of the CHF with pressure for saturated boiling of water. Zuber's (1958) CHF correlation is used widely and is also the basis for a number of corrections that account for effects discussed below.

**4.1.2 Effect of subcooling** The CHF increases linearly with subcooling as found by several investigators, including Kutateladze (1952), Zuber et al. (1961), and Ivey and Morris (1962).

The correction for the subcooling effect has the general form of the equation proposed by Ivey and Morris (1962):

$$\frac{q''_{\text{crit, sub}}}{q''_{\text{crit, sat}}} = 1 + 0.1 \left(\frac{\rho_L}{\rho_G}\right)^{3/4} \frac{c_L(T_{\text{sat}} - T_L)}{h_{LG}} \tag{12}$$

although the group of variables multiplying the subcooling and the coefficients may differ.

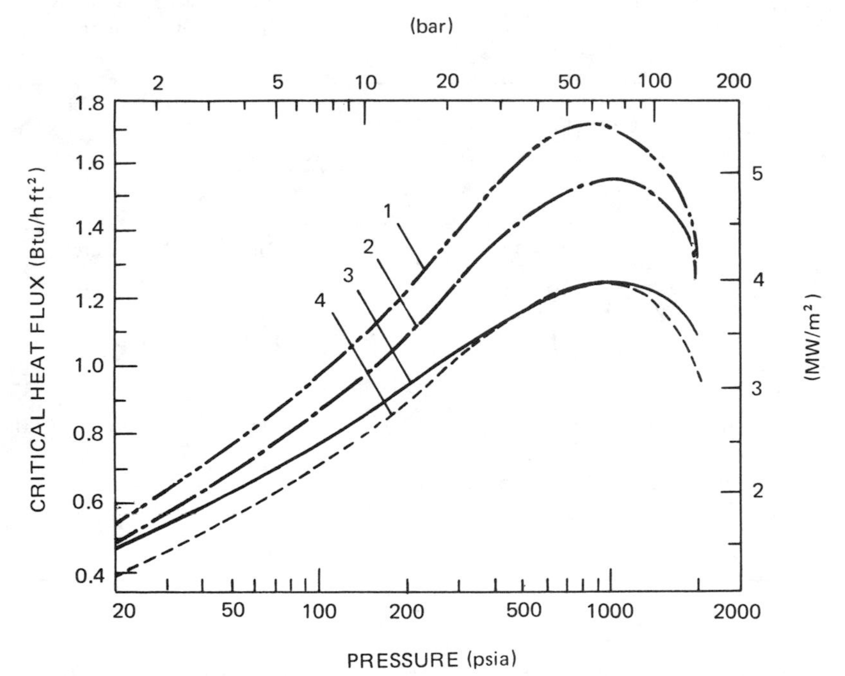


**Figure 7** Variation of the pool-boiling critical heat flux with pressure for saturated water (Lahey and Moody, 1977). Line 1, Zuber (1958), with a recommended value of 0.18 of the coefficient; line 2, Kutateladze (1952); line 3, Rohsenow and Griffith (1956); and line 4, Zuber (1958), with the original value of 0.131 of the coefficient.

**4.1.3 Other parametric effects** The effect of heater geometry on the CHF has been investigated extensively by Lienhard and his coworkers (Sun and Lienhard, 1970; Lienhard and Dhir, 1973). Lienhard and Dhir presented a unified treatment of critical heat fluxes from bodies of various shapes based on the hydrodynamic instability theory. They essentially correct the value of the critical heat flux given by Zuber's (1958) prediction in terms of a characteristic dimensionless length of the heater:

$$L' = \frac{L}{\lambda}$$

where $\lambda$ is the wavelength of the instability given by Eq. (10).

Contrary to what happens in nucleate boiling, surface finish has a minor effect on CHF, in agreement with the hydrodynamic-instability explanation of the phenomena. Ivey and Morris (1962) reported variations of $q''_{\text{crit}}$ of less than 20% for a wide variety of surface materials and finishes.

Other parametric effects are discussed in the reviews of pool-boiling CHF by Gambill (1966, 1968a) and Bergles (1975).

## 4.2 Burnout in Flow Boiling

Our understanding of flow-boiling burnout phenomena is not complete and our ability to predict burnout in convective boiling is far inferior to our ability to predict DNB in pool boiling. This is because burnout in flow boiling is a more complicated phenomenon. The larger number of independent variables and the several possible physical mechanisms that may lead to burnout also considerably complicate the problem.

The occurrence of burnout may be precipitated in a system experiencing unstable flow. Burnout in the presence of flow oscillations is not discussed, however, since it is not possible to account for the effect of flow oscillations on burnout heat flux.

**4.2.1 Physical mechanisms and models** In a typical experimental situation and for a given test section and fluid, critical heat flux data are obtained by fixing system operating conditions such as exit pressure, test-section-inlet subcooling, and mass flux, and by increasing the heat flux until the heated wall temperature exhibits oscillations closely followed by a temperature excursion. For uniformly heated test sections the location where the temperature excursion appears first is with very few exceptions (Groeneveld, 1974) the end of the heated length. Upstream burnout can be observed in test sections that are not uniformly heated and/or in the presence of local flow perturbating devices, such as spacers in rod bunbles (Lahey and Moody, 1977).

The burnout mechanisms are quite different in subcooled or low-quality flow where heat transfer takes place by nucleation and in high-quality flow, where the likely heat transfer mode is forced convection evaporation with an annular-flow pattern. In subcooled flow we can properly speak of DNB, while the burnout in the high-quality flow occurs by dryout of the liquid film on the wall.

DNB in flow boiling resembles DNB in pool boiling since in both types of boiling the boiling crisis results from blanketing of the wall with vapor. It is evident then that burnout when the wall is blanketed with vapor will depend mostly on local conditions and on surface-proximity parameters such as the local void distribution, the thickness and superheat of the boundary layer, and possibly, the surface conditions and materials.

Burnout in high-quality flow is associated with the removal of the liquid film from the wall by dryout or some other disruptive cause. Since the film characteristics depend in an integral fashion on the flow history from the beginning of the heated length up to the dryout point, burnout in this case will also depend on flow history.

The DNB transition in subcooled and low-quality flow is generally very fast (fast burnout) and dramatic, often leading to physical destruction of the heater. On the contrary, the dryout transition is generally slower (slow burnout) and the resulting change in wall temperature less pronounced.

The various possible burnout mechanisms have been reviewed by Tong (1972), Tong and Hewitt (1972), and more recently by Hewitt (1977) and Bergles (1977a, 1977b). According to Tong and Hewitt, burnout in subcooled bubbly or slug flow may be caused by

1. Overheating of the wall surface at a nucleation site resulting from dewetting of the surface under a bubble prior to detachment; this mechanism is postulated to occur at high subcooling and mass fluxes.
2. Crowding of the bubbles near the surface as a result of insufficient heat exchange with the subcooled liquid core leading to growth of the boundary layer and vapor blanketing. This phenomenon is thought to be controlled either by a critical value of the superheat in the liquid layer near the wall (Tong et al., 1965) or by boundary-layer separation or blowoff as a result of the injection of vapor bubbles from the wall into the main stream. The heat exchanges between the boundary layer and the liquid core have also been modeled on the basis of the Reynolds flux concept: Vapor injection into the flow stream reduces the velocity gradient near the wall. As the vapor injection increases, the boundary layer thickens, the bubbles become almost stagnant, and DNB occurs.
3. Formation of a dry spot during the passage of a large slug of vapor in slug flow. This mechanism was observed by Fiori and Bergles (1970) and is responsible for dryout in subcooled, low-pressure flows where the formation of large slugs of vapor is possible.

Models of the burnout process based on the boundary layer separation mechanism were proposed by Kutateladze and Leont'ev (1966). Tong (1968, 1975) obtained fairly successful correlations of burnout data starting from boundary-layer separation considerations. Purcupile and Gouse (1972) employed the Reynolds flux concept to derive a DNB correlation.

Postulated burnout mechanisms in saturated annular flow are discussed by Tong and Hewitt (1972) and Hewitt (1977). These mechanisms include the formation of a vapor film under the liquid film, sudden disruption of the liquid film as a result of some interface instability, and dry-patch formation. The fact, however, that burnout in high-quality annular flow is a result of dryout of the liquid film on the wall (Isbin et al., 1961; Hewitt et al., 1965–1966) is now quite well established: Dryout is controlled by the competing processes of liquid loss from the film because of evaporation and liquid entrainment and liquid addition from droplet deposition. Dryout occurs near the point where, as a result of these processes, the thickness of the liquid film is reduced to zero. A significant amount of liquid may remain present in the tube at the burnout point as entrained droplets.

According to the dryout model outlined above, burnout in annular flow would be entirely controlled by hydrodynamic entrainment and redeposition considerations. This seems indeed to be the case, at least as a first approximation, since experiments by Hewitt (1970) have shown that the value of the heat flux does not affect significantly the entrainment and deposition mechanisms. The key then to the prediction of burnout in annular flow is an adequate description of the liquid entrainment and deposition rates.

Information on entrainment in annular flow can be found in books by Wallis (1969) and Hewitt and Hall-Taylor (1970), and in articles by Hutchinson and Whalley (1973) and others. Hewitt (1977), summarizing the ongoing development work in this area at Harwell, showed that fairly reliable predictions of the dryout point in tubes, as well as in bundles and annuli, under steady-state and transient conditions can be obtained using predictive methods based on the physical models outlined above and empirical information on liquid entrainment.

Under certain accident conditions, the flow rate through the core of LWRs may be very low. As a result, the core is largely voided, and burnout may occur because of the absence of a sufficient quantity of liquid in the core. In BWRs, for example, at a sufficiently low inlet flow rate, most of the liquid entering the bundles is vaporized. Even if water is available in the upper plenum, it may be prevented from flowing down in the bundle if the velocity of the exiting steam is sufficiently large. This countercurrent-flow limiting condition (CCFL) may lead to CCFL-limited burnout, as discussed by Lahey and Moody (1977).

Most of the available CHF correlations use the critical quality and mass flux as correlation parameters. In quasi-stagnant-flow situations resembling to some extent pool boiling, these correlations are not applicable. Moreover, under the countercurrent-flow conditions prevailing in these situations even the definition of the quality loses its meaning. A correlation of CHF data in terms of the local void fraction seems more appropriate. Walkush (1974) proposed to use the pool-boiling CHF value given by Zuber's (1958) correlation, corrected as a function of the local void fraction. The correction factor is 0.9 for void fractions inferior to 0.2 and decreases linearly with the void fraction, reaching zero at a void fraction of one. Work is needed to validate analytical methods in this area. Practically all the available CHF correlations are based on data obtained in cocurrent upward flow. Worley and Griffith (1973), adopting the assumption of CHF dependence on local void fraction (instead of local flow quality), proposed a method for converting upward flow correlations to downward flow situations on the basis of equal critical void fraction at the burnout point.

**4.2.2 Parametric effects** In general the critical heat flux $q''_{\text{crit}}$ will be a function of system geometry (tube, annulus, bundle, with or without spacers, etc.) and orientation, heat flux distribution (for example, axial and radial heat flux distribution in a rod bundle), fluid, operating pressure, inlet subcooling, and mass flux, and could also depend on secondary variables that are difficult to control, such as heated-surface characteristics and condition, and gas content of the fluid.

Disregarding the effect of secondary variables and for a given geometry and fluid, the critical heat flux will be a function of the hydraulic diameter $D$, test section length $L$, inlet subcooling $\Delta h_{\text{in}}$, mass flux $G$, and pressure $p$:

$$q''_{\text{crit}} = q''_{\text{crit}}(D, L, \Delta h_{\text{in}}, G, p)$$

Using the heat balance relationship, the inlet subcooling may be replaced by the critical enthalpy or quality $x_{\text{crit}}$ at the burnout point,

$$q''_{\text{crit}} = q''_{\text{crit}}(D, L, x_{\text{crit}}, G, p) \tag{13}$$

The six variables of Eq. (13) are not independent, and any change in operating conditions results in changes in at least two variables. Thus it is not possible to examine parametric effects by varying one parameter at a time; this inherent limitation obscures the discussion of parametric effects if sufficient care is not taken to specify which variables are held constant.

A detailed discussion of parametric effects in burnout can be found in Collier (1972). These can be summarized as follows (Hewitt, 1977):

1. Everything else being constant, the CHF usually increases linearly with inlet subcooling, as shown in Fig. 8.
2. For constant $L$, $D$, and $\Delta h_{\text{in}}$, the CHF increases with the mass flux, as also shown in Fig. 8. The mass flux effect is stronger at low mass fluxes.
3. For constant $G$, $D$, and $\Delta h_{\text{in}}$, the CHF decreases with increasing channel length; the critical power required to achieve burnout, however, increases with increasing length.
4. For constant $G$, $\Delta h_{\text{in}}$, and $L$, the CHF increases with the hydraulic diameter; the effect is most pronounced at small diameters.

Burnout data obtained at constant $G$, $L$, $D$, and $p$ are often represented by a straight line in the $(q''_{\text{crit}}, x_{\text{crit}})$ plane, as shown in Fig. 9. If the occurrence of the CHF condition depended only on local variables $(G, x_{\text{crit}}, D, p)$, data obtained with different heat flux distributions and with test sections of various lengths would have fallen on the same line in this plane. This would have shown that the occurrence of CHF is independent of the flow history from channel inlet up to the burnout point.

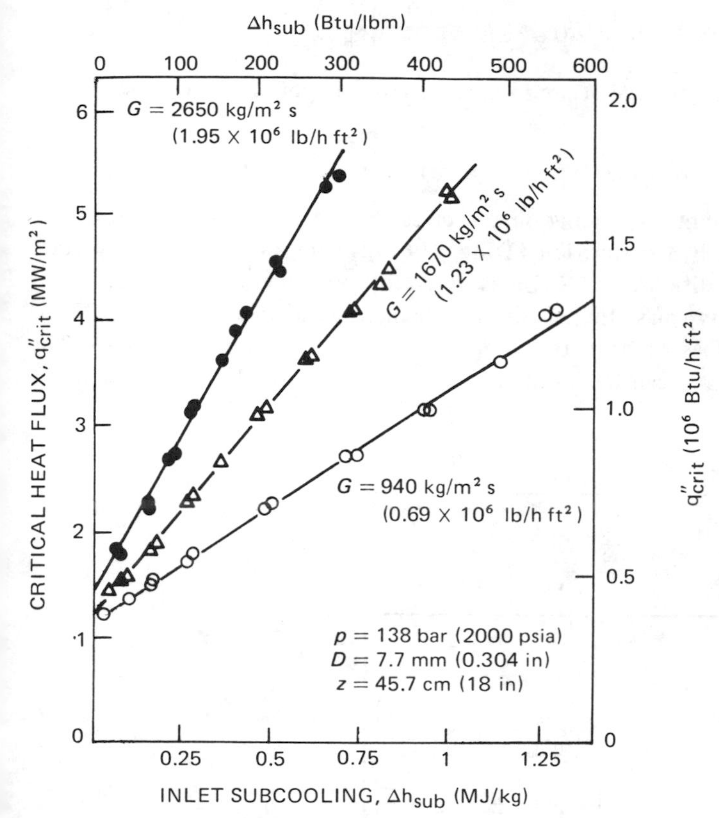


**Figure 8** Effect of inlet subcooling and mass flux on critical heat flux (Weatherhead, 1963).

The assumption of absence of flow memory or independence from past flow history is referred to as the local-conditions hypothesis, and it may be true only for subcooled or low-quality burnout, as mentioned above. Figure 9 shows that this assumption is certainly not true when burnout occurs at relatively high quality.

**4.2.3 Empirical correlations** In the absence of comprehensive analytical models of the burnout phenomena, empirical correlations or CHF data must be used for design purposes. While the form of some correlations is based on analytical developments, some of the most successful and widely used correlations were obtained as numerical fits of a large number of data points. Comprehensive compilations of available CHF correlations and data bases were prepared by Gambill (1968b), Tong (1972), and Collier (1972).

One of the later, most general and widely used CHF correlations for uniformly heated round tubes was developed by Thompson and Macbeth (1964). This correlation is based on the local-conditions hypothesis and assumes linearity in the $(q''_{crit}, x_{crit})$ plane. If the relationship between inlet subcooling and exit quality is used, the correlation can be put in either of the two following forms:

$$q''_{\text{crit}} = A - C\frac{D}{4}Gh_{LG}x_{\text{crit}} = \frac{A + C(D/4)G\,\Delta h_{\text{in}}}{1 + CL}$$

The empirical coefficients $C$ and $A$ are in essence given as power functions of $D$ and $G$:

$$A = y_0 D^{y_1} G^{y_2} \qquad C = y_3 D^{y_4} G^{y_5}$$

where the $y$'s are constants depending on the pressure range.

The Westinghouse *W*-3 correlation (Tong, 1967) is being widely used to predict CHF under PWR conditions; the CHF is given by a numerical fit in terms of pressure, critical quality, mass flux, hydraulic diameter, and inlet subcooling. Thus the *W*-3 correlation does not use the simplifying local-conditions hypothesis. The effect of axial flux shape can be taken into account using the flux shape factor $F_c$ of Tong et al. (1965):

$$F_c \equiv \frac{q''_{\text{crit},U}(z_{\text{crit}})}{q''_{\text{crit},NU}(z_{\text{crit}})} = \frac{C}{1 - e^{-Cz_{\text{crit}}}} \int_0^{z_{\text{crit}}} \frac{q''(z)}{q''(z_{\text{crit}})} e^{-C(z_{\text{crit}}-z)}\,dz$$

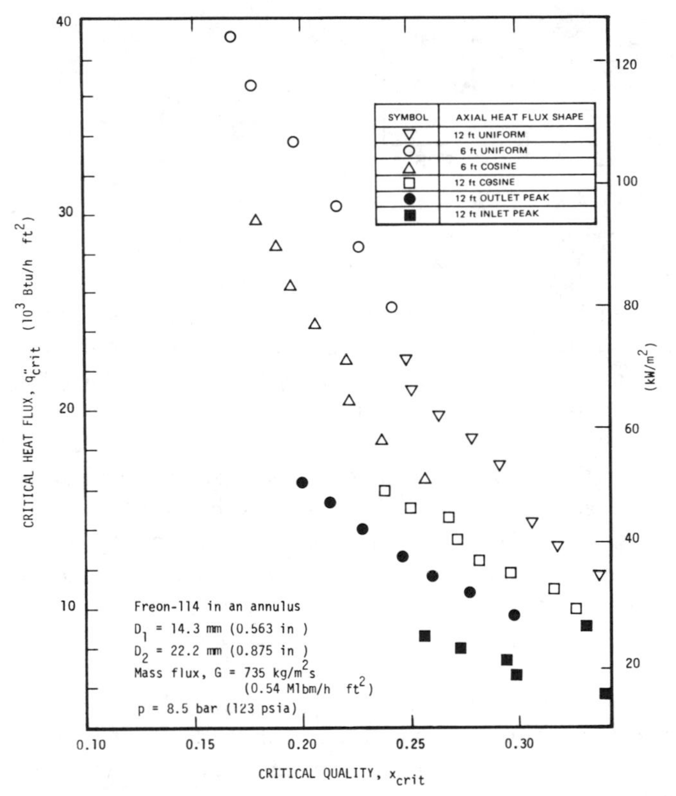


**Figure 9** Critical heat flux data represented in the quality–heat flux plane. Various axial heat flux shapes (Shiralkar, 1972).

where the subscripts $U$ and $NU$ refer to the uniform and nonuniform heat flux distributions, respectively. The local critical nonuniform-distribution heat flux is calculated by dividing the $q''_{crit,U}$ value provided by the $W$-3 correlation by $F_c$. The constant $C$ is given empirically in terms of $x_{crit}$ and $G$.

For BWR design, limit lines that constitute a lower envelope to all available CHF data have been used in the past. The Janssen and Levy (1962) lines provided the CHF values as simple linear functions of mass flux and critical quality. The Janssen-Levy limit lines were subsequently replaced by the Hench-Levy limit lines (Healzer et al., 1966). To eliminate some undesirable features inherent in the limit-line approach, a new correlation, known as the General Electric Company critical quality–boiling length correlation (GEXL) was developed (General Electric Company, 1973). The critical quality against boiling length correlation concept was introduced by workers at CISE in Italy (Bertoletti et al., 1965) and has resulted in a number of quite successful CHF correlations (Gaspari et al., 1974) capable of correlating nonuniform heat flux data. The upstream history is accounted by the dependence on the critical boiling length, $L_B$, i.e., the distance from the point where the fluid reaches saturation to the burnout point. The CISE correlations have the following functional form:

$$x_{crit} = \frac{a(p, G)L_B}{b(p, G, D) + L_B}$$

Figure 10 shows the data of Fig. 9 plotted now in the ($x_{crit}$, $L_B$) plane, confirming the adequacy of the boiling length concept in correlating data obtained under axially nonuniform heat flux conditions.

The effect of radial nonuniformities in the heat flux distribution in a rod bundle can be taken into account by the use of empirical factors correcting the results obtained using bundle-average parameters, such as the "generalized local peaking factor $R$" used with the GEXL correlation. An approach that, in principle, is more satisfactory consists of performing first a subchannel analysis to determine the distributions of mass flux and enthalpy in the bundle. A CHF correlation can then be applied to the hot subchannel.

This approach is indeed followed to predict CHFs in PWRs using the $W$-3 correlation (Tong, 1969). The subchannel analysis approach has not been very successful yet in BWR applications, mainly because of the difficulties inherent in the formulation of high-quality, two-phase flow mixing and subchannel analysis. Lahey and Moody (1977) and Guarino et al. (1974) discussed these problems.

## 5 REWETTING

During the last few years, a large number of experimental and theoretical studies have been devoted to the investigation of the physical mechanisms involved in the rewetting of a hot, dry surface. Our present physical understanding of the rewetting phenomena, the various proposed analytical descriptions of the rewetting process, and some information on available semiempirical rewetting correlations are discussed in this section. Additional details can be found in recent reviews of the subject by

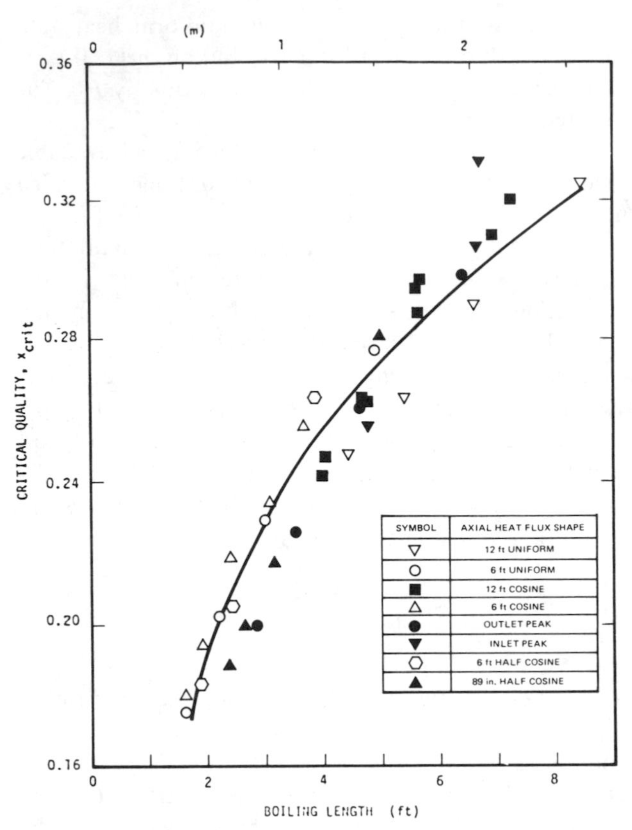


**Figure 10** Critical heat flux data of Fig. 9 represented in the critical quality–critical boiling length plane with additional data from shorter lengths (Shiralkar, 1972).

Sawan and Carbon (1975), Butterworth and Owen (1975), Elias and Yadigaroglu (1977b, 1978), and McAssey and Bonilla (1977).

## 5.1 Rewetting Mechanism of the Nuclear Fuel Rods

Rewetting, or quenching, of a hot surface can be defined as the transition from a regime in which heat transfer takes place predominantly through a vapor layer covering the surface (film boiling) to a regime in which the liquid is in direct contact with a large fraction of the wall (nucleate boiling and forced convection vaporization). However, droplets may make temporary contact with the hot wall in film boiling, and small vapor patches may develop for short times on the wall in nucleate boiling. In transition boiling, portions of the surface alternate between these two regimes.

Rewetting occurs in emergency cooling by both top spraying and by bottom reflooding. During rewetting by top spraying, a liquid film attaches to the upper front of the rods; rewetting and sputtering take place at the edge of this film as it slowly moves downward. During rewetting by bottom reflooding a two-phase mixture rises around the fuel rods, and the quenching front moves upward; at any time this front might be well below the froth level. During both rewetting conditions the velocity of the quench front is either constant or varies slowly.

Figure 11 shows the axial wall temperature profiles recorded during a recent series of bottom-reflooding experiments. The quenching, or rewetting, region is defined as the narrow region where the wall temperature drops from values characteristic of film boiling to temperatures characteristic of nucleate boiling. The exact shape of the temperature profile in this region cannot be easily determined, as discussed below. Figure 12 shows a thermocouple temperature trace obtained during the same bottom-reflooding experiments. We notice that the transition from one heat transfer regime to another, as the quench front moves over the location of the thermocouple, is very fast.

Several definitions of a rewetting temperature are possible, as discussed in Sec. 5.2. We will define the rewetting temperature loosely as the wall temperature at which the dramatic transition from film-boiling-like to nucleate-boiling-like heat transfer takes place. If we plot the heat transfer coefficient distribution relating to the wall temperature along the quench front region, as shown in Fig. 13, the

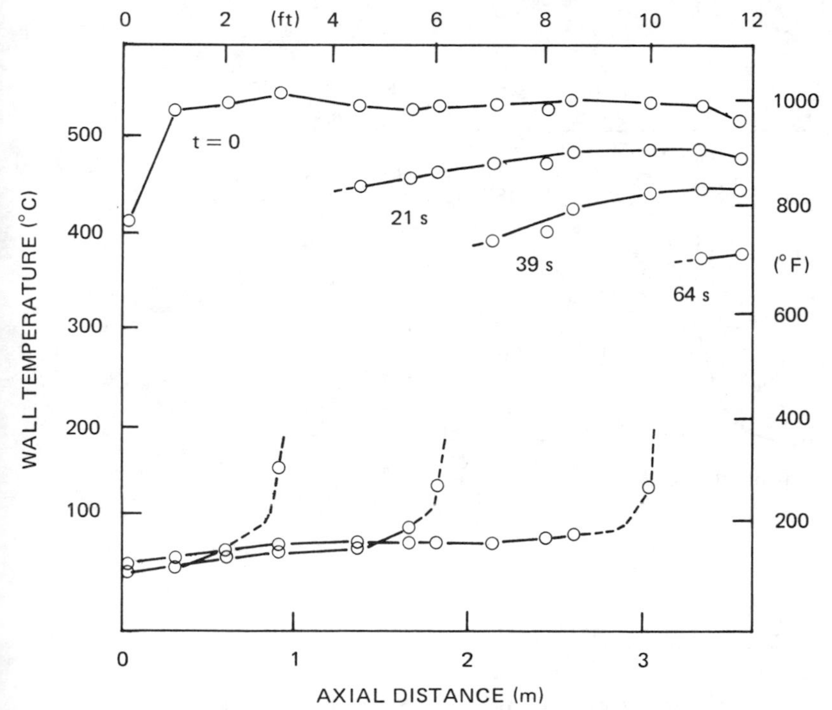


**Figure 11** Variation of the axial temperature profile in time during a bottom-reflooding experiment (run 128, $V_{in} = 124$ mm/s, $T_{in} = 25°C$, atm. pressure) (Yu, 1978).

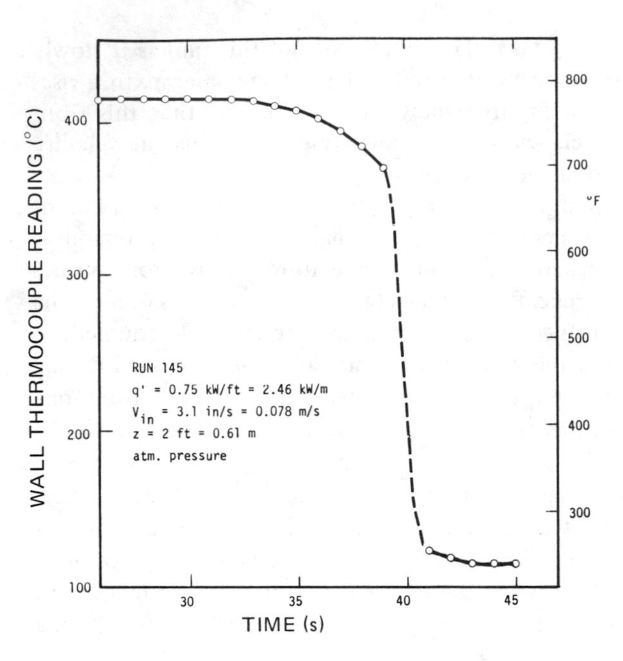


**Figure 12** External wall thermocouple reading during bottom-reflooding experiment (Yu, 1978).

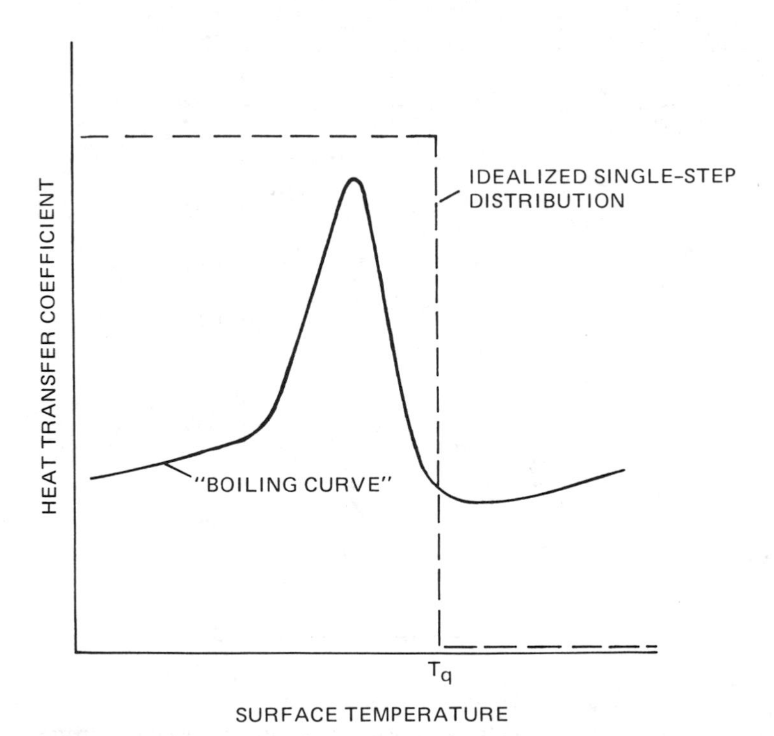


**Figure 13** Variation of the heat transfer coefficient in the quench-front region.

rewetting temperature can be associated with some discontinuity, maximum, or minimum of this distribution. The quench front is the location where the rewetting temperature is reached.

The exact definition of a rewetting temperature is only important when idealized heat transfer coefficient distributions are assumed to exist, as shown in Fig. 13. When a complete description of a boiling curve [that is, of the variation of the heat transfer coefficient with wall temperature and possibly with local flow conditions, $h = h(T, G, x, p, \ldots)$] along the quench region is provided, there is no need to identify a particular rewetting temperature. In this sense the rewetting point and the quench front are situated somewhere on the steep-gradient part of the quenching profiles of Figs. 11 and 12.

According to the physical picture used in rewetting models, the temperature profile in the quench-front region moves along the rod without changing its shape. This quench front moves by cooling the surface ahead of it to a rewetting temperature below which the rod quenches. It is often assumed that below the rewetting temperature, heat is removed by transition boiling, nucleate boiling, and convective heat transfer to the liquid. The hot portion of the rod in front of the quench front experiences, in turn, free and forced convective cooling by steam, dispersed-flow film boiling, and continuous film boiling.

**5.1.1 Importance of axial conduction** Axial conduction (especially in the cladding) plays an important role in the rewetting process. Heat from the hot portion of the rod downstream of the quench front, where the heat transfer coefficient is low, is transported by axial conduction toward the colder portion of the rod where good heat transfer to the coolant takes place. Heat from the hot portion of the rod is also evacuated by direct convection to the coolant. This heat transfer mechanism is often referred to as precursory cooling.

The relative importance of these two cooling mechanisms depends on the physical situation. With top-spray cooling, very poor heat transfer may exist ahead of the quench front, and axial conduction will clearly be the dominant cooling mechanism. With bottom reflooding, the cooling of the hot surface ahead of the wet front may be significant, in which case both axial conduction and precursory cooling may be important.

In some models of the rewetting process during bottom reflooding, such as those of Martini and Premoli (1972) and Kirchner (1976), the effects of axial conduction are ignored, and it is postulated that the rod rewets when its surface temperature, calculated by taking into account only radial conduction, falls below a rewetting temperature, chosen to be an elevated temperature. Kirchner obtained satisfactory agreement with experimental data by this procedure. The choice of a high rewetting temperature artificially increases convection to the coolant in the vicinity of the quench front (since the heat transfer coefficient increases dramatically below the quench temperature) and thus compensates for the absence of axial conduction in these models. It seems that axial conduction may be neglected only if no sharp discontinuities in the heat transfer coefficient distribution produce, in turn, steep axial temperature gradients. Complete resolution of these questions requires measurements of the heat transfer coefficient distribution along the quench front region, a difficult task, as discussed in Sec. 5.2.

Some investigators have defined as rewetting temperature the temperature at the downward knee of the temperature traces of Figs. 11 or 12, since quenching follows closely in time and in space the attainment of this temperature. If, following Kirchner's (1976) approach, we ignore any effects of axial conduction and assume that rewetting occurs practically as soon as this rewetting temperature is reached, the progression of the quench front becomes completely determined by the wall temperature history downstream of the quench front.

It seems, however, more reasonable to consider the temperature at the knee of the traces as the "initial" wall temperature (or "temperature at infinity") for the axial-conduction dominated phenomena taking place in the narrow quench region. In this approach cooling of the wall ahead of the quench front provides the boundary conditions for the phenomena taking place in the quench region. The velocity of the quench front is determined by the phenomena in the quench region itself and not only by the temperature history of the wall downstream. The cooling of the wall downstream of the quench front remains, however, a very important consideration since it sets the conditions along the path of the quench front. These conditions, that is, the initial wall temperature, in turn influence to a considerable extent the velocity of the quench front.

Support for the second approach comes from the experimental fact that the temperature at the knee of the temperature trace varies widely. Moreover there has been considerable success in explaining and predicting quench velocity trends using axial-conduction models of the quench-front region.

**5.1.2 Hydrodynamically controlled rewetting** When the rods are cooled by a spray entering from the top of the bundle, a countercurrent flow of rising steam and falling droplets and liquid film is created. If the velocity of the rising steam exceeds a certain critical velocity, the progression of the liquid film is halted, and the droplets may be prevented from falling downward (a hydrodynamically controlled "flooding" situation). In this case, the progression of the quench front will also be hydrodynamically controlled. This situation has been examined by Chan and Grolmes (1975).

All investigators assumed that rewetting is conduction-controlled and are not concerned with the hydrodynamics and stability of the liquid film spreading on the hot surface. The stability of hot patches was examined by Zuber and Staub (1966) and by Shiralkar and Lahey (1973) in a different context.

## 5.2 Rewetting Temperature

In flow boiling, a universally accepted definition of a rewetting temperature, often referred to also as sputtering, quenching, calefaction, minimum film boiling, or Leidenfrost temperature, does not exist. These terms are not exactly synonymous. Various approaches taken in attempts to define a rewetting temperature are reviewed in this section.

The Leidenfrost temperature is generally defined as the maximum temperature at which a small droplet floating on a vapor film over a hot surface eventually collapses and touches the wall. In pool boiling, the minimum of the classical pool-boiling curve is referred to as the minimum film boiling temperature.

Spiegler et al. (1963) considered the rewetting temperature as a thermodynamic property and related it directly to the thermodynamic critical temperature. In the analyses of Henry (1974) and Baumeister and Simon (1973), however, the rewetting temperature is obtained from models in which both the hydrodynamics of film boiling and transient conduction effects in the wall are considered. In this case, properties of the wall enter into the expression for the rewetting temperature. Moreover, it appears that the rewetting temperature in rod-quenching experiments also depends on the surface condition, as noted by Shires et al. (1964).

In dispersed-flow film boiling, the wall may be cooled both by convection to steam and by the droplets entering the vapor boundary layer and possibly contacting the wall. As the wall temperature increases, heat transfer by these two modes varies in opposite directions, and a minimum in the boiling curve is produced. According to these considerations by Iloeje et al. (1975), there is no fundamental change in the heat transfer mechanisms above and below this minimum film boiling temperature. However, the part of the curve with a negative slope is dynamically unstable, and in heat flux-controlled cases, a temperature excursion down to the nucleate boiling wall temperature is observed.

This limited review of the phenomena involved shows that our level of understanding of the basic rewetting mechanisms is still uncertain. Moreover, making accurate experimental measurements of the rewetting temperature and of the heat transfer coefficient distribution in the quench-front region is extremely difficult, since the phenomena are highly transient and involve very steep temperature gradients. The temperature trace of Fig. 12 was obtained with a thermocouple welded to the outside surface of a relatively thin-wall (0.76 mm), internally cooled test section. The transition from film boiling to nucleate boiling temperatures shown by this trace is extremely fast, and it could be even faster at the inside surface. Indeed for a rewetting-front velocity range of 3-50 mm/s, an estimated width of the quench-front region of, say, 10 mm, and temperature differences of as much as several hundred degrees, the average rate of change of the wall temperature is of the order of 100°C/s. The average axial temperature gradients are of the order of maybe 20-50°C/mm, and in sufficiently thick test sections, there could be appreciable radial temperature gradients also. It becomes evident that the rewetting temperature transient and the rewetting velocity cannot be recorded accurately.

This uncertainty is reflected in the large variation of available data on rewetting temperatures and on the values of the heat transfer coefficient near the quench front.

## 5.3 Analytical Solutions of the Axial-conduction-controlled Rewetting Problem

The following simplifications are generally made to solve analytically the conduction equation for a wall undergoing quenching:

1. The wall is a homogenous lamina, or tube, of infinite length and uniform thickness $\delta$ having constant physical and thermal properties.
2. The quench front velocity ($U$) is constant and the temperature ($T$) distribution is invariant in a coordinate system that moves along with the wet front. Axial coordinate ($z$) and time ($t$) variations are then related by

$$\frac{\partial T}{\partial t} = -U \frac{\partial T}{\partial z}$$

3. One side of the wall is insulated while the other (the wet side) is cooled.
4. There is no heat generation in the wall.
5. The reference coolant temperature $T_c$ is constant and equal to the saturation temperature $T_s$.

The last three assumptions can sometimes be easily relaxed. The second assumption is usually justified since Bukur and Isbin (1972) have shown that the quenching front reaches an asymptotic constant velocity very rapidly. With the above assumptions, the governing equation in Cartesian coordinates* reduces to

$$\frac{\partial^2 T}{\partial y^2} + \frac{\partial^2 T}{\partial z^2} + \frac{\rho c U}{k} \frac{\partial T}{\partial z} = 0 \tag{14}$$

where $y$ is the radial coordinate, $c$ is the specific heat, $k$ is the thermal conductivity, and $\rho$ is the density of the wall. The variation of the heat transfer coefficient $h$ with surface temperature must be specified and used as a boundary condition.

Dimensional analysis shows that the nondimensional temperature

$$\theta \equiv \frac{T - T_s}{T_q - T_s} \tag{15}$$

will depend on the Biot and Péclet numbers

$$\mathrm{Bi} \equiv \frac{h\delta}{k} \tag{16}$$

$$\mathrm{Pe} \equiv \frac{U \delta \rho c}{k} \tag{17}$$

and on a temperature ratio, often defined as

$$\theta_w \equiv \frac{T_w - T_s}{T_q - T_s} \tag{18}$$

involving the hot wall temperature at infinity, or initial wall temperature $T_w$. The rewetting temperature $T_q$ enters into the solution of the problem as the temperature at which a discontinuity in the variation of $h$ occurs.

One- and two-dimensional analytical solutions available in the literature were

*This equation can also be written and solved in cylindrical coordinates; see, for example, Blair (1975) and Yeh (1975).

reviewed in detail by Elias and Yadigaroglu (1977b, 1978). Equation (14) determines the complete two-dimensional problem. One-dimensional solutions are obtained by ignoring any transverse temperature gradients in the wall and assuming that the temperature is uniform throughout the wall thickness. This assumption holds only for thin walls and low values of the heat transfer coefficient and of the quench velocity, that is, for small Péclet and Biot numbers ($\mathrm{Bi} < 1$, $\mathrm{Pe} < 1$).

All the analytical solutions presented in the literature consider only the cladding (or a solid rod having uniform composition) and ignore heat storage and transient conduction in the fuel pellet as well as in the gap between the pellet and the cladding. Numerical, finite-difference methods must be used to treat the complete fuel-rod problem realistically. Note that the analytical solution for a solid rod having the properties of the cladding does not represent an actual fuel rod better, since the quench-front propagation is largely a surface phenomenon, and the conductivities of the cladding and the fuel pellet are quite dissimilar.

The resistance to heat transfer in the gap decouples, to some extent, transient conduction in the cladding from conduction in the pellet. Hein et al. (1974) and Pearson et al. (1977) used two-dimensional conduction codes to show that a fuel rod with a significant gap resistance rewets at a faster rate than a rod with no gap resistance. This is explained by the fact that a lesser amount of heat is transferred to the fluid right at the quench front and a large fraction of the heat stored in the pellet is transferred to the liquid after quenching. The presence of filler material inside the fuel rod always slows down the quench-front progression.

Elias et al. (1976) used a simple hybrid model to investigate the effects of gap conductance. The model accounted analytically for axial conduction in the cladding and numerically for radial conduction in the pellet and the gap and showed that the rewetting velocity can be doubled if the cladding is completely isolated from the pellet.

**5.3.1 One-dimensional solutions** The wall is generally divided axially into a number of regions. The heat transfer coefficient is specified, and the conduction equation is solved in each region. The solutions are then matched by requiring continuity of temperature and axial heat flux at the boundaries of the regions. The only differences among the various available solutions stem from the assumed variation of the heat transfer coefficient with temperature and the number of regions used.

Séméria and Martinet (1965) and Yamanouchi (1968) were the first to propose one-dimensional solutions. The wall was divided axially into a wet and a dry region. The heat transfer coefficient was considered to be constant in the wet region and zero on the hot, dry surface ahead of the quench front. With this assumption, the rewetting velocity can be analytically determined as

$$U^{-1} = \frac{\rho c}{2}\sqrt{\frac{\delta}{hk}}\sqrt{\left[\frac{2(T_w - T_q)}{T_q - T_s} + 1\right]^2 - 1} \tag{19}$$

or in nondimensional notation

$$\frac{\sqrt{\mathrm{Bi}}}{\mathrm{Pe}} = \sqrt{\theta_w(\theta_w - 1)}$$

Several authors, including Sun et al. (1974a), Chun and Chon (1975), Thompson (1972), and Andréoni (1975) presented one-dimensional solutions varying only with respect to the number of axial regions defined. Thompson (1972) and Sun et al. (1974b) used exponential variations of the heat transfer coefficient in some regions. Ishii (1975) attempted to account for two-dimensional effects by using the penetration depth as an effective wall thickness in the sputtering region. Yao (1976) analyzed approximately the effect of internal heat generation. Thompson (1972, 1974a) presented one- and two-dimensional numerical solutions of the problem for a variation of the heat transfer coefficient with the third power of the wall superheat.

Elias and Yadigaroglu (1977a) developed a generalized approach by subdividing the wall axially into an arbitrary number of segments of specified thermal properties and heat transfer coefficient. In this "boiling curve" approach the wall temperature at the limits of all the segments is specified; no particular value of *a* rewetting temperature has to be specified. Heat generation in the wall and/or external heating of the wall by a variable heat flux coming from the fuel pellets are also included. The procedure is quite general and can be adapted to a variety of situations, provided that the behavior of the cladding remains essentially one dimensional.

**5.3.2 Two-dimensional solutions** Séméria and Martinet (1965) formulated the two-dimensional problem several years before the first solutions became available. The mathematical complications involved in an exact two-dimensional solution have forced all the investigators to deal with no more than two axial heat transfer regions.

Duffey and Porthouse (1973) solved Eq. (14) by separation of variables for $h = 0$ in the hot region and $h = \text{const}$ in the wetted region. The temperature distribution in each region is given by an infinite series. The two distributions must be matched at the boundary between the regions, and this match determines the values of the coefficients in the series expansion. Duffey and Porthouse obtained the rewetting velocity analytically by neglecting all terms but the first in the series. Blair (1975) has shown, however, that this approximation leads to inaccuracies, since the series converges slowly.

An accurate solution by separation of variables is presented by Coney (1974). The constants of the series expansion were calculated by matching the temperature and axial heat flux distributions in the wet and dry regions at a large number (up to 150) of discrete interface points, using a computer program. Coney's results are plotted in Fig. 14.

The two-dimensional solution for a cylindrical rod is given by Blair (1975) and later by Yeh (1975).

Explicit solutions for the limiting cases of large and small Péclet numbers were obtained by Tien and Yao (1974), who utilized the Wiener-Hopf technique and the kernel-substitution method. Functional relationships among the three dimensionless parameters are presented for the two-region model with no precursory cooling.

Recently Dua and Tien (1977), following the earlier work of Andersen and Hansen (1974), found approximate relationships between a modified Biot number,

$$\overline{\mathrm{Bi}} = \frac{\mathrm{Bi}}{\theta_w(\theta_w - 1)}$$

and the Péclet number. Using these relationships, they have fitted the theoretical results of Tien and Yao (1974) by an expression that is valid over the entire range of Biot numbers:

$$\mathrm{Pe} = [\overline{\mathrm{Bi}}(1 + 0.40\,\overline{\mathrm{Bi}})]^{1/2}$$

This simple formula agrees with the theoretical results at the limits $\mathrm{Bi} \gg 1$ and $\mathrm{Bi} \ll 1$ and provides an approximate interpolation for intermediate values of Bi that could be very useful for practical applications.

All the solutions discussed above ignore the effects of both precursory cooling and of the variation of the heat transfer coefficient in the wetted zone near the quench front. Piggott and Porthouse (1974) suggested, however, that an effective heat transfer coefficient be defined as the ratio of the average heat flux to the average temperature on the wet side, both averages being taken with respect to temperature.

Edwards and Mather (1973) considered exponential variation of the heat fluxes both upstream and downstream of the wet front. This assumption completely decouples the heat transfer inside the wall from heat transfer to the coolant, but it permits an exact solution. Their study demonstrates an increase in the wet-front velocity resulting from heat transfer immediately ahead of the quench front.

Dua and Tien (1976) also accounted for the effect of precursory cooling by

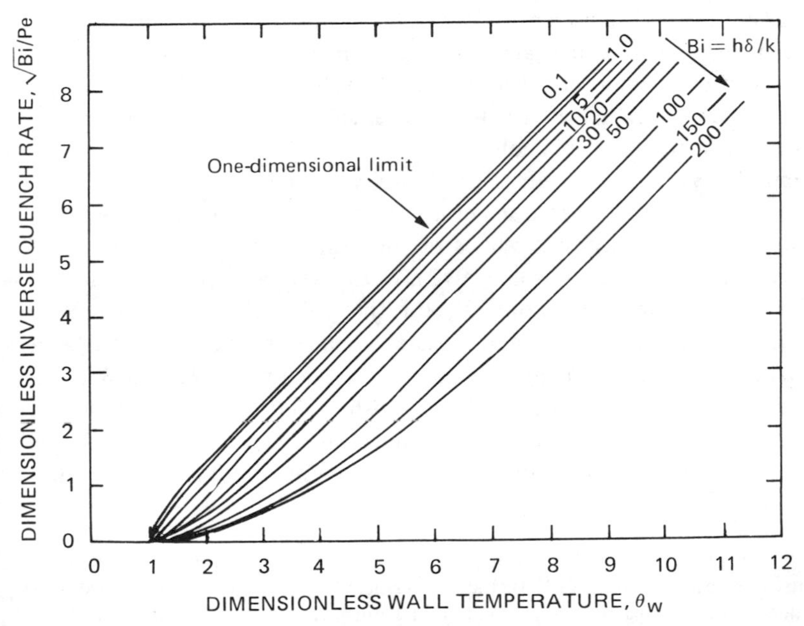


**Figure 14** The variation of $\sqrt{\mathrm{Bi}}/\mathrm{Pe}$ with dimensionless temperature $\theta_w$ as calculated by Coney (1974).

assuming an exponentially decaying heat flux downstream of the quench front. Their mathematical technique is identical to that of Tien and Yao (1974).

Thompson (1972) solved the two-dimensional conduction equation numerically and presented an extensive study of parametric trends (1974a). He showed that at high rewetting velocities, beyond a certain wall thickness, the wall thickness does not affect the rewetting rate. This wall thickness depends on thermal conductivity, initial wall temperature, and saturation temperature (but for practical purposes is approximately 0.5 mm). At a low rewetting rate the problem is of course one-dimensional, and the entire thickness of the wall, as well as the pellets, participate.

## 5.4 Parametric Dependence and Empirical Rewetting Correlations

Experimental data for spray cooling and bottom reflooding show a complicated dependence of the quench-front velocity on the system variables, including wall temperature, coolant flow rate and subcooling, pressure, wall material and geometry, and surface condition. In the analytical work described above the effects of wall geometry, material, and temperature are considered in the solution of the conduction equations. The effect of pressure can be included in the analysis by using the saturation temperature as the coolant temperature. The effects of coolant flow rate and subcooling are either incorporated in the wet-region heat transfer coefficient or are accounted for by precursory cooling.

Many experimental results have been reported in terms of inlet subcooling. The rewetting velocity depends, however, in general, on the local conditions at the position of the quench front, which are, in turn, an integral function of the system geometry and operational conditions. Yu et al. (1977) showed that some apparently conflicting data trends are explained when the local subcooling at the quench front is considered instead of the inlet subcooling.

Yamanouchi (1968) measured values of the velocity and rewetting temperature and, using Eq. (19), calculated heat transfer coefficients $h$ as high as $10^6$ W/m$^2$ °C for the wetted region.* Equally high values of the heat transfer coefficient in the wet region were derived by Thompson (1972). Both these authors used relatively low values of the rewetting temperature $T_q$ (of the order of 150–200°C). When higher values of the rewetting temperature are assumed (typically 260°C), the wet-side heat transfer coefficient acquires values in the vicinity of 20,000 W/m$^2$ °C, values of the order of magnitude of the heat transfer coefficient near burnout in pool boiling. Numerical exercises show that rewetting data can be correlated indifferently with a multiplicity of pairs of ($h$, $T_q$) values: When one of these

*Such values are not encountered in ordinary steady-boiling situations. However, there are indications, as discussed by Thompson (1972), that the critical heat flux could be increased by an order of magnitude by suppressing the hydrodynamic instability responsible for the disruption of nucleate boiling. Heat transfer would then be controlled by evaporation of a microlayer on the surface and could potentially result in extremely high heat fluxes.

parameters is specified, an average correlating value for the other can be extracted from available data. Chambré and Elias (1977) have indeed shown that the solution depends on the maximum value of the heat flux at the quench front, $q''_q = h(T_q - T_c)$. Yu et al. (1977) used the product $(T_q - T_c)\sqrt{h}$ as a correlating parameter and showed that the accuracy of the resulting correlation is insensitive to the particular value of $T_q$ used in this expression. $T_c$ is the coolant temperature at the quench front.

It is evident that the lack of direct, reliable measurements of the rewetting temperature and of the heat transfer coefficient in the immediate vicinity of the quench front leaves these questions unresolved. Attempts to include precursory cooling effects in the analysis add new parameters, parameters that are not directly measurable and that must also be extracted from available data. In the absence of direct measurements, the values obtained from simplified analytical models will have to be considered as fitting parameters with uncertain physical significance.

Several authors, including Yamanouchi (1968), Elliot and Rose (1970, 1971), and Thompson (1974b) correlated experimental data and obtained simple formulas for the rewetting velocity or the heat transfer coefficient in terms of system parameters. A review of these correlations and of the parametric trends was made by Elias and Yadigaroglu (1977b, 1978).

Yu et al. (1977) considered practically all the available falling-film data and a number of atmospheric-pressure bottom-reflooding data (a total of some 3750 points) and derived correlations of fairly broad validity. Their method is summarized below.

Yu et al. started from the two-dimensional solution of the axial conduction problem presented by Coney (1974), who had assumed a constant value of the heat transfer coefficient $h$ in the wet region and no heat transfer in the hot region. In essence the method consists of correlating the value of $h$ in terms of system pressure, mass flux, and local fluid temperature at the quench front, $T_c$. Instead of $h$, Yu et al. correlated the value of the product

$$F_c = (T_q - T_c)\sqrt{h}$$

There are two parameters in the correlation, $h$ and $T_q$. The authors found that in spite of the fact that the value of the rewetting superheat $T_q - T_s$ (optimized individually for different sets of data) varied between slightly negative values and ~200°C, overall correlations of all the data obtained by using an average value of 80°C were capable of predicting the data well. With this value of $T_q - T_s$, for falling films, the correlation takes the form

$$F_c = 4.52 \times 10^4 (1 + 0.036\,\Delta T_c G_p)(1 + 1.216 \log p)^{0.5} G_p^{0.0765/p} \tag{20}$$

where $\Delta T_c = T_s - T_c$ = subcooling at the quench front in degrees Celsius (0–90°C)
$G_p$ = flow rate per unit perimeter in kilograms per meter per second (0.06–1 kg/m s)
$p$ = pressure in bar (1–69 bar)

Atmospheric-pressure bottom reflooding data were correlated with $T_q - T_s$ = 80°C and

$$F_c = 4.24 \times 10^4 A V^{0.15} (1 + V\,\Delta T_c^2)^a \tag{21}$$

where
$$A = 1 \qquad a = 0.13 \qquad \text{if } (1 + V\,\Delta T_c^2) \leqslant 40$$
$$A = 0.4839 \qquad a = 0.346 \qquad \text{if } (1 + V\,\Delta T_c^2) \geqslant 40$$

Velocity $V$ is the inlet coolant velocity in meters per second (0.005–1.5 m/s). Equation (21) developed for bottom reflooding correlates also falling film data if $V$ is the average velocity in the falling film.

Arrieta and Yadigaroglu (1978) were able, using Eq. (21), to correlate well bottom-reflooding data obtained with an internally cooled tubular test section over a wide range of subcoolings and mass fluxes.

Figure 15 shows the variation of the heat transfer coefficient predicted by Eq. (21). It is clear that the effects of subcooling and mass flux are small at low subcoolings and low mass fluxes and become increasingly important at high mass fluxes. Further analysis of the University of California, Berkeley, bottom-reflooding data by Yu (1978) indicates also a small effect of local quality and mass flux when saturated conditions prevail at the quench front. A similar effect was reported by Andréoni (1975).

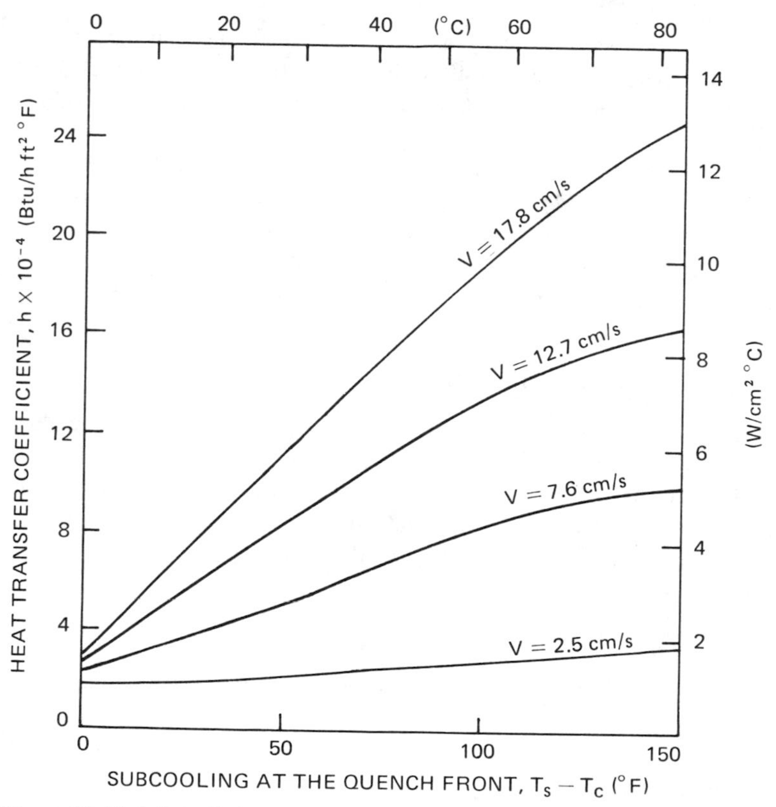


**Figure 15** Variation of the wet-side heat transfer coefficient as predicted by the correlation of Yu et al. (1977) during atmospheric-pressure, bottom-reflooding experiments.

The trends shown in Fig. 15 explain why it is not possible to correlate rewetting data using a unique value of the heat transfer coefficient $h$ at high mass fluxes. At high mass fluxes, the effects of mass flux and local subcooling must be correlated either by a method similar to the one proposed by Yu et al., or by the inclusion of the effect of precursory cooling in the analysis as proposed by Sun et al. (1974b) and Dua and Tien (1976).

Regarding the first alternative, the idealization of the heat transfer coefficient distribution by a step is not satisfactory on physical grounds. Indeed one would not expect the heat transfer coefficient to vary in a steplike manner since small local random fluctuations would always tend to smooth the distribution. Attempts to correlate by the method of Yu et al. data obtained with heater elements other than thin-walled tubes have been less successful, probably because the method lumps all the effects of the various system parameters into the stepwise variable $h$. For significantly different heater geometries, e.g., fuel rods, the stepwise variation of $h$ does not seem to be acceptable and suggests that the effect of system variables should have been "distributed" along a boiling curve. Thus, a boiling curve approach to the description of the variation of the heat transfer coefficient seems more appropriate, but has not been tried yet because of the large number of correlating parameters involved.

## 6 ONSET OF DISPERSED-FLOW FILM BOILING

The dispersed-flow regime is characterized by the fact that the vapor phase is continuous, while the liquid is dispersed in the form of drops. At sufficiently high wall temperatures the heat transfer mechanism in such a flow regime is DFFB. DFFB starts either downstream of the dryout point in annular flow (low-flooding-rate case of Fig. 2) or at the onset of liquid carryover (OC) in the presence of an upstream IAFB region (high-flooding-rate case of Fig. 2).

In the low-flooding-rate case the onset of DFFB can be identified with the dryout point. Consideration of the hydrodynamic stability of the inverted-annular flow pattern leads to criteria for the onset of DFFB following an IAFB region.

The various entrainment mechanisms responsible for the separation of liquid drops from a liquid film (annular flow) or a liquid core (inverted-annular flow) were discussed by Elias and Yadigaroglu (1978). When the steam velocity exceeds a certain critical value, entrained droplets can be "carried over." This situation marks the OC point.

Very little information is available on the mechanisms of liquid entrainment and carryover in the inverted-annular flow regime. Recent visual observations by Goodman and Elias (1978) have confirmed that the disruption of the liquid core and the creation of dispersed flow is caused by a hydrodynamic (Helmholtz) type instability that appears at the interface when the relative velocity between the two phases reaches a critical value. The visual observations show that a stationary wave is created at the liquid–vapor interface. After a few wavelengths the amplitude of the wave becomes comparable to the radial dimensions of the liquid core, and at this point large chunks of liquid become separated from the liquid core. These large drops are unstable and rapidly break into smaller droplets that can be carried over by the vapor.

A criterion for the onset of entrainment (OE) in inverted-annular flow was recently proposed by Arrieta and Yadigaroglu (1978). The onset of entrainment is related to a critical Weber number based on the diameter of the liquid core (approximately equal to the channel diameter $D$) and the relative velocity between the phases at the point of OE,

$$\mathrm{We}_{\mathrm{OE}} = \frac{\rho_G (u_G - u_L)^2 D}{\sigma}$$

Jensen (1972) studied the stability of the interface in inverted-annular flow. His analysis shows that the Weber number based on the radial channel dimension is indeed the controlling parameter. Jensen's parametric studies show the following dependence of the Weber number on the wavelength $\lambda_{max}$ that results in the maximum growth rate of the instability:

$$\frac{D}{\lambda_{max}} \approx 0.1 \ \mathrm{We}$$

Assuming, in agreement with the observations of Jensen and Goodman and Elias (1978), that $D/\lambda_{max}$ is of the order of unit, we obtain

$$\mathrm{We}_{\mathrm{OE}} \approx 10$$

According to Arrieta and Yadigaroglu (1978), this criterion seems to correlate well experimental OE data obtained during reflooding experiments.

## NOMENCLATURE

| | |
|---|---|
| Bi | Biot number [Eq. (16)] |
| $c$ | heat capacity |
| $D$ | diameter |
| $f$ | friction factor |
| $g$ | acceleration due to gravity |
| $G$ | mass flux |
| $h$ | heat transfer coefficient |
| $h_{LG}$ | latent heat of vaporization |
| $k$ | thermal conductivity |
| $L$ | length |
| $L_B$ | boiling length |
| Nu | Nusselt number |
| $p$ | pressure |
| Pe | Péclet number |
| Pr | Prandtl number |

| | |
|---|---|
| $q''$ | heat flux |
| Re | Reynolds number |
| $r$ | radial distance |
| St | Stanton number |
| $T$ | temperature |
| $U$ | velocity |
| $v$ | specific volume |
| $v_{LG}$ | $v_G - v_L$ |
| $V$ | velocity |
| We | Weber number |
| $x$ | flowing quality |
| $y$ | transverse coordinate |
| $z$ | axial coordinate |
| $\delta$ | thickness |
| $\theta$ | nondimensional temperature [Eq. (15)] |
| $\lambda$ | wavelength |
| $\rho$ | density |
| $\sigma$ | surface tension |

**Subscripts**

| | |
|---|---|
| $c$ | coolant at the quench front |
| crit | critical |
| in | inlet |
| $g$ | gassy |
| $G$ | gaseous phase |
| $L$ | liquid phase |
| $q$ | quench |
| $s$ | coolant |
| sat | saturation condition |
| sub | subcooled |
| $w$ | wall |

**Abbreviations**

| | |
|---|---|
| BB | bulk boiling boundary |
| BD | bubble detachment point |
| BE | bubble ejection point |
| CCFL | countercurrent-flow limiting condition |
| CHF | critical heat flux |
| DFFB | dispersed-flow film boiling |
| DNB | departure from nucleate boiling |
| DO | dryout |
| EQ | thermodynamic equilibrium |
| IAFB | inverted-annular film boiling |
| IB | incipience of boiling |
| NVG | net vapor generation |

OC onset-of-carryover point
OE onset of entrainment
QG quench front
SAT slug-to-annular transition point

## REFERENCES

Ahmad, S. Y., Axial Distribution of Bulk Temperature and Void Fraction in a Heated Channel with Inlet Subcooling, *J. Heat Transfer, Trans. ASME*, vol. 92, pp. 595–609, 1970.

Andersen, J. G. M. and Hansen, P. Two-dimensional Heat Conduction in Rewetting Phenomenon, Danish Atomic Energy Commission Research Establishment Riso rept. NORHAV-D-6, 1974.

Andréoni, D., Echanges Thermiques lors du Renoyage d'un Coeur de Réacteur à Eau, Thèse de docteur ingénieur, Université Scientifique et Médicale, Institut National Polytechnique, Grenoble, 1975.

Arrieta, L. A. and Yadigaroglu, G., Analytical Model for Bottom Reflooding Heat Transfer in Light Water Reactors (The UCFLOOD Code), EPRI NP-756, 1978.

Baumeister, K. J. and Simon, F. F., Leidenfrost Temperature–Its Correlation for Liquid Metals, Cryogens, Hydrocarbons, and Water, *J. Heat Transfer, Trans. ASME*, vol. 95, pp. 66–173, 1973.

Béhar, M., Courtaud, M., Ricque, R., and Séméria, R., Fundamental Aspects of Subcooled Boiling with and without Dissolved Gases, *Proc. 3d Int. Heat Transfer Conf., Chicago, Aug. 1966*, AIChE-ASME, vol. 4, pp. 1-11, 1966.

Bergles, A. E., Burnout in Boiling Heat Transfer. Part I: Pool-Boiling Systems, *Nucl. Saf.*, vol. 16, pp. 29–42, 1975.

Bergles, A. E., Burnout in Boiling Heat Transfer. Part II: Subcooled and Low Quality Forced Convection Sustems, in *Two-Phase Flows and Heat Transfer*, eds. S. Kakaç and T. N. Veziroğlu, vol. 2, pp. 693–720, Hemisphere, Washington, D.C., 1977a.

Bergles, A. E., Burnout in Boiling Heat Transfer. Part II: Subcooled and Low-Quality Forced Convection Systems, *Nucl. Saf.*, vol. 18, no. 2, pp. 154–167, 1977b.

Bergles, A. E. and Rohsenow, W. M., The Determination of Forced-Convection Surface-Boiling Heat Transfer, *J. Heat Transfer, Trans. ASME*, vol. 86, pp. 365–372, 1964.

Bertoletti, S., Gaspari, G. P., Lombardi, C., Peterlongo, G., Silvestri, M., and Tacconi, F. A., Heat Transfer Crisis with Steam-Water Mixtures, *Energia Nucleare*, vol. 12, no. 3, pp. 121–172, 1965.

Betten, P. R. and Paul, F. W., Determination of the Point of Net Vapor Generation in Forced Convection Subcooled Boiling, ASME paper 76-WA/HT-86, 1976.

Blair, J. M., An Analytical Solution to a Two-dimensional Model of the Rewetting of a Hot Dry Rod, *Nucl. Eng. Des.*, vol. 32, pp. 159–170, 1975.

Borishanskii, V. M., An Equation Generalizing Experimental Data on the Cessation of Bubble Boiling in a Large Volume of Liquid, *Sov. Phys. Tech. Phys.*, vol. 1, no. 1, pp. 438–442, 1956.

Bowring, R. W., Physical Model, Based on Bubble Detachment, and Calculation of Steam Voidage in the Subcooled Region of a Heated Channel, Institut for Atomenergi, Halden rept. HPR-10, 1962.

Brown, W., Study of Flow Surface Boiling, Ph.D. thesis, Massachusetts Institute of Technology, Cambridge, 1967.

Bukur, D. B. and Isbin, H. S., Numerical Solution of the Yamanouchi Model for Core Spray Cooling, *Nucl. Eng. Des.*, vol. 23, pp. 195–197, 1972.

Butterworth, D. and Owen, R. G., The Quench of Hot Surfaces by Top and Bottom Flooding–A Review, UKAEA rept. AERE-R-7922, 1975.

Chambré, P. and Elias, E., Rewetting Model Using a Generalized Boiling Curve, EPRI NP-571, 1977.

Chan, S. H. and Grolmes, M. A., Hydrodynamically Controlled Rewetting, ASME paper 75-HT-70, 1975.

Chang, Y. P. and Snyder, N. W., Heat Transfer in Saturated Boiling, *Chem. Eng. Prog. Symp. Ser.*, vol. 56, no. 30, pp. 25–28, 1960.

Chun, M. H. and Chon, W. Y., Analysis of Rewetting in Water Reactor Emergency Core Cooling Inclusive of Heat Transfer in the Unwetted Region, ASME paper 75-WA/HT-32, 1975.

Cobb, C. B. and Park, E. L., Jr., Nucleate Boiling: A Maximum Heat Flux Correlation for Corresponding States Liquids, *Chem. Eng. Prog. Symp. Ser.*, vol. 65, no. 92, pp. 188–193, 1969.

Collier, J. G., *Convective Boiling and Condensation*, McGraw-Hill, London, 1972.

Coney, M. W. E., Calculations on the Rewetting of Hot Surfaces, *Nucl. Eng. Des.*, vol. 31, pp. 246–259, 1974.

Costa, J., Mesure de la Perte de Pression par Accélération et Etude de l'Apparition du Taux de Vide en Ebullition Locale à Basse Pression, CENG note TT 244, 1967.

Davis, E. J. and Anderson, G. H., The Incipience of Nucleate Boiling in Forced Convection, *AIChE J.*, vol. 12, pp. 774–780, 1966.

Dix, G. E., Vapor Void Fractions for Forced Convection with Subcooled Boiling at Low Flow Rates, Ph.D. thesis, Department of Mechanical Engineering, University of California, Berkeley, 1971.

Dua, S. S. and Tien, C. L., Two-dimensional Analysis of Conduction-controlled Rewetting with Precursory Cooling, *J. Heat Transfer, Trans. ASME*, vol. 98, no. 3, pp. 407–413, 1976.

Dua, S. and Tien, C. L., A Generalized Two-Parameter Relation for Conduction-controlled Rewetting of a Hot Vertical Surface, *Int. J. Heat Mass Transfer*, vol. 20, no. 2, pp. 174–176, 1977.

Duffey, R. B. and Porthouse, D. T. C., The Physics of Rewetting in Water Reactor Emergency Cooling, *Nucl. Eng. Des.*, vol. 25, pp. 379–394, 1973.

Edwards, A. R. and Mather, D. J., Some U. K. Studies Related to the Loss-of-Coolant Accident, *Proc. Topical Meet. on Water Reactor Safety, Salt Lake City*, CONF-730304, pp. 720–739, 1973.

Elias, E. and Yadigaroglu, G., A General One-dimensional Model for Conduction-controlled Rewetting of a Surface, *Nucl. Eng. Des.*, vol. 42, pp. 185–194, 1977a.

Elias, E. and Yadigaroglu, G., The Reflooding Phase of the LOCA–State of the Art. II. Rewetting and Liquid Entrainment, in *Two-Phase Flows and Heat Transfer*, eds. S. Kakaç and T. N. Veziroğlu, vol. 2, Hemisphere, Washington, D.C., 1977b.

Elias, E. and Yadigaroglu, G., The Reflooding Phase of the LOCA in PWR's. Part II: Rewetting and Liquid Entrainment, *Nucl. Saf.*, vol. 19, no. 2, pp. 160–175, 1978.

Elias, E., Arrieta, L., and Yadigaroglu, G., An Improved Model for the Rewetting of a Hot Fuel Rod, *Trans. ANS*, vol. 24, pp. 299–300, 1976.

Elliott, D. F. and Rose, P. W., The Quenching of a Heated Surface by a Film of Water in a Steam Environment at Pressure up to 53 Bar, AEEW-M-976, 1970.

Elliott, D. F. and Rose, P. W., The Quenching of a Heated Zircaloy Surface by a Film of Water in a Steam Environment at Pressure up to 53 Bar, AEEW-M-1027, 1971.

Fiori, M. P. and Bergles, A. E., Model of Critical Heat Flux in Subcooled Flow Boiling, *Heat Transfer 1970, Proc. 4th Int. Heat Transfer Conf., Paris*, vol. 6, paper B6.3, 1970.

Frost, W. and Dzakowic, G. S., An Extension of the Method of Predicting Incipient Boiling on Commercially Finished Surfaces, ASME paper 67-HT-61, 1967.

Gambill, W. R., Burnout in Boiling Heat Transfer, *Nucl. Saf.*, vol. 7, no. 4, pp. 436–443, 1966.

Gambill, W. R., Burnout in Boiling Heat Transfer–Part I: Pool Boiling Systems, *Nucl. Saf.*, vol. 9, no. 5, pp. 351–362, 1968a.

Gambill, W. R., Burnout in Boiling Heat Transfer–Part II: Subcooled Forced-Convection Systems, *Nucl. Saf.*, vol. 9, no. 6, pp. 467–480, 1968b.

Gaspari, G. P., Hassid, A., and Lucchini, F., A Rod-centered Subchannel Analysis with Turbulent (Enthalpy) Mixing for Critical Heat Flux Prediction in Rod Clusters Cooled by Boiling Water, *Heat Transfer 1974, Proc. 5th Int. Heat Transfer Conf., Tokyo, Sept. 1974*, vol. 4, paper B6.12, pp. 295–299, 1974.

General Electric Company, General Electric BWR Thermal Analysis Basis (GETAB): Data, Correlation and Design Application, NEDO-10958, 1973.

Goodman, J. and Elias E., Heat Transfer in the Inverted Annular Flow Regime during Reflooding, *Trans. ANS*, vol. 28, pp. 397–399, 1978.

Griffith, P., Clark, J. A., and Rohsenow, W. M., Void Volumes in Subcooled Boiling Systems, ASME paper 58-TH-19, 1958.

Griffith, P., Schumann, W. A., and Neustal, A. D., Flooding and Burnout in Closed-End Vertical Tubes, Two-Phase Flow Symp., IME, London, paper 5, 1962.

Groeneveld, D. C., The Occurrence of Upstream Dryout in Uniformly Heated Channels, *Heat Transfer 1974, 5th Int. Heat Transfer Conf., Tokyo, Sept. 1974*, vol. 4, paper B.6, pp. 265–269, 1974.

Guarino, D., Marinelli, V., and Pastori. L., State-of-the-art in BWR Rod Bundle Burnout Predictions, *Nucl. Technol.*, vol. 23, p. 38, 1974.

Healzer, J. M., Hench, J. E., Jannsen, S., and Levy, S., Design Basis for Critical Heat Flux Condition in Boiling Water Reactors, APED-5286, 1966.

Hein, D., Köhler, W., and Riedle, K., Analysis of Mono-Tube and 340-Rod-Bundle Experiments with a Rewetting Model, European Two-Phase Flow Group Meeting, Harwell, June 1974.

Henry, R. E., A Correlation for the Minimum Film Boiling Temperature, *Chem. Eng. Prog. Symp. Ser.*, vol. 70, no. 138, pp. 81–90, 1974.

Henry, R. E., Singer, R. M., Quinn, D. J., and Jeans, W. C., Incipient Superheat in a Convective Sodium System, *Heat Transfer 1974, Proc. 5th Int. Heat Transfer Conf., Tokyo, Sept. 1974*, vol. 4, paper B7.1, pp. 305–309, 1974.

Hewitt, G. F., Experimental Studies on the Mechanisms of Burnout in Heat Transfer to Steam-Water Mixtures, *Heat Transfer 1970, Proc. 4th Int. Heat Transfer Conf., Paris*, vol. 6, paper B6.6, 1970.

Hewitt, G. F., Mechanism and Prediction of Burnout, in *Two-Phase Flows and Heat Transfer*, eds. S. Kakaç and T. N. Veziroğlu, vol. 2, pp. 721–745, Hemisphere, Washington, D.C., 1977.

Hewitt, G. F. and Hall-Taylor, N., *Annular Two-Phase Flow*, Pergamon, New York, 1970.

Hewitt, G. R., Kearsey, H. A., Lacey, P. M. C., and Pulling, D. J., Burnout and Film Flow in the Evaporation of Water in Tubes, *Proc. IME*, vol. 180, part 3C, pp. 206–215, 1965–1966.

Hsu, Y. Y., On the Size Range of Active Nucleation Cavities on a Heating Surface, *J. Heat Transfer, Trans. ASME*, vol. 84, no. 3, pp. 207–216, 1962.

Hsu, Y. Y. and Graham, R. W., An Analytical and Experimental Study of the Thermal Boundary Layer and Ebullition Cycle in Nucleate Boiling, NASA TN-D-594, 1961.

Hutchinson, P. and Whalley, P. B., A Possible Characterization of Entrainment in Annular Flow, *Chem. Eng. Sci.*, vol. 28, pp. 974–975, 1973.

Iloeje, O. C., Rohsenow, W. M., and Griffith, P., Three-step Model of Dispersed Flow Heat Transfer (Post-CHF Vertical Flow), ASME paper 75-WA/HT-1, 1975.

Isbin, H., Vanderwater, R., Fauske, H., and Singh, S., A Model for Correlation Two-Phase Steam-Water, Burnout Heat Fluxes, *J. Heat Transfer, Trans. ASME*, vol. 83, pp. 149–157, 1961.

Ishii, M., Study on Emergency Core Cooling, *J. Br. Nucl. Energy Soc.*, vol. 14, pp. 237–242, 1975.

Ivey, H. J. and Morris, D. J., On the Relevance of the Vapour Liquid Exchange Mechanism for Subcooled Boiling Heat Transfer at High Pressure, AEEW-R137, 1962.

Janssen, E. and Levy, S., Burnout Limit Curves for Boiling Water Reactors, APED-3892, 1962.

Jensen, R., Inception of Liquid Entrainment during Emergency Cooling of Pressurized Water Reactors, Ph.D. thesis, Utah State University, 1972.

Kirchner, W. L., Reflood Heat Transfer in a Light-Water Reactor, Ph.D. thesis, Department of Nuclear Engineering, Massachusetts Institute of Technology, NUREG-0106 (V.1-2), 1976.

Kutateladze, S. S., Heat Transfer in Condensation and Boiling, AEC-TR-3700, 1952.

Kutateladze, S. S. and Leont'ev, A. I., Some Applications of the Asymptotic Theory of the Turbulent Boundary Layer, *Proc. 3d Int. Heat Transfer Conf., Chicago*, vol. 3, pp. 1–6, 1966.

Lahey, R. T., Jr., and Moody, F. J., *The Thermal-Hydraulics of a Boiling Water Nuclear Reactor*, American Nuclear Society, Hinsdale, Ill., 1977.

Levy, S., Forced Convection Subcooled Boiling–Prediction of the Vapor Volumetric Fraction, *Int. J. Heat Mass Transfer*, vol. 10, pp. 951–965, 1967.

Lienhard, J. H. and Dhir, V. K., Hydrodynamic Prediction of Peak Boiling Heat Fluxes from Finite Bodies, *J. Heat Transfer, Trans. ASME*, vol. 95, pp. 152–158, 1973.

Lienhard, J. H. and Schrock, V. E., The Effect of Pressure, Geometry, and the Equation of State upon the Peak and Minimum Boiling Heat Flux, *J. Heat Transfer, Trans. ASME*, vol. 85, pp. 261–272, 1963.

Lienhard, J. H. and Watanabe, K., On Correlating the Peak and Minimum Boiling Heat Fluxes with Pressure and Heater Configuration, *J. Heat Transfer, Trans. ASME*, vol. 88, pp. 94–100, 1966.

Madejski, J., Vapour Bubble Departure Conditions in Flow Boiling, *Trans. Inst. Fluid Flow Machinery*, Polish Acad. Sciences, vol. 49, pp. 3–22, 1970.

Maitra, D. and Subba Raju, K., Vapour Void Fraction in Subcooled Flow Boiling, *Nucl. Eng. Des.*, vol. 32, pp. 20–28, 1975.

Martini, R. and Premoli, A., A Simple Model for Predicting ECC Transients in Bottom Flooding Conditions, *Proc. CREST Meet. on Emergency Core Cooling for Light Water Reactors, Garching/München, Oct. 18–20, 1972*, MRR 115, vol. 2, 1972.

McAssey, E. V. and Bonilla, C. F., A Survey of Rewetting Following a Postulated LOCA, ASME paper 77-HT-93, 1977.

Moissis, R. and Berenson, P. J., On the Hydrodynamic Transitions in Nucleate Boiling, *J. Heat Transfer, Trans. ASME*, vol. 85, pp. 221–229, 1963.

Murphy, R. W. and Bergles, A. E., Subcooled Flow Boiling of Fluorocarbons—Hysteresis and Dissolved Gas Effects on Heat Transfer, *Proc. 1972 Heat Transfer and Fluid Mechanics Inst.*, Stanford University Press, Stanford, Calif., 1972.

Noyes, R. C., An Experimental Study of Sodium Pool Boiling Heat Transfer, *J. Heat Transfer, Trans. ASME*, vol. 85, pp. 125–131, 1963.

Noyes, R. C. and Lurie, H., Boiling Sodium Heat Transfer, *Proc. 3d Int. Heat Transfer Conf., Chicago, Aug. 1966*, AIChE-ASME, vol. 5, pp. 92–100, 1966.

Pearson, K. G., Piggott, B. D. C., and Duffey, R. B., The Effect of Thermal Diffusion from Fuel Pellets on Rewetting of Over-heated Water Reactor Pins, *Nucl. Eng. Des.*, vol. 41, pp. 165–173, 1977.

Piggott, B. D. G. and Porthouse, D. T. C., A Correlation of Rewetting Data, *Nucl. Eng. Des.*, vol. 32, pp. 171–181, 1974.

Purcupile, J. C. and Gouse, S. W., Jr., Reynolds Flux Model of Critical Heat Flux in Subcooled Forced Convection Boiling, ASME paper 72-HT-4, 1972.

Rohsenow, W. M., Boiling, in *Handbook of Heat Transfer*, eds. W. M. Rohsenow and J. P. Hartnett, sec. 13, pp. 13.1–13.75, McGraw-Hill, New York, 1973.

Rohsenow, W. M. and Griffith, P., Correlation of Maximum Heat Transfer Data for Boiling of Saturated Liquids, *Chem. Eng. Prog. Symp. Ser.*, vol. 52, no. 18, pp. 47–49, 1956.

Rouhani, S. Z., Calculation of Steam Volume Fraction in Subcooled Boiling, *J. Heat Transfer, Trans. ASME*, vol. 90, pp. 158–164, 1968.

Roy, R. P. and Yadigaroglu, G., An Investigation of Heat Transport in Oscillatory Turbulent Subcooled Flow, *J. Heat Transfer, Trans. ASME*, vol. 98, pp. 630–637, 1976.

Saha, P. and Zuber, N., Point of Net Vapor Generation and Vapor Void Fraction in Subcooled Boiling, *Heat Transfer 1974, Proc. 5th Int. Heat Transfer Conf., Tokyo, Sept. 1974*, vol. 4, paper B4.7, pp. 175–179, 1974.

Sawan, M. E. and Carbon, M. W., A Review of Spray-Cooling and Bottom-Flooding Work on LWR Cores, *Nucl. Eng. Des.*, vol. 32, pp. 191–207, 1975.

Sekoguchi, K., Nishikawa, K., and Nakatasomi, M., Flow Boiling in Subcooled and Low Quality Regions—Heat Transfer and Local Void Fraction, *Heat Transfer 1974, Proc. 5th Int. Heat Transfer Conf., Tokyo, Sept. 1974*, vol. 4, paper B4.8, pp. 180–184, 1974.

Séméria, R. and Martinet, B., Calefaction Spots on a Heating Wall: Temperature Distribution and Resorption, *Proc. IME*, vol. 180, part 3C, pp. 192–205, 1965.

Shiralkar, B. S., Analysis of Non-uniform Flux CHF Data in Simple Geometries, NEDM-13279, 1972.

Shiralkar, B. S. and Lahey, R. T., Jr., The Effect of Obstacles on a Liquid Film, *J. Heat Transfer, Trans. ASME*, vol. 95, pp. 528–533, 1973.

Shires, G. L., Pickering, A. R., and Blacker, P. T., Film Cooling of Vertical Fuel Rods, AEEW-R-343, 1964.

Spiegler, P., Hopenfeld, J., Silberberg, M., Bumpus, C. F., Jr., and Norman, A., Onset of Stable Film Boiling and the Foam Limit, *Int. J. Heat Mass Transfer*, vol. 6, pp. 987–994, 1963.

Staub, R. W., The Void Fraction in Subcooled Boiling–Prediction of the Initial Point of Net Vapor Generation, *J. Heat Transfer, Trans. ASME*, vol. 90, no. 1, pp. 151–158, 1968.

Sun, K. H. and Lienhard, J. H., The Peak Pool Boiling Heat Flux on Horizontal Cylinders, *Int. J. Heat Mass Transfer*, vol. 13, no. 9, pp. 1425–1439, 1970.

Sun, K. H., Dix, G. E., and Tien, C. L., Cooling of a Very Hot Vertical Surface by a Falling Liquid Film, *J. Heat Transfer, Trans. ASME*, vol. 96, p. 151, 1974a.

Sun, K. H., Dix, G. E., and Tien, C. L., Effect of Precursory Cooling on Falling-Film Rewetting, ASME paper 74-WA/HT-52, 1974b.

Thompson, T. S., An Analysis of the Wet Side Heat Transfer Coefficient during Rewetting of a Hot Dry Patch, *Nucl. Eng. Des.*, vol. 22, pp. 212–224, 1972.

Thompson, T. S., On the Process of Rewetting a Hot Surface by a Falling Liquid Film, *Nucl. Eng. Des.*, vol. 31, pp. 234–235, 1974a.

Thompson, T. S., Rewetting of a Hot Surface, *Heat Transfer 1974, Proc. 5th Int. Heat Transfer Conf., Tokyo, Sept. 1974*, vol. 4, paper B.13, pp. 139–143, 1974b.

Thompson, B. and Macbeth, R. V., Boiling Water Heat Transfer–Burnout in Uniformly Heated Round Tubes: A Compilation of World Data with Accurate Correlations, AEEW-R-356, 1964.

Tien, C. L. and Yao, L. S., Analysis of Conduction Controlled Rewetting of a Vertical Surface, ASME paper 74-WA/HT-49, 1974.

Tong, L. S., Prediction of Departure from Nucleate Boiling for an Axially Non-uniform Heat Flux Distribution, *J. Nucl. Energy*, vol. 21, pp. 241–248, 1967.

Tong, L. S., Boundary Layer Analysis of the Flow Boiling Crisis, *Int. J. Heat Mass Transfer*, vol. 11, pp. 1208–1211, 1968.

Tong, L. S., Critical Heat Fluxes in Rod Bundles, *Two-Phase Flow and Heat Transfer in Rod Bundles*, pp. 31–46, ASME, New York, 1969.

Tong, L. S., *Boiling Crisis and Critical Heat Flux*, AEC Critical Rev. Ser., TID-25887, 1972.

Tong, L. S., A Phenomenological Study of Critical Heat Flux, ASME paper 75-HT-68, 1975.

Tong, L. S. and Hewitt, G. F., Overall Viewpoint of Flow Boiling CHF Mechanisms, ASME paper 72-HT-54, 1972.

Tong, L. S., Currin, H. B., Larsen, P. S., and Smith, O. G., Influence of Axially Non-uniform Heat Flux on DNB, *Chem. Eng. Prog. Symp. Ser.*, vol. 62, no. 64, pp. 35–40, 1965.

Ünal, H. C., Determination of the Initial Point of Net Vapor Generation in Flow Boiling Systems, *Int. J. Heat Mass Transfer*, vol. 18, pp. 1095–1099, 1975.

Walkush, J. P., High Pressure Counterflow CHF, Master's thesis, Department of Mechanical Engineering, Massachusetts Institute of Technology, Cambridge, 1974.

Wallis, G. B., Two-Phase Flow Aspects of Pool Boiling from a Horizontal Surface, AEEW-R103; also Symp. on Two-Phase Fluid Flow, IME, London, paper 3, 1961.

Wallis, G. B., *One-Dimensional Two-Phase Flow*, McGraw-Hill, New York, 1969.

Weatherhead, R. J., Nucleate Boiling Characteristics and Critical Heat Flux Occurrence in Sub-cooled Axial Flow Systems, ANL-6675, 1963.

Worley, L. C., III, and Griffith, P., Downflow Void Fraction, Pressure Drop and Critical Heat Flux, MIT Heat Transfer Lab. rept. DSR 80620-83, 1973.

Yamanouchi, A., Effect of Core Spray Cooling in Transient State after Loss-of-Coolant Accident, *J. Nucl. Sci. Technol.*, vol. 5, pp. 547–558, 1968.

Yao, L. S., Rewetting of a Vertical Surface with Internal Heat Generation, 16th Natl. Heat Transfer Conf., St. Louis, Mo., Aug. 8–13, AIChE paper 9, 1976.

Yeh, H. C., An Analysis of Rewetting of a Nuclear Fuel Rod in Water Reactor Emergency Core Cooling, *Nucl. Eng. Des.*, vol. 34, pp. 317–322, 1975.

Yin, S. T. and Abdelmessih, A. H., Prediction of Incipient Flow Boiling from a Uniformly Heated Surface, 16th Natl. Heat Transfer Conf., St. Louis, Mo., Aug. 8–13, AIChE paper, 1976.

Yu, K.-P., An Experimental Investigation of Reflooding of a Bare Tubular Test Section, Ph.D. thesis, Department of Nuclear Engineering, University of California, Berkeley, 1978.

Yu, S. K. W., Farmer, P. R., and Coney, M. W. E., Methods and Correlations for the Prediction of Quenching Rates on Hot Surfaces, *Int. J. Multiphase Flow*, vol. 3, pp. 415–443, 1977.

Zuber, N., On Stability of Boiling Heat Transfer, *Trans. ASME*, vol. 80, pp. 711–720, 1958.

Zuber, N., Discussion of paper by P. J. Berenson, Film-Boiling Heat Transfer from a Horizontal Surface, *J. Heat Transfer, Trans. ASME*, vol. 83, no. 3, pp. 351–358, 1961.

Zuber, N. and Staub, F. W., Stability of Dry Patches Forming in Liquid Films Flowing over Heated Surfaces, *Int. J. Heat Mass Transfer*, vol. 9, pp. 897–905, 1966.

Zuber, N., Tribus, M., and Westwater, J. W., The Hydrodynamic Crisis in Pool Boiling of Saturated and Subcooled Liquids, *International Developments in Heat Transfer, Proc. Int. Heat Transfer Conf., Boulder*, pt. 2, ASME, pp. 230–236, 1961.

CHAPTER

# SEVENTEEN

# TWO-PHASE FLOW INSTABILITIES AND PROPAGATION PHENOMENA

**G. Yadigaroglu**

The study of instabilities in two-phase flow systems was initiated 40 years ago with the publication of the pioneering article by Ledinegg (1938). Greatly increased attention was given to these problems approximately 25 years ago with the introduction of high-power-density boilers, space systems, and the boiling-water reactor (BWR).

The first comprehensive analyses of two-phase flow instabilities were published in the early 1960s. A period of relative confusion followed, with many authors attempting to explain various widely different observations of unstable behavior by the same physical instability mechanism. It was only during the late 1960s that publications on this subject finally started bringing some order, clarity, and classification of instability mechanisms. At the same time stability data were being gathered, a number of predictive analytical models were published, and fair success was obtained regarding the prediction of the threshold and other characteristics of unstable behavior.

Our understanding of most instability mechanisms is fairly complete now, but our analytical modeling of these phenomena remains limited for reasons that will become apparent in Sec. 3.1. Excellent reviews of this field have been published by Bouré et al. (1973), Ishi (1976), and Bergles (1977).

In this chapter the various known types of instabilities are reviewed, and the most common types, namely, the instabilities resulting from the particular form of the pressure drop–flow rate characteristic of the channel and the density wave oscillations, are discussed rather extensively. After this general review, specific stability topics related to nuclear-power systems are briefly treated.

Instability mechanisms are linked to perturbation propagation phenomena in two-phase flow because feedback mechanisms involve delays created by the finite propagation time of perturbations in the system.

Pressure perturbations create dynamic waves that propagate at the "sonic" velocity of the mixture. Temperature or enthalpy perturbations travel at the flow velocity or create in turn void-fraction perturbations that propagate at the kinematic-wave velocity of the mixture as density waves. Kinematic waves have been studied by Lighthill and Whitham (1955) and are discussed by Wallis (1969) and Zuber and Staub (1966, 1967). Consideration of the continuity and energy equations is sufficient for the formulation of kinematic-wave propagation problems, while the momentum equation must be used in formulating dynamic-wave propagation phenomena.

Most two-phase flow instabilities, with the exception of the so-called acoustic oscillations, are connected with the propagation of kinematic waves. In this case pressure variations are relatively slow, and the phenomena can be studied by assuming that pressure perturbations propagate instantaneously throughout the system. It is only for acoustic oscillations that the finite propagation time of the pressure perturbations must be taken into account. Alternatively speaking, in the presence of acoustic oscillations the time constants of the various important phenomena are of the order of magnitude of the dynamic-wave transit time through the system; for instabilities linked to kinematic-wave propagation, the period of the oscillations and the time constants of important contributing phenomena must be of the order of the mixture transit time through the system.

Acoustic oscillations will be only briefly reviewed in Sec. 4.3. Dynamic-wave propagation phenomena find a number of important applications, such as critical flow of two-phase mixtures, discussed in another chapter of this book.

Kinematic-wave propagation phenomena are considered to some extent in Sec. 3.1 in relation to density wave oscillations. The discussion of that section applies in general to all two-phase flow dynamics problems.

## 1 TYPES OF INSTABILITIES

It is first necessary to distinguish between microscopic instabilities that occur locally at the liquid–gas interface (such as the well-known Helmholtz and Taylor instabilities, bubble collapse, and film-boiling instability) and macroscopic instabilities, which involve the entire two-phase flow system. Microscopic instabilities can be treated by methods of classical fluid mechanics and are outside the scope of this review. Several classifications of macroscopic instabilities are possible.

The main classification proposed by Bouré et al. (1973) is based on the distinction between the static and dynamic character of the conservation laws needed to explain the dynamics of unstable equilibrium states. Thus the threshold for static instabilities can be predicted from steady-state considerations only, while the triggering of dynamic instabilities involves transient inertia and dynamic feedback effects. Known types of static instabilities include

1. Flow excursions (Leginegg instability)
2. Relaxation instabilities, such as flow pattern transitions, nucleation instabilities, bumping, chugging, and geysering

The most common dynamic instabilities are characterized as

1. Density wave oscillations
2. Pressure drop oscillations
3. Acoustic oscillations

Since several elementary physical instability mechanisms often contribute to the overall system behavior (for example, nucleation, heat transfer from the walls, flow pattern changes, and void propagation), instabilities can be qualified as compound when several mechanisms interact simultaneously. For example, instabilities in BWRs involve both thermohydrodynamic considerations as well as the void-reactivity coupling. Flow excursions as well as density wave oscillations occurring in an array of parallel channels (such as BWR bundles or steam generator tubes) can be compounded by secondary flow distribution instabilities among the parallel channels. Static flow excursions may lead to compound pressure drop oscillations. Destabilizing mechanisms are also often provided by peculiarities of components of the two-phase flow system; instabilities related to the presence of a pressurizer or accumulator or to some particular operating characteristic of a system component (direct-contact condenser, defective control valve, etc.) are such examples.

Instabilities that have been postulated or observed to occur in nuclear-power systems, such as steam condensation instabilities in BWR suppression pools, steam–water mixing instabilities during emergency core cooling system (ECCS) operation, oscillations during the reflooding phase of the loss of coolant accident (LOCA) in pressurized-water reactors (PWR), and surface temperature oscillations at the dryout point of steam generator tubes are briefly discussed in Sec. 5. Countercurrent two-phase flow phenomena, such as those that may occur in the downcomer of PWRs during refill of the vessel and in BWRs during spray cooling of the bundles, seem to be connected with stability considerations; these are, however, outside the scope of this review.

Thermohydrodynamic instabilities can be classified in alternate categories. According to the circulation mode, for example, instabilities occurring with natural and forced convection can be distinguished. If the similarity of the fundamental physical mechanisms is considered, however, this distinction seems unimportant. Only the range of certain parameters will differ in the two cases.

An important consideration in studying the stability of a two-phase flow system is the correct specification of the boundary conditions. Boundary conditions generally determine which parts of the system may participate in the transients and which components will operate in a steady mode.

Boundary conditions are seldom exact; they usually represent an idealization of complex physical situations that is useful in simplifying the problem. For example, a pressurizer connected to a particular point of a loop may, under some conditions, be capable of keeping the pressure approximately constant at that point. Under different conditions, however, the pressurizer dynamics may contribute to the overall behavior of the system; in this case the constant-pressure boundary condition must be removed and replaced by a model of the behavior of the pressurizer. As another example, consider a BWR core. Study of the redistribution of flow among the subassemblies resulting from a large external disturbance, requires examination of the entire primary loop. On the contrary, when the stability of the hot channel alone is investigated, it is sufficient to specify the pressure at the inlet and exit plenums since flow changes in a single channel will not affect significantly the total flow, and consequently the pressure drop, across the core.

These considerations seem obvious but there has been considerable confusion in the past because many stability experiments were designed without making an effort to clearly define the limits of the unstable part of the system. In particular, system-induced instabilities have often interfered with the basic oscillation mechanisms of the heated channel. Furthermore, the term "parallel channels" has been used frequently to denote a constant-pressure-drop boundary condition, while it should have been used more properly to denote flow distribution phenomena.

## 2 INSTABILITIES RESULTING FROM THE PRESSURE DROP–FLOW RATE CHARACTERISTICS OF THE SYSTEM

Two-phase flow channels occasionally exhibit the particular S-shaped steady-state pressure drop–flow rate characteristic shown in Fig. 1*a*. We will show that operation in the negative-slope part of the characteristic is unstable and generally leads to a flow excursion [the Ledinegg (1938) instability]. Maulbetsch and Griffith (1965) and Stenning et al. (1967) have shown that this can also lead to pressure drop oscillations if appropriate dynamic feedback mechanisms exist.

Let us first examine the shape of the characteristic for a boiling channel with constant heat addition and consider the effects of the frictional, gravitational, and accelerational components of the pressure drop separately. An influence that modifies the slope of the flow rate–pressure drop characteristic will be called destabilizing if it makes the slope more negative and stabilizing if it tends to make the slope more positive.

The flow quality always decreases with increasing flow rate. If we assume that at high mass fluxes the void fraction depends almost exclusively on the quality, increases in flow rate reduce the quality and make the gravitational pressure gradient always more negative in upward flow and more positive in downward flow. Thus at high mass fluxes, the gravity effect is always stabilizing for upward flow

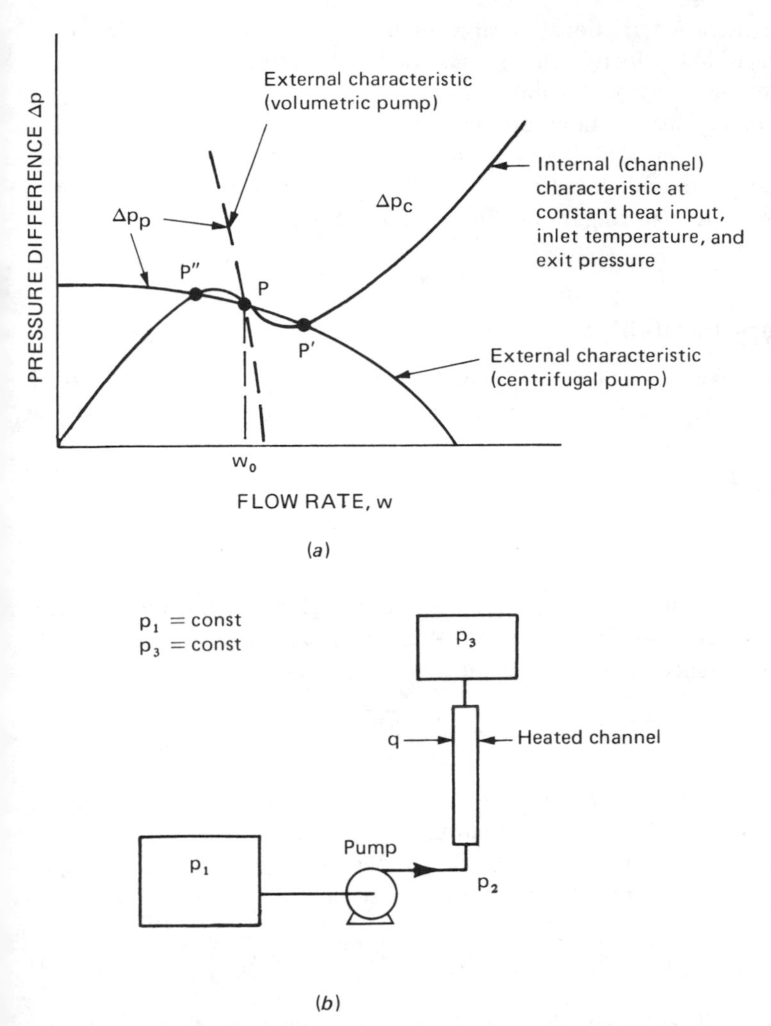


**Figure 1** Two-phase flow system. (*a*) Internal and external pressure difference–flow rate characteristics. (*b*) Flow system.

and destabilizing for downward flow. At very low mass fluxes, the drift velocity of the vapor (Zuber and Findlay, 1965) starts playing a role, making the void fraction mass flux dependent: at the same quality, increasing the mass flux tends to increase the void fraction in upward flow and decrease it in downward flow. Thus, at very low mass fluxes the effect of the drift velocity of the vapor is always destabilizing and tends to diminish the stabilizing effect of the gravity in upward flow and enhance its destabilizing effect in downward flow.

In bulk boiling, the frictional pressure drop decreases with increasing flow rate if the boiling boundary is sufficiently downstream in the channel. For subcooled boiling at high fluxes small reductions in the flow rate may produce local boiling near the wall and increased frictional shear without a corresponding significant decrease in the average flow velocity effect. Thus frictional pressure drop in both saturated and subcooled flow may be destabilizing.

When small reductions in flow produce small increases in exit quality but relatively large increases in void fraction, the acceleration pressure drop increases with decreasing flow rate and thus has a destabilizing effect. The acceleration pressure drop may also be destabilizing for condensing flows in tubes.

## 2.1 The Ledinegg Instability

Consider the simple two-phase flow system of Fig. 1*b*. A typical pump characteristic is shown in Fig. 1*a*; it is referred to as the external characteristic with respect to the channel.

The pressure difference between points 1 and 2 is given by

$$p_2 - p_1 = \Delta p_p - I_p \frac{dw}{dt} \tag{1}$$

where $I_p$ represents the inertia of this leg. We have neglected for simplicity any frictional and gravitational pressure drops in this external part of the system.

The pressure difference across the heated channel is given by

$$p_2 - p_3 = \Delta p_c + I_c \frac{dw}{dt} \tag{2}$$

where $\Delta p_c$ is the steady-state internal characteristic of the channel shown in Fig. 1*a* and $w$ is the flow rate. We have approximately represented the temporal acceleration effects by the last term of Eq. (2), where $I_c$ represents the inertia of the "internal" leg of the system. At any time we must have

$$p_3 - p_1 = \Delta p_p - \Delta p_c - (I_p + I_c) \frac{dw}{dt} = \text{const} \tag{3}$$

Consider now small perturbations around an equilibrium point denoted by the superscript $^0$:

$$\Delta p_p(t) = \Delta p_p^0 + \delta\, \Delta p_p(t) \cong \Delta p_p^0 + \left.\frac{\partial\, \Delta p_p}{\partial w}\right|_0 \delta w(t) \tag{4}$$

$$\Delta p_c(t) = \Delta p_c^0 + \delta\, \Delta p_c(t) \cong \Delta p_c^0 + \left.\frac{\partial\, \Delta p_c}{\partial w}\right|_0 \delta w(t) \tag{5}$$

$$w(t) = w^0 + \delta w(t) \tag{6}$$

Substituting these expressions into Eq. (3) and eliminating the steady-state terms, we obtain the governing equation for the flow rate perturbations:

$$I \frac{d\,\delta w}{dt} = A\, \delta w \tag{7}$$

where

$$I \equiv I_p + I_c$$

$$A \equiv \frac{\partial}{\partial w} \Delta p_p \bigg|_0 - \frac{\partial}{\partial w} \Delta p_c \bigg|_0$$

The solution of Eq. (7) for an initial flow perturbation $\delta w_0$

$$\delta w = \delta w_0 e^{(A/I)t}$$

shows that the perturbation will grow if $A$ is positive. Thus we conclude that the system will be unstable if the slope of the channel characteristic is more negative than the slope of the external characteristic. Figure 1*a* also shows that we can stabilize an unstable channel by using a volumetric pump with a steep negative characteristic; for this condition the flow rate is practically fixed, and the system cannot but operate at that flow rate. We also see that since pump characteristics always have negative slopes, a nonnegative-slope channel characteristic is sufficient for stability.

We have used the time-dependent, momentum conservation equation to derive the stability criterion; however, the criterion itself (negative value of $A$) is based on steady-state considerations only. Thus the Ledinegg instability is classified as a static instability.

A stability map can be drawn by obtaining analytical expressions for the maximum and minimum points of the ($w$, $\Delta p$) characteristic. Chilton (1957) and Hands (1974) used simple analytical expressions for the three components of the pressure drop to produce such a map. A system is unconditionally stable if the ($w$, $\Delta p$) curve becomes monotonic.

Ledinegg-type flow excursions have been observed in subcooled-boiling situations by Daleas and Bergles (1965) and Maulbetsch and Griffith (1965), in once-through boilers having N-shaped tubes in which the flow was partly downward by Krasyakova and Glusker (1965), and in U-shaped tubes of superheaters in which the flow was initially flowing downward by Giphshman and Levinzon (1966).

For steam generators the situation is complicated since heat input to the secondary side depends on the secondary-side flow rate. The ($w$, $\Delta p$) characteristic can be generated either at constant load (by adjusting primary-side inlet conditions to compensate for the effect of the secondary flow rate) or at constant primary-side inlet conditions (and, consequently, variable heat exchange rates). Characteristics obtained in this fashion often exhibit negative slopes. Waszink and Efferding (1973) and Llory (1975) reported that liquid metal-cooled fast-breeder reactor (LMFBR) steam generators did not exhibit any flow excursions in spite of having negative-slope characteristics. Thus the thermal coupling between the primary and secondary sides might be stabilizing. Derivation of stability criteria for steam-generator operation will require consideration of this coupling.

If the external characteristic of a channel is essentially flat, a flow excursion will generally occur if the channel originally operating at a stable point is led into operation in the negative-slope part of its characteristics. Such a transient can happen, for example, in a gradual loss of flow through the core of an LMFBR as a result of a pump trip. For this condition the external characteristic of the channel will change in time, and the operating point will move as shown in Fig. 2. An excursion will occur when a position near the minimum of the characteristic is reached, and the operating point will move along the dashed line of Fig. 2. For the LMFBR transient, the speed of the flow transition from $A_3$ to $A_6$ is important since it is controlling the thermal behavior of the fuel pins. In a study by Schmitt (1974) the flow excursion from $A_3$ to $A_6$ was relatively very slow (10–30 s) and largely controlled by the thermal inertia of the system and the slow progression of a boiling front along the channel. Points along the operating locus during the excursion correspond to operation at variable heat transfer to the coolant. Akagawa et al. (1971) have also observed an extremely long Ledinegg flow excursion occurring in a long evaporator tube (time period of the order of hours). Thus it seems that if operation at a variable heat input rate during the excursion is possible, the excursion may be largely controlled by thermal input considerations; if on the contrary no steady operation points at variable heat input exist along the external characteristic, the excursion may be very fast and controlled only by the hydraulic inertia of the flow.

## 2.2 Flow Distribution Instabilities

When several channels having multivalued (negative-slope) ($w$, $\Delta p$) characteristics are operated in parallel, severe flow maldistributions may result. Such phenomena can occur during the LOCA in light-water reactors (LWRs), in steam generators, and in chemical process equipment. Eselgroth and Griffith (1967, 1968) examined flow inversions in parallel LWR channels. The distribution of flow rates in long parallel tubes of steam generators was studied by Akagawa et al. (1971) and Hayama (1963, 1964, 1967). Criteria for the stability of parallel-channel systems were derived by Gerliga and Dulevskiy (1970).

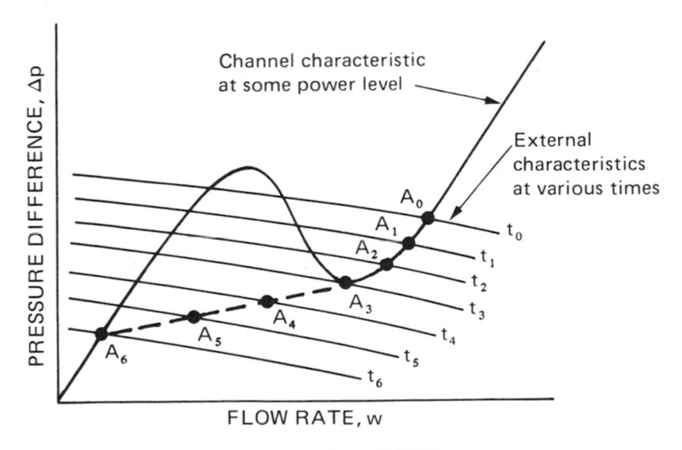


**Figure 2** Loss of flow transient in a LMFBR.

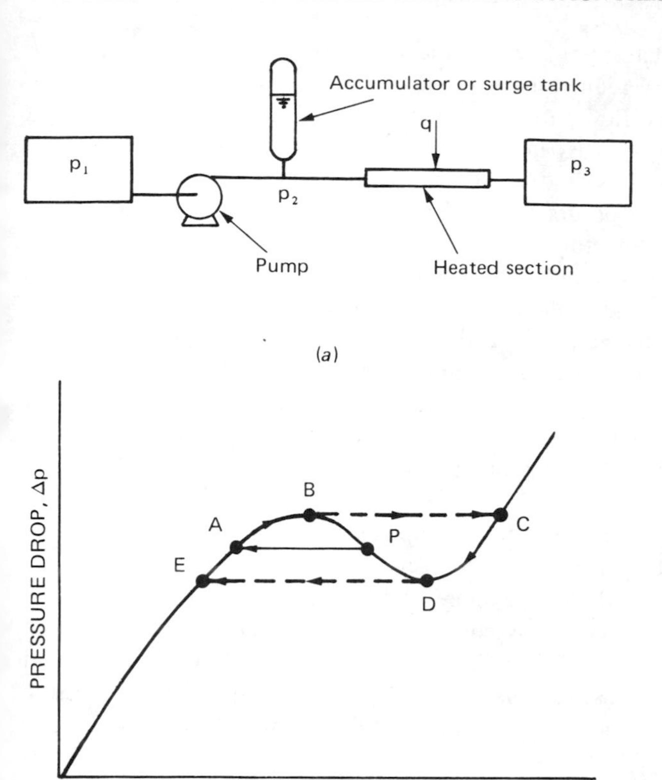


(b)

**Figure 3** Pressure drop oscillations. (*a*) System capable of sustaining pressure drop oscillations. (*b*) Limit cycle shown on channel internal characteristic.

## 2.3 Pressure Drop Oscillations

A multivalued ($w$, $\Delta p$) channel characteristic may also lead to flow oscillations (rather than an excursion) if there is sufficient interaction and delayed feedbacks between the inertia of the flow and compressibility of the two-phase mixture. The required compressible volume may be situated outside the heated section as reported by Stenning and Veziroğlu (1965, 1967), Maulbetsch and Griffith (1965), and Stenning et al. (1967), but it might also be provided by the internal compressibility of long test sections as indicated by Maulbetsch and Griffith (1965) and Andoh (1964). With reference to their cause, such oscillations have been referred to as pressure drop–flow rate oscillations, or simply as pressure drop oscillations by Stenning and Veziroğlu (1967).

Consider, for example, a compressible volume (accumulator, surge tank, etc.) situated upstream of the heated section, as shown in Fig. 3*a*. Operation of the heated channel at an unstable point $P$ on the $(w, \Delta p)$ characteristic of Fig. 3$b$ results in an initial excursive reduction of the flow rate through the heated section, along the line $PA$. Since the flow rate through the pump is not affected initially, a flow imbalance and storage of mass in the compressible volume result; the pressure in the compressible volume rises, and the operating point moves along the segment $AB$. A second flow excursion occurs at $B$ and the operating point moves now along $BC$. The increased flow rate through the heated section now results in net evacuation of mass from the compressible volume, reduced pressure in that volume, and movement of the operating point along the segment $CD$. From point $D$ a flow excursion to point $E$ occurs again, and the limit cycle $ABCDEA$ repeats itself thereafter. That delayed feedback mechanisms are responsible for the cyclic operation should be clear from this idealized explanation. In reality the flow excursions are not instantaneous, and the "corners" of the limit cycle are rounded, as shown in the recording of the locus of the operating points in the $(w, \Delta p)$ plane by Stenning and Veziroğlu (1967) in Fig. 4.

If the compressible volume is situated upstream of the test section, the oscillations can be damped by throttling the inlet of the heated channel. Throttling may of course not be effective when the compressibility is internal to the channel.

The frequency of the oscillations is controlled by the time constant associated with the compressible volume and not by the transit time through the heated channel as for density wave oscillations, discussed next. The periods of pressure drop oscillations are generally longer than the periods of density wave oscillations.

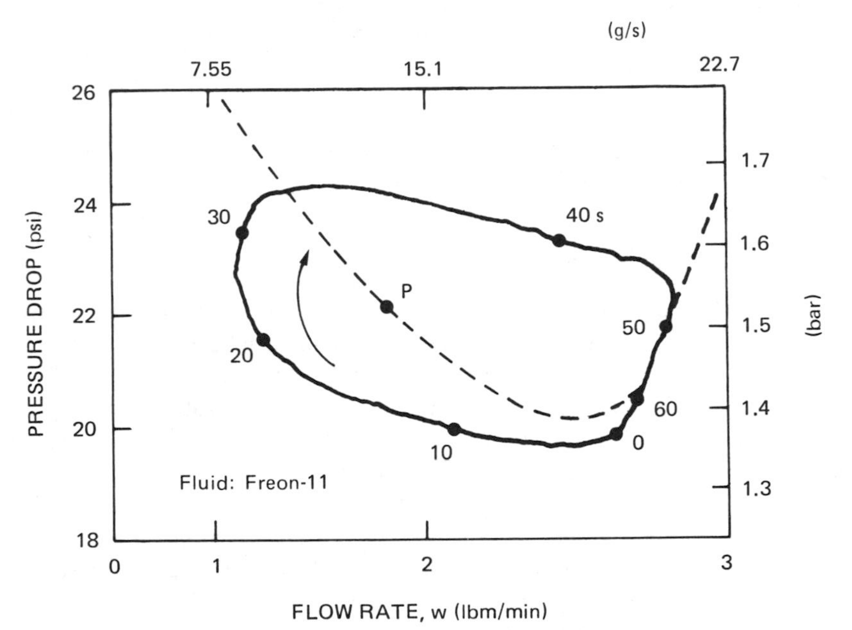


**Figure 4** Limit cycle for pressure drop oscillation in the flow rate–pressure drop plane. $P$ denotes the steady-state operating point on the channel internal characteristic shown by dotted lines (Stenning and Veziroğlu, 1967).

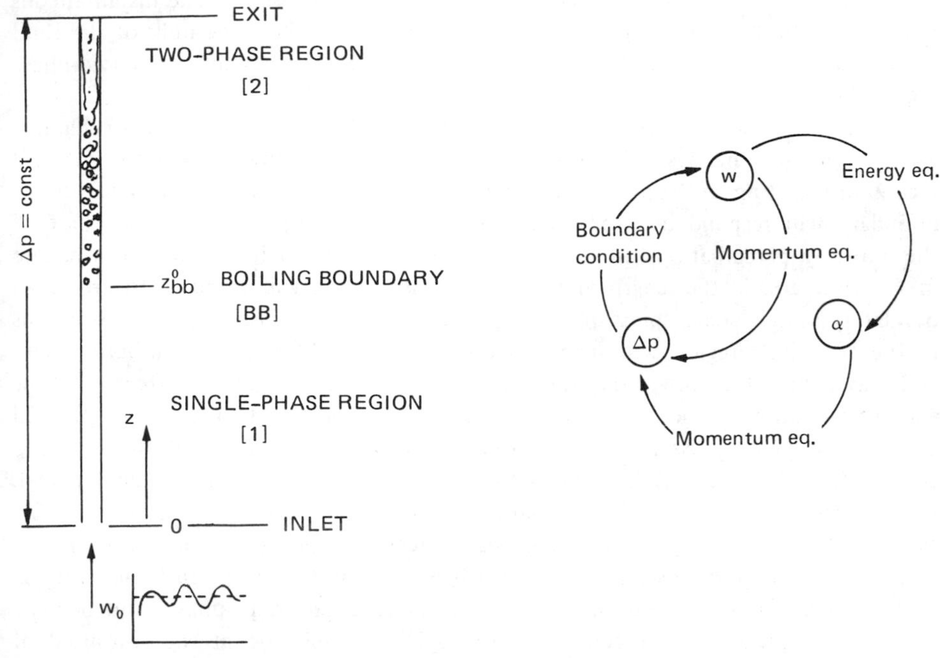


**Figure 5** Boiling channel and various feedbacks.

With a relatively stiff system, however, it might be possible to lower the time constant of the compressible volume to the point where the period of pressure drop oscillations becomes comparable to the period of density wave oscillations. Daleas and Bergles (1965) have indeed shown that the amount of upstream compressibility needed to produce pressure drop oscillations may be surprisingly small. When the heat flux is sufficiently high, Ledinegg flow excursions and pressure drop oscillations result in "premature" burnout of the test section. Such situations occur often in subcooled boiling (Daleas and Bergles, 1965).

Simple analytical models that could be used to predict the threshold and the period of pressure drop oscillations have been proposed by Stenning and Veziroğlu (1965, 1967), Maulbetsch and Griffith (1965), and Andoh (1964).

## 3 DENSITY WAVE OSCILLATIONS

Density wave oscillations are probably the most common type of instabilities encountered in two-phase flow systems and result from the multiple feedbacks between the flow rate, the vapor generation rate, and the pressure drop in a boiling channel (see Fig. 5). The physical mechanism leading to density wave oscillations is now clearly understood and can be described in equivalent ways. One such description by Yadigaroglu and Bergles (1972) follows.

Consider a heated channel such as the one depicted in Fig. 5. The instantaneous position of the boiling boundary BB, that is, the point where the bulk of the fluid reaches saturation, divides the channel into a single-phase region and a two-phase region. Subcooled boiling will be considered later.

Consider now an oscillatory subcooled inlet flow entering this channel. Yadigaroglu and Bergles (1972) showed that the inlet flow oscillations $\delta w$ will create propagating enthalpy perturbations in the single-phase region. The boiling boundary will respond by oscillating according to the amplitude and the phase of the enthalpy perturbations at the point where the flow reaches saturation. Changes in the flow and in the length of the single-phase region will combine to create an oscillatory, single-phase, pressure drop perturbation $\delta\,\Delta p_1$. Enthalpy perturbations in the two-phase region will appear as quality and void-fraction perturbations and will travel with the flow along the heated channel. The combined effects of flow and void-fraction perturbations and the variations of the two-phase length will create a two-phase pressure drop perturbation $\delta\,\Delta p_2$. Since the total pressure drop across the boiling channel is imposed by the external characteristic of the channel, the two-phase pressure drop perturbations will create feedback pressure perturbations of the opposite sign in the single-phase region, which can either reenforce or attenuate the imposed oscillation by creating a feedback flow perturbation $\delta w_{fdb}$. With correct timing, the perturbations can acquire appropriate phases and become self-sustained. Under these conditions the system would be at the threshold of stability since it would have the capability of oscillating without externally imposed perturbations, while satisfying the boundary conditions; this situation is illustrated in Fig. 6*b*.

Alternatively one can say that the system crosses the threshold of instability when for a set of operating conditions and for a given inlet flow oscillation frequency $\omega$, the total pressure drop perturbation vanishes:

$$\delta\,\Delta p(\omega) = \delta\,\Delta p_1(\omega) + \delta\,\Delta p_2(\omega) = 0$$

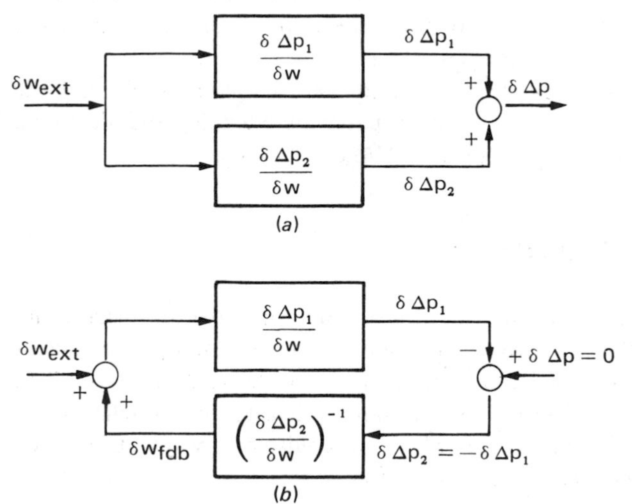


**Figure 6** Block diagrams of a boiling channel.

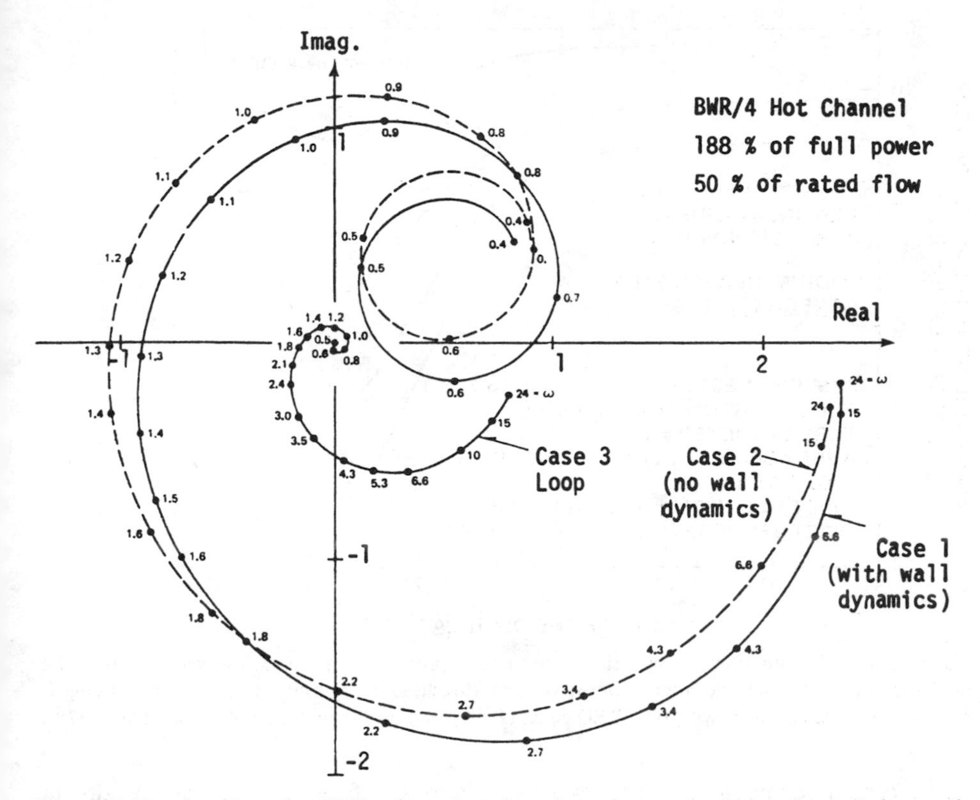


**Figure 7** Nyquist plots for BWR channel and entire primary circulation loop (Lahey and Yadigaroglu, 1974). Cases 1 and 2 refer to the stability of a core channel. Case 3 refers to the stability of the entire primary loop including the downcomer and the jet pumps. (The loci are marked with the period of the oscillation; the system becomes unstable with a period of approximately 1.3 s.)

Under those conditions, since the boundary conditions imposed on the system are satisfied with oscillatory flow, the system tends to be unstable. This second stability criterion is also illustrated in Fig. 6*a*.

The boiling channel can be modeled by transfer functions as a closed-loop feedback system. This type of modeling readily allows a frequency-domain interpretation of the stability margins; Nyquist plots can be used, for example, to investigate system stability. Figure 7 shows such a Nyquist plot, and Fig. 8 shows the variation of the magnitude of $\delta \Delta p$ as the stability limit is approached.

The preceding description of the instability mechanism reveals that transportation delays in the channel are of paramount importance regarding stability of the system, although inertia effects are also responsible for the generation of phase shifts. The presence of transportation and inertial delays led to the suggestion for naming these instabilities time-delay oscillations. Emphasizing the feedback mechanism, Neal and Zivi (1967) referred to these as flow-void feedback instabilities. The

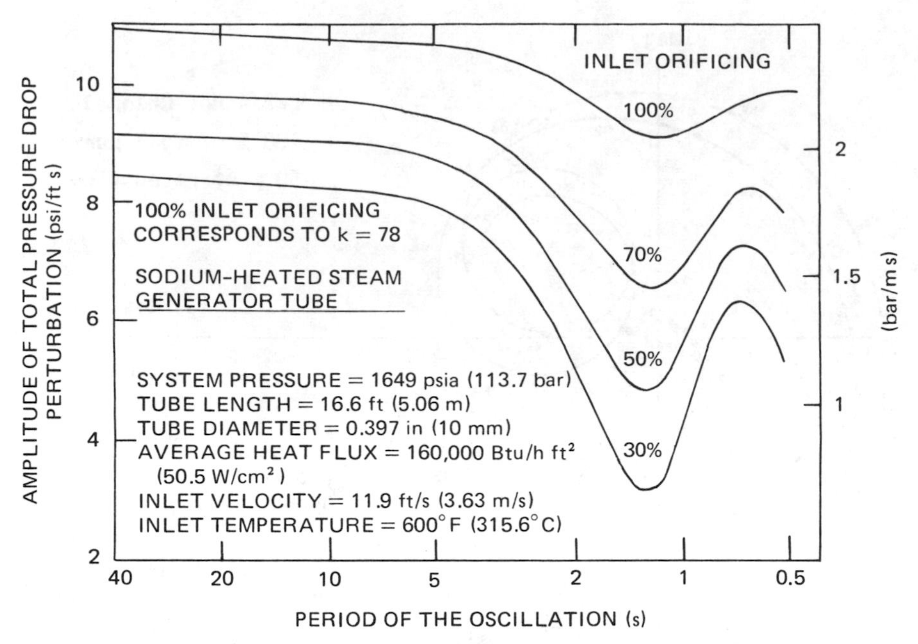


**Figure 8** Transfer functions of inlet flow rate to total pressure drop for various values of the inlet orificing. Figure shows the approach toward the threshold of stability as the inlet orificing is reduced. Results obtained with NUFREQ code. (Lahey and Yadigaroglu, 1974; Chen et al., 1976).

well-accepted terminology density wave oscillations is from Stenning and Veziroğlu (1965) and illustrates that waves of alternatively more and less dense mixture travel along the channel. Zuber (1966) referred to these as thermally induced oscillations.

A number of auxiliary considerations, including transient heat storage in the channel walls, variation of the saturation temperature with pressure, compressibility of the vapor phase, thermal nonequilibrium, variable heat transfer, and changes in the relative velocity between the phases, contribute to the fundamental feedback mechanism described above by altering the phase and amplitude of the various perturbations. Although none of these is the cause of density wave oscillations, they might acquire under certain conditions a controlling role.

Density wave oscillations can be manifested in all kinds of two-phase flow systems, including BWRs (Kirchenmayer, 1960; Fleck, 1961; Nahavandi and Von Hallen, 1964; Sanathanan et al., 1965), although modern BWRs are not limited from stability considerations (Lahey and Moody, 1977). Two-phase flow loops (for example, Fabréga, 1965; Jain et al., 1966; Schuster and Berenson, 1967; Mathisen, 1967; Masini et al., 1968; Dijkman, 1969), cryogenic equipment (Jones and Peterson, 1974; Friedly et al., 1967), reboilers and other chemical process equipment (Blumenkrantz and Taborek, 1972), and steam generators (for example, Waszink and Efferding, 1973; Llory, 1975; Moxon, 1973; Sano et al., 1973) can also exhibit density wave oscillations. Instabilities in steam generators are complicated by coupling with the primary fluid dynamics. In BWRs there is nuclear coupling owing to the void-reactivity-neutron flux feedbacks. Density wave oscillations can also appear in fluids operating in the near-critical and the supercritical thermodynamic region. These were investigated by Zuber (1966) and Friedly et al. (1967). A pertinent bibliography on density wave oscillations is given in Table 1.

## 3.1 Analytical Investigation

All analyses of the dynamics of two-phase flows start from the analytical expression of the conservation laws and from a specification of the boundary conditions. The system of resulting equations must be closed by specifying an appropriate number of constitutive laws. Stability may be investigated in either the time or the frequency domains as discussed below.

**3.1.1 Two-phase flow descriptions** Sophisticated two-phase flow models including a set of conservation equations for each phase have become available recently (Ishii, 1975). The phase interaction laws required for coupling the two sets of conservation equations at the interfaces are, however, not always well-established yet and therefore limit the usefulness of such "six-equation" models. All two-phase stability analyses have been conducted in the past using the three conventional mixture conservation equations in various forms (Yadigaroglu and Lahey, 1976). All these analyses are limited by the fact that macroscopic, one-dimensional constitutive laws obtained from steady-state experiments are used to describe system behavior during transients when local multidimensional effects might be controlling.

The period of density wave oscillations is of the order of the transit time of the fluid in the heated channel. Thus density wave oscillations involve relatively slow transients. For high-pressure systems it is often acceptable to assume constant properties along the channel, functions only of a system pressure. In this way the local density of the mixture becomes a function of the mixture enthalpy only and a decoupling of the mass and energy conservation equations from the momentum equation is achieved (Meyer and Rose, 1963). At the same time the effects of compressibility of the phases and acoustic wave propagation effects are eliminated as discussed in the beginning of this chapter. For relatively slow transients these are, however, not important.

**3.1.2 Kinematic-wave propagation** Density wave and some other types of two-phase flow instabilities are linked to kinematic-wave propagation phenomena, as discussed in the beginning of this chapter. A brief review of relevant work in this area is given below.

Kinematic or continuity waves occur whenever the steady-state values of flow rate and concentration are related. A continuity wave is created whenever a system propagates from a given state into another state. There are no effects of inertia or momentum, and in fluid flow problems continuity waves can be described by the mass and energy conservation equations alone. On the contrary dynamic waves

**Table 1 Selected bibliography on density wave oscillations**

| Author | Linearized stability models | Nonlinear stability models | Successful simple models | Models considering BWR dynamics | Stability data | Data with nonuniform power distribution | Several parallel channels |
|---|---|---|---|---|---|---|---|
| Anderson (1970) | x | | | | x | | |
| Becker et al. (1964) | | x | | | x | | |
| Biancone et al. (1965) | | | | | x | | |
| Bogaardt et al. (1966) | x | | | | x | | |
| Bouré and Mihaila (1967) | x | | x | | | | |
| Carver (1970) | x | | | | | | |
| Collins and Gacesa (1970) | | | | | x | | |
| Crowley et al. (1967) | | | | | x | | x |
| Davis and Potter (1967) | x | | x | | | | |
| Dean and Murray (1976) | x | | | | | | |
| Dijkman (1969) | | x | | | x | x | |
| Efferding (1968) | x | | | | | | |
| Fabréga (1965) | | | | | x | | |
| Fleck (1961) | | x | | x | | | |
| Friedly et al. (1967) | x | | | | | | |
| Ishii (1971) | x | | x | | x | | |
| Jain et al. (1966) | | | | | x | | |
| Jones (1961–1964) | x | | | | | | |
| Jones and Peterson (1974) | | x | | | | | |
| Jones and Yarbrough (1964, 1965) | x | | | x | | | |
| Kanai et al. (1961) | | | | x | | | |
| Kakaç (1977) | | | | | x | | |
| Kirchenmayer (1960) | | | | x | | | |
| Koshelev et al. (1970) | | | | | x | | x |
| Lahey and Yadigaroglu (1974) | x | | x | x | | | |
| Masini et al. (1968) | | | | | x | x | x |
| Mathisen (1967) | | | | | x | | |
| Meyer and Rose (1963) | | x | | | | | |
| Moxon (1969) | | x | | | | | |
| Nahavandi and Von Hallen (1964) | | x | | x | | | |
| Neal and Zivi (1967) | | | | | x | | |
| Quandt (1961) | x | | | | x | | |
| Saha (1974) | x | | x | | x | | |
| Sanathanan et al. (1965) | x | | | x | | | |
| Schuster and Berenson (1967) | | | | | x | | x |
| Shotkin (1967) | x | | | | | | |
| Stenning (1964) | x | | | | | | |
| Stenning and Veziroğlu (1965, 1967) | x | | | | x | | |
| Takahashi and Futami (1977) | x | | | | | | |
| Wallis and Heasley (1961) | x | | x | | | | |
| Yadigaroglu and Bergles (1972) | x | | x | | x | x | x |
| Zuber (1966) | x | | | | | | |

depend for their existence on acceleration forces that result from concentration gradients.

Examples of continuity waves can be found in the flow of rivers, where the flow rate depends on the depth of the water; on the flow rate of vehicles on highways, where speed depends on vehicle concentration (Lighthill and Whitham, 1955); and in two-phase flow problems, where the mixture density depends on the flow rate.

At least two distinct mechanisms are responsible for the relationship between density and flow rate in two-phase flow. The first mechanism present with or without heat addition, is the dependence of the void fraction on flow rate. Within the framework of Zuber and Findlay's (1965) drift-flux description of two-phase flow, this dependence can be explained by the relative effect of the drift velocity of one phase (largely independent of flow rate) with respect to the mixture velocity.

The second mechanism is a result of the dependence of the enthalpy of the mixture (and consequently of its density) on flow rate through the energy equation: If the flow rate is decreased, the two-phase mixture density decreases. A theory and examples of continuity waves are given by Wallis (1969).

Zuber and Staub (1967), starting from the mass and energy conservation equations, developed a general expression for the transient response of the void fraction in a two-phase system with change of phase:

$$\frac{\partial \alpha}{\partial t} + C_k \frac{\partial \alpha}{\partial z} = \frac{\rho}{\rho_L - \rho_G} \Omega \tag{8}$$

where $C_k$ is the velocity of the kinematic wave, given in terms of drift-flux model variables by

$$C_k = j + V_{Gj} + \alpha \frac{\partial V_{Gj}}{\partial \alpha} \tag{9}$$

In Eq. (9) $j$ is the volumetric flux of the mixture and $V_{Gj}$ is the drift velocity of the gas, $V_{Gj} = u_G - j$. The characteristic reaction frequency $\Omega$ is related to the vapor generation rate. If the properties of the fluid are constant, $\Omega$ is given by

$$\Omega = v_{LG} \Gamma_G$$

where $\Gamma_G$ is the rate of vapor generation and $v_{LG}$ is the volume change resulting from vaporization, $v_{LG} = v_G - v_L$. If all the heat input is used to generate vapor

$$\Gamma_G = \frac{q'''}{h_{LG}} \qquad \Omega = \frac{v_{LG}}{h_{LG}} q'''$$

where $q'''$ is the heat input per unit volume of fluid and $h_{LG}$ is the heat of vaporization. We see that $\Omega$ describes the relationship between energy input and volume change of the mixture.

The mass and energy conservation equations can also be combined to give

$$\frac{\partial j}{\partial z} = \Omega \tag{10}$$

which can be integrated into

$$j = j_0 + \int_0^z \Omega \, dz$$

Substituting this expression into Eq. (9), we obtain

$$C_k = j_0 + \int_0^z \Omega \, dz + V_{Gj} + \alpha \frac{\partial V_{Gj}}{\partial \alpha} \tag{11}$$

Equation (11) shows clearly that the kinematic-wave velocity depends on the integral relationship between heat input and flow rate, on the drift velocity of the gas, and on the variation of the drift velocity with concentration.

Note that Eq. (9) can also be written in terms of the liquid fraction $(1 - \alpha)$ as

$$C_k = j + V_{Lj} + (1 - \alpha) \frac{\partial V_{Lj}}{\partial(1 - \alpha)} \tag{12}$$

where the liquid drift velocity, $V_{Lj}$ is defined as $u_L - j$. The choice between Eqs. (9) and (12) will depend on the flow regime being described. For example, it is natural to use the drift velocity of the bubbles in bubbly flow and the drift velocity of the droplets in dispersed-droplet flow.

Zuber and Staub (1966) and Staub and Zuber (1967) used the formulation just outlined to study the propagation of void fraction perturbations in oscillatory flow. This formulation is also the basis of the two-phase flow stability analyses of Ishii and Zuber (1970) and Saha (1974). Shiralkar et al. (1973) and Hancox and Nicoll (1971) extended the analysis to account for the effects of a subcooled inlet condition and of the variations of the radial distributions of void, velocity, and enthalpy.

Gonzalez-Santalo and Lahey (1973) applied a simplified, homogeneous-flow version of the theory of kinematic-wave propagation to study flow decay transients in BWRs. The propagation of density disturbances in air–water flow has also been studied by Nassos and Bankoff (1966).

Although the kinematic behavior of the mixture, that is, the relationships between phase velocities and void fraction, can be described fairly well at steady state, there is unfortunately very little basic information (Nassos and Bankoff, 1967; Sakaguchi et al., 1973) available on this behavior during transients.

**3.1.3 Approaches to stability investigation** Two-phase flow instability models result in space- and time-dependent partial differential equations. In early analyses attempts were made to eliminate the space dependence by integrating the equations along the channel. This is possible if an arbitrary, a priori assumption on the space dependence of the various perturbations is made. Such lumped-parameter stability models cannot be successful in general and have been abandoned in favor of distributed-parameter models in which a space and time dependent solution is performed before any integration.

Two general approaches are generally possible for the thermohydrodynamic stability analyses:

1. Frequency-domain, linearized models
2. Time-domain, nonlinear finite-difference models

In frequency-domain, linearized models the conservation equations and the necessary constitutive laws governing fluid flow and heat transfer are linearized about an operating point; the resulting linear differential equations are Laplace-transformed, and the transfer functions that can be obtained in this manner are used to arrive at a stability evaluation of the system using classic control-theory techniques. This method can predict only the threshold of instability since the characteristics of the limit-cycle oscillation appearing beyond this threshold can be obtained only from nonlinear models of the system. The method is inexpensive with respect to computer time, relatively straightforward to implement, yields results that are easy to interpret, and is free of the numerical stability problems of finite-difference techniques.

Alternatively, one can use a finite-difference, nonlinear model of the system for stability evaluations. In this model the steady state is perturbed with small stepwise changes of some operating parameter simulating an actual transient, such as power escalation in a real system. The stability threshold is reached when undamped or diverging oscillations appear following such a small change [see, for example, Meyer and Rose (1963)]. Stability can also be investigated by momentarily perturbing the steady state, for example, with a delta-function variation of a boundary condition.

Time-domain finite-difference techniques are very time consuming when used for stability analyses, since a large number of cases must be run to produce a stability map, and each run is itself time consuming because of the limits on the allowable time step. It is also sometimes difficult to clearly distinguish between the true hydrothermodynamic oscillations and numerical instabilities.

**3.1.4 The threshold of stability and the limit cycle** In theory, a system becomes unstable when infinitesimal oscillations around an equilibrium point diverge or are indefinitely sustained. In practice, it is occasionally difficult to define experimentally the threshold of stability since the system often becomes increasingly noisy as the stability limit is approached. This is especially true at low inlet subcoolings and when certain flow regimes, such as slug flow, dominate. Therefore, some arbitrary criterion must be applied in defining the experimental threshold of stability. Experimental thresholds are often defined by simple visual examination of traces. An extrapolation toward zero of the oscillation amplitude plotted against the power input can also be used to define the threshold power level, but the amplitude of the oscillations does not necessarily grow linearly.

Various attempts have been made to systematize the procedure. When the inception point is readily distinguishable, the knee of the amplitude curve plotted as a function of some independent variable can be used. When a clear-cut threshold is not observed, statistical methods involving use of an rms value (Bogaardt et al.,

1966) or of the flow-noise variance (Kjaerheim and Rolstad, 1967) or some other noise function (Mathisen, 1967) become necessary.

Time-domain analyses are in principle capable of predicting the limit cycle of the oscillations although such predictions might be difficult and unreliable. Predictions of the limit cycle could be very useful since, near the threshold of instability, the amplitude of the oscillations might be small and acceptable. Furthermore, comparisons of experimental and computed limit-cycle characteristics could provide valuable information on the adequacy of two-phase flow dynamics codes.

**3.1.5 Stability models** Several distributed-parameter stability models, both linear and nonlinear, have been published, as shown in Table 1. Although models became increasingly more complex and realistic in their description of the systems, they remained limited in their ability to predict system stability, as further discussed in Sec. 3.1.9.

The features of complex computer codes developed by national laboratories and industrial organizations for two-phase flow dynamics studies are reviewed by Bouré et al. (1973). Probably some of the best available models are proprietary.

The tendency in recent years has been toward formulation of simpler models (see Table 1) that retain the essential physics, provide valuable physical insight, and permit rapid and inexpensive evaluation of parametric trends.

In all the recent simpler models in Table 1 the conservation equations are linearized and solved in the frequency domain since the capability of linearized models to predict the threshold and the period of the oscillations is well accepted now.

Early work in the stability field that led to the present formulation of the problem is reviewed by Ishii (1976). It appears that the first correct formulation of the relationship between the variation of the two-phase mixture velocity and the heat flux is by Serov (1953). This analysis and several subsequent analyses were, however, incomplete since some effects were either ignored or modeled in a very approximate fashion.

The recent linearized models are in essence refinements and improvements of the pioneering analysis of Wallis and Heasley (1961); in all these models simple linearized forms of the mass and energy conservation equations are integrated analytically along the channel to determine the time-dependent distributions of two-phase density and velocity. These distributions are then used in the decoupled momentum equation to calculate the pressure drop perturbations. The basic features of five recent models are compared in Table 2.

Inspection of Table 2 shows that no single model incorporates all the desirable features needed for a detailed and complete stability analysis, although there is no fundamental reason for not combining the desirable features of several models in a single code.

**3.1.6 Modeling of subcooled boiling** When subcooled boiling is not neglected, the net vapor generation (NVG) boundary, that is, the point in the channel beyond

**Table 2 Comparison of the basic features of recent frequency-domain models**

| Feature | Author | | | | |
|---|---|---|---|---|---|
| | Davies and Potter (1967) (LOOP code) | Bouré and Mihaila (1967) | Lahey and Yadigaroglu (1974) (NUFREQ code) | Ishii and Zuber (1970) | Saha (1974) |
| Heat input distribution in single-phase region | Uniform | | Arbitrary variation | | Uniform |
| Heater dynamics in single-phase region | Constant heat input | | Heater dynamics considered[a] | Constant heat input | |
| Subcooled boiling | Not considered | Considered through density–enthalpy relationship | Not considered | | Considered by profile-fit method |
| Heat input in two-phase region | Constant and uniform | | | | |
| Two-phase flow model | Homogeneous | Particular relationship between enthalpy and density | Homogeneous | Drift-flux model | |
| Superheated steam region | Included assuming constant steam density | Not included[a] | | | |
| Other features | | | Nuclear feedback effects considered | | |

[a]The more detailed description of heater dynamics by Yadigaroglu and Bergles (1972) and a superheated steam region are now included in NUFREQ.

which the subcooled void fraction increases rapidly, must replace the boiling boundary as the limit between the single-phase and the two-phase regions. The oscillation of the NVG point will be determined by the amplitude and phase of the temperature and velocity perturbations near the heated wall. Roy and Yadigaroglu (1976) have shown that there are significant departures from one-dimensional (axial) behavior regarding these perturbations near the wall, even at very low frequencies. Thus the usual one-dimensional treatments of the dynamics of the single-phase region cannot describe adequately the conditions near the heated wall. Two-dimensional (axial-radial), time-dependent, turbulent-flow analyses of the single-phase region must depend on empirical treatments of the turbulence. The reliability of such turbulence models under transient conditions has not been adequately tested. Furthermore, when the time constants associated with large eddies approach the period of the oscillations, the conventional turbulence models may lose their validity, and other approaches would have to be found. The flow oscillations could, for example, alter the fundamental characteristics of the turbulence. Thus there appear to be serious difficulties in modeling the behavior of the NVG point and consequently of subcooled boiling in oscillatory flow.

Saha (1974) and Saha et al. (1976) have recently included the effects of subcooled boiling in Ishii and Zuber's (1970) stability model by accounting for the oscillations of the NVG point as determined by quasi-steady-state considerations. Saha's nonequilibrium analysis improved all predictions of the frequency of the oscillations and of the experimental stability boundaries at low subcooling. Ishii and Zuber's equilibrium model, however, gave better predictions of stability boundaries at high subcooling.

**3.1.7 Heat capacity of the heater** Heat capacity and transient heat conduction effects in the heater influence stability by providing the necessary degree of freedom for variation of the heat flux to the coolant. Flow rate and local temperature or quality perturbations influence the value of the heat transfer coefficient and, together with bulk temperature variations, are responsible for the perturbations of the heat flux to the coolant.

The effect is probably most important in the single-phase region because of the relative insensitivity of boiling heat transfer coefficients to flow rate variations. A detailed analysis of the effects of heater capacity and conductivity in the single-phase regions was presented by Yadigaroglu and Bergles (1972) and shows the importance of these effects with respect to stability. This analysis considers only the case of constant (but not necessarily uniform) heat generation rate. In a steam generator, the situation is complicated by coupling of the dynamics of the primary and secondary fluids. The analysis of Yadigaroglu and Bergles is being extended by Chan (1979) to account for the variations of the temperature of the primary fluid resulting from secondary-side flow rate variations.

**3.1.8 Void-power and Doppler coupling in BWRs** In BWRs the heat generation rate is perturbed by the void-reactivity-power coupling. The Doppler effect also introduces an additional space- and time-dependent coupling between fuel rod temperature and power generation rate. There was considerable concern about combined thermohydraulic and nuclear instabilities during the early days of BWR development, as evidenced by the references of Table 1. Such instabilities are not, however, of serious concern for modern BWRs operating at pressures around 70 bar, as shown by Lahey and Yadigaroglu (1974). At high pressure the void coefficient of reactivity is small and void fluctuations create only small changes in reactivity. Moreover, the uranium dioxide fuel rods used in modern BWRs have a thermal time constant of the order of 10 s that tends to strongly damp perturbations of the heat generation rate (Lahey and Moody, 1977).

An exact analysis of the void-reactivity-power coupling effects is rather difficult since the heat generation-rate perturbations in the "hot" bundle depend on the void distribution perturbations in that particular bundle as well as on the relatively unperturbed void distributions in neighboring bundles. Thus a space- and time-dependent coupled, neutronic, thermohydraulic analysis of the core is needed. Only simplified analyses of this situation have been published (Table 1).

There are also some early, relatively crude studies of the effects of oscillatory changes in gravity on the stability of a boiling channel (Beckjord, 1960) with application to naval BWRs.

**3.1.9 Adequacy and verification of analytical methods** None of the various stability models proposed in the past provided consistently good predictions of experimental stability thresholds and oscillation frequencies. This is not really surprising given that even steady-state, two-phase flow predictions are often inaccurate.

The available advanced analytical models were reviewed and tested against experimental data by Neal and Zivi (1967) and Grumbach (1969). Neal and Zivi concluded that the STABLE code (Jones, 1961-1964; Jones and Yarbrough, 1964, 1965) based on a sectionalized, linear, frequency-domain model best described the data trends. In fact the STABLE code and related codes (Efferding, 1968) have been used extensively in the United States for stability investigations of BWRs and steam generators, respectively.

The simpler frequency-domain models discussed in Sec. 3.1.4 provided good predictions of parametric trends and fairly accurate predictions of stability thresholds and oscillation frequencies.

There is always some difficulty in defining the experimental stability thresholds, as discussed in Sec. 3.1.3. In view of this difficulty it seems appropriate to compare model predictions of various system transfer functions with experimental measurements of these transfer functions. Such comparisons are of a more fundamental level and therefore can provide more detailed information on the adequacy of the various thermohydraulic models embedded in stability codes. Measurements of flow rate-to-void (Anderson, 1970; Kresja et al., 1967; Paul and Riedle, 1972); flow rate-to-pressure drop (Anderson, 1970; Kresja et al., 1967); and power-to-void (Christensen, 1961; Zuber et al., 1967; St.-Pierre, 1965) or power-to-pressure drop, or power-to-flow (Space Technology Laboratories, 1962) transfer functions are available and can be used to test the adequacy of analytical methods.

The results of such a test are plotted in Fig. 9. The results are qualitatively

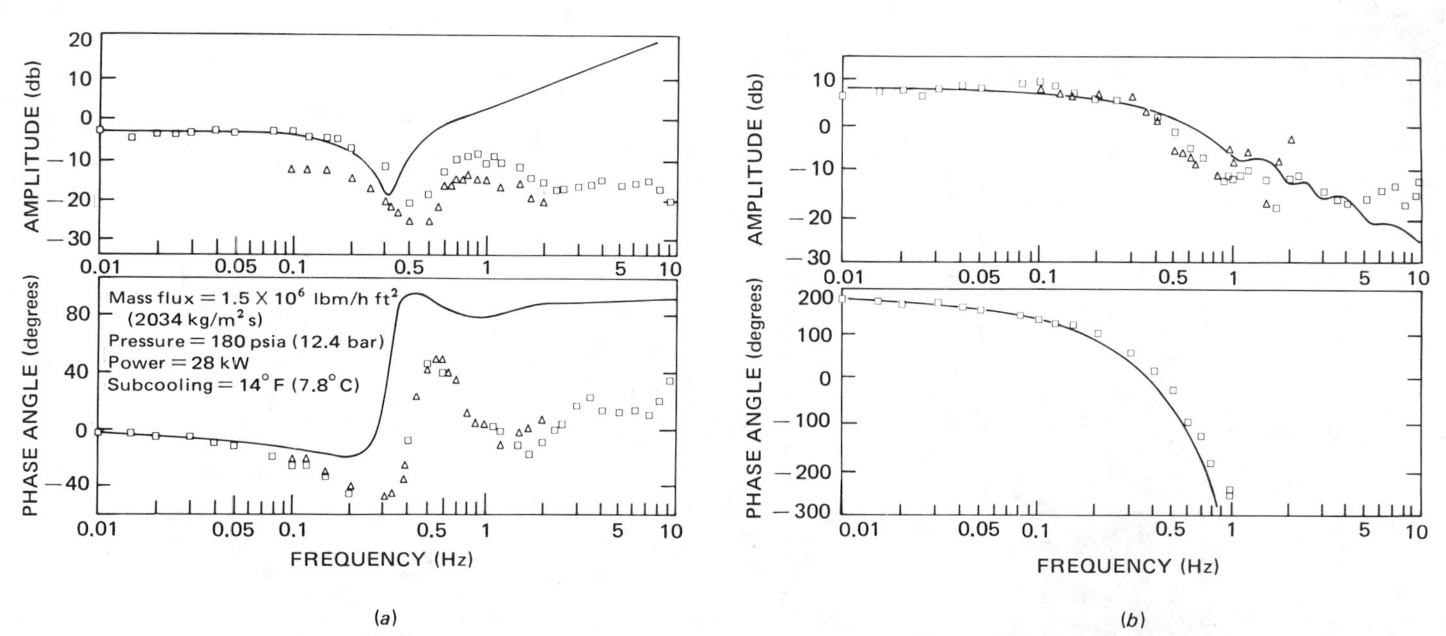


**Figure 9** Comparison of NUFREQ frequency-response results (lines) with data of Paul and Riedle (1972) (points) (Lyon, 1976). (*a*) Inlet flow rate-to-inlet pressure transfer function. (*b*) Inlet flow rate-to-exit void fraction transfer function.

excellent although at high frequencies there are phase shifts that are not accounted for properly by the analysis.

## 3.2 Parametric Effects–Stability Maps

The operating state of a boiling channel is generally specified by the following parameters:

1. Fluid
2. Channel geometry (length, hydraulic diameter, heated perimeter, frictional characteristics, local variations of geometry, etc.)
3. Operating pressure level (for example, channel exit pressure)
4. Axial linear heat input distribution $q'(z)$, and total heat input $q$
5. Flow rate $w$, or alternatively channel pressure drop $\Delta p$
6. Inlet coolant temperature $T_{\text{in}}$, or inlet subcooling $\Delta h_{\text{in}}$ (in enthalpy units)

For stability investigations, the first three items in the list and the axial heat input distribution are generally specified. The task of the analyst is then to find the regions of stable and unstable operation in the three-dimensional space ($w$, $q$, $\Delta h_{\text{in}}$). A mapping of these regions in two dimensions is referred to as the stability map of the system. In natural-circulation loops the dimensions of the problem are reduced from three to two since the heat input, flow rate, and subcooling are related.

It has become clear now that there is no universal stability map. Furthermore, the stability boundary in the ($w$, $q$, $\Delta h_{\text{in}}$) space is a surface and can be represented only by a family of curves in any two-dimensional map. Several such alternative representations are possible. Although no unique pair of nondimensional parameters describes completely the behavior of a given system, a judicious choice of nondimensional parameters can lead to a clustering of the family of stability lines useful in understanding system behavior.

Stability maps were drawn in the nondimensional enthalpy difference–subcooling plane by Yadigaroglu and Bergles (1972) and Lahey and Yadigaroglu (1974). These two parameters are defined as

$$\frac{q}{wh_{LG}} \qquad \frac{\Delta h_{\text{in}}}{h_{LG}}$$

where $h_{LG}$ is the latent heat of vaporization.

Ishii and Zuber (1970) used a phase change number $N_{pch}$ and a subcooling number $N_{\text{sub}}$. The phase change number is the ratio of the characteristic frequency of phase change $\Omega$ defined in Sec. 3.1.2 to the inverse of a single-phase transit time in the system. Equation (10) shows that the gradient of the volumetric flux of the mixture $j$ is equal to $\Omega$. Thus $\Omega$ is controlling the two-phase mixture velocity distribution.

The phase change number and the subcooling number are given by

$$N_{pch} = \frac{\Omega}{V_{\text{in}}/L} = \frac{v_{LG}}{v_L}\frac{q}{wh_{LG}} \qquad N_{\text{sub}} = \frac{\Delta h_{\text{in}}}{h_{LG}}\frac{v_{LG}}{v_L}$$

where $V_{\text{in}}$ is the inlet velocity and $L$ is the heated length. Thus at a given pressure level, the $(q/wh_{LG}, \Delta h_{\text{in}}/h_{LG})$ and $(N_{pch}, N_{\text{sub}})$ maps are identical. The advantage of Ishii's parameters is that they include the effects of pressure variation. Ishii (1976) and Yadigaroglu and Bergles (1969) listed and discussed other nondimensional parameters that influence stability.

Figure 10 shows a stability map for a typical BWR bundle geometry. The constant-mass-flux and the constant-heat-flux threshold lines are clustered together. At sufficiently high subcooling, these lines are constant-quality lines, approximately. Note that a modern BWR channel becomes unstable only when operating well beyond its nominal state.

Mapping of the stability domain in other alternative planes seemed to offer some advantages. For example, the ratio of single-phase to total pressure drop, $\Delta p_1/\Delta p$ may be substituted for the dimensionless subcooling as the ordinate of the stability plane. Experimental data obtained with a Freon-113 system clustered fairly well around a single stability threshold line when plotted in this fashion (Yadigaroglu and Bergles, 1969). Figure 11 shows the same BWR stability predictions in this plane; the clustering is less evident. In conclusion, the $(N_{pch}, N_{\text{sub}})$ plane seems to provide the best means of illustrating stability trends. This plane also correlates well with the pressure level effect as noted in Sec. 3.2.2.

Ishii (1976) presented detailed parametric stability investigations. His trends

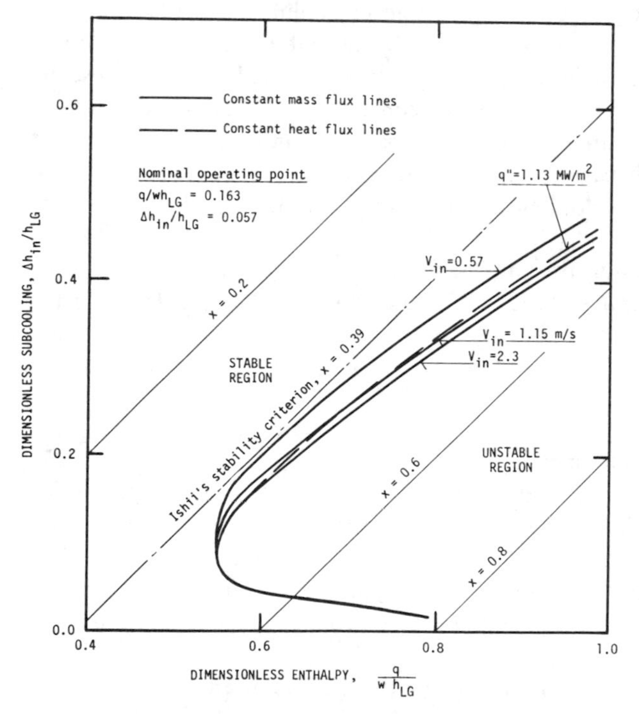


**Figure 10** A stability map for a typical BWR channel. The map includes constant-quality lines and Ishii's simplified stability criterion. The calculations are with NUFREQ code.

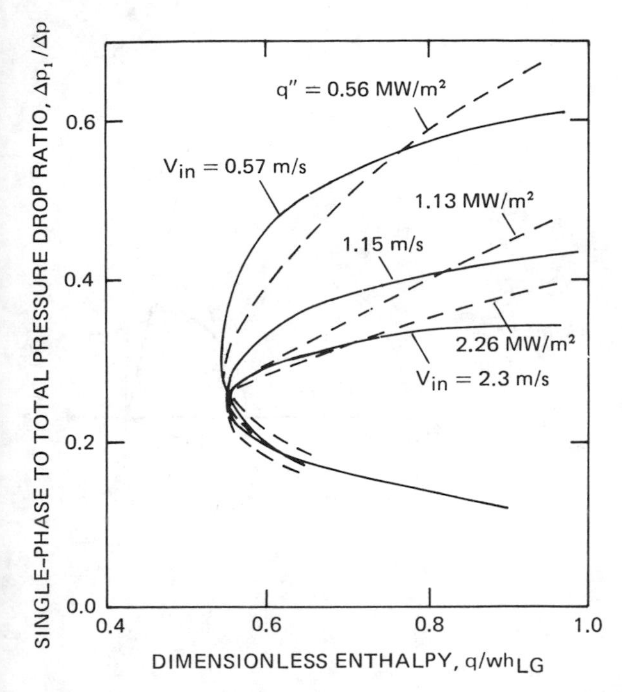


**Figure 11** A stability map for a typical BWR channel in the dimensionless enthalpy versus pressure ratio plane. The calculations are with NUFREQ code.

agree with the generally accepted parametric trends summarized below. In the following discussion the influence of a change in a certain parameter is said to be stabilizing if it tends to bring the operating point from the unstable into the stable region. In practice, this means that, everything else being equal, the system will be capable of operation at a higher power level without experiencing oscillations.

**3.2.1 Effects of heat input, flow rate, and exit quality** A stable system is brought into the unstable operating region by increases in heat input or decreases in flow rate; both effects increase the exit quality. Figure 12 shows the various perturbations in the complex plane during an increase of heat flux at constant mass flux and constant inlet temperature. One can clearly see the rotation of the perturbation vectors as the heat flux is increased. The single-phase pressure drop perturbation does not depend much on the heat flux since it is essentially controlled by inlet throttling. The two-phase pressure drop perturbation grows and rotates with increases in the heat flux, and at the threshold becomes equal but opposite in direction to the single-phase perturbation.

The destabilizing effect of increasing the ratio $q/w$ seems to be universally accepted, although islands of instability have been observed to occur by Yadigaroglu and Bergles (1969) within the stable operating region. Figure 13 shows the inlet flow rate trace recorded during a stability experiment at constant power input and

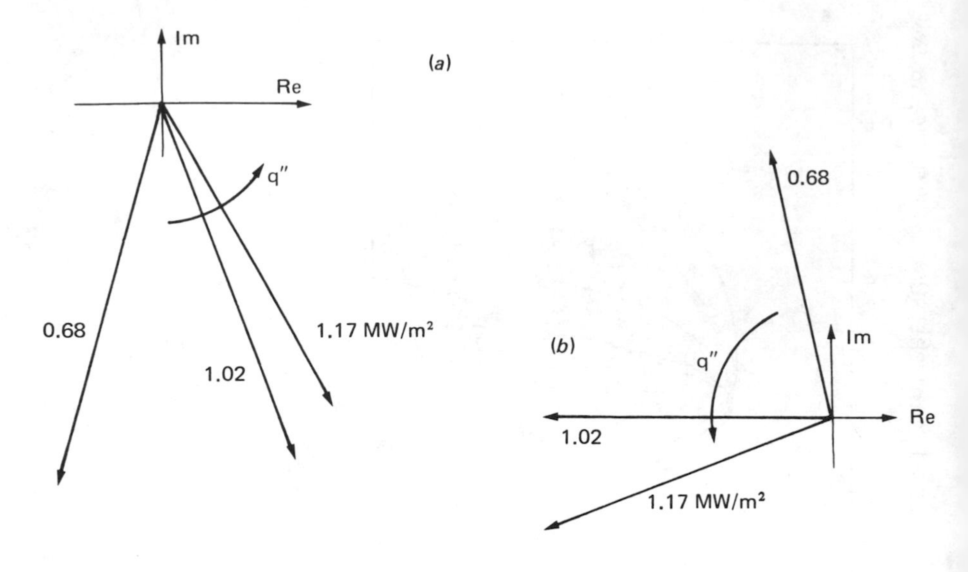


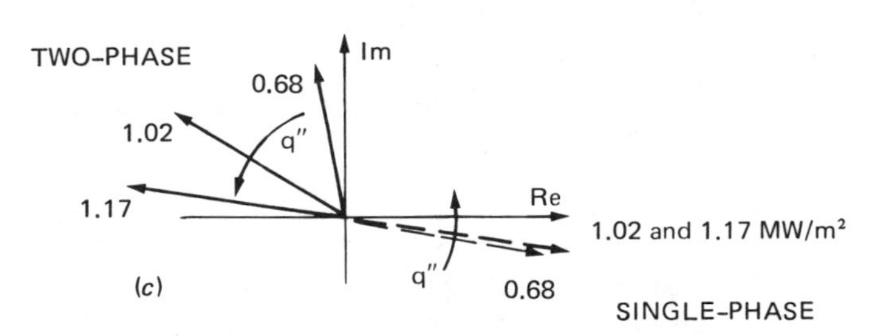


**Figure 12** Variation of the perturbations in the complex plane as the threshold power is approached (at constant mass flux). (*a*) Perturbation of the boiling boundary $\delta z_{bb}/\delta w$. (*b*) Perturbation of the exit flow rate $\delta w_e/\delta w$. (*c*) Pressure drops in single-phase and two-phase regions, $\delta \, \Delta p/\delta w$.

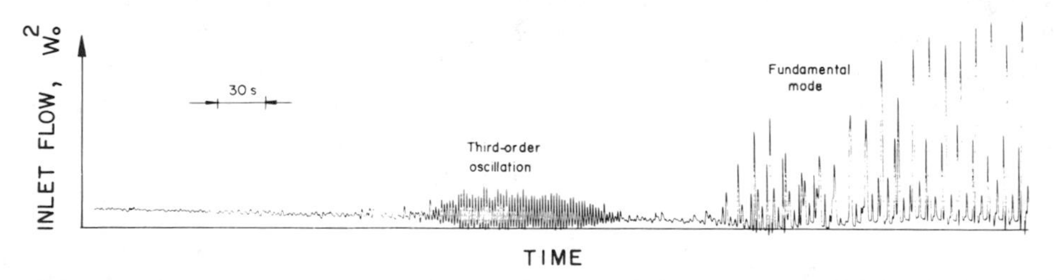


**Figure 13** Appearance of higher-mode oscillations before the threshold of the fundamental-mode oscillation. The experiment was conducted by slowly varying the inlet flow rate (Yadigaroglu and Bergles, 1969).

constant inlet subcooling. This experiment was conducted by varying slowly and continuously the inlet flow rate. Figure 13 shows that oscillations appeared at some value of the flow rate only to disappear a little later. Further reduction of the flow rate brought the system into the unstable region again. The frequency of the oscillations was different in the two unstable regions leading to the conclusion that the early oscillations were of higher mode, as discussed in Sec. 3.3.1.

**3.2.2 Effects of pressure level** An increase in pressure level has always been found to be stabilizing, although one must be careful in stating which system parameters were kept constant while the pressure level was increased. At constant values of the dimensionless enthalpy difference and subcooling, that is, at constant dimensionless inlet subcooling and exit quality, the pressure effect is made apparent by the density ratio $v_{LG}/v_L$ appearing in the definition of Ishii and Zuber's (1970) phase change and subcooling numbers. Although these two dimensionless numbers do not describe completely all the effects of the pressure level on the stability of the system, Ishii (1976) noted that calculated stability boundaries at three different pressure levels could not be differentiated in the $(N_{pch}, N_{\text{sub}})$ plane. Thus the density ratio $v_{LG}/v_L$ (approximately equal to $\rho_L/\rho_G$) could be used to extrapolate data obtained at a given pressure level and produce a map for a different pressure level.

**3.2.3 Effect of inlet subcooling** The effect of inlet subcooling is stabilizing at high subcoolings and destabilizing at low subcoolings as shown in Fig. 10. Intuitively this effect may be explained by the fact that, as we increase or decrease the inlet subcooling, the two-phase channel tends towards stable single-phase liquid and vapor operation, respectively. Thus both increases and decreases of subcooling tend to pull the system out of the unstable two-phase operating mode.

The value of the inlet subcooling determines the location of the boiling boundary. In turn, the phase and the amplitude of the oscillations of the boiling boundary depend on the length of the single-phase region. The location and the behavior of the boiling boundary control to a large extent the phase of the exit flow rate perturbation and the stability of the system, as noted by Shotkin (1967) and Yadigaroglu and Bergles (1969, 1972), who gave a detailed derivation of the transfer function between the inlet flow rate and the position of the boiling boundary. The importance of the movements of the boiling boundary was investigated by Lahey and Yadigaroglu (1974), who compared the pressure drop perturbations with and without a moving boiling boundary.

**3.2.4 Effect of nonuniform, axial, heat input distribution** The axial heat input distribution in BWRs is generally peaked near the center of the channel and is often assumed to be a cosine distribution. In steam generators the heat flux peaks generally near the inlet of the channel.

The effect of a cosine distribution on stability has been investigated experimentally, as shown in Table 1. These observations, which may be apparently conflicting, have been analyzed and were discussed by Yadigaroglu and Bergles

(1969, 1972). The effect of a cosine distribution is to move the boiling boundary toward the center of the channel; this has an effect on the length of the single-phase region and on the phase of the oscillation of the boiling boundary. The effect seems to be largest and most destabilizing when the boiling boundary is high in the channel. Thus it was concluded that cosine distributions could be moderately stabilizing at low subcoolings and destabilizing at high subcoolings. This effect is illustrated in Fig. 14 and can also be explained by analogy with the well-established effect of inlet subcooling. Indeed, lowering a high subcooling brings the boiling boundary down toward the center of the channel, much the same way that the cosine distribution does at high subcooling. The effect of nonuniform distributions on pressure drop is generally minor (Dijkman, 1969; Yadigaroglu and Bergles, 1969), since most of the two-phase pressure drop takes place near the exit of the channel, where the quality (at constant power) is independent of the heat flux distribution.

**3.2.5 Effects of inlet and exit throttling** The effect of inlet (single-phase) throttling is always strongly stabilizing and is used to assure the stability of otherwise unstable channels. On the contrary, the effect of flow resistances near the exit of the channel (two-phase flow region) is strongly destabilizing. For example, stable channels can become unstable if an orifice is added at the exit.

These effects can be understood by examining the pressure drop perturbations

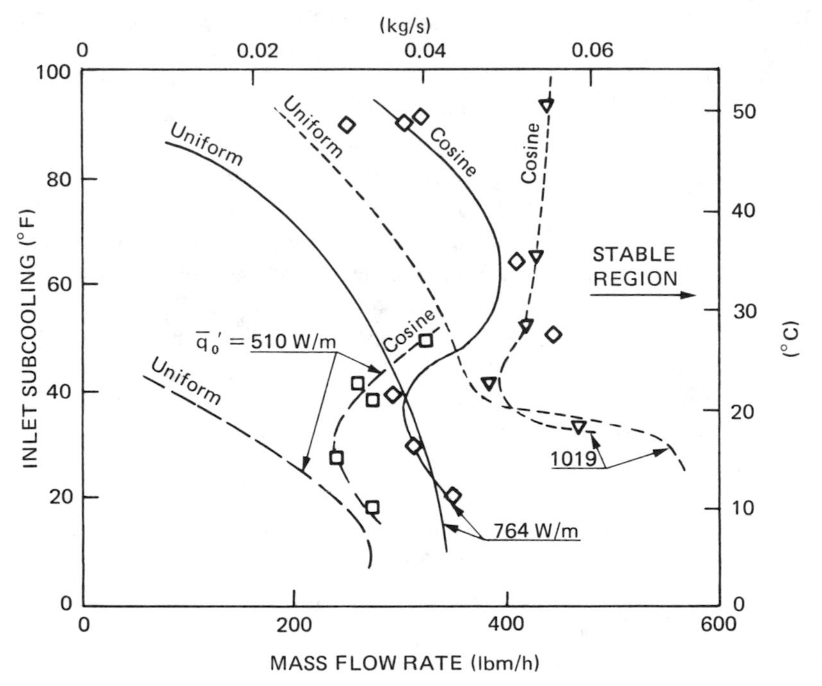


**Figure 14** Comparison of the stability boundaries obtained with a uniform and a cosine heat input distribution at three different power levels (Yadigaroglu and Bergles, 1969).

of Fig. 12. Starting from a low power level, the threshold of instability is crossed when the single-phase and two-phase pressure perturbations become equal but diametrically opposed in the complex plane. Figure 12 shows how the two-phase pressure drop perturbation grows progressively and finally matches the single-phase pressure drop perturbation as the power level is increased. It is evident that any increase of the single-phase term will retard, while increases of the two-phase term will precipitate crossing of the threshold.

**3.2.6 Effect of gravity** Gravity may affect the stability of the system only when the gravitational pressure drop is relatively important as compared with the frictional and acceleration pressure drops. This condition happens at low quality and low mass flux. The effect of gravity on the threshold of stability was negligible for the BWR channel of Figs. 7, 10, and 11.

**3.2.7 System effects** In single-channel loops all the loop components influence the stability of the system to some extent. In general, loop components in the single-phase part of the loop tend to stabilize the boiling channel by adding inlet resistance. On the contrary, an unheated riser for the two-phase mixture has a destabilizing effect.

In natural-circulation loops the stability trends are often obscured by the intrinsic relationships between heat input, flow rate, and inlet subcooling. Bouré et al. (1973) noted that in closed loops enthalpy perturbations at the channel exit travel around the loop and may reappear as inlet enthalpy perturbations with a delay controlled by the transit time around the loop. This additional feedback mechanism should be accounted for in stability analyses of such systems.

**3.2.8 Simplified stability criteria** A detailed stability investigation can only be conducted through the approaches discussed in Sec. 3.1.2, which dictate use of a computer code. Simple approximate stability criteria are, however, highly desirable for preliminary stability appraisals. Such a criterion was proposed by Ishii (1971, 1976) and later modified by Saha (1974) to include the effects of subcooled boiling. According to Ishii the stability boundary in the $(N_{pch}, N_{\text{sub}})$ plane can be conservatively approximated by a constant-quality line given by

$$N_{pch} = N_{\text{sub}} + \frac{k_i + 4f_{2\phi}L/D + k_e}{1 + \frac{1}{2}\left[\frac{1}{2}(4f_{2\phi}L/D) + k_e\right]}$$

where $k_i$ and $k_e$ are the loss coefficients for the inlet and exit local pressure drops $(\Delta p = k\rho V^2/2)$, $f_{2\phi}$ is the Fanning friction factor in the two-phase region, and $L$ and $D$ are the length and hydraulic diameter of the heated channel, respectively. Ishii's criterion expresses the relationship between the channel operating conditions ($N_{pch}$ and $N_{\text{sub}}$) and the frictional characteristics of the channel. Gravitational and relative velocity effects can be included in more complex forms of this criterion given by Ishii (1971). At sufficiently high subcooling, the stability boundary follows a constant-quality line, as shown in Fig. 10. Ishii's criterion essentially expresses the relationship between the value of the quality at the threshold and the frictional characteristics of the channel. This criterion predicted well the stability boundaries of Fig. 10.

## 3.3 Frequency of the Oscillations

**3.3.1 Fundamental oscillation period** The period of the oscillations is closely related to transportation delays in the channel, as stated earlier. In the single-phase region enthalpy perturbations propagate with the flow velocity (Yadigaroglu and Bergles, 1972). In the two-phase region the enthalpy perturbations create void fraction disturbances that, according to Zuber and Staub (1966, 1967), propagate with kinematic-wave velocity, which is of the order of magnitude of the velocity of the vapor phase. Since a cycle is completed by the passage of two perturbations through the channel (one positive and one negative), the period of the oscillation should be of the order of twice the transit time of the mixture. Figure 15 shows the ratio of the predicted period of the oscillation to twice the transit time of the mixture (assuming homogeneous flow) at points located on the threshold boundaries of the stability map of Fig. 10. The ratio is not exactly equal to 1 since the transportation delays do not take place in strict succession.

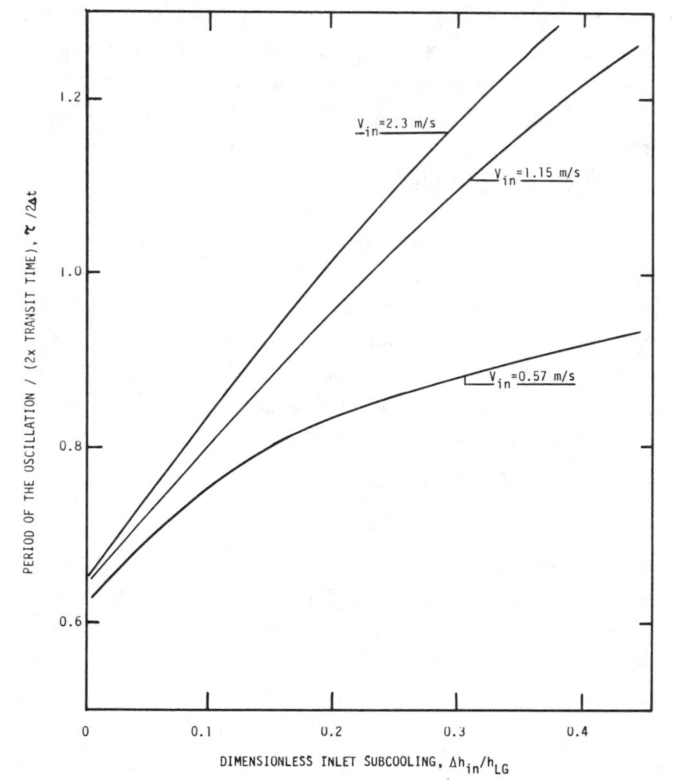


**Figure 15** Variation of the period of the oscillation on the stability boundary. Typical BWR channel (stability map of Fig. 10).

The transit time of the mixture in homogeneous flow is given by

$$\Delta t = \Delta t_1 + \Delta t_2 = \frac{\rho_L \, \Delta h_{\text{in}}}{q'''} + \frac{h_{LG}}{q''' v_{LG}} \ln \left(1 + \frac{v_{LG}}{v_L} x_e\right)$$

Since the exit quality, $x_e$, is approximately constant on the threshold lines, this expression shows that the period of the oscillation increases with subcooling and decreases with heat input. These trends were evident in the threshold data of Yadigaroglu and Bergles (1969) and Ishii (1971).

Dynamic heat storage in the heater walls and the nonuniformity of the axial heat input in the single-phase region influence the phase of the enthalpy perturbations in the single-phase region. Stability data obtained with a cosine distribution confirm the theoretical trends (Yadigaroglu and Bergles, 1969, 1972).

**3.3.2 Fundamental and higher-mode oscillations** Frequency-domain stability analyses, such as that of Ishii (1971), show that there are several stability boundaries, each corresponding to a different range of oscillation periods. As the system is brought and operated into the unstable region, sudden changes in oscillation frequency might occur. Such transitions from mode to mode are usually not observed since operation is generally discontinued as soon as the system becomes unstable (generally at the fundamental mode).

Yadigaroglu and Bergles (1969, 1972) observed higher-mode oscillations in a Freon-113 system that was operated well within the unstable region, confirming experimentally the theoretical possibility for the occurrence of such oscillations. Figure 13 shows this system oscillating in two different modes; in this rather unusual case, a stable-operation region happened to separate the two unstable regions.

Experimental stability maps showing the various higher-mode regions are given by Yadigaroglu and Bergles (1969, 1972), who presented a detailed discussion. Ishii (1971, 1976) and Saha (1974) presented theoretical stability maps including the higher-mode boundaries. For higher-mode oscillations the period becomes a simple fraction of the transit time in the channel. Higher-mode oscillations are likely to appear at high inlet subcooling and low flow rates and power inputs, that is, when the transit time in the channel becomes very long.

## 3.4 Parallel Oscillating Channels

A number of experiments were conducted to examine the behavior of several parallel channels connected to common plena (see Table 1). It is evident, however, that parallel channels cannot influence one another unless they are at least weakly coupled through the common plena; this coupling is difficult to observe and quantify. In other words, the stability of a given channel is determined by the boundary conditions imposed on it; the stability of parallel channels will not be affected unless their boundary conditions are perturbed by the instabilities occurring in neighboring channels.

These remarks are confirmed by several experiments. For example, a large bypass is often used to impose a constant-pressure-drop boundary condition on the boiling channel. This boundary condition is maintained only for a sufficiently large bypass ratio, as confirmed experimentally by Collins and Gacesa (1970) and theoretically by Carver (1970), who showed that stability is independent of the bypass ratio for sufficiently large values of this ratio. These authors also showed that a stiff boundary condition (high bypass ratio) was destabilizing.

When parallel channels are fed through a common plenum preceded by a common supply path [for example, BWR bundles, Lahey and Yadigaroglu (1974); steam generator parallel heated tubes, Chen et al. (1976)], it is generally the heated channel alone, rather than the entire loop, that is most unstable since any upstream resistance tends to stabilize the system. Thus in experiments where a small number of channels (two to five) were fed through a common inlet plenum, the channels oscillated while the flow in the remaining parts of the loop was approximately constant. This condition is possible since the phases of the oscillations in the different channels adjust in such a way as to produce a zero loop-flow perturbation. For example, two channels may oscillate out of phase (Masini et al., 1968), while three channels may oscillate with 120° phase differences (Crowley et al., 1967) or may otherwise adjust their phases and amplitudes (Akagawa et al., 1971; Koshelev et al., 1970).

Kakaç et al. (1974), Lee et al. (1977), and Kakaç (1977) investigated extensively the behavior of one channel and two and four parallel channels with and without cross connections at axial locations. One conclusion is that the cross connections stabilize the system. This conclusion is explained by the observation that cross-connected channels cannot oscillate out of phase and therefore bring into action the entire flow loop, which is generally more stable than a single channel.

## 4 OTHER TYPES OF OSCILLATIONS

The major types of instabilities were discussed in Secs. 2 and 3. Several other types of two-phase flow instabilities have been observed (Bouré et al., 1973; Bergles, 1977) and are briefly reviewed below.

### 4.1 Periodic Expulsion or Chugging and Geysering

Chugging and geysering are relaxation instabilities characterized by periodic expulsion of coolant from the channel. The resulting transient behavior may be very mild or may lead to violent ejection of the mixture out of the heated channel from one or both ends. Chugging seems to be the term used to describe periodic expulsion phenomena in a flow situation, while geysering often refers to no net-flow situations.

The expulsion of the fluid results from sudden vaporization. The bulk of the fluid must therefore be superheated for this vaporization to occur. It is clear then that bubble formation, growth, and collapse phenomena are of importance to chugging and geysering. Hawtin (1970) presented a review of these phenomena with relation to chugging. It appears that liquid-metal and low-pressure water systems are most susceptible to chugging because of the possibility of erratic boiling (bumping) under those conditions. Core voiding by coolant expulsion is a serious concern regarding the safety of LMFBRs. Schlechtendahl (1970) listed references on liquid-metal boiling problems.

Griffith (1962) noted that two mechanisms can be responsible for the generation of large liquid superheats. The first one is the delay in nucleation because of, for example, the absence of nucleation sites on the wall. The second mechanism is dynamic and results from the relationship between gravitational pressure and voids: When voids are created in a liquid column, the hydrostatic pressure decreases; this reduction in pressure results in superheat of the liquid and further void generation. Under certain conditions the phenomenon might diverge rapidly and result in a violent expulsion of the coolant. This second mechanism is believed to be the cause of geysering.

Chugging and geysering cycles can be divided into an incubation, an expulsion, and a refill period. The incubation period is needed to produce the superheat required for violent nucleation. Since the delay to nucleation may vary randomly, the period of chugging or geysering phenomena may be irregular.

Mild chugging oscillations were observed by Chexal and Bergles (1973); these preceded the appearance of density wave oscillations in their system and could have been confused with the latter. Geysering has been studied experimentally by Griffith (1962).

Since geysering and chugging depend to a large extent on the nucleation characteristics of the fluid and the walls, and these are difficult to quantify, no basic analysis of these phenomena has been published. There is, however, a large body of literature dealing with core voiding in LMFBRs.

## 4.2 Heat Transfer and Two-Phase Flow Regime Instabilities

Two-phase flow regime changes are triggered by subtle mechanisms not always well understood. The boiling heat transfer regime at the heated wall may also change suddenly and dramatically as a result of small changes in some parameter (boiling crisis or burnout, rewetting, transition from boiling to single-phase heat transfer, etc.). Since the flow and heat transfer regimes influence sometimes strongly the overall behavior of the boiling channel, such changes can lead to flow instabilities.

Stenning and Veziroğlu (1967), for example, reported the occurrence of thermal oscillations in their experimental apparatus and postulated a mechanism for these based on a relatively rare combination of flow and heat transfer characteristics. Flow regime changes have been observed by Fabréga (1965) and, during subcooled boiling, by Jeglic and Grace (1965). The latter type of oscillations are reminiscent of chugging since nucleation instabilities and bubble growth in subcooled flow near the wall seem to be the controlling phenomenon. In fact, chugging could have been included under the general heading of this section.

Nakajima et al. (1975) recently described unstable behavior of a two-phase loop induced by periodic discharge and coalescence of steam bubbles in a segment of horizontal piping. Cho et al. (1971) described unexplained changes, called flips, occurring in the behavior of a once-through steam generator. These may well have been the consequence of changes in the heat transfer regime in the tubes.

Since the prediction of two-phase flow regimes is difficult even at steady state, the analysis of heat transfer and flow regime-related instabilities seems difficult in general.

## 4.3 Acoustic Oscillations

Acoustic oscillations are associated with dynamic pressure waves, and their periods are of the order of magnitude of the transit time of such waves through the system. Indeed commonly observed frequencies are in the range 10-100 Hz.

Acoustic oscillations have been observed in subcooled boiling, bulk boiling, and film boiling (Bouré et al., 1973). Although generally their amplitude is small, instances of relatively high pressure variations have been reported (Cornelius, 1965). Acoustic oscillations often seem to be triggered by or coupled to heat transfer mechanisms such as subcooled boiling (collapse of the bubbles) (Bergles et al., 1967), and film boiling (variations of the film thickness and vapor generation rate owing to changes in pressure) (Edeskuty and Thurston, 1967; Thurston et al., 1967). Acoustic oscillations have also been observed in cryogenic systems by Cornelius (1965), Edeskuty and Thurston (1967), and, at supercritical pressures, by Bishop et al. (1964). Acoustic oscillations have been analyzed by Bergles et al. (1967).

## 4.4 Condensation Instabilities

The possibility of flow excursions related to the negative slope of the $(w, \Delta p)$ characteristic in condensing systems has been mentioned in Sec. 2.

Oscillatory instabilities in condensing tubes have been observed by Soliman and Berenson (1970). According to these authors, in upward condensing flow the liquid film thickens to where the vapor-induced shear is no longer sufficient for moving it upward. Falling portions of the film periodically plug the tube and result in oscillatory behavior. Thus, a relaxation mechanism and flow regime changes seem to be responsible for this type of oscillation. In parallel-tube configurations these conditions lead to flow distribution instabilities. Instabilities in downflow condensation of steam are also reported by Goodykoontz and Dorsh (1967).

Soliman and Berenson (1970) also reported the occurrence of pressure fluctuations because of interfacial waves in annular condensing flow striking the vapor-liquid interface at the point of complete condensation. Instabilities created by wave motion in boiling flow are also discussed in Sec. 5.3.1.

Oscillatory instabilities have also been observed in direct-contact condensers where vapor is brought into direct contact with subcooled liquid. Westendorf and Brown (1966) observed both high-frequency (50-200 Hz) pressure oscillations and low-frequency (1-10 Hz) pressure and condensing-interface oscillations in a system where vapor and liquid were injected concentrically at one end of a tube (liquid on the wall, vapor jet in the center). The high-frequency oscillations increased with subcooling of the liquid and are believed to be caused by the collapse of vapor voids. The low-frequency oscillations were associated with movements of the condensing length that alternatively grew to some maximum and then diminished to zero. These variations were related to changes in the flow pattern from bubbly (short condensing length) to annular (long condensing length). Condensing length excursions were also observed. A detailed explanation of these phenomena and an analysis are presented by Westendorf (1970).

Pressure oscillations that under certain conditions can reach large amplitudes have been observed both in small-scale experimental setups with immersed steam jets and during condensation of steam jets in BWR vapor-suppression pools. These phenomena and oscillations occurring during mixing of steam and water are discussed in Sec. 5.1.

## 5 CURRENT INSTABILITY CONCERNS IN NUCLEAR SYSTEMS

The stability problems of BWRs that seemed to be limiting the operation of prototypes of this reactor type were eliminated by the increase of system pressure and implementation of forced circulation in the BWR vessel. Modern BWRs are normally stable, but may approach the threshold of instability during abnormal operational transients that could result in unusual flow rate and power level combinations.

Other classes of stability problems have, however, surfaced recently. These are related to the performance of the emergency systems of LWRs, and include steam condensation instabilities in the vapor-suppression pools of BWRs (Lahey, 1977), steam-water mixing oscillations during emergency coolant injection in PWRs (Rothe et al., 1977), oscillations during the reflooding phase of the LOCA in PWRs, and complex phase interaction problems in countercurrent flow of steam and water in PWR downcomers (Block and Schrock, 1977), as well as in BWR fuel rod bundles (Lahey, 1977). There are also flow stability problems in steam generators for LMFBRs or other types of plants. Some of these instability phenomena are discussed in this section.

### 5.1 Condensation-related Instabilities

Steam-water mixing and condensation of vapor jets are discussed by Hsu in Chap. 15. Information is presented here on instabilities observed during the condensation of immersed steam jets in BWR suppression pools and during the mixing and subsequent condensation of steam with cold water in PWR primary-system pipes.

**5.1.1 Condensation of immersed steam jets** Immersed-jet condensation instabilities have been observed in BWR vapor-suppression pools during normal high-pressure steam relief operations (U.S. NRC, 1975) and could also occur during low-pressure containment relief transients following a hypothetical LOCA (Lahey, 1977).

Relief of high-pressure steam by injection as a jet and condensation in the vapor-suppression pool of BWRs has produced strong pressure pulsations that resulted in damage to the pool, when the temperature of the pool water exceeded approximately 70°C (U.S. NRC, 1975). This has prompted investigation of these immersed-jet condensation phenomena both in Germany and in the United States.

Immersed-jet condensation instabilities were observed by Kerney (1970) in a small experimental setup. A small-scale immersed-jet condensation experiment was also conducted at the University of California, Berkeley. In this experiment, 6–7 bar steam was injected from a horizontal 12.7-mm pipe into an atmospheric-pressure tank of modest dimensions ($0.3 \times 0.6 \times 1.0$ m). Large-amplitude (to ±1–2 bar) and short-duration (less than 0.1 ms) pressure pulses were recorded by transducers immersed in the tank. These pulses were observable at pool temperatures as low as approximately 45°C; they peaked in amplitude in the vicinity of 80°C and disappeared as the water temperature approached saturation. Pulsations were present with both subsonic and sonic flow rates of steam. The pulsation frequency was of the order of a few hundred hertz and decreased with pool temperature. Cinematographic observations of the condensing jet revealed that when pulsations were present, the jet was pinched to a point; the large bubble that formed near its tip in this manner was oscillating and condensing extremely rapidly. Probably the condensation of these bubbles that were separated from the jet produced the pressure pulses. This was the condition in a different experiment by Wright and Albrecht (1975). Instabilities occurring during vapor suppression were also investigated experimentally by Saito et al. (1974).

The actual phenomena in BWR suppression pools seem to be complicated by interaction of the pressure pulsations created by individual jets with the walls of the pools and with adjoining jets, by the presence of air in the steam, by the capacitance of the steam lines, by nonuniformities in the pool water temperature, by pool geometry and scale, etc. Detailed studies of these phenomena are being made but have not been reported in the open literature yet. In practice, the safety problems are solved by replacing single large jets by smaller multiple jets.

**5.1.2 Steam–water mixing during ECC injection** During the LOCA in a PWR, when the primary system pressure falls below typically 40 bar (600 psia), check valves open and the large volume of cold water stored under compressed nitrogen in the accumulators is injected through the side branch of a tee into the primary system piping, where steam generated in the reactor core is flowing, as shown in Fig. 16*a*. Under some conditions a liquid plug can form in the main pipe. The steam condensation rate is high when the ECC discharge port is not covered by the liquid plug and the water is injected directly into a steam environment. On the contrary, the condensation rate is low when the discharge of the ECC line is covered and the water is injected into the liquid plug. The steam-volume pressure increases during this low-condensation phase of the cycle and pushes the liquid plug downstream;

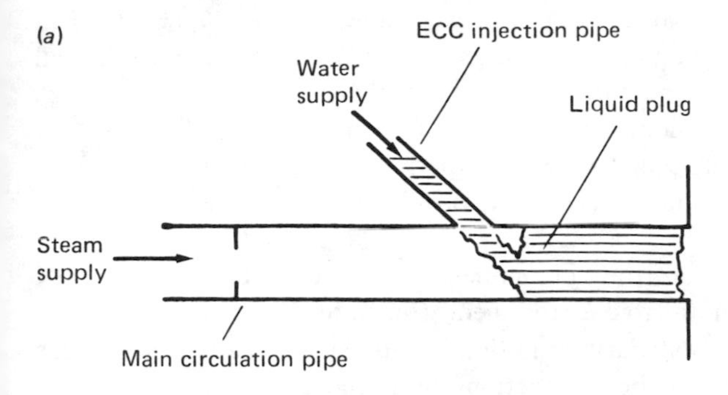


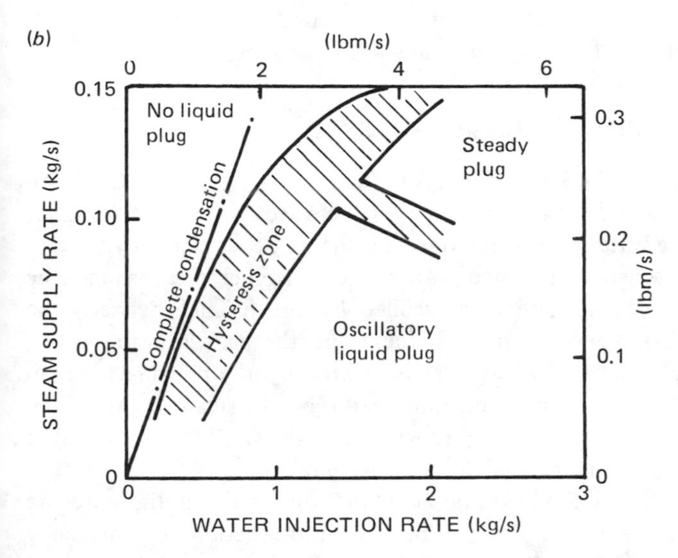


**Figure 16** Experimental flow regime map for steam-water mixing experiments (Rothe et al., 1977).

the ECC injection point is then uncovered, the condensation rate increases, the steam pressure is reduced, and the liquid plug is pulled back upstream. Such cyclic operation has been observed in scale-model tests of the ECC injection region and can result in pressure oscillations having a frequency of a few hertz and amplitudes as high as the absolute pressure.

Wallis and Rothe (1976) studied the conditions necessary for the formation of a water plug and arrived at a Froude number criterion. Since the conditions for plug formation are generally satisfied, Rothe et al. (1976, 1977) analyzed the behavior of this plug using a simple one-dimensional analysis that was successful in predicting the regions of unstable operation in a stability map such as that of Fig. 16*b*, and the frequency of the oscillations. Their analysis shows that the phenomena are dominated by inertial interactions of the fluids, provided the condensation rate varies between sufficiently high and low values.

Information on the available experimental data from tests at various geometric scales can be found in Rothe et al. (1977). The data show that no liquid plug and oscillations can exist when the liquid flow rate is not sufficient for condensing all the steam on a thermodynamic equilibrium basis. For higher liquid flow rates, Fig. 16*b* shows a region where the liquid plug cannot be maintained in the pipe, and regions of oscillatory and steady plug behavior. Experimentally, the transitions from one type of behavior to another take place with considerable hysteresis, as shown in Fig. 16*b*.

The tests and analyses of the phenomena just described consider the steam–water mixing tee as an isolated component (separate-effects tests), that is, they apply essentially constant boundary conditions to its limits. In an actual reactor configuration, however, there will be interactions with the adjoining components (downcomer, piping, pump) that could result in much more complex integral system behavior. The dynamics of such couplings have not been analyzed in detail yet.

## 5.2 Oscillations during Bottom Reflooding of PWRs

Bottom reflooding of the core of PWRs will take place in a postulated LOCA following the blowdown phase, during which the pressure vessel loses its coolant from the break. Reflooding water is then provided by the ECCS to recover the core. This emergency water is generally pumped into the downcomer area around the core. The reflooding rate of the core is controlled by the balance between the driving gravity head forcing the water into the core and the back pressure in the upper plenum. The upper-plenum back pressure is created by the pressure losses in the coolant lines leading from the upper plenum to the break; steam exiting from the core escapes into the containment through these lines. This back-pressure limitation of the reflooding data is referred to as steam binding.

At the beginning of the reflooding transient the fuel rods in the core are overheated, and heat transfer is by film boiling. The introduction of emergency coolant lowers the rod temperature; the film boiling regime is progressively replaced by nucleate boiling, following the passage of a quench front that propagates essentially from the bottom to the top of the core.

The FLECHT-SET reflooding simulation tests (Blaisdell et al., 1973; Waring et al., 1974) conducted in the United States showed that when flooding is driven by a gravity head in a downcomer connected in a U-tube-manometer fashion to a heated test section, the flow rate oscillates. These oscillations are especially violent during the initial phase of the transient. The effect of the oscillations is also felt as heated surface-temperature fluctuations. No comparable oscillations were observed during reflooding tests in which the test section flooding rate was controlled externally.

Figure 17 shows the measured downcomer head during a FLECHT-SET run. For this particular case the average reflooding rate was approximately 0.055 m/s, while peak values of the reflooding velocity were as high as ±1–2 m/s. The period of the oscillation during the initial part of the transient was of the order of 3 s, decreasing later as the water level in the bundle and the downcomer increased. The oscillations died before the quench front reached the midplane elevation.

Oscillations of a similar nature were observed during the Siemens AG

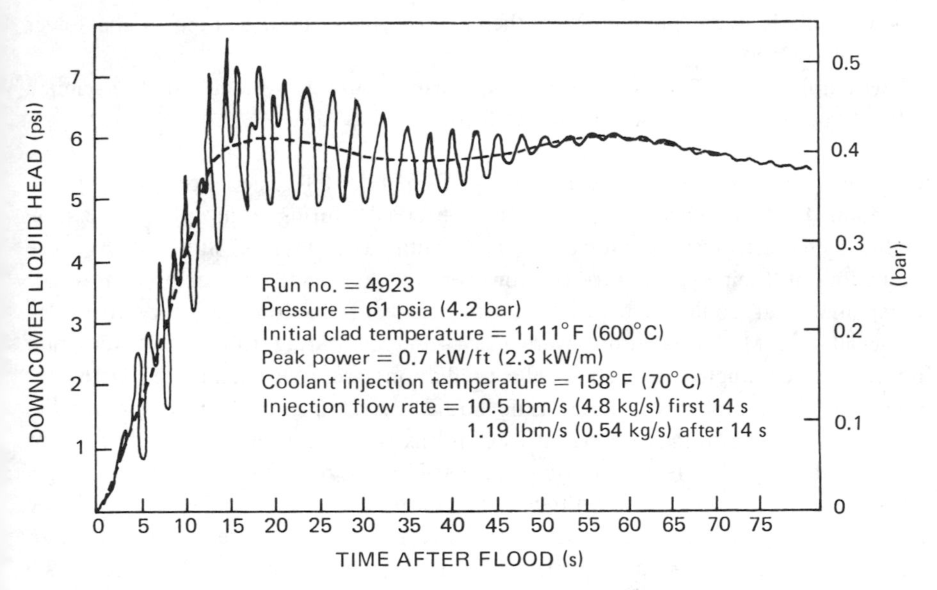


**Figure 17** Oscillation of the downcomer level during a gravity-feed FLECHT-SET run (Blaisdell et al., 1973).

340-pin-bundle, gravity-feed reflooding tests reported by Riedle and Winkler (1972) and Riedle et al. (1976) and during BWR bottom-reflooding tests conducted by Hitachi Ltd. in Japan.

Oscillations having periods of 1–2 s were also produced by White and Duffey (1974) in a small-scale single-rod mockup of the core, downcomer, upper plenum, and discharge loop of a PWR. Steam binding was simulated by a variable test section outlet orifice. Different ranges of downcomer fill-up rates and a broad range of test section outlet resistances were investigated. The various apparatus configurations tested included a different lower plenum with approximately 10 times the volume of the initial one, and two alternative downcomers, one having 10 times the original cross-sectional area and the other twice the height of the reference downcomer.

White and Duffey (1974) reported that as the water level started moving up the heated channel, vigorous boiling occurred at the quench front. The steam and entrained-water mixture leaving the test section produced a back pressure that forced the liquid level downward, leaving a liquid film on the rod, which then dried out. Reduction of the back pressure because of venting of the contents of the test section again increased the liquid level. This process was repeated in a series of violent oscillations of gradually decaying amplitude. The oscillations were found to damp out in one-half the time when the lower plenum volume was increased by an order of magnitude; their amplitude and frequency remained unchanged, however, probably because only a small fraction of the liquid in the lower plenum participated in the oscillations. The two alternative downcomers also had little effect on the oscillation frequency. The experiments confirmed, however, that

1. The oscillations are present when the quench front is progressing in the lower part of the rod.
2. The amplitude of the oscillations is increasing with the velocity of the quench front and is strongest when the downcomer level is low.

These observations are in agreement with the FLECHT-SET experience.

Flooding-rate oscillations were also observed during atmospheric-pressure, forced-flooding-rate tests conducted by Martini and Premoli (1973). A close examination of their system reveals, however, the existence of a damper near the channel inlet that could have provided the degree of freedom necessary to create the oscillations. Mild oscillations were also observed during forced-reflooding-rate flooding tests conducted in France. The oscillations seem to enhance heat transfer during reflooding, probably by increasing liquid carryover higher up in the bundle.

All evidence regarding reflooding oscillations was obtained with single rods or single bundles. The effects of flow redistribution on the stability of large, open-lattice cores with a large number of communicating channels is practically unknown. Also practically unknown is the stability of large cores with respect to more conventional modes of oscillation, such as density waves. In both cases the stability of the most unstable "hot" channel is influenced by radial flow distribution and mixing effects. One can intuitively claim that as radial resistance to flow decreases, stability of the channel increases. No published experimental or detailed analytical study of these effects is known to the author.

**5.2.1 Theoretical analysis and mechanism of the oscillations** The oscillations encountered during gravity-feed reflooding have been often labeled as manometer-type oscillations. The experimental observations of White and Duffey (1974) and the appearance of oscillations during some of the forced-reflooding-rate tests suggest the possibility of their existence even if the level in the downcomer remained constant. Thus the oscillations appear to be thermohydrodynamic and are probably a result of the feedbacks among the flooding rate, the vapor generation rate, and the back pressure created by the mixture leaving from the top of the test section. With respect to the presence of such feedbacks they resemble density wave oscillations. It should be noted, however, that the steam-generation dynamics of reflooding are strongly influenced by the propagation of the quench front, a rather unique phenomenon.

In a simple model developed by White and Duffey (1974), the mass of liquid in the downcomer and the two-phase mixture around the rod are regarded as a piston trapping a volume of steam leaking through the test section outlet orifice. This model correctly predicts an exponential decay in oscillation amplitude for an exponential rise of the quench front and for steam generation proportional to the immersed length. The predicted frequencies of the oscillations were in approximate agreement with experimental values and depended on the steam-generation rate.

Thus it appears that the reflooding oscillations are not simple manometer oscillations owing to sloshing of the liquid in the U-tube-manometer reactor downcomer and core configuration. If these were indeed simple manometer oscillations, their frequency $f$ would have been given by

$$f = \frac{1}{2\pi}\sqrt{\frac{g}{I}\left(\frac{1}{A_1} + \frac{1}{A_2}\right)}$$

where $I = \int [dz/A(z)]$ is the inverse of the characteristic system inertial length, and $A_1$ and $A_2$ are the flow cross-sectional areas at the location of the water level in the two legs. Indeed, one observes that this frequency is only geometry dependent.

## 5.3 Stability Considerations in Steam Generators

Steam generators are subject to Ledinegg excursions and to density wave oscillations, as discussed in Secs. 2 and 3. Sodium- or helium-heated steam generators for nuclear applications often must be stabilized by orificing the flow at the inlet of the tubes, at the expense of pumping power. Since the tubes of such generators are several tens of meters long, the transit time of the fluid, and consequently the period of potential density wave oscillations become very long.

The stability of flow in the secondary side of steam generators is influenced by the coupling with the primary side. According to Moxon (1973), this coupling has a stabilizing influence. Although a detailed stability analysis is needed to fully investigate the effects of variable heat flux from the primary side, the enhancement of stability can be explained as follows: the heat flux varies most in the economizer region of the tubes since two-phase heat transfer coefficients are less sensitive to flow rate and quality variations. An increase of the flow rate in the secondary side tends to move the boiling boundary higher in the channel, but it also tends to increase the heat flux in the economizer. The two effects tend to cancel each other, making a fluid-heated system more stable than a system with externally imposed heat flux, such as a nuclear reactor or an electrically heated test loop.

The thermohydrodynamic stability of steam generators was investigated experimentally by Waszink and Efferding (1973), Llory (1975), and Sano et al. (1973) and theoretically by Moxon (1973), Efferding (1968), Dean and Murray (1976), and Takahashi and Futami (1977).

**5.3.1 Oscillations of the dryout point in steam generators** Many sodium- or gas-heated steam generators for nuclear applications are designed to operate with part of their tubes in the postdryout regime. For these generators there are concerns regarding fatigue of the tube metal from thermal stresses created by small oscillations of the dryout point (Chu et al., 1978; Chiang et al., 1977). Thermal stresses could also result in exfoliation of the oxide layer on the tube surface, leading to accelerated corrosion (Chiang et al., 1977).

There is experimental evidence discussed by the authors just mentioned indicating that the dryout point is not stationary, even if all operating parameters such as inlet temperature and heat flux are kept constant to the extent practically possible. In recent tests in the United States of a mockup LMFBR steam generator, a wall-temperature thermocouple embedded into the tube wall to a distance of 1.75 mm from the inside (water) surface recorded maximum temperature oscillations of about 18°C with a representative period of about 3 s. The amplitude of the temperature oscillations was certainly higher at the inside surface of the tube. For 10-mm-diameter tubes, the dryout point moved axially approximately four tube diameters (Chu et al., 1977).

The cause of these oscillations is not clearly understood. The oscillations have a random character and may be a result of the arrival at the dryout location of interfacial disturbance waves created upstream in the annular flow region. The occurrence of such disturbance waves has been investigated by several authors and is discussed, for example, by Hewitt and Hall-Taylor (1970). Wedekind and Stoecker (1966), and Wedekind (1971) have studied similar movements of the point of complete vaporization in two-phase flow and adopt the same explanation.

Chu et al. (1978) adopted an alternative but not necessarily conflicting explanation based on the well-known appearance of liquid rivulets at the dryout point. They presented a thermal-stress analysis of the tube wall based on a model postulating the existence of three or four rivulets around the periphery of the tube. These rivulets are rotating in the circumferential direction exposing the tube surface alternatively to high and low heat transfer coefficients. The parallel analysis of Chiang et al. (1977) was based on an axial movement of the dryout front and concluded that the observed frequency and amplitude of the dryout temperature oscillations could create rapid fatigue of the tube material. Since it is difficult to simulate exactly the actual steam-generator operating conditions in mockups, full resolution of these considerations might only be possible through long-term operating experience with the actual equipment.

## NOMENCLATURE

| | |
|---|---|
| $C_k$ | kinematic-wave velocity |
| $D$ | diameter |
| $f$ | friction factor |
| $g$ | acceleration of gravity |
| $h_{LG}$ | heat of vaporization |
| $I$ | inertia of hydraulic leg |
| $j$ | volumetric flux |
| $k$ | loss coefficient |
| $L$ | length |
| $N_{pch}$ | phase change number, Sec. 3.2 |
| $N_{\text{sub}}$ | subcooling number, Sec. 3.2 |
| $p$ | pressure |
| $q$ | heat input |
| $q'$ | heat input per unit length |
| $q''$ | heat flux |
| $q'''$ | heat input per unit fluid volume |
| $t$ | time |
| $u$ | phase velocity |
| $v$ | specific volume |
| $v_{LG}$ | $v_G - v_L$ |

$V$ velocity
$V_{Gj}$ drift velocity of the gas, $u_G - j$
$V_{Lj}$ drift velocity of the liquid, $u_L - j$
$w$ flow rate
$x$ quality
$z$ axial coordinate
$\alpha$ void fraction
$\Gamma_G$ rate of vaporization
$\Delta h_{in}$ inlet subcooling (in enthalpy units)
$\tau$ period of the oscillation
$\Omega$ $v_{LG}\Gamma_G$, Sec. 3.1.2
$\omega$ radial oscillation frequency

**Subscripts**

$bb$ boiling boundary
$c$ heated channel
$e$ channel exit
ext external
fdb feedback
$G$ gaseous phase
$i$ channel inlet
$L$ liquid phase
$p$ pump
0 steady state
1 single-phase region
2 two-phase region

**Abbreviations**

BB boiling boundary
BWR boiling-water reactor
ECC emergency core coolant
ECCS emergency core cooling system
LOCA loss of coolant accident
LWR light-water reactor (PWR or BWR)
NVG net vapor generation
PWR pressurized-water reactor

## REFERENCES

Akagawa, K., Sakaguchi, T., Kono, M., and Nishimura, M., Study on Distribution of Flow Rates and Flow Stabilities in Parallel Long Evaporators, *Bull. JSME*, vol. 14, p. 837, 1971.

Anderson, T. T., Hydraulic Impedance: A Tool for Predicting Boiling Loop Stability, *Nucl. Appl. Technol.*, vol. 9, p. 422, 1970.

Andoh, H., Discharged Flow Oscillation in a Long Heated Tube, Master's thesis, Department of Mechanical Engineering, Massachusetts Institute of Technology, Cambridge, 1964.

Becker, K. M., Jahnberg, S., Haga, I., Hansson, P. T., and Mathisen, R. P., Hydrodynamic Instability and Dynamic Burn-out in Natural Circulation Two-Phase Flow—An Experimental and Theoretical Study, *Proc. 1964 Geneva Conf. on the Peaceful Uses of the Atomic Energy*, P/607, vol. 8, p. 325, United Nations, New York, 1964.

Beckjord, E. S., The Stability of Two-Phase Flow Loops and Response to Ship's Motion, GEAP-3493, Rev. 1, 1960.

Bergles, A. E., Review of Instabilities in Two-Phase Systems, in *Two-Phase Flows and Heat Transfer*, eds. S. Kakaç and F. Mayinger, vol. 1, pp. 383–422, Hemisphere, Washington, D.C., 1977.

Bergles, A. E., Goldberg, P., and Maulbetsch, J. S., Acoustic Oscillations in High Pressure Single Channel Boiling System, Symp. on Two-Phase Flow Dynamics, EURATOM, EUR 4288e, The Technological University of Eindhoven, vol. 1, pp. 525–550, 1967.

Biancone, F., Campanile, A., Galini, G., and Coffi, M., Forced Convection Burnout and Hydrodynamic Instability Experiments for Water at High Pressures, Part I, Presentation of Data for Round Tubes with Uniform and Non Uniform Heat Flux, EUR-2490e, 1965.

Bishop, A. A., Sandberg, R. O., and Tong, L. S., Forced Convection Heat Transfer to Water at Near-critical Temperature and Super-critical Pressures, WCAP-2056, 1964.

Blaisdell, J. A., Hochreiter, L. E., and Waring, J. P., PWR FLECHT Phase A Report, WCAP-8238, 1973.

Block, J. A. and Schrock, V. E., Emergency Cooling Water Delivery to the Core Inlet of PWR's during LOCA, in *Thermal and Hydraulic Aspects of Nuclear Reactor Safety*, vol. 1. *Light Water Reactors*, eds. O. C. Jones and S. G. Bankoff, pp. 109–132, ASME, New York, 1977.

Blumenkrantz, A. and Taborek, J., Application of Stability Analysis for Design of Natural Circulation Boiling Systems and Comparison with Experimental Data, *AIChE Symp. Ser.*, vol. 68, no. 118, pp. 136–146, 1972.

Bogaardt, M., Spigt, C. L., Dijkman, F. J. M., and Verheugen, A. N. J., On the Heat Transfer and Fluid Flow Characteristics in a Boiling Channel under Conditions of Natural Convection, in *Boiling Heat Transfer in Steam Generating Units and Heat Exchangers, Proc. Inst. Mech. Eng. 1965–1966*, vol. 180, pt. 3C, pp. 77–87, 1966.

Bouré, J. A. and Mihaila, A., The Oscillatory Behavior of Heated Channels, Symp. on Two-Phase Flow Dynamics, EURATOM, EUR 4288e, The Technological University of Eindhoven, vol. 1, pp. 695–720, 1967.

Bouré, J. A., Bergles, A. E., and Tong, L. S., Review of Two-Phase Flow Instability, *Nucl. Eng. Des.*, vol. 25, pp. 165–192, 1973.

Carver, M. B., Effect of By-pass Characteristics on Parallel-Channel Flow Instabilities, in *Fluid Mechanics and Measurements in Two-Phase Flow Systems, Proc. Inst. Mech. Eng. 1969–1970*, vol. 184, pt. 3C, pp. 84–92, 1970.

Chan, K. C., Thermal-hydraulic Stability Analysis of Steam Generators, Ph.D. thesis, Department of Nuclear Engineering, University of California, Berkeley, 1979.

Chen, K., Yadigaroglu, G., and Wolf, S., Dynamic Stability of the CRBRP Prototype Steam Generator Test Loop and the CRBRP Evaporator Loop under Low-Power, Low-Flow Rate Natural Circulation Conditions, NEDM-14157, 1976.

Chexal, V. K. and Bergles, A. E., Two-Phase Instabilities in a Low Pressure Natural Circulation Loop, *AIChE Symp. Ser.*, vol. 69, no. 31, pp. 37–45, 1973.

Chiang, T., France, D. M., and Bump, T. R., Calculation of Tube Degradation Induced by Dryout Instability in Sodium-heated Steam Generators, *Nucl. Eng. Des.*, vol. 41, pp. 181–191, 1977.

Chilton, H., A Theoretical Study of Stability in Water Flow through Heated Passages, *J. Nucl. Energy*, vol. 5, pp. 273–284, 1957.

Cho, S. M., Ange, L. J., Fenton, R. E., and Gardner, K. A., Performance Changes of a Sodium-heated Steam Generator, ASME paper 71-HT-15, 1971.

Christensen, H., Power-to-Void Transfer Functions, ANL-6385, 1961.

Chu, C. L., Roberts, J. M., and Dalcher, A. W., DNB Oscillatory Temperature and Thermal Stress Responses for Evaporator Tubes Based on Rivulet Model, *J. Eng. Power, Trans. ASME*, vol. 100, no. 3, pp. 424–431, 1978.

Collins, D. B. and Gacesa, M., Hydrodynamic Instability in a Full Scale Simulated Reactor Channel, in *Fluid Mechanics and Measurements in Two-Phase Flow Systems, Proc. Inst. Mech. Eng. 1969-1970*, vol. 184, pt. 3C, pp. 115–126, 1970.

Cornelius, A. J., An Investigation of Instabilities Encountered during Heat Transfer to a Supercritical Fluid, ANL-7032, 1965.

Crowley, J. D., Deane, C., and Gouse, S. W., Two-Phase Flow Oscillations in Vertical, Parallel, Heated Channels, Symp. on Two-Phase Flow Dynamics, EURATOM, EUR 4288e, The Technological University of Eindhoven, vol. 2, pp. 1131–1171, 1967.

Daleas, R. S. and Bergles, A. E., Effect of Upstream Compressibility on Subcooled Critical Heat Flux, ASME paper 65-HT-67, 1965.

Davies, A. L. and Potter, R. Hydraulic Stability: An Analysis of the Causes of Unstable Flow in Parallel Chennels, Symp. on Two-Phase Flow Dynamics, EURATOM, EUR 4288e, The Technological University of Eindhoven, vol. 2, pp. 1225–1266, 1967.

Dean, R. and Murray, J., The Prediction of Dynamic Stability Limits in Once-through Boilers Using DYMEL, European Two-Phase Flow Group Meet., Erlangen, West Germany, 1976.

Dijkman, F. J. M., Some Hydrodynamic Aspects of a Boiling Water Channel, WW-R 144, Doctor of engineering science thesis, Laboratory for Heat Transfer and Reactor Engineering, The Technological University of Eindhoven, 1969.

Edeskuty, F. J. and Thurston, R. S., Similarity of Flow Oscillations Induced by Heat Transfer in Cryogenic Systems, Symp. on Two-Phase Flow Dynamics, EURATOM, EUR 4288e, The Technological University of Eindhoven, vol. 1, pp. 551–568, 1967.

Efferding, L. E., DYNAM—A Digital Computer Program for Study of the Dynamic Stability of Once-through Boiling Flow with Steam Superheat, GAMD-8656, 1968.

Eselgroth, P. W. and Griffith, P., Natural Convection Flows in Parallel Connected Vertical Channels with Boiling, MIT, EPL rept. 70318-49, 1967.

Eselgroth, P. W. and Griffith, P., The Prediction of Multiple Heated Channel Flow Patterns from Single Channel Pressure Drop Data, MIT EPL rept. 70318-57, 1968.

Fabréga, S., Instabilités Hydrodynamiques Limitant la Puissance des Réacteurs à Eau Bouillante, Eur-1509f, 1965.

Fleck, J. A., Jr., The Influence of Pressure on Boiling Water Reactor Dynamic Behavior at Atmospheric Pressure, *Nucl. Sci. Eng.*, vol. 9, p. 271, 1961.

Friedly, J. C., Kroeger, P. G., and Manganaro, J. L., Stability Investigation of Thermally Induced Flow Oscillations in Cryogenic Heat Exchangers, NAS 8-21014, final rept., 1967.

Gerliga, V. A. and Dulevskiy, R. A., The Thermohydraulic Stability of Multi-Channel Steam-generating Systems, *Heat Transfer Sov. Res.*, vol. 2, no. 2, pp. 63–72, 1970.

Giphshman, I. N. and Levinzon, V. M., Disturbances in Flow Stability in the Pendent Superheater of the TPP-110 Boiler, *Thermal Eng. (USSR)* (transl. of *Teploenergetika*), vol. 13, no. 5, pp. 45–49, 1966.

Gonzalez-Santalo, J. M. and Lahey, R. T., Jr., An Exact Solution of Flow Decay Transients in Two-Phase Systems by the Method of Characteristics, *J. Heat Transfer, Trans. ASME*, vol. 95, pp. 470–476, 1973.

Goodykoontz, J. H. and Dorsch, R. G., Local Heat Transfer Coefficients and Static Pressures for Condensation of High-Velocity Steam within a Tube, NASA Lewis Res. Ctr. rept. E 3498, 1967.

Griffith, P., Geysering in Liquid Filled Lanes, ASME paper 62-HT-39, 1962.

Grumbach, R., A Systematic Comparison of Different Hydrodynamics Models, *Nucl. Sci. Eng.*, vol. 36, pp. 429–433, 1969.

Hancox, W. T. and Nicoll, W. B., A General Technique for the Prediction of Void Distributions in Nonsteady Two-Phase Forced Convection, *Int. J. Heat Mass Transfer*, vol. 14, p. 1377, 1971.

Hands, B. A., A Re-examination of the Ledinegg Instability Criterion and Its Application to Two-Phase Helium Systems, in *Multi-Phase-Flow Systems, IChE Symp. Ser. 38*, vol. 1, paper E2, 1974.

Hawtin, R., Chugging Flow, AERE-R 6661, 1970.

Hayama, S., A Study of Hydrodynamic Instability in Boiling Channels, *Bull. JSME*, vol. 6, no. 23, pp. 549–56, 1963; vol. 7, no. 25, pp. 129–135, 1964; vol. 10, no. 38, pp. 308–319, 320–327, 1967.

Hewitt, G. F. and Hall-Taylor, N., *Annular Two-Phase Flow*, Pergamon, New York, 1970.

Ishii, M., Thermally Induced Flow Instabilities in Two-Phase Mixtures in Thermal Equilibrium, Ph. D. thesis, School of Mechanical Engineering, Georgia Institute of Technology, Atlanta, 1971.

Ishii, M., *Thermo-Fluid Dynamic Theory of Two-Phase Flow*, Eyrolles, Paris, 1975.

Ishii, M., Study of Flow Instabilities in Two-Phase Mixtures, ANL-76-23, 1976.

Ishii, M. and Zuber, N., Thermally Induced Flow Instabilities in Two Phase Mixtures, *Heat Transfer 1970, Proc. 4th Int. Heat Transfer Conf., Paris*, vol. 5, paper B5.11, 1970.

Jain, K. C., Petrick, M., Miller, D., and Bankoff, S. G., Self Sustained Hydrodynamic Oscillation in a Natural-Circulation Boiling Water Loop, *Nucl. Eng. Des.*, vol. 4, pp. 233–252, 1966.

Jeglic, F. A. and Grace, T. M., Onset of Flow Oscillations in Forced Flow Subcooled Boiling, NASA TN-D-2821, 1965.

Jones, A. B., Hydrodynamic Stability of a Boiling Channel, Part I: KAPL-2170, 1961; Part II: KAPL-2208 (with D. G. Dight), 1962; Part III: KAPL-2290 (with D. G. Dight), 1963; Part IV: KAPL-3070, 1964.

Jones, A. B. and Yarbrough, W. M., Reactivity Stability of a Boiling Reactor, Part I: KAPL-3072, 1964; Part II: KAPL-3093, 1965.

Jones, M. C. and Peterson, R. G., A Study of Flow Stability in Helium Cooling Systems, ASME paper 74-WA/HT-24, 1974.

Kakaç, S., Boiling Flow Instabilities in a Multi-Channel Upflow System, in *Two-Phase Flows and Heat Transfer*, eds. S. Kakaç and F. Mayinger, vol. 1, pp. 511–547, Hemisphere, Washington, D.C., 1977.

Kakaç, S., Veziroğlu, T. N., Akyuzlu, K., and Berkol, O., Sustained and Transient Boiling Flow Instabilities in a Cross-connected Four-parallel-channel Upflow System, *Heat Transfer 1974, Proc. 5th Int. Heat Transfer Conf., Tokyo*, vol. 4, paper B5.11, pp. 235–239 (see also discussion in vol. 8), 1974.

Kanai, T., Kawai, T., and Aoko, R., Void Reactivity Response in Boiling Water Reactors, *J. At. Energy Soc. Jpn.*, vol. 3, no. 3, pp. 168–178, 1961.

Kerney, P. J., Characteristics of a Submerged Steam Jet, Ph.D. thesis, Department of Mechanical Engineering, Pennsylvania State University, University Park, 1970.

Kirchenmayer, A., On the Kinetics of Boiling Water Reactors, *J. Nucl. Energy Part A*, vol. 12, pp. 155–161, 1960.

Kjaerheim, G. and Rolstad, E. In-pile Hydraulic Instability Experiments with a 7-Rod Natural Circulation Channel, Symp. on Two-Phase Flow Dynamics, EURATOM, EUR 4288e, The Technological University of Eindhoven, vol. 1, pp. 231–276, 1967.

Koshelev, I. I., Surnov, A. V., and Nikitina, L. V., Inception of Pulsations Using a Model of Vertical Water-Wall Tubes, *Heat Transfer Sov. Res.*, vol. 2, no. 3, pp. 111–115, 1970.

Krasyakova, L. Y. and Glusker, B. N., Hydraulic Study of Three-Pass Panels with Bottom Inlet Headers for Once-through Boilers, *Thermal Eng.*, vol. 12, no. 8, pp. 17–23, 1965.

Krejsa, E. A., Goodykoontz, J. H., and Stevens, G. H., Frequency Response of a Forced Flow Single Tube Boiler, NASA TN-D-4039, 1967.

Lahey, R. T., Jr., The Status of Boiling Water Nuclear Reactor Safety Technology, in *Thermal and Hydraulic Aspects of Nuclear Reactor Safety*, vol. 1. *Light Water Reactors*, eds. O. C. Jones and S. G. Bankoff, pp. 151–172, ASME, New York, 1977.

Lahey, R. T., Jr. and Moody, F. J., *The Thermal-Hydraulics of a Boiling Water Nuclear Reactor*, American Nuclear Society, Hinsdale, Ill., 1977.

Lahey, R. J., Jr. and Yadigaroglu, G., A Lagrangian Analysis of Two-Phase Hydrodynamic and Nuclear-coupled Density-Wave Oscillations, *Heat Transfer 1974, Proc. 5th Int. Heat Transfer Conf., Tokyo*, vol. 4, paper B5.9, pp. 225–229, 1974.

Ledinegg, M., Instability of Flow during Natural and Forced Circulation, *Waerme*, vol. 61, no. 8, pp. 891–898, 1938; AEC-TR-1861, 1954.

Lee, S. S., Veziroğlu, T. N., and Kakaç, S., Sustained and Transient Boiling Flow Instabilities in Two Parallel Channel Systems, in *Two-Phase Flows and Heat Transfer*, eds. S. Kakaç and F. Mayinger, vol. 1, pp. 467–510, Hemisphere, Washington, D.C., 1977.

Lighthill, M. J. and Whitham, G. B., On Kinematic Waves. I. Flood Movement in Long Rivers. II. A Theory of Traffic Flow on Long Crowded Roads, *Proc. R. Soc. London Ser. A*, vol. 229, pp. 281–345, 1955.

Llory, M. A., Experimental and Theoretical Study of Static and Dynamic Stability on Two Scale Model Steam Generator Mock-ups, AIChE Meet., Los Angeles, Nov. 1975.

Lyon, W. Y., Comparison of Two-Phase Flow Instability Analytical Model (NUFREQ) against Experimental Data, M. S. Project, Department of Mechanical Engineering, University of California, Berkeley, 1976.

Martini, R. and Premoli, A., Bottom Flooding Experiments with Simple Geometries under Different ECC Conditions, *Energia Nucleare*, vol. 20, pp. 540–553, 1973.

Masini, G., Possa, G., and Tacconi, F. A., Flow Instability Thresholds in Parallel Heated Channels, *Energia Nucleare*, vol. 15, p. 777, 1968.

Mathisen, R. P., Out of Pile Instability in the Loop Skälvan, Symp. on Two-Phase Flow Dynamics, EURATOM, EUR 4288e, The Technological University of Eindhoven, vol. 1, pp. 19–64, 1967.

Maulbetsch, J. S. and Griffith, P., A Study of System-induced Instabilities in Forced Flow Convection with Subcooled Boiling, MIT, EPL rept. 5382-35, 1965.

Meyer, J. E. and Rose, R. P., Application of a Momentum Integral Model to the Study of Parallel Channel Boiling Flow Oscillations, *J. Heat Transfer, Trans. ASME*, vol. 85, pp. 1–9, 1963.

Moxon, D. The SUPERSLIP Programme for the Steady State and Dynamics Performance of Once-through Boilers, AEEW-M 870, 1969.

Moxon, D. Stability of Once-through Steam Generators, ASME paper 73-WA/HT-24, 1973.

Nahavandi, A. N. and Von Hallen, R. F., A Space Dependent Dynamic Analysis of Boiling Water Reactor Systems, *Nucl. Sci. Eng.*, vol. 20, p. 392, 1964.

Nakajima, I., Kawada, A., Fukuda, K., and Kobori, T., An Experimental Study on the Instability Induced by Voiding from a Horizontal Pipe Line, ASME paper 75-WA/HT-20, 1975.

Nassos, G. P. and Bankoff, S. G., Propagation of Density Disturbances in an Air-Water Flow, *Proc. 3rd Int. Heat Transfer Conf., Chicago*, vol. 4, pp. 234–246, AIChE, New York, 1966.

Nassos, G. P. and Bankoff, S. G. Slip Velocity Ratios in an Air-Water System under Steady-state and Transient Conditions, *Chem. Eng. Sci.*, vol. 22, p. 4, 1967.

Neal, L. G. and Zivi, S. M., Stability of Boiling-Water Reactors and Loops, *Nucl. Sci. Eng.*, vol. 30, pp. 25–38, 1967.

Paul, F. W. and Riedle, K. J., Experimental and Analytical Investigation of the Dynamic Behavior of Diabatic Two-Phase Flow in a Vertical Monotube Vapor Generator, Part I, ASME paper 72-WA/HT-46, 1972.

Quandt, E. R., Analysis and Measurement of Flow Oscillations, *Heat Transfer Buffalo, Chem. Eng. Prog. Symp. Ser.*, vol. 57, no. 32, pp. 111–126, 1961.

Riedle, K. and Winkler, F., EEC-Reflooding Experiments with a 340-Rod Bundle, CREST Specialist Meetg on Emergency Core Cooling for Light Water Reactors, Garching/München, rept. MRR-115, Oct. 18–20, 1972.

Riedle, K., Gaul, H. P., Ruthrof, K., and Sarkar, J., Reflood and Spray Cooling Heat Transfer in PWR and BWR Bundles, ASME paper 76-HT-10, 1976.

Rothe, P. A., Wallis, G. B., and Block, J. A., Cold Leg ECC Flow Oscillations in *Thermal and Hydraulic Aspects of Nuclear Reactor Safety*, vol. 1. *Light Water Reactors*, eds. O. C. Jones and S. G. Bankoff, pp. 133–150, ASME, New York, 1977.

Rothe, R. P., Wallis, G. B., and Thrall, D. E., Cold Leg ECC Flow Oscillations, EPRI topical rept. NP-282, 1976.

Roy, R. P. and Yadigaroglu, G., An Investigation of Heat Transport in Oscillatory Turbulent Subcooled Flow, *J. Heat Transfer, Trans. ASME*, vol. 98, pp. 630–637, 1976.

Saha, P., Thermally Induced Two-Phase Flow Instabilities, Including the Effect of Thermal Non-equilibrium between the Phases, Ph.D. thesis, School of Mechanical Engineering, Georgia Institute of Technology, Atlanta, 1974.

Saha, P., Ishii, M., and Zuber, N., An Experimental Investigation of the Thermally Induced Flow Oscillations in Two-Phase Systems, *J. Heat Transfer, Trans. ASME*, vol. 98, pp. 616–622, 1976.

St.-Pierre, C. C., Frequency-Response Analysis of Steam Voids to Sinusoidal Power Modulation in Thin-walled Boiling Water Coolant Channel, ANL-7041, 1965.

Saito, T., Uchida, H., Morita, T., Oishi, T., and Saito, S., On the Unsteady Phenomena Relating to Vapor Suppression, ASME paper 74-WA/HT-47, 1974.

Sakaguchi, T., Akagawa, K., and Hamaguchi, H., Transient Behavior of Air-Water Two-Phase Flow in a Horizontal Tube, ASME paper 73-WA/HT-21, 1973.

Sanathanan, C. K., Carter, J. C., and Miraldi, F., Dynamic Analysis of Coolant Circulation in Boiling Water Nuclear Reactors, Part I and Part II, *Nucl. Sci. Eng.*, vol. 23, pp. 119–129, 130–137, 1965.

Sano, A., Kanamori, A., Tsuchiya, T., and Yamashita, H., 1 MWt Steam Generator Operating Experience, ASME paper 73-HT-53, 1973.

Schlechtendahl, E. G., Theoretical Investigations on Sodium Boiling in Fast Reactors, *Nucl. Sci. Eng.*, vol. 41, p. 99, 1970.

Schmitt, F., Contribution Expérimentale et Théorique à l'Etude d'un Type Particulier d'Ecoulement Transitoire de Sodium en Ebullition: La Redistribution de Débit, Thèse de docteur-ingénieur, Université Scientifique et Médicale de Grenoble, 1974.

Schuster, J. R. and Berenson, P. J., Flow Stability of a Five-Tube Forced Convection Boiler, ASME paper 67-WA/HT-20, 1967.

Serov, E. P., The Operation of Once-through Boilers in Variable Regimes, *Tr. Mosk. Energ. Inst.*, vol. 11, 1953.

Shiralkar, B. S., Schnebly, L. E., and Lahey, R. T., Jr., Variation of the Vapor Volumetric Fraction during Flow and Power Transients, *Nucl. Eng. Des.*, vol. 25, pp. 350–368, 1973.

Shotkin, L. M., Stability Considerations in Two-Phase Flow, *Nucl. Sci. Eng.*, vol. 28, pp. 317–324, 1967.

Soliman, M. and Berenson, P. J., Flow Stability and Gravitational Effects in Condenser Tubes, *Heat Transfer 1970, Proc. 4th Int. Heat Transfer Conf., Paris*, Sept. 1970, vol. 6, paper Cs 1.8, Elsevier, Amsterdam, 1970.

Space Technology Laboratories, Kinetic Studies of Heterogeneous Water Reactors, STL-6212, Annual Summary rept., 1962.

Staub, F. W. and Zuber, N., Void Response to Flow and Power Oscillations in a Forced-Convection Boiling System with Axially Non-uniform Power Input, *Nucl. Sci. Eng.*, vol. 30, pp. 296–303, 1967.

Stenning, A. H., Instabilities in the Flow of a Boiling Liquid, *J. Basic Eng., Trans. ASME*, vol. 86, pp. 213–217, 1964.

Stenning, A. H. and Veziroğlu, T. N., Flow Oscillation Modes in Forced-Convection Boiling, *Proc. 1965 Heat Transfer and Fluid Mechanics Inst.*, pp. 301–316, Stanford University Press, Stanford, Calif., 1965.

Stenning, A. H. and Veziroğlu, T. N., Oscillations in Two-Component Two Phase Flow, vol. 1, NASA CR-72121; and Flow Oscillation in Forced Convection Boiling, vol. 2, NASA CR-72122, 1967.

Stenning, A. H., Veziroğlu, T. N., and Callahan, G. M., Pressure-Drop Oscillations in Forced Convection Flow with Boiling, Symp. on Two-Phase Flow Dynamics, EURATOM, EUR 4288e, The Technological University of Eindhoven, vol. 1, pp. 405–427, 1967.

Takahashi, R. and Futami, T., Theoretical Study of Flow Instability of a Sodium-heated Steam Generator, *Nucl. Eng. Des.*, vol. 41, pp. 193–204, 1977.

Thurston, R. S., Rogers, J. D., and Skoglund, V. J., Pressure Oscillations Induced by Forced Convection Heating of Dense Hydrogen, *Adv. Cryog. Eng.*, vol. 12, p. 438, 1967.

U.S. Nuclear Regulatory Commission, Relief Valve Discharge to Suppression Pool, *Operating Experience–Information on Inspection and Enforcement, Bulletins and Replies*, Apr. 8, 1975.

Wallis, G. B., *One-dimensional Two-Phase Flow*, McGraw-Hill, New York, 1969.

Wallis, G. B. and Heasley, J. H., Oscillations in Two-Phase Flow Systems, *J. Heat Transfer, Trans. ASME*, vol. 83, p. 363, 1961.

Wallis, G. B. and Rothe, P. H., Water Plug Formation in Subscale PWR Cold Leg Models during Simulating ECC Injection, Creare TN-237, 1976.

Waring, J. P., Rosal, E. R., and Hochreiter, L. E., PWR FLECHT-SET Phase B1 Data Report, WCAP-8431, 1974.

Waszink, R. P. and Efferding, L. E., Hydrodynamic Stability and Thermal Performance Tests of a 1-MWt Sodium-heated Once-through Steam Generator Model, ASME paper 73-Pwr-16, 1973.

Wedekind, G. L., An Experimental Investigation into the Oscillatory Motion of the Mixture-Vapor Transition Point in Horizontal Evaporating Flow, *J. Heat Transfer, Trans. ASME*, vol. 93, pp. 47–54, 1971.

Wedekind, G. L. and Stoecker, W. F., Transient Response of the Mixture-Vapor Transition Point in Horizontal Evaporating Flow, *Trans. ASHRAE*, vol. 72, pt. 1, IV.2, pp. 1–15, 1966.

Westendorf, W. H., A Model for Predicting the Onset of Oscillatory Instability Occurring with the Intermixing of High-Velocity Vapor with Its Subcooled Liquid in Cocurrent Streams, *Heat Transfer 1970, Proc. 4th Int. Heat Transfer Conf., Paris*, Sept. 1970, vol. 5, paper B5.4, Elsevier, Amsterdam, 1970.

Westendorf, W. H. and Brown, W. F., Stability of Intermixing High-Velocity Vapor with Its Subcooled Liquid in Cocurrent Streams, NASA TN-D-353, 1966.

White, E. P. and Duffey, R. B., A Study of the Unsteady Flow and Heat Transfer in the Reflooding of Water Reactor Cores, CEGB rept. RD/B/N3134, 1974.

Wright, S. A. and Albrecht, R. W., Sodium Boiling Detection in LMFBR's by Acoustic-neutronic Cross Correlation, *Trans. ANS*, vol. 21, p. 478, 1975.

Yadigaroglu, G. and Bergles, A. E., An Experimental and Theoretical Study of Density-Wave Oscillations in Two-Phase Flow, MIT rept. DSR 74629-3 (HTL 74629-67), 1969.

Yadigaroglu, G. and Bergles, A. E., Fundamental and Higher-Mode Density Wave Oscillations in Two-Phase Flow, *J. Heat Transfer, Trans. ASME*, vol. 94, pp. 189–195, 1972.

Yadigaroglu, G. and Lahey, R. T., Jr., On the Various Forms of the Conservation Equations in Two-Phase Flow, *Int. J. Multiphase Flow*, vol. 2, pp. 477–494, 1976.

Zuber, N., An Analysis of Thermally Induced Flow Oscillations in the Near-Critical and Super-critical Thermodynamic Region, NAS 8-11422, final rept., 1966.

Zuber, N. and Findlay, J. A., Average Volumetric Concentration in Two-Phase Flow Systems, *J. Heat Transfer, Trans. ASME*, vol. 87, pp. 453–468, 1965.

Zuber, N. and Staub, F. W., The Propagation and the Wave Form of the Vapor Volumetric Concentration in Boiling, Forced Convection System under Oscillatory Conditions, *Int. J. Heat Mass Transfer*, vol. 9, pp. 871–895, 1966.

Zuber, N. and Staub, F. W., An Analytic Investigation of the Transient Response of the Volumetric Concentration in a Boiling Forced-Flow System, *Nucl. Sci. Eng.*, vol. 30, pp. 268–278, 1967.

Zuber, N., Staub, F. W., Bijwaard, G., and Kroeger, P. G., Steady State and Transient Void Fraction in Two-Phase Flow Systems—Final Report for the Program of Two-Phase Flow Investigation, GEAP-5417, EURAEC 1949, 1967.

CHAPTER

# EIGHTEEN

# CRITICAL FLOWS

**M. Giot**

The calculation of critical flow rates, that is, of choking flow rates or maximum flow rates, and the understanding of the phenomena related to wave propagation are important for the design of equipment involving two-phase flows. Understanding these factors is important because critical flows are easily reached in such equipment as a result of the small values of the corresponding flow velocities. Even a relatively small pressure difference applied to a flow restriction (for example, a valve or an orifice) can lead to sonic blockage.

Further, in practice, critical flows are often associated with a phase change in steady-state flow. This phase change in steady-state flow occurs for example in refrigerating and desalination plants as a result of flashing.

The incidence of critical flows in the safety analysis of nuclear reactors is well known: the critical flow rate calculations are of paramount importance for loss of coolant accident (LOCA) studies. The rarefaction wave propagation after the failure must be described with sufficient accuracy.

Finally, from the theoretical point of view, two-phase critical flows are a subject of major interest since their investigation requires a deep insight into the mass, momentum, and heat transfers between the phases.

This chapter starts with the definitions of critical flow and sonic velocity from the physical and mathematical points of view. A review of the experiments is then presented, with an analysis of the testing-facilities, fluids, test sections, and instrumentation. The main results and observations are reported. Finally, several theoretical models from the literature are discussed. The models are derived in a logical way from a general set of equations; the necessary assumptions are shown and calculation results are compared with experimental data.

# 1 DEFINITIONS

## 1.1 Critical Flow

**1.1.1 Physical standpoint** Consider a single- or two-phase flow in a pipe of any geometry between two vessels. Assume the state of the fluid is kept constant in the upstream vessel. The flow rate is called *critical* when one refers to the maximum flow rate, which can be obtained under such conditions by lowering the pressure in the downstream vessel.

When the critical flow rate is reached, if one measures along the pipe the axial profiles of the variables used to describe the flow, the following peculiarity is noticed: any perturbation of the state of the fluid in the downstream vessel, if the pressure is maintained at a value lower than that giving the critical flow rate, does not affect the profiles upstream from a certain cross section of the pipe. This cross section is called the critical section.

This peculiarity of critical flows, which is well-known for single-phase flows, was observed by numerous investigators in two-phase flow experiments. We may refer to the studies of Réocreux (1974) concerning the flashing of water due to the discharge through a vertical, cylindrical tube ended by a conical diffuser with an angle of 7°. The flow was directed upward.

Figure 1 shows an example of pressure and void-fraction profiles for such a critical flow. It can be seen that the critical section is downstream from the throat.

**1.1.2 Mathematical standpoint** Consider the set of differential equations describing the steady-state flow of a fluid in a pipe.

For a single-phase flow, there are three equations: a mass, a momentum, and an energy equation. They can be written in such a way that the unknowns are the derivatives with respect to the length $z$ of the channel of time- and space-averaged variables, such as the pressure, the temperature, and the velocity of the fluid.

For a two-phase flow, there are six equations; in this chapter we will consider the set of equations given in the Appendix and derived from the two-fluid model.

Here, the averaged variables, whose derivatives are the unknowns of the set of differential equations, are denoted by $x_j$ ($j = 1, \ldots, n$). This set is written in matrix form as

$$b_{ij} \frac{dx_j}{dz} = c_i \qquad i, j = 1, \ldots, n \tag{1}$$

where elements $b_{ij}$ of the system matrix, as well as elements $c_i$ of the right-hand side (rhs) column vector are functions of the averaged variables $x_j$. The integration of this set of equations involves a discretization of the distance $z$ in steps of length $\delta z$ and the calculation of the corresponding increments $\delta x_j$.

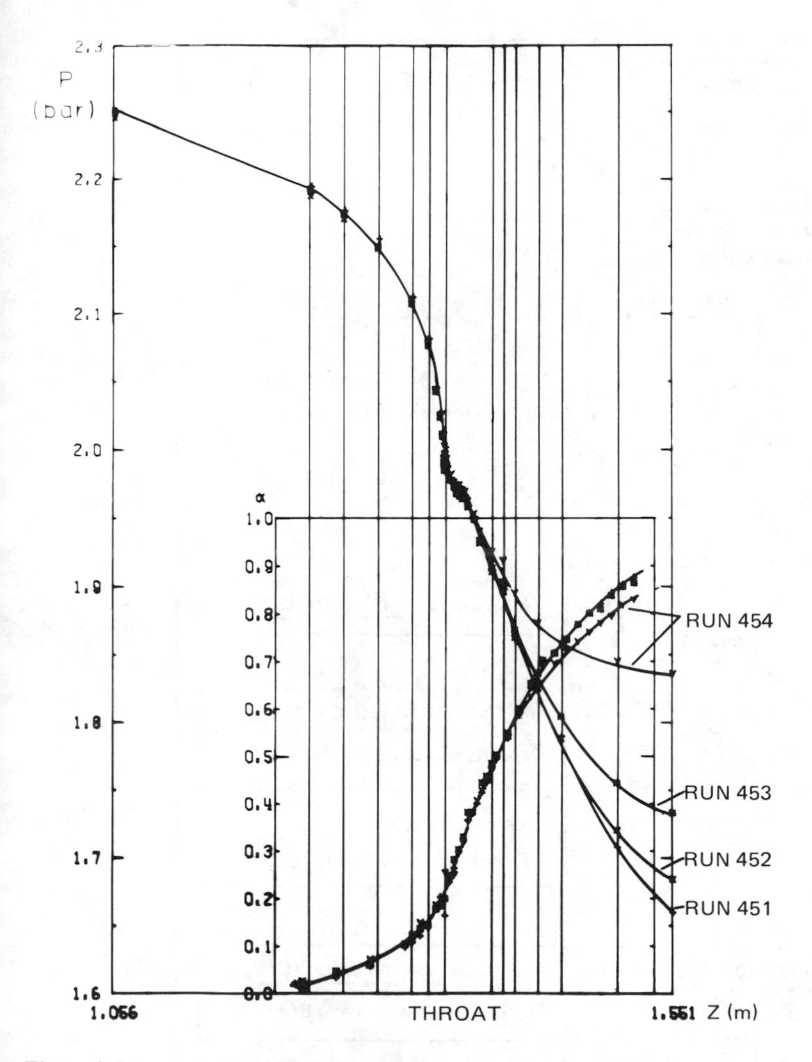


**Figure 1** Pressure and void-fraction profiles in a cylindrical tube ended by a diffuser with an angle of 7° (Réocreux, 1974). For runs 451 (+), 452 (×), 453 (∗), and 454 (Y), the mass velocities (kg/m² s) and inlet temperatures (°C) are, respectively: 8520, 125.1; 8531, 125.2; 8533, 125.2; and 8541, 125.2.

Let us follow the calculation procedure illustrated by the flow chart in Fig. 2 proposed by Bouré et al. (1975, 1976).

We start with a set of values $x_j(0)$ at the inlet of the channel. This set of inlet values $x_j(0)$ consists of data (pressure, temperature, etc.) and trial quantities (void fraction, velocities, etc.). The set of equations is solved by means of evaluation at each step of the ratios:

$$\frac{dx_j}{dz} = \frac{N_j}{\det(b_{ij})}$$

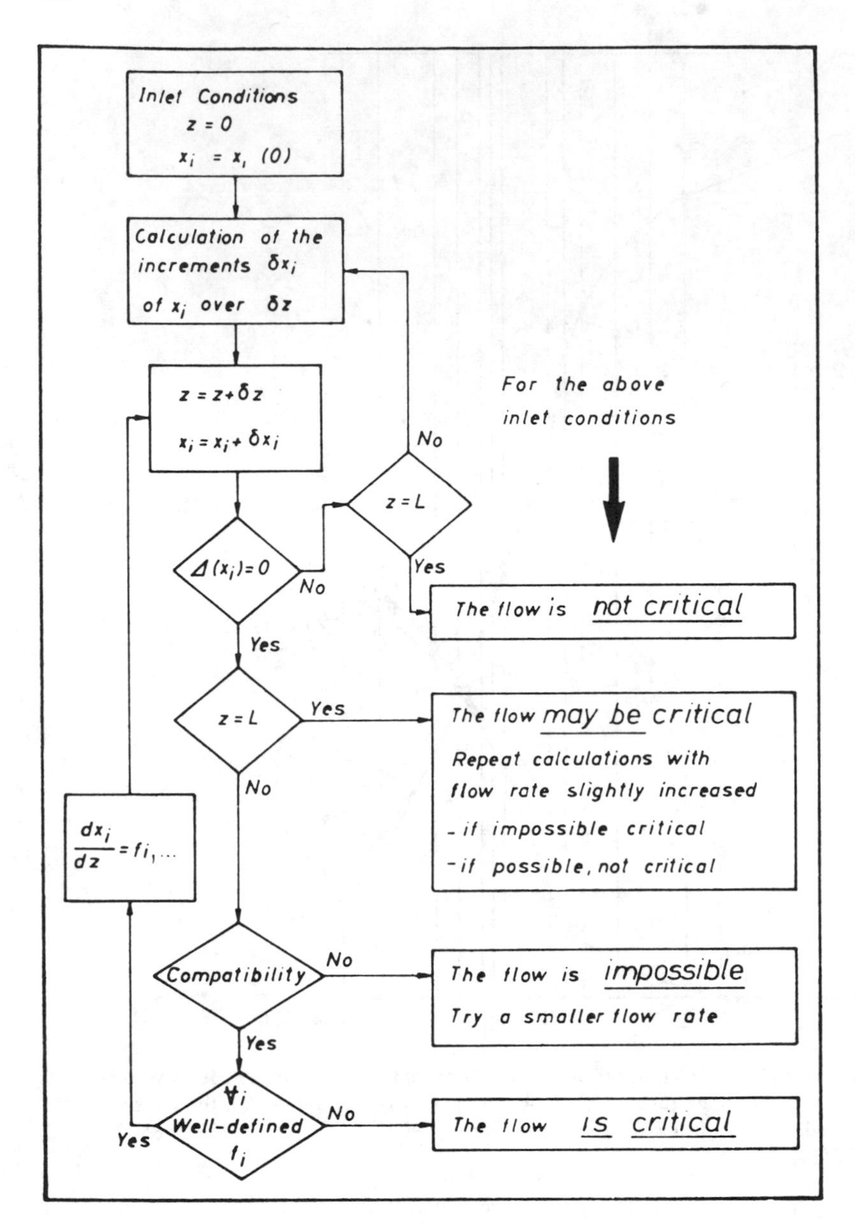


**Figure 2** Flow chart for the calculation of a critical flow (Bouré et al., 1975, 1976).

where $N_j$ stands for the determinant of the system matrix in which the $j$th column is replaced by the elements $c_i$ of the column vector.

The calculation can proceed downstream to the channel outlet ($z = L$) if the singularity det $(b_{ij}) = 0$ does not appear for $z \leqslant L$. The axial profiles of the $x_j$'s are found; the solution corresponds to a possible flow within the validity limits of the chosen model, that is, a set of equations and inlet conditions. This flow is not critical.

On the contrary, if $\det(b_{ij}) = 0$ in any cross section after the inlet, the problem is impossible or its solution is indeterminate. Here, it must be emphasized that some trivial cases are a priori excluded, for example,

$$x_j = \text{const} \qquad \text{for } 0 \leqslant z \leqslant L$$

$$\det(b_{ij}) = 0 \qquad \text{for } z = 0$$

The problem is impossible if $N_j \neq 0$, because $dx_j/dz = \infty$ does not correspond to any physical reality. The calculation must be resumed from the inlet with another set of inlet values $x_j(0)$.

Indetermination occurs when all the determinants $N_j$ vanish at the same abscissa as $\det(b_{ij})$ if $\det(b_{ij})$ is not a factor of $N_j$. As a consequence of the exclusion of the trivial case $x_j = \text{const}$ one cannot have the $N_j$'s all identically equal to zero, and the necessary and sufficient condition for indetermination at the section where $\det(b_{ij}) = 0$, is the vanishing condition $N_j = 0$ of one of any $N_j$ that is not identically zero. This condition $N_j = 0$ is called the compatibility condition.

If the compatibility condition $N_j = 0$ is nonidentically satisfied at the cross section where $\det(b_{ij}) = 0$, the flow is said to be critical. For a critical flow, it can indeed be numerically verified that a slight variation of one of the inlet values $x_j(0)$ results in a choice of

1. Shifting the sections where $\det(b_{ij}) = 0$; the problem is then impossible.
2. Eliminating the indetermination; the condition $\det(b_{ij}) = 0$ is then no longer satisfied.

If one excepts the condition where the derivatives $dx_j/dz$ have well-defined values in any section, even where $\det(b_{ij}) = 0$, the flow is critical only if we have

$$\det(b_{ij}) = 0 \qquad \text{and} \qquad N_j = 0$$

at a section of the channel, which for that reason is called the critical section.

It must be noticed that the compatibility condition, and only the compatibility condition, involves the terms $c_i$ of the column vector, such as $dA/dz$, $\tau_{kC}$, $q_{kC}$, $q$, and $\phi$ and the interfacial transfer terms $M_{kI}$, $(MV)_{kI}$, $\tau_{kI}$, $(MH)_{kI}$, $(EC)_{kT}$, $\chi_{kI}$, and $q_{kI}$ (see the appendix).

## 1.2 Sound Velocity

Let us first recall the distinction between the displacement velocity and propagation velocity of a wave in a flowing fluid: The displacement velocity is measured in a frame of reference fixed with respect to the channel, whereas the propagation velocity is measured in a frame of reference moving at the average velocity of the fluid.

**1.2.1 Physical standpoint** The sound velocity is the propagation velocity of a small amplitude pressure wave. According to the medium in which the propagation takes place, the wave can be attenuated or dispersed or it can produce shocks. These aspects are not studied here; only the incidence of the two-phase system parameters (thermodynamic state of the phases, flow pattern, and velocities of the flow) on the propagation velocities of small disturbances is considered.

**1.2.2 Mathematical standpoint** The quasilinear set of partial differential equations describing the unsteady single- or two-phase flow can be written in a matrix form:

$$a_{ij} \frac{\partial x_j}{\partial t} + b_{ij} \frac{\partial x_j}{\partial z} = c_i \tag{2}$$

Several methods have been used to determine the propagation velocities of small amplitude waves. The method of characteristics and the method of small perturbations are considered below.

The method of characteristics (Courant and Friedrichs, 1961) consists of looking for directions $V_k^*$ $(k = 1, \ldots, n)$ of the plane $(z, t)$ along which the derivatives of the variables $x_j$ are indeterminate. By crossing these directions, one meets regions of the plane where the variables $x_j$ are submitted to small amplitude variations, which are called, therefore, small perturbations. Setting

$$\frac{dx_j}{dt} = \frac{\partial x_j}{\partial t} + V \frac{\partial x_j}{\partial z} \tag{3}$$

and replacing $\partial x_j/\partial t$ in Eq. (2) by its expression obtained from the definition in Eq. (3) of the total derivative $dx_j/dt$ yields

$$(b_{ij} - a_{ij}V) \frac{\partial x_j}{\partial z} = c_i - a_{ij} \frac{dx_j}{dt} \tag{4}$$

The solution of this quasilinear set of equations is obtained by calculating at each section of the channel the $n$ ratios:

$$\frac{\partial x_j}{\partial z} = \frac{(N_V)_j}{\det (b_{ij} - a_{ij}V)}$$

where $(N_V)_j$ denotes the determinant of Eq. (4) where the $j$th column is replaced by the column vector

$$c_i - a_{ij} \frac{dx_j}{dt}$$

The $n$ characteristics $V_k^*$ are the roots of

$$\det (b_{ij} - a_{ij}V) = 0 \tag{5}$$

If the set of partial differential equations in Eq. (2) is hyperbolic, all the roots are real. In order that the derivatives $\partial x_j/\partial z$ be indeterminate along the characteristics, the following compatibility condition must be nonidentically satisfied:

$$(N_V)_j = 0 \qquad \text{for each of the } n \text{ characteristics } V_k^*$$

Consequently, there are $n$ conditions:

$$(N_{V_k^*})_j = 0 \qquad k = 1, \ldots, n \tag{6}$$

By developing these compatibility conditions, one gets (Fritte, 1974)

$$\sum_j d_{jk} \frac{dx_j}{dt} = f_k \tag{7}$$

where $\qquad d_{jk} = d_{jk}(a_{ij}, b_{ij}, V_k^*)$

and $\qquad f_k = f_k(c_i, a_{ij}, b_{ij}, V_k^*)$

The left-hand side (lhs) member of Eq. (5) is an $n$th degree polynomial with respect to $V$ and may be written

$$\det(b_{ij} - a_{ij}V) = \sum_{l=1}^{n} V^l \phi_l(a_{ij}, b_{ij}) + \det(b_{ij})$$

Consequently, the necessary condition of critical flow, $\det(b_{ij}) = 0$, is the vanishing condition of at least one root: At the critical section, the roots of $\det(b_{ij} - a_{ij}V)$ take the values

$$V_k^* = V_{k,c}^*$$

and one of them vanishes. It can be concluded that at least one perturbation is stationary at the critical section. If the set of equations (2) is hyperbolic, all the roots $V_{k,c}^*$ are real. It is then sufficient that all of them be positive or zero to prevent a small perturbation from propagating toward the upstream direction.

Information about waves propagating along the characteristics can be obtained from the analysis of the compatiblity conditions as shown by Fritte (1974).

For the method of small perturbations we set

$$x_j = \bar{x}_j + x'_j \qquad \text{with} \qquad x'_j \ll x_j$$

where $\bar{x}_j$ is the undisturbed value of variable $x_j$, and $x'_j$ is a small perturbation of $x_j$. The set of partial differential equations (2) may be written

$$(\bar{a}_{ij} + a'_{ij})\left(\frac{\partial \bar{x}_j}{\partial t} + \frac{\partial x'_j}{\partial t}\right) + (\bar{b}_{ij} + b'_{ij})\left(\frac{\partial \bar{x}_j}{\partial z} + \frac{\partial x'_j}{\partial z}\right) = \bar{c}_i + c'_i$$

or, if only the terms having the same order of magnitude are kept,

$$\bar{a}_{ij}\frac{\partial \bar{x}_j}{\partial t} + \bar{b}_{ij}\frac{\partial \bar{x}_j}{\partial z} + \bar{a}_{ij}\frac{\partial x'_j}{\partial t} + \bar{b}_{ij}\frac{\partial x'_j}{\partial z} = \bar{c}_i \tag{8}$$

If the set of equations written in terms of mean values of the $x_j$'s is satisfied, one has

$$\bar{a}_{ij}\frac{\partial x'_j}{\partial t} + \bar{b}_{ij}\frac{\partial x'_j}{\partial z} = 0 \tag{9}$$

Looking for progressive waves, one can set

$$x'_j = x_{jo} e^{j(\omega t - kz)}$$

where $x_{jo}$ is the amplitude of the wave, $\omega$ is its pulsation, and $k$ is the wave number. The homogeneous set of equations (9) becomes

$$jk \left( \frac{\omega}{k} \bar{a}_{ij} - \bar{b}_{ij} \right) x'_j = 0 \tag{10}$$

The solution of this set is different from zero if

$$\det \left( \frac{\omega}{k} \bar{a}_{ij} - \bar{b}_{ij} \right) = 0 \tag{11}$$

It can be seen that Eq. (11) is very similar to Eq. (5) as deduced from the method of characteristics; its lhs is a polynomial of the $n$th degree with respect to $\omega/k$, which may be written

$$\det \left( \frac{\omega}{k} \bar{a}_{ij} - \bar{b}_{ij} \right) = \sum_{l=1}^{n} \left( \frac{\omega}{k} \right)^l \psi_l(\bar{a}_{ij}, \bar{b}_{ij}) - \det (\bar{b}_{ij})$$

Consequently, the necessary condition of critical flow, $\det (\bar{b}_{ij}) = 0$, leads to the vanishing condition of at least one root: at the critical section, the roots of

$$\det \left( \frac{\omega}{k} \bar{a}_{ij} - \bar{b}_{ij} \right) = 0$$

take the values

$$\left( \frac{\omega}{k} \right)^* = \left( \frac{\omega}{k} \right)_c^*$$

and one of them is zero. All the conclusions of the method of characteristics are again reached.

## 1.3 Links between the Critical Flow and the Sound Velocity

The above-mentioned necessary condition of critical flow, $\det (\bar{b}_{ij}) = 0$, involves the vanishing condition of a characteristic direction $V_k^*$ or the vanishing condition of a wave displacement velocity $(\omega/k)^*$. By definition, the wave propagation velocity $\mathbf{a}_k$ is given by the vectorial relationship:

$$\mathbf{V}_k^* \triangleq \mathbf{w}_k + \mathbf{a}_k \qquad k = 1, \ldots, n \tag{12}$$

where the entrainment velocity $\mathbf{w}_k$ is an average velocity of the fluid. For example, for a single-phase compressible fluid, we have ($n = 3$):

$$w_1 = w_2 = w_3 = w$$

$$a_1 = 0 \qquad a_2 = a_s \qquad a_3 = -a_s$$

with $a_s^2 \triangleq p'_{\rho,s}$, and hence

$V_1^* = w$, a continuity wave, moving downstream
$V_2^* = w + a_s$, a wave moving downstream
$V_3^* = w - a_s$, a wave moving upstream if $w < a_s$

If the flow is critical, that is, if

$$w = w_c = a_s$$

one has

$$V_{1c}^* = w_c \qquad V_{2c}^* = 2a_s \qquad V_{3c}^* = 0$$

One wave is stationary, whereas the other two move downstream. A disturbance generated below the critical section (a pressure or temperature disturbance) can affect neither the state of the fluid nor its velocity above the critical section. This property explains why the blockage of the flow rate occurs even if the pressure below the critical section is lowered.

For a two-phase fluid, these principles remain true; however, it is not a priori evident that the application of these principles to two-phase systems would produce identical results or even results consisting in a simple transposition of the single-phase flow results. Consider some examples:

1. It is not a priori established that any model suggested in the literature provides displacement velocities that are all real and positive when the critical flow condition is verified. Nevertheless, the existence of the critical flow has been confirmed by numerous experiments. Consequently, any model that does not involve the physical basis of this property, namely the independence of the conditions upstream of the critical section with respect to perturbations generated downstream, would be irrelevant.

2. Some critical flow models involve two velocities, for example, an average velocity for each phase ($w_L$, $w_G$), or the average velocity of a phase, and a relative velocity between both phases. Then several velocities $w_k$ can be expected in Eq. (12); they would be the velocities chosen for the model or functions of these velocities.

3. The rhs of the set of partial differential equations of the single-phase flows involve, among others, wall friction and heat transfer terms. It has been shown that these terms do not play any part either in the necessary condition of critical flow, or in Eqs. (5) and (11) used to calculate the wave-displacement velocities.

Moreover, the rhs of the set of partial differential equations of two-phase flows involve mass ($M$), momentum ($MV$), and energy ($MH$) interfacial transfer terms. The expressions of these terms still remain largely unknown; however, they might be given by the following expansions:

$$\begin{aligned} M &= M_o - \sum \lambda_{x_j}^t \frac{\partial x_j}{\partial t} - \sum \lambda_{x_j}^z \frac{\partial x_j}{\partial z} - \cdots \\ (MV) &= (MV)_o - \sum \mu_{x_j}^t \frac{\partial x_j}{\partial t} - \sum \mu_{x_j}^z \frac{\partial x_j}{\partial z} - \cdots \end{aligned} \tag{13}$$

$$(MH) = (MH)_o - \sum \nu_{x_j}^t \frac{\partial x_j}{\partial t} - \sum \nu_{x_j}^z \frac{\partial x_j}{\partial z} - \cdots \qquad (13\ cont.)$$

where $M_o$, $(MV)_o$, $(MH)_o$, $\lambda$, $\mu$, $\nu$, . . . depend on $z$, $t$, and the $x_j$'s.

If Eq. (13) is restricted to the first order, the set of partial differential equations remains quasilinear, but matrices $a_{ij}$ and $b_{ij}$ are then functions of the interfacial transfer laws; the same holds for the necessary condition of critical flows and expression of the wave-displacement velocities.

From a physical standpoint, one can accept, for example, that the expressions of the wave velocities be different, depending on whether these waves propagate in a liquid-vapor system at equilibrium, or isentropically in each phase separately, or in a two-component liquid-gas system.

4. Bauer et al. (1976) using the CLYSTERE code have established that a steam-water flow rate can reach a blockage value without the fluid flowing at the speed of sound: it is therefore sufficient, when the downstream pressure is lowered, for the resulting vaporization to increase the acceleration pressure drop by an amount such that the pressure remains unchanged upstream from the potential critical section. This phenomenon, named pseudo-critical, is still unexplained.

These four examples illustrate the complexity of two-phase flow dynamics. We will return to these analyses after examining the results of the experimental studies.

## 2 EXPERIMENTAL RESULTS

Despite the difficulty in analyzing experimental results dealing with many different facilities (about which detailed descriptions are seldom available in the literature), we will try to give a survey of the main tests performed up to now. Table 1 helps the reader locate the studies in their chronological and technological contexts.

### 2.1 Test Facilities

Two kinds of test facilities can be distinguished according to the stationary or transient character of the phenomena to be studied:

1. Steady-state operating loops, which allow the flow of the fluid(s) in stationary conditions through an expansion device, where the critical section is located. Provided the indispensable conditions for stable operation are present, these loops allow critical flow rate, pressure, and void-fraction profiles to be measured in the easiest way.
2. Blowdown testing facilities, where an initially pressurized capacity is discharged through the expansion device. These facilities ensure the best simulation of conditions for possible nuclear reactor accidents, but they require sophisticated instrumentation. Some blowdown testing facilities allow a long discharge, whereas in many facilities the discharge takes place in a few milliseconds (shock tubes). Usually the aim of such experiments is of the benchmark type for code validation. The analysis of their results is outside the scope of this survey.

| Reference | Fluid (direction of flow[a]) | Test sections | Range of pressures[b] (bar) | Range of mass velocities (kg/m² s) | Range of qualities |
|---|---|---|---|---|---|
| Silver (1940) | Saturated water (VD) | Nozzles with rounded inlets: $L/D = 1.5 \dots 12$ $D = 4.76$ mm | $p_{up} = 1 \dots 3.5$ $p_{down} = 1$ | 10,000 . . . 17,000 | – |
| Bailey (1951) | Saturated water (H) | Orifices, converging nozzles, tubes: $L/D = 0.87 \dots 20$ $D = 3.2$ and 6.4 mm | $p_{up} = 0.5 \dots 3$ $\Delta p = 0 \dots 1$ | 4,000 . . . 18,000 | – |
| Isbin et al. (1957) | Steam–water mixtures (H) | Tubes with circular or annular cross section: $L/D = 24 \dots 144$ $D = 6.4 \dots 25.4$ mm | $p_c = 0.28 \dots 3$ $p_{down} = \dots 0.07$ | 100 . . . 4,400 | 0.01 . . . 1 |
| Zaloudek (1961) | Steam–water mixtures (H) | $L/D = 46$ and 69 $D = 13.2$ mm | $p_c = 2.8 \dots 7.6$ | 500 . . . 5,000 | 0.004 . . . 0.99 |
| Faletti and Moulton (1963) | Steam–water mixtures (H) | Annular cross section: $L = 13.5 \dots 892.5$ mm $D_{rod} = 4.75$ and 9.5 mm $D_{tube} = 14.6$ mm | $p_c = 1.8 \dots 7.3$ $p_{down} = 1.2 \dots 4.0$ | 1,000 . . . 30,000 | 0.001 . . . 1 |
| Fauske (1963b) | Steam–water mixtures (H) | $L/D = 408$ and 450 $D = 3.2$ and 6.8 mm | $p_c = 2.8 \dots 25$ | 2,500 . . . 20,000 | 0.01 . . . 0.7 |
| Faukse and Min (1963) | Refrigerant 11 saturated or subcooled (H) | Orifices, tubes: $L/D = 2 \dots 55$ | $p_{up} = 1$ $\Delta p = 0.1 \dots 0.7$ | 3,000 . . . 9,000 | – |
| Zaloudek (1963) | Saturated or subcooled water (H) | $L/D = 0.06 \dots 6$ $D = 6.4 \dots 15.8$ mm | $p_{up} = 7 \dots 28$ $\Delta p = 0.07 \dots 7$ | 3,000 . . . 30,000 | – |

See footnotes on page 418.

**Table 1 Experimental studies (*Continued*)**

| Reference | Fluid (direction of flow[a]) | Test sections | Range of pressures[b] (bar) | Range of mass velocities (kg/m² s¹) | Range of qualities |
|---|---|---|---|---|---|
| Zaloudek (1964) | Saturated or subcooled water (H) | $L/D = 20$<br>$D = 12.8$ mm | $p_{up} = 140$ | 10,000 . . . 90,000 | – |
| Edmonds and Smith (1965) | Saturated refrigerant 11 (H) | Long and short converging nozzles, short tube:<br>$L/D = 21$<br>$D = 8.4$ mm | $p_{up} = 3.7$<br>$\Delta p = 0 \ldots 2.1$ | 3,200 . . . 7,000 | – |
| Fauske (1965) | Air–water (VD) | Rectangular section:<br>3.1 × 25.4 mm<br>$L/D = 11.5, 15.5$ and 30<br>$D = 7.7$ and 11.6 mm | $p_c = 1.17$ | 3,100 . . . 22,000 | 0.006 . . . 0.08 |
| Muir and Eichhorn (1966) | Air–water (VU) | Converging (60°) to diverging (8°) nozzle, rectangular cross section | $p_{up} = 1.4 \ldots 6.3$ | 3,800 . . . 31,000 | $1.5 \times 10^{-4} \ldots 5 \times 10^{-3}$ |
| Chen and Isbin (1966) | Saturated water (H) | Tubes and diverging nozzles:<br>$L/D = 2 \ldots 4.6$<br>$D_{inlet} = 6.35, 12.7,$ and 25.4 mm | $p_{up} = 2.7 \ldots 14.8$ | 8,000 . . . 20,000 | – |
| Fauske et al. (1967) | Saturated sodium | $L/D = 60$ | $p_c = 0.069 \ldots 0.48$ | 250 . . . 5,000 | 0.005 . . . 0.065 |
| Hammitt et al. (1967) | Water<br>Mercury | Venturi | $p_{up} = 1$ | $w_c = 19.7$ m/s<br>$w_c = 10.1$ m/s | $10^{-9} \ldots 10^{-3}$ |
| Smith et al. (1967, 1968)<br>Smith (1972) | Air–water (VU) | Annular movable venturi | $p_c = 1 \ldots 2$ | 300 . . . 10,000 | 0.01 . . . 1 |

| | | | | | |
|---|---|---|---|---|---|
| Henry (1968) | Subcooled water (H) | Circular and rectangular tube ended by a divergent (7° or 120°): $L/D > 100$ | $p_c = 3.75 \ldots 10.3$ | 2,500 . . . 32,000 | 0.002 . . . 0.216 |
| Ogasawara (1969a) | Steam–water mixtures (H) | $L/D = 2 \ldots 190$<br>$D = 10 \ldots 50$ mm | $p_c = 4 \ldots 48$ | 5,000 . . . 30,000 | 0.02 . . . 0.14 |
| Ogasawara (1969b) | Saturated water (blowdown) (H) | Orifices: $D = 10 \ldots 70$ mm | $p_{up} = 10 \ldots 70$<br>$\Delta p = 2 \ldots 70$ | 10,000 . . . 55,000 | – |
| Klingebiel and Moulton (1971) | Steam–water mixtures (H) | $L/D = 44$<br>$D = 12.8$ mm | $p_c = 1.9 \ldots 5.3$ | 700 . . . 45,000 | 0.01 . . . 0.99 |
| Wallis and Sullivan (1972) | Air–water (H) | Nozzles with an upper converging–diverging wall | $p_{up} = 1.1 \ldots 3.85$ | – | 0.02 . . . 0.35 |
| Costa and Charlety (1973) | Boiling sodium (VU) | Tube ($D = 4$ mm) downstream of a heating rod | $p_c = 1.1 \ldots 1.3$ | 1,200 . . . 3,200 | – |
| Flinta et al. (1973) | Saturated water (blowdown) (H) | Orifices with straight or round edges, short tubes:<br>$L/D = 1 \ldots 100$<br>$D = 35$ mm | $p_{up} = 5 \ldots 15$ | 4,000 . . . 20,000 | – |
| Katto and Sudo (1973) | Air–water (VD) | $L/D = 1 \ldots 5$<br>$D = 7.6 \ldots 10$ mm | $p_c = 1.1 \ldots 4.5$ | 7,400 . . . 17,600 | 0.024 . . . 0.028 |
| Muncaster and Thomson (1973) | Saturated or subcooled water | Nozzles, various lengths and inlet shapes:<br>$L/D = 3 \ldots 5$<br>$D = 25$ mm | $p_c = 1 \ldots 2$ | 18,000 . . . 24,000 | – |
| Réocreux (1974, 1976) | Subcooled water (VU) | Tubes ($D = 20$ mm) ended by a divergent (7°) | $p_c = 1.5 \ldots 2$ | 4,000 . . . 11,000 | < 0.01 |

See footnotes on page 418.

**Table 1 Experimental studies (*Continued*)**

| Reference | Fluid (direction of flow[a]) | Test sections | Range of pressures[b] (bar) | Range of mass velocities (kg/m² s¹) | Range of qualities |
|---|---|---|---|---|---|
| Réocreux and Seynhaeve (1974) | Subcooled water (VU) | Tubes ($D = 20$ mm) ended by a divergent (7°) | $p_c = 1.5 \ldots 3.4$ | 3,700 . . . 9,000 | 0.002 . . . 0.014 |
| Flinta (1975) | Subcooled water + dissolved air (blowdown) (H) | Orifices: $D = 35.48$ and 60 mm | $p_{up} = 6.5 \ldots 12$ | 7,000 . . . 27,000 | – |
| Burgess (1976) | Subcooled water | Vortex diode | $p_{up} = 1.13$ | – | – |
| Duggins (1976) | Subcooled water | $L/D = 10 \ldots 160$<br>$D = 12.7$ and 15.85 mm | $p_{up} = 3 \ldots 9$ | 15,000 . . . 25,000 | – |
| Seynhaeve et al. (1976) | Subcooled water (VU) | $L/D = 17.7$ and 24.5<br>$D = 12.5$ mm | $p_c = 1.5 \ldots 7$ | 10,000 . . . 20,000 | 0.002 . . . 0.015 |
| Mauget (1977) | Subcooled water (blowdown) (H) | Orifices: $D = 6.1$, 6.5 and 7.04 mm | $p_{up} = 60 \ldots 200$ (initial) | . . . 50,000 . . . | – |
| Seynhaeve (1977) | Subcooled water (VU) | Orifices: $D = 12.75$ and 16.25 mm | $p_{up} = 6.5$<br>$\Delta p = 1.1 \ldots 3.3$ | 1,600 . . . 3,300 | 0.01 . . . 0.05 |

[a] VU = vertical upward; VD = vertical downward; H = horizontal.

[b] $p_c$ = critical pressure; $p_{up}$ = upstream pressure; $p_{down}$ = downstream pressure; $\Delta p = p_{up} - p_{down}$.

The steady-state operating loops are closed or open according to whether the fluid(s) is (are) recycled after the expansion.

## 2.2 Fluids

The two-phase critical-flow experiments deal with systems with or without phase change. For systems without phase change, air–water flows were mainly studied. Let us point out the work performed by Smith, Cousins, and Hewitt (1967, 1968), who observed annular flow patterns and were able to adjust and measure the liquid film thickness of these flows.

In flows with phase change, either of two characteristics is distinguished:

1. The fluid is a subcooled liquid at the inlet of the test section
2. Both phases are mixed prior to expansion

It can be expected that experiments of the first kind will give much smaller critical qualities ($10^{-4} \ldots 10^{-2}$ compared with $\ldots 10^{-2} \ldots 1$) and the critical flow rates will be larger than those of the second kind. Furthermore, the nucleation mechanisms play a major role. The experimenter must carefully note all phenomena that could influence the incipience of the second phase during expansion, for example,

1. Wall roughness, such as at the junctions between mechanical parts and at the pressure taps of the test section (Réocreux, 1974)
2. Presence of dissolved gas (Henry, 1968; Flinta, 1975), and micro-bubbles
3. Presence of a suspension of solid particles (Réocreux and Seynhaeve, 1974)

To illustrate the importance of these phenomena, Fig. 3 shows the critical flow rates obtained by experiments performed at the Centre d'Etudes Nucléaires de Grenoble (C.E.N.G.) on a stainless steel loop, and at the Université Catholique de Louvain (U.C.L.) on a mild steel loop, using the same test section; critical flow rates are related to fluid inlet temperature. The results obtained at the U.C.L. are similar to those obtained at the C.E.N.G. when cavitation is increased at the inlet of the test section by placing a grid at this location.

## 2.3 Test Sections

Four common kinds of geometries are distinguished: orifices, nozzles, short tubes, and long tubes. The name orifice is supposed to indicate tubes with no length, and, in practice, tubes where the length-to-diameter ($L/D$) ratio is less than 0.25. Short tubes have a cylindrical or annular shape, with $0.25 < L/D < \ldots 40 \ldots$; long tubes have the same geometric shapes, but with $L/D > \ldots 40 \ldots$. Finally, nozzles are characterized by their variable section, converging, diverging, or converging-diverging. Very large pressure and velocity gradients, the features of critical flows, are apparent at the outlets of the tubes and at the throats of the nozzles.

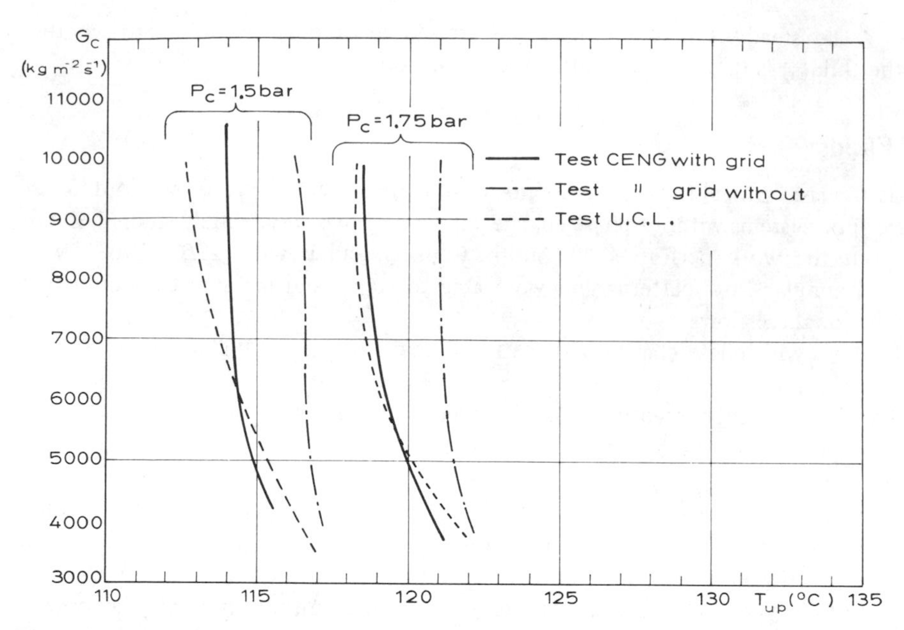


**Figure 3** Comparative tests (Réocreux and Seynhaeve, 1974).

No study involving a heating test section has been found in the literature.

One must point out the small values of the diameters of the test sections for steady-state operating loops: this limitation can be explained by the limited range of flow rates and power values that can be tested.

**2.3.1 Length-to-diameter ($L/D$) ratio** The experimenters are unanimous in establishing that the critical flow rate increases when the $L/D$ ratio decreases. This increase is illustrated in Fig. 4, which is taken from a figure published by Fauske and Min (1963) in a report concerning saturated refrigerant-11 flows. Similar results for water were presented by Diggins (1976).

Bailey (1951) interpreted his own experiments with short tubes by suggesting that, if the pressure difference ($\Delta p \triangleq p_{up} - p_{down}$) increases (Fig. 5), the liquid flow first fills the whole tube; as the downstream pressure decreases, however, the liquid jet contracts, and the pressure at the vena contracta reaches saturation value $p_{sat}$. Then vapor starts to flow around the contracted jet, which remains saturated. The relationship between the mass velocity $G$ and $\Delta p$ is similar to that of incompressible flows, that is,

$$G = C_1 C_2 \sqrt{2\rho\, \Delta p} \tag{14}$$

where $C_1$ is the classic contraction coefficient ($C_1 = 0.61, \ldots, 0.64$) and $C_2$ is a second contraction coefficient, taking into account the annular vaporization, if present. Beyond a sufficient value of the pressure difference, the flow rate is blocked at a critical value; the higher this value, the lower the $L/D$ ratio.

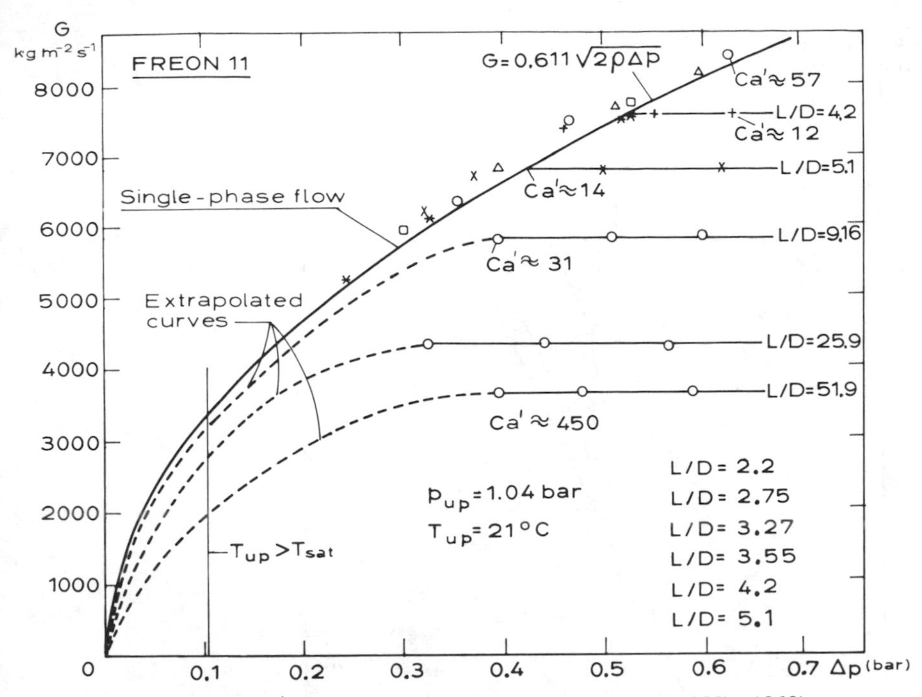


**Figure 4** Influence of the $L/D$ ratio on the critical flow rate (Fauske and Min, 1963).

Zaloudek (1963) published some results of similar investigations except that he used a transparent nozzle. His $G$-$\Delta p$ diagrams (Fig. 6) clearly exhibit two different and successive shock phenomena. The first shock phenomenon appears at the inlet of the tubes, near the vena contracta, when pressure reaches its saturation values at that point; it appears at a stage of constant $G$ of the characteristic curves of these tubes, and seems not to depend on the $L/D$ ratio. The second shock phenomenon occurs at the outlet of the short tubes for much smaller values of the back pressure and appears at a second stage of the characteristic curves. Visualizations have shown that, for the first shock phenomenon cavitation appears in a collar between the vena contracta and the wall, without any vapor escaping downstream. With increasing $\Delta p$, the vapor collar is no longer stagnant and surrounds the metastable liquid jet. When the critical flow is reached, vapor bubbles appear in the jet near to the outlet section. Subsequent tests with high pressure performed by Zaloudek (1964) confirm

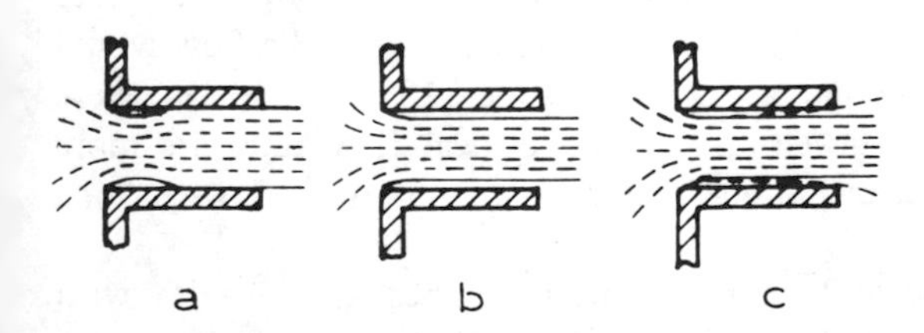


**Figure 5** Flow pattern in a short tube, according to Bailey (1951). (a) The liquid flow fills the whole tube. (b) Saturation pressure at the vena contracta. (c) Vapor flowing around the contracted jet.

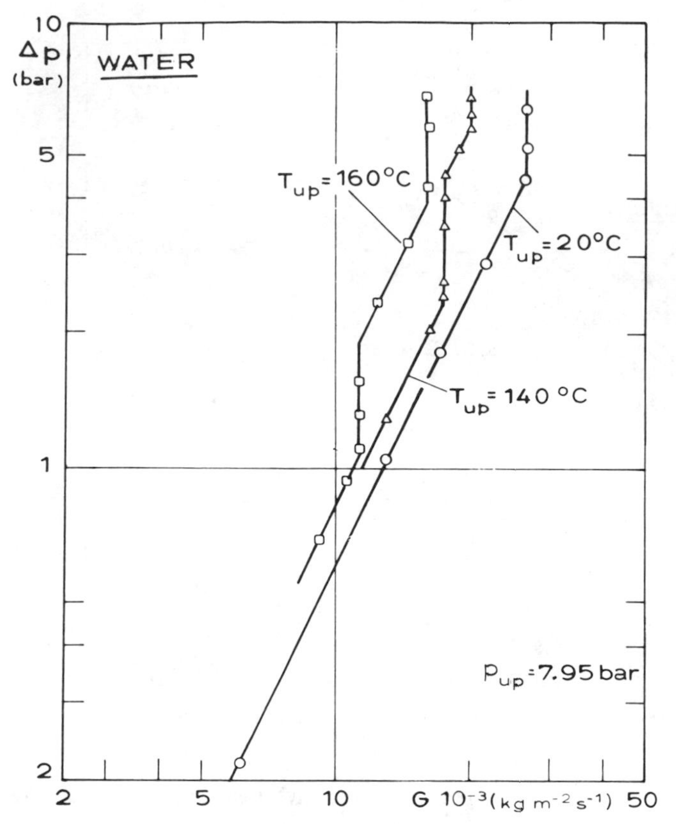


**Figure 6** Characteristic curves of a short tube, $D = 12.7$ mm, $L = 38.1$ mm, according to Zaloudek (1963).

the existence of a shock at the inlet when the upstream pressure is higher than the saturation pressure.

Fauske and Min (1963) suggested using a nondimensional quantity, called the modified cavitation number, to calculate the boundaries between single-phase metastable flow and two-phase flow:

$$\mathrm{Ca}' \triangleq \frac{2\,\Delta p}{\rho_{\mathrm{up}} w_{\mathrm{up}}^2} \frac{L}{D}$$

where $\rho_{\mathrm{up}}$ and $w_{\mathrm{up}}$ are, respectively, the density and the velocity of the liquid at the inlet of the test section. For values of $\mathrm{Ca}' < 10$, which correspond to the case of orifices and short tubes, no critical flow would be obtained, but the flow would be single-phase and metastable. Beyond a transition zone ($10 < \mathrm{Ca}' < 14$), where two-phase phenomena progressively develop, recorded critical flow rates are more accurately predicted by the models proposed for thermal equilibrium flows in long tubes.

For the calculation of critical flow rates through tubes and nozzles with saturated liquid at the inlet, Flinta et al. (1973) suggested the following correlation:

$$G_c = 1400(p_{sat} - 0.02)^{0.75}$$

where $G_c$ is measured in kilograms per square meter per second and $p_{sat}$, given in bars, is the saturation pressure just upstream of the critical section. This correlation has been improved quite recently (Flinta, 1975) to calculate the critical mass velocity $G_c$ of expansion of subcooled water containing or not containing a dissolved gas.

**2.3.2 Inlet and outlet profiles** The results depend on the inlet and outlet profiles of the tubes and nozzles.

The inlet shape determines the inlet pressure drop, and, consequently, the value of the critical pressure, for a given upstream pressure. Moreover, if the fluid is a subcooled liquid at the inlet of the channel, the inlet pressure drop determines the location of the section where the vaporization starts, and consequently, influences the critical quality. For short tubes with rounded inlet profiles, Flinta et al. (1973) measured critical flow rates 25% in excess of those obtained with sharp inlet profiles.

For a 3-m tube, this increase would still be of the order of 11%. Muncaster and Thomson (1973) obtained similar results. Any disturbance able to induce cavitation at the inlet of the channel affects the value of the critical flow rate significantly. Réocreux (1974) checked this feature by placing grids at the inlet of a test section. Some results are presented in Fig. 7, where critical mass velocity is related to void fraction $\bar{R}_{G1}$ measured at the critical section. The results are compared with those obtained by Henry (1968).

Outlet profiles also play a role, as shown in Fig. 8, which is drawn from Henry's report. Experimental results obtained by three test sections are given: C7 and R7 are tubes with, respectively, circular and rectangular cross sections, ended by a 7° divergent, whereas C120 is a cylindrical tube having the same diameter as C7, but ended by a 120° divergent. For the C120 tube, the outlet pressure of the test section depends on the back pressure. This condition can be explained, according to Henry, by the existence of a radial pressure profile in the outlet plane of the C120 test section. The critical flow rate is compared with the equilibrium quality $x_{eq}$ calculated for the outlet section of the channel.

We note that the critical flow rate is affected not only by the outlet profile of the test section but also by the diameter of the pipe into which the test section emerges. This has been established for orifices by Ogasawara (1969b) and Seynhaeve (1977). Seynhaeve showed that when a subcooled liquid flows through an orifice in a tube, the critical section is in the tube, downstream of the orifice, beyond a section where the metastable liquid jet surrounded by a saturated vapor jet is broken. This is illustrated in Fig. 9. The orifice is placed between pressure taps $p_4$ and $p_3$. The void-fraction profile measurements show that the liquid jet breaks at section 3′.

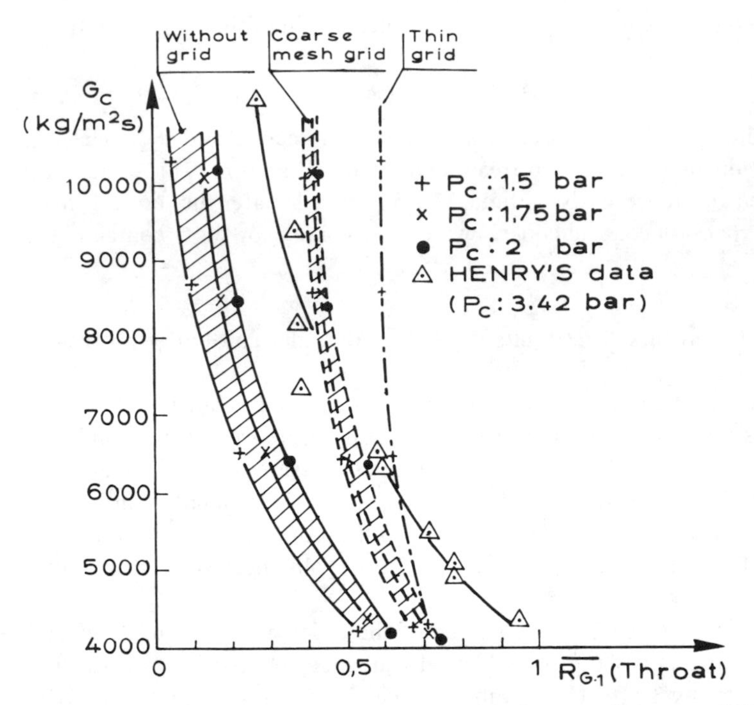


**Figure 7** Influence of inlet geometry on the critical flow rate (Réocreux, 1974).

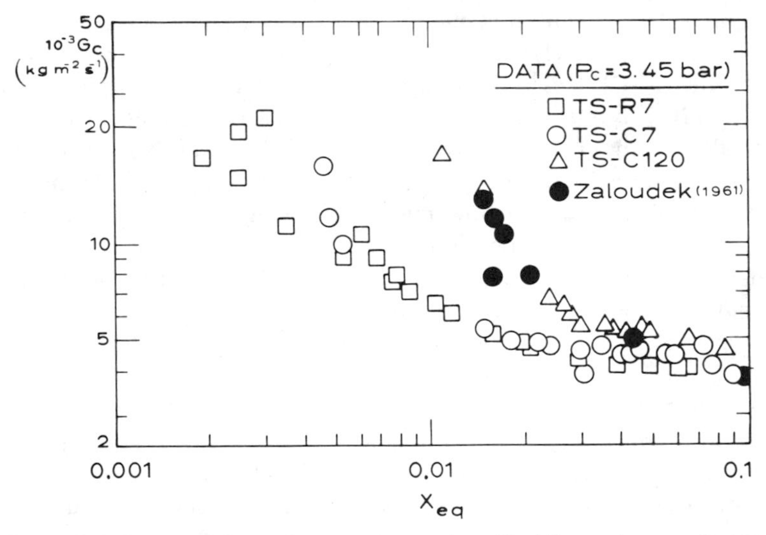


**Figure 8** Influence of the outlet geometry on the critical flow rate, according to Henry (1968).

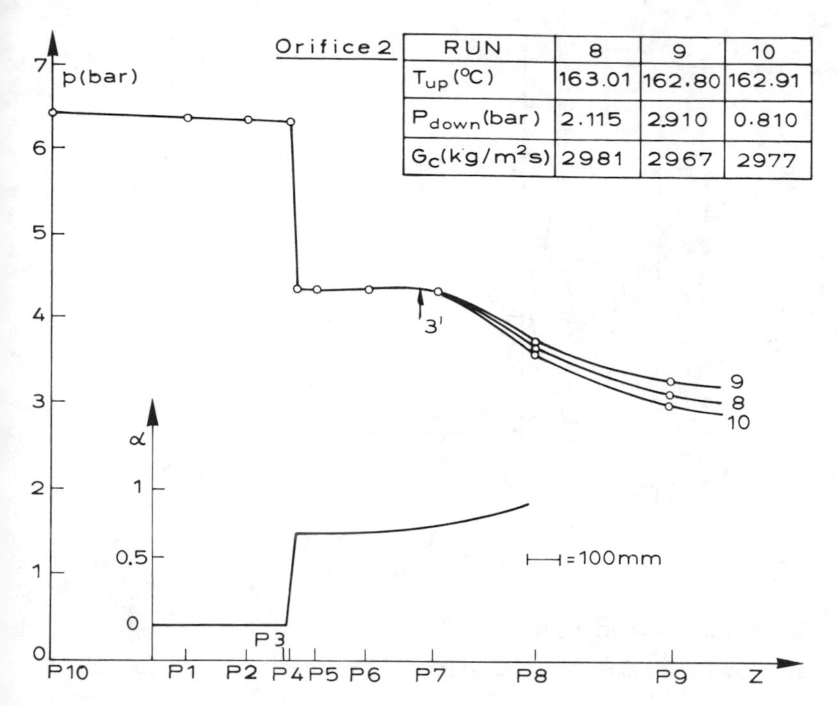


**Figure 9** Critical flows through an orifice (Seynhaeve, 1977).

## 2.4 Measurements

We will now describe the means used to determine the pressure, the void fraction, the quality, the slip ratio, and the thermal nonequilibriums in the works mentioned in Table 1. Only the first two parameters, and sometimes the third, have been the object of direct measurements.

**2.4.1 Pressure** Experimenters have used three methods to measure axial pressure profiles:

1. Wall pressure taps to the shortest possible distance from the critical section in a channel with a fixed geometry
2. Wall pressure taps along the cylindrical wall of a pipe having the shape of an annular venturi, the internal movable wall consisting of two cones with a common base (Fig. 10)
3. Probes movable in the direction of the flow in a channel with fixed geometry

Silver (1948) was the first to establish that a pressure tap disturbs the critical flow: in his experiments he observed that the critical flow rate decreased by 17% as a result of the hole of a pressure tap. In the most favorable cases the critical

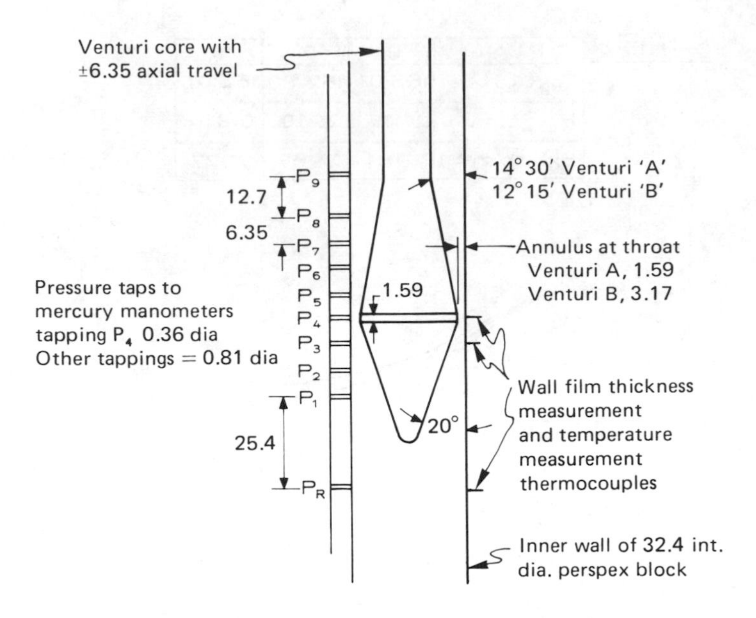


**Figure 10** Annular venturi with a movable nose cone (Smith et al., 1968).

pressure value is deduced from interpolation, otherwise by extrapolation of the measurement points. If the test section consists of a tube emerging into a vessel or a tube with larger diameter, or even, as for some of Henry's tests (1969), if it is ended by a large-angle divergent, two-dimensional effects can make the extrapolation meaningless. On the contrary, if the tube is ended by a 7° divergent, it is possible to measure the pressure profile precisely (Fig. 1) (Réocreux, 1974).

The use of movable probes is reported by Faletti and Moulton (1963), Edmonds and Smith (1965), Klingebiel and Moulton (1971), and Muncaster and Thomson (1973). Edmonds and Smith (1965) pointed out that because of the presence of the probe, the critical mass velocity is changed in tubes but not in nozzles. Klingebiel and Moulton used a movable probe to determine the radial pressure profiles near the outlet of a tube, and the axial profile in the free jet emerging from the tube.

**2.4.2 Void fraction** The tests that enabled void fraction to be determined are reported in Table 2.

**2.4.3 Quality** The quality of two-component (air–water) flows is determined by flow rate measurements. For flow with phase change, the estimation of quality at the critical section is difficult. Let us consider, for example, the adiabatic expansion of a liquid near to but below saturation at the inlet (subscript $i$) of a horizontal pipe of uniform cross section. In steady-state flow, the energy equation of the mixture may be written, in the two-phase part of the flow, as

$$\frac{d}{dz}\left[\alpha\rho_G w_G\left(h_G + \frac{w_G^2}{2}\right) + (1-\alpha)\rho_L w_L\left(h_L + \frac{w_L^2}{2}\right)\right] = 0$$

Integration of this equation and the corresponding single-phase flow equation between the inlet and the critical section (subscript $c$) gives

$$h_L(p_i, T_i) + \tfrac{1}{2}w_{L,i}^2 = x_c h_G(p_c, T_{G,c}) + (1-x_c)h_L(p_c, T_{L,c}) + \tfrac{1}{2}[x_c w_{G,c}^2 + (1-x_c)w_{L,c}^2] \tag{15}$$

where quality is defined by

$$x \triangleq \frac{\alpha\rho_G w_G}{G}$$

The energy equation [Eq. (15)] may also be written as

$$x = \frac{h_L(p_i, T_i) - h_L(p_c, T_{L,c}) + \frac{1}{2}(w_{L,i}^2 - w_{L,c}^2)}{h_G(p_c, T_{G,c}) - h_L(p_c, T_{L,c}) + \frac{1}{2}(w_{G,c}^2 - w_{L,c}^2)} \tag{16}$$

and since the differences between the enthalpies and their saturation values are defined by

$$\Delta h_L \triangleq h_L(p, T_L) - h_L^{\text{sat}}(p)$$

and

$$\Delta h_G \triangleq h_G(p, T_G) - h_G^{\text{sat}}(p)$$

Eq. (16) can be written as

$$x = \frac{h_L(p_i, T_i) - h_L^{\text{sat}}(p_c) - \Delta h_{L,c} + \frac{1}{2}(w_{L,i}^2 - w_{L,c}^2)}{\mathcal{L}(p_c) + \Delta h_{G,c} - \Delta h_{L,c} + \frac{1}{2}(w_{G,c}^2 - w_{L,c}^2)} \tag{17}$$

**Table 2 Void-fraction measurements in critical flows**

| Author | Flow | Method | Type |
|---|---|---|---|
| Fauske (1965) | Air-water, rectangular channel | $\gamma$ attenuation (thulium-170) | "One shot" near the critical section $\bar{R}_{G_2}$ |
| Hammitt et al. (1967) | Steam-water, vapor mercury, venturi | $\gamma$ attenuation (cobalt-60) | Axial profile, and chord by chord $\bar{R}_{G_1}$ |
| Henry (1968) | Steam-water, tube ended by a divergent | $\gamma$ attenuation (thulium-170) | "One shot," axial profiles $\bar{R}_{G_2}$ |
| Ogasawara (1969a) | Steam-water, long tubes | Electric impedance | Electrode fixed on the wall near the critical section $\bar{R}_{G_3}$ |
| Réocreux (1974) | Steam-water, tube ended by a 7° divergent | $X$ attenuation | Axial profiles, and chord by chord $\bar{R}_{G_1}$ |
| Seynhaeve (1977) | Steam-water, orifice | $\gamma$ attenuation (cesium-137) | Diameter of a liquid jet $\bar{R}_{G_1}$ |

In Eq. (17) $h_L(p_i, T_i)$ can be replaced by $h_L^{\text{sat}}[p_{\text{sat}}(T_i)]$. Most often three assumptions are made to evaluate the quality used for the presentation of experimental data:

First assumption: No difference between the temperatures and their saturation value; Eq. (17) is then written as

$$x = \frac{h_L(p_i, T_i) - h_L^{\text{sat}}(p_c) + \frac{1}{2}(w_{L,i}^2 - w_{L,c}^2)}{\mathfrak{L}(p_c) + \frac{1}{2}(w_{G,c}^2 - w_{L,c}^2)} \tag{18}$$

Second assumption: No drift ($w_{G,c} = w_{L,c}$); when the second assumption is added to the first assumption, the expression of quality is simplified to

$$x = \frac{h_L(p_i, T_i) - h_L^{\text{sat}}(p_c) + \frac{1}{2}(w_{L,i}^2 - w_{L,c}^2)}{\mathfrak{L}(p_c)} \tag{19}$$

Third assumption: The term resulting from the kinetic energy per unit mass variation is negligible with respect to the drop of enthalpy per unit mass. When the three assumptions are made together, quality is given by a very simple expression, which is called the equilibrium quality $x_{\text{eq}}$ (or the thermodynamic quality):

$$x_{\text{eq}} \triangleq \frac{h_L(p_i, T_i) - h_L^{\text{sat}}(p_c)}{\mathfrak{L}(p_c)} \tag{20}$$

This version of quality is the one most often used to correlate experimental data.

**2.4.4 Slip ratio** An estimation of the slip ratio of two-component, two-phase (air–water) flows can be obtained from the measurements of quality and void fraction (Fauske, 1965) by using the relationship

$$k = \frac{x}{1-x}\,\frac{1-\alpha}{\alpha}\,\frac{\rho_L}{\rho_G} \tag{21}$$

For single-component flows, the determination of the true quality $x$ is no longer possible, and the slip ratio is calculated by using Eq. (21), where the true quality is replaced by thy equilibrium quality $x_{\text{eq}}$ of Eq. (20), and where $p_i$, $p_c$, and $T_i$ are measured values.

Klingebiel and Moulton (1971) determined the slip ratio at the critical section by measuring the thrust $F$ of a two-phase jet (Fig. 11) emerging from a test section

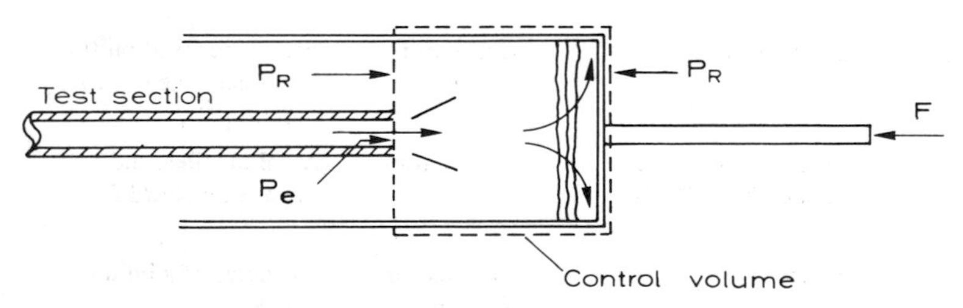


**Figure 11** Measuring device of the thrust of a two-phase jet (Klingebiel and Moulton, 1971).

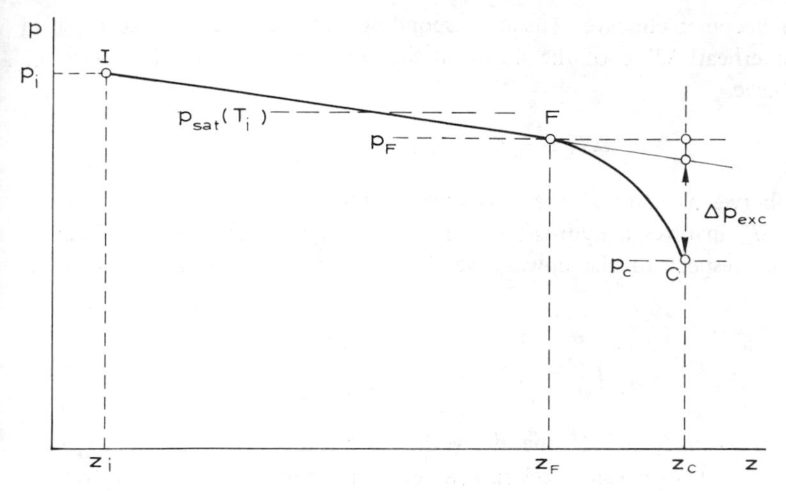


**Figure 12** Pressure profile of critical flows with flashing.

into a momentum cage. The momentum equation in the direction of the flow, applied to the control volume defined by the dashed lines in Fig. 11, can be written as

$$\alpha A_c \rho_G w_G^2 + (1-\alpha)A_c \rho_L w_L^2 + p_c A_c + p_R(A_R - A_c) = F + p_R A_R$$

or, with the use of Eq. (21) to eliminate $\alpha$,

$$A_c(p_R - p_c) = A_c G^2 \left[ \frac{x^2}{\rho_G} + x(1-x)\left(\frac{k}{\rho_L} + \frac{1}{k\rho_G}\right) + \frac{(1-x)^2}{\rho_L} \right] \tag{22}$$

The true quality $x$ and the slip ratio $k$ can be calculated by Eqs. (17) and (22), considered as a set of equations whose unknowns are $x$ and $k$. Klingebiel and Moulton pointed out that for $x_{eq}$ between 0.01-1, the maximum observed slip ratios agree with the values predicted by the correction of Vance and Moulton (1965) for subcritical flows:

$$k = 1.465 \left( \frac{1 - x_{eq}}{x_{eq}} \frac{G}{G_c} \right)^{0.3021} \tag{23}$$

The discrepancies between the values predicted by Eq. (23) and the correlation of the experimental data of Klingebiel and Moulton are small:

$$k_c = 1.339 x_{eq}^{-0.3352} \tag{24}$$

Two-phase jet-thrust measurements have also been made by Chen and Isbin (1966).

Another method for determining slip ratio has been proposed by Réocreux et al. (1973) and developed by Seynhaeve (1976). It is based on the use of the axial pressure profile for critical flows with flashing (Fig. 12). The flashing onset position is well defined on this profile: It is the abscissa $z_F$ of the point $F$ where the pressure profile becomes concave. The corresponding superheat can be expressed in terms of the superheat $\Delta T_{s,F}$ of the liquid at the position of the onset of flashing. If $T_F = T_i$, we have

$$\Delta T_{s,F} \triangleq T_i - T_{\text{sat}}(p_F)$$

At each point between $i$ and $F$, the derivative of the pressure of the single-phase flow (subscript *SP*) involves a hydrostatic term ($\phi$ being the angle of the direction of the flow with respect to the upward vertical) and a friction term (subscript *f, SP*):

$$-\left(\frac{dp}{dz}\right)_{SP} = \rho_L g \cos\phi - \left(\frac{dp}{dz}\right)_{f,SP} \tag{25}$$

At each point between $F$ and $C$, the derivative of the pressure of the two-phase flow (subscript *TP*) involves an acceleration term, a gravity term, and a friction term:

$$-\left(\frac{dp}{dz}\right)_{TP} = \frac{d}{dz}\,[(1-\alpha)\rho_L w_L^2 + \alpha\rho_G w_G^2] + [(1-\alpha)\rho_L + \alpha\rho_G]g\cos\phi - \left(\frac{dp}{dz}\right)_{f,SP} \tag{26}$$

By subtracting, side by side, Eqs. (25) and (26), we obtain

$$-\left[\left(\frac{dp}{dz}\right)_{TP} - \left(\frac{dp}{dz}\right)_{SP}\right] = \frac{d}{dz}\,[(1-\alpha)\rho_L w_L^2 + \alpha\rho_G w_G^2] - \alpha(\rho_L - \rho_G)g\cos\phi - \left(\frac{dp}{dz}\right)_{f,SP}(\Phi_{L,o}^2 - 1) \tag{27}$$

where $\Phi_{L,o}$ is one of the parameters of the Martinelli-Nelson method for the evaluation of friction pressure drops. The integration of Eq. (27) between $F$ and $C$ gives the excess pressure drop ($\Delta p_{\text{exc}}$), defined in Fig. 12:

$$\Delta p_{\text{exc}} = (1-\alpha_c)\rho_L w_{L,c}^2 + \alpha_c\rho_{G,c}w_{G,c}^2 - \rho_L w_{L,F}^2 - g\cos\phi\int_F^C \alpha(\rho_L - \rho_G)\,dz - \left(\frac{dp}{dz}\right)_{f,SP}\int_F^C (\Phi_{L,o}^2 - 1)\,dz \tag{28}$$

One can verify that the gravity term in Eq. (28) is negligible with respect to the two other terms. Further, one can write

$$\Phi_{L,o}^2 = (1-x)^{1.75}\,\Phi_L^2$$

and for flows with small qualities

$$\Phi_{L,o}^2 \cong \Phi_L^2 \cong (1-\alpha)^{-1.75}$$

With these approximations, Eq. (28) becomes

$$\Delta p_{\text{exc}} \cong (1-\alpha_c)\rho_L w_{L,c}^2 + \alpha_c \rho_{G,c} w_{G,c}^2 - \rho_L w_{L,F}^2 - \left(\frac{dp}{dz}\right)_{f,SP} \int_F^C [(1-\alpha)^{-1.75} - 1]\, dz \tag{29}$$

The second term of the rhs can be calculated if the void-fraction profile is known and is small compared with $\Delta p_{\text{exc}}$. Thus, the first term is the leading one; it can be written as

$$(1-\alpha_c)\rho_L w_{L,c}^2 + \alpha_c \rho_{G,c} w_{G,c}^2 - \rho_L w_{L,F}^2 = G^2 \left[\frac{(1-x_c)^2}{(1-\alpha_c)\rho_L} + \frac{x_c^2}{\alpha_c \rho_{G,c}} - \frac{1}{\rho_L}\right]$$

or approximately

$$G^2 \left[\frac{1}{(1-\alpha_c)\rho_L} + \frac{x_c^2}{\alpha_c \rho_{G,c}}\right]$$

One can deduce the value of $x_c$, or, by using Eq. (21), the critical value $k_c$ of the slip ratio. To apply this method, it is necessary to have the pressure profile and the value of the void fraction at the throat.

**2.4.5 Temperatures** The direct measurement of the temperature nonequilibriums of critical flows is difficult for the two following reasons:

1. To measure temperature fluctuations within each phase with a thermocouple its response time would need to be small compared with transit times; for high-speed flows it has not yet been possible to satisfy this condition.
2. The thermocouple holders inside the flow could disturb this flow by generating cavitation.

For the temperature differences between the phases to be evaluated, pressure measurements must suffice along with assumptions concerning temperature evolutions (isothermal liquid phase, saturated vapor phase, etc).

# 3 PREDICTION MODELS

## 3.1 Two-Component, Two-Phase Flows

**3.1.1 Literature survey** The air–water critical-flow experiments carried out by Fauske (1965) give useful data for a large range of qualities (Fig. 13). The values of the flow rates are significantly higher than the values predicted by the homogeneous model, those from the homogeneous model resulting from the vanishing condition of the determinant det $(b_{ij})$ of a set of three equations, namely the two-phase mass, momentum, and energy balances. This vanishing condition gives

$$G^2 = \rho^2 \left(\frac{\partial p}{\partial \rho}\right)_s \tag{30}$$

where
$$\frac{1}{\rho} \triangleq \frac{x}{\rho_G} + \frac{1-x}{\rho_L}$$

Fauske suggested using the same expression as Eq. (30), but with

$$\rho \triangleq \alpha\rho_G + (1-\alpha)\rho_L$$

where $\alpha$ is calculated by means of the slip ratio $k$ and the quality $x$ of the mixture, using Eq. (21) and the empirical relationship

$$k = 0.17x^{0.18}\left(\frac{\rho_L}{\rho_G}\right)^{1/2} \tag{31}$$

Further, he assumed $x = \text{const}$, $\rho_L = \text{const}$, and $\rho_G = p/RT$.

Smith et al. (1967, 1968) and Smith (1972) presented experimental data in surprisingly good agreement with Fauske's data since neither the test section geometry (venturi instead of tubes) nor the flow patterns (annular with adjustable liquid film thickness instead of bubbly or dispersed flow patterns) are similar in the two studies. The authors claimed that the flow of the mixture is critical when the gas flow is critical. They found that the pressure profile is approximated by the pressure profile of the isentropic gas flow when the liquid flow rate is small and that it is approximated by the pressure profile of the isothermal gas flow when the liquid flow rate increases: Thus the liquid stabilizes the temperature of the gas phase. Unfortunately, this model implies the use of an adjustable parameter, that is, the effective gas cross-sectional area, which takes into account the interfacial waves and two-dimensional effects.

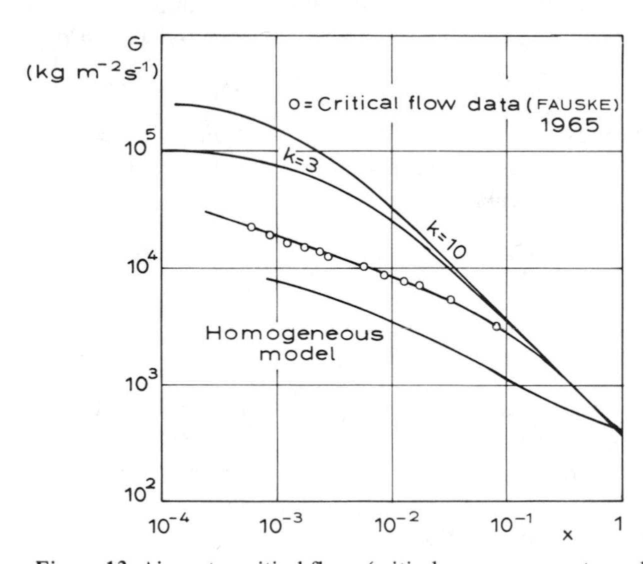


**Figure 13** Air–water critical flows (critical pressure near atmospheric pressure).

Smith (1973) used the same ideas in a theoretical study on helium critical flow.

Chisholm (1968) recommended an expression of the critical flow rate based on the assumptions of an incompressible liquid phase and polytropic expansion of the gas phase. He found good agreement with the experimental data of Fauske and other authors by choosing the value $n = 1.2$ for the exponent of the polytropic expansion. The proposed expression is

$$\frac{1}{G^2} = \frac{x}{\rho_G n p_c}\left(x + \frac{1-x}{k^2}\right) \tag{32}$$

where the slip ratio $k$ is given by an improved version of Eq. (31):

$$k = 1 + (4.46x^{0.18} - 1)\frac{\rho_L/\rho_G}{(\rho_L/\rho_G)_p} \tag{33}$$

where subscript $p = 1$ bar.

Finally, Katto and Sudo (1973) assumed that the phases are separated, that the liquid flow is isothermal and incompressible, and that the gas phase is submitted to an isentropic expansion. They obtained the following expression for the critical condition:

$$\mathrm{Ma}_G^2 = 1 + K\,\mathrm{Ma}_G^3 \tag{34}$$

where $\mathrm{Ma}_G$ is the Mach number of the gas phase, and

$$K \triangleq \frac{1-x}{x}\left(\frac{\rho_G}{\rho_L}\right)^2\left(\frac{a_{s,G}}{w_L}\right)^3$$

where $a_{s,G}$ is the isentropic sound velocity of the gas phase. In Eq. (34), one has

$$1 \leqslant \mathrm{Ma}_G \leqslant \sqrt{3} \quad \text{and} \quad 0 \leqslant K \leqslant \frac{2}{\sqrt{27}}$$

However, the mechanism of the critical phenomenon does not appear clearly in Katto and Sudo's publication. In addition, these authors claimed that pressure disturbances generated downstream of the critical section can be transmitted upstream through the liquid flow.

**3.1.2 Set of equations** The set of appendix equations (A.1) to (A.7) is equivalent to the following set ($k = L$ or $G$):

Mass:

$$\frac{d}{dz}(\alpha_k \rho_k w_k) = -\alpha_k \rho_k w_k \frac{1}{A}\frac{dA}{dz} \tag{35}$$

Momentum: Subtracting Eq. (A.4) from Eq. (35) multiplied by $w_k$ and dividing the resulting equation by $\alpha_k$ (the cases where $\alpha_k = 0$ are excluded), give

$$\rho_k w_k \frac{dw_k}{dz} + \frac{dp}{dz} = -\frac{1}{\alpha_k}(\tau_{kI} + \tau_{kC}) - \frac{1}{\alpha_k A}\phi_k - \rho_k g \cos\phi \tag{36}$$

Energy: Subtracting Eq. (A.7) from Eq. (35) multiplied by $h_k + w_k^2/2$, and dividing the resulting equation by $\alpha_k \rho_k w_k$ (the cases where $w_k = 0$ are excluded) give

$$\frac{d}{dz}\left(h_k + \frac{w_k^2}{2}\right) = \frac{1}{\alpha_k \rho_k w_k}(q_{kI} + q_{kC} + \chi_{kI}) + \frac{1}{\alpha_k A \rho_k w_k}\psi_k - g\cos\phi \qquad (37)$$

The state variables on the lhs of Eqs. (35)-(37) are $\rho_k$, $p$, and $h_k$. One can eliminate $\rho_k$ and $h_k$ and keep $p$ and $T_k$ as variables, since

$$d\rho_k = \left(\frac{\partial \rho_k}{\partial p}\right)_{T_k} dp + \left(\frac{\partial \rho_k}{\partial T_k}\right) dT_k \triangleq \rho'_{k,p}\, dp + \rho'_{k,T_k}\, dT_k \qquad (38)$$

$$dh_k = \left(\frac{\partial h_k}{\partial p}\right)_{T_k} dp + \left(\frac{\partial h_k}{\partial T_k}\right) dT_k \triangleq h'_{k,p}\, dp + h'_{k,T_k}\, dT_k \qquad (39)$$

**3.1.3 Critical-flow condition** The necessary condition of critical flow is the vanishing condition of the determinant of the set of six equations [Eqs. (35)-(37)]. It gives

$$\sum_{k=L,G} \frac{1}{\alpha_k}\rho_k^2 w_k^2 h'_{k,T_k}[\rho_k w_k^2(\rho'_{k,p}h'_{k,T_k} - h'_{k,p}\rho'_{k,T_k}) + (w_k^2\rho'_{k,T_k} - \rho_k h'_{k,T_k})] = 0 \qquad (40)$$

Giot and Fritte (1972) showed that this expression can be written as

$$G^2 = \frac{\{[(1-\alpha)k^2/\rho_L] + \alpha/\rho_G\}\{[(1-\alpha)\rho_L/k] + \alpha\rho_G\}^2}{[(1-\alpha)/\rho_L]\,[\rho'_{L,p} - \rho'_{L,T_L}(s'_{L,p}/s'_{L,T_L})] + (\alpha/\rho_G)[\rho'_{G,p} - \rho'_{G,T_G}(s'_{G,p}/s'_{G,T_G})]} \qquad (41)$$

where $s'_{k,p}$ and $s'_{k,T_k}$ are the partial derivatives of the entropy per unit mass, with respect to the pressure and temperature of phase $k$.

In Fig. 13, one can see the curves corresponding to Eq. (41), where $T_G = T_L$, $k = 3$, and $k = 10$. It is clear that the six-equation model leads to values of the critical flow rate that are much higher than those in reality.

An approximate expression of the critical-flow rate can be found by introducing the assumptions of incompressible liquid and ideal gas, either in Eq. (40) or in Eq. (41), that is,

$$\rho'_{L,p} = 0 \quad \rho'_{L,T_L} = 0 \quad \rho'_{G,p} = \frac{1}{RT_G} \quad \rho'_{G,T_G} = -\frac{\rho_G}{T_G}$$

$$h'_{G,p} = 0 \quad h'_{G,T_G} = c_{pG} \quad s'_{G,p} = -\frac{R}{p} \quad s'_{G,T_G} = \frac{c_{pG}}{T_G}$$

One finds

$$G^2 = \gamma R T_G \left(1 + \frac{1-\alpha}{\alpha}\frac{\rho_G}{\rho_L}k^2\right)\left[\frac{(1-\alpha)\rho_L}{k} + \alpha\rho_G\right]^2 \qquad (42)$$

where $\gamma = c_{pG}/c_{vG}$, or

$$\mathrm{Ma}_G^2 = 1 + \frac{1-\alpha}{\alpha}\frac{\rho_G}{\rho_L}\frac{w_G^2}{w_L^2} \tag{43}$$

To calculate the rhs of Eq. (43), let us eliminate the void fraction $\alpha$ and introduce the quality $x$; then we have

$$\mathrm{Ma}_G^2 = 1 + \frac{1-x}{x}\left(\frac{\rho_G}{\rho_L}\right)^2\left(\frac{w_G}{w_L}\right)^3 \tag{44}$$

This expression of the critical condition is identical to the condition in Eq. (34) established by Katto and Sudo (1973). If we calculate the slip ratio $k$ by means of Fauske's correlation, we find

$$\mathrm{Ma}_G \simeq 1.002 \qquad \text{for } x = 0.001$$

$$\mathrm{Ma}_G \simeq 1.04 \qquad \text{for } x = 0.1$$

The critical condition of the six-equation model is thus approximately

$$\mathrm{Ma} \simeq 1$$

which corresponds to values of the critical flow rate much too high for small qualities.

**3.1.4 Particular models** ***Uniform cooling*** If the thermal equilibrium of both phases is reached at every section of the flow, the following condition must be satisfied everywhere, including at the critical section:

$$\frac{dT_L}{dz} = \frac{dT_G}{dz}$$

This peculiarity of the axial temperature profiles must come from the solution of the set of Eqs. (35)-(37), and this implies some particular expression of the interfacial energy fluxes. The two phasic energy equations can then be replaced by the condition $dT_L/dz = dT_G/dz$, and a two-phase energy equation obtained by summing the liquid and gas expressions of Eq. (37).

The vanishing condition of the determinant of the new set of equations gives values of the critical flow rates near those we obtained from the vanishing condition of the initial set, that is, Eq. (44).

The flows with thermal equilibrium ($T_L = T_G$) are a particular case of the uniform-cooling model.

***Isothermal flow*** If the energy equations (37) are equivalent to the relationships

$$\frac{dT_G}{dz} = 0 \qquad \text{and} \qquad \frac{dT_L}{dz} = 0$$

which implies some particular expressions for the rhs of these equations (especially the presence of derivatives with suitable coefficients), then the initial set of equations is equivalent to the subsystem composed of the four equations (35) and (36). The necessary condition for critical flow is then

$$\begin{vmatrix} \rho_G w_G & \alpha\rho'_G w_G & \alpha\rho_G & 0 \\ -\rho_L w_L & (1-\alpha)\rho'_L w_L & 0 & (1-\alpha)\rho_L \\ 0 & 1 & \rho_G w_G & 0 \\ 0 & 1 & 0 & \rho_L w_L \end{vmatrix} = 0$$

If we develop the determinant, we obtain

$$(1-\alpha)\rho_G w_G^2 \left[ w_L^2 \left( \frac{\partial \rho_L}{\partial p} \right)_T - 1 \right] + \alpha\rho_L w_L^2 \left[ w_G^2 \left( \frac{\partial \rho_G}{\partial p} \right)_T - 1 \right] = 0 \quad (45)$$

With the approximations of incompressible liquid and ideal gas, this gives

$$\mathrm{Ma}_G = \frac{1}{\sqrt{\gamma}} \quad (46)$$

and corresponds to the limiting state of the gas isothermal flow. This condition is not in better agreement with the data than Eq. (44) because the factor $1/\sqrt{\gamma}$ does not decrease the theoretical values significantly.

***Uniform cooling and acceleration*** If the following conditions at the critical section are assumed,

$$\frac{dT_G}{dz} = \frac{dT_L}{dz}$$

$$\frac{dw_G}{dz} = \frac{dw_L}{dz}$$

the set of six equations is equivalent to the set comprising two phasic mass balance equations, a mixture momentum equation, and a mixture energy equation; this set of four equations is complemented by the two conditions just given. The necessary condition of critical flow is written as

$$\begin{vmatrix} \rho_G w_G & \alpha\rho'_{G,p} w_G & \alpha\rho_G & \alpha\rho'_{G,T} w_G \\ -\rho_L w_L & 0 & (1-\alpha)\rho_L & 0 \\ 0 & 1 & G & 0 \\ 0 & 0 & x w_G + (1-x) w_L & x c_{pG} + (1-x) c_{pL} \end{vmatrix} = 0$$

where the assumption of liquid incompressible flow has been introduced. The development of the determinant gives, for the ideal gas,

$$\frac{xG^2}{\rho_G RT} \left[ 1 - \frac{\rho_G}{\rho_L} R \frac{1-x}{1-\alpha} \frac{kx + 1 - x}{x c_{pG} + (1-x) c_{pL}} \right] = \rho_G [\alpha + k(1-\alpha)] \quad (47)$$

For low pressures the second term between the brackets in the lhs is small compared with unity, and the critical condition is approximately

$$G^2 \cong \frac{\rho_G^2 RT}{x} [\alpha + k(1-\alpha)]$$

or

$$\mathrm{Ma}_G^2 \cong \frac{x}{\gamma\alpha}\left(1 + k\,\frac{1-\alpha}{\alpha}\right) \tag{48}$$

or finally,

$$\mathrm{Ma}_G^2 \cong \frac{1}{\gamma}\,x\left(1 + k\,\frac{1-x}{x}\,\frac{\rho_G}{\rho_L}\right)\left(1 + k^2\,\frac{1-x}{x}\,\frac{\rho_G}{\rho_L}\right) \tag{49}$$

Comparing Eq. (49) with Eq. (46), one observes a factor near $\sqrt{x}$ at low pressure. Figure 14 illustrates this result. The critical flow rate is related to quality for three values of the slip ratio: $k = 5$, 10, and 15. The experimental points are data given in Fauske (1965).

A slip ratio value of one ($k = 1$, no drift between the phases) would be a particular case of the model for uniform acceleration, that is the homogeneous model. Anyhow, the exact value of the slip ratio at the critical section can be deduced only

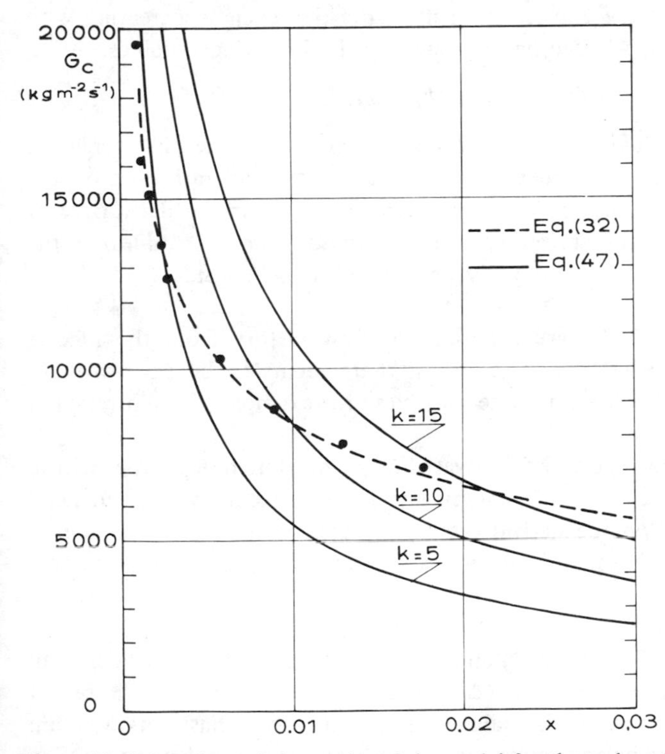


**Figure 14** Comparison of the results of the model for thermal equilibrium and uniform acceleration with the experimental data given by Fauske (1965).

from the complete integration of the set of equations from the mixing section down to the critical section. It is unlikely that a universal law would give the value of the critical-flow rate whatever the length and diameter of the channel, the kind of mixing, the pressure, etc.

## 3.2 Liquid–Vapor Flows at High Qualities ($> 5 \times 10^{-2}$)

By high qualities we mean flows for which the quality at the critical section is higher than about $5 \times 10^{-2}$. In two-phase flow experiments, this kind of flow is usually produced by mixing both phases upstream of a test section exhibiting a large $L/D$ value (see Table 1). Numerous models in the literature (Giot and Fritte, 1972; Fritte, 1974; Réocreux, 1974) predict the critical flow rate. An easy way to present these models briefly consists in classifying them according to the number of equations.

**3.2.1 Six-equation models** The critical condition of the six-equation model is the vanishing condition of the determinant of Eqs. (A.1), (A.4), and (A.7) written for steady flow, assuming that the terms of the rhs members do not involve derivatives of the variables $dx_j/dz$. Equation (41) gives the critical condition. As for two-component, two-phase flows, the mass velocities calculated by this model have higher values than those of the experimental data, the discrepancy increasing with decreasing quality. Equation (41) can be expressed by the following variables:

$$G^2 = G^2(p, x, k, \Delta h_G, \Delta h_L)$$

where $\Delta h_G$ and $\Delta h_L$ are defined in Sec. 2.4.3. Calculations show that for given pressure and quality the flow rate depends very little on the slip ratio and on the thermal nonequilibriums. At the critical section, restrictions on the derivatives of some variables have to be found, rather than restrictions on the values of the variables themselves. This point can be easily understood if one considers

1. Local values of the variables are functions of flow history from the mixing section to the critical section and, a priori, there seems to be no reason to obtain some prescribed value for these variables (for example, the slip ratio) at the critical section.
2. Experimental data show that the derivatives are maximum near the critical section; the mass, momentum, and energy interfacial transfer laws may depend under these conditions on the derivatives $dx_j/dz$, the result of this dependence being that particular evolutions of the flow are obtained in this region of the test sections.

**3.2.2 Five-equation models** Katto (1968) and Sudo and Katto (1974) considered an equilibrium adiabatic liquid–vapor flow ($\Delta h_L = \Delta h_G = 0$) described by a set of equations involving a mixture mass balance equation, two phasic momentum equations, a mixture energy balance equation, and a vapor energy balance equation.

According to these authors the critical condition corresponds to the limiting condition of the variation of state that satisfies the fundamental equations mentioned above.

The limit is determined by zeroing the cause of the variation of state, that is, considering that all the terms of the rhs member of the equation are zero. Then the necessary conditions for the coexistence of the five homogeneous equations with four unknowns ($p$, $x$, $w_L$, $w_G$) are the vanishing conditions of two determinants of the fourth order. These conditions can be solved to give $w_L$ and $w_G$ as functions of $p$ and $x$ at the critical section. Comparing the results with various data for $0.02 \leqslant x \leqslant 1$, these authors find a rather good agreement.

**3.2.3 Four-equation models** Several models are based on the use of four equations; all of them assume thermal equilibrium.

***Fauske's model*** Fauske's model uses three mixture balance equations, namely, mass, momentum, and entropy (Fauske, 1963b, 1964, 1965). The entropy equation is a result of a combination of the energy equation with the two other mixture balance equations. In the energy equation, the enthalpies are eliminated because of the relationship

$$\frac{ds_k}{dz} = \frac{1}{T}\frac{dh_k}{dz} - \frac{1}{\rho_k T}\frac{dp}{dz}$$

The four variables of the model are $p$, $x$, $G$, and $k$. The critical condition corresponds to a maximum of $G$, calculated by the vanishing condition of the derivative $\partial G/\partial p$. The value of the slip ratio at the critical section is given by an additional relationship:

$$\left(\frac{\partial G}{\partial k}\right)_p = 0$$

which leads to

$$k_C = \left(\frac{\rho_L}{\rho_G}\right)^{1/2}$$

Giot and Meunier (1968) showed that if the set of three equations is completed by

$$\frac{dk}{dz} + k_1\frac{dp}{dz} + k_2\frac{dx}{dz} + k_3\frac{dw_G}{dz} = 0 \tag{50}$$

Fauske's critical condition is identical to the vanishing condition of the determinant of the resulting set of four equations. The critical value of the slip ratio given by Fauske corresponds to the maximum flow rate available for given momentum, pressure, and quality at the critical section.

A comparison with data from the literature is given for steam–water flows in Figs. 15 and 16 (Giot, 1970).

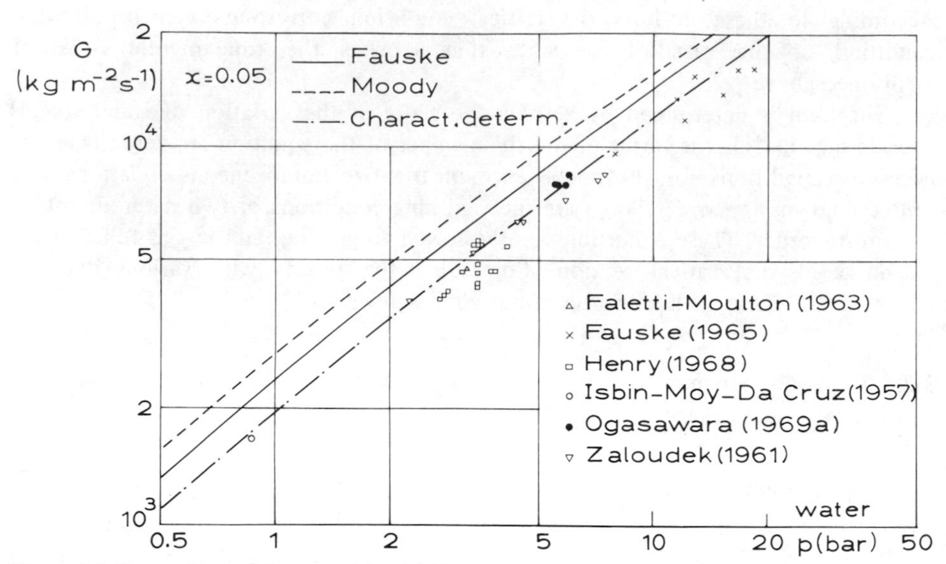


Figure 15 Comparison of three models with data.

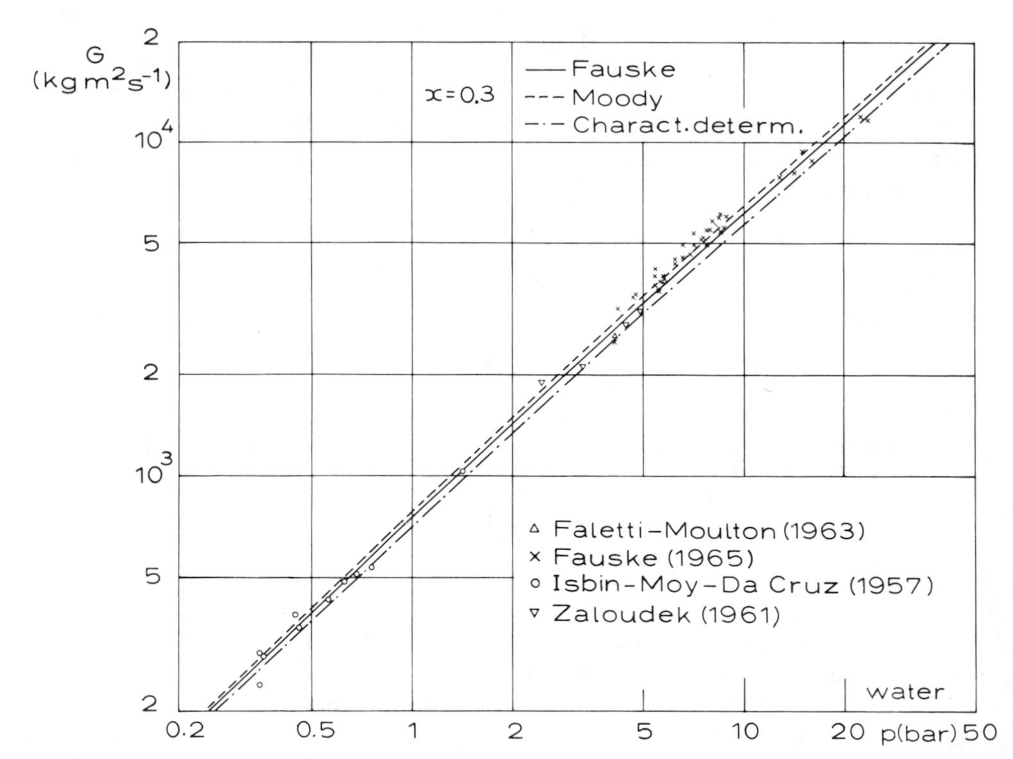


Figure 16 Comparison of three models with data.

***Moody's models*** One model uses three mixture balance equations, namely, those of mass, energy, and entropy (Moody, 1965, 1966). The four variables of the model are $p$, $x$, $G$, and $k$. The critical condition corresponds to a maximum of $G$, calculated by the vanishing condition of the derivative $\partial G/\partial p$. The value of the slip ratio at the critical section is given by an additional relationship:

$$\left(\frac{\partial G}{\partial k}\right)_p = 0$$

which leads to

$$k_C = \left(\frac{\rho_L}{\rho_G}\right)^{1/3}$$

Giot and Meunier (1968) have shown that on completing the set of three equations by Eq. (50), Moody's critical condition becomes identical to the vanishing condition of the determinant of the resulting set of four equations. The critical value of the slip ratio given by Moody corresponds to the maximum flow rate available for given energy, pressure, and quality at the critical section.

A comparison with data from the literature is given for steam-water flows in Figs. 15 and 16 (Giot, 1970). A similar approach has been developed by Cruver and Moulton (1967).

A second model, called the pressure pulse model, was proposed by Moody in 1969. All the balance equations are written in a frame of reference moving upward with a pressure pulse. The critical condition is obtained when the pressure pulse is at rest. The homogeneous and two-fluid models appear to be particular cases of the pressure-pulse model, depending on the equations chosen.

***Method of the characteristic determinant*** A method analogous to those of Fauske and Moody was tried by Giot and Meunier (1968). The Giot and Meunier method consists in writing the vanishing condition of the set of the mass, momentum, and energy mixture balance equations, the four variables being $p$, $x$, $G$, and $k$. If one takes Fauske's critical slip ratio, the resulting flow rates are very similar to those obtained by Fauske. On the contrary, when Moody's critical slip ratio is taken, the resulting flow rates are lower, as can be seen in Figs. 15 and 16.

***Ogasawara's model*** Several models have been published by Ogasawara (1967, 1969, 1969a). One of them is called the energy model, another is called the entropy model, according to whether the energy or entropy equation is used. Each model gives two or three solutions; the author chose the value of the critical slip ratio that leads to a double solution for the critical flow rate.

***Levy's model*** The same set of equations and the same critical condition as in the method of the characteristic determinant are used in the model of Levy (1965), but the two one-phase momentum equations are used neglecting friction and head losses to derive a relationship between quality and void fraction.

***Variable slip model*** Meunier and Fritte (1969), Giot (1970), and Giot and Fritte (1972) studied a model taking into account a mass mixture balance equation, two one-phase momentum equations, and an energy mixture equation. The critical-flow condition deduced from this set of four equations enables one to calculate the critical mass velocity as a function of the slip ratio for given pressure and quality. The result is illustrated by curve 1 in Fig. 17 for Freon-12. One sees that the critical mass velocity increases as the slip ratio decreases; this result agrees with experimental data showing an increase of mass velocity with a decrease of the $L/D$ ratio. Figure 17 shows results obtained by replacing the energy equation by a simplified entropy equation (curve 2) as Fauske did, and by replacing, as Moody did, the momentum mixture equation by a simplified entropy equation (curve 3). For the value of the critical slip ratio recommended by Fauske, curve 2 intersects the curve given by Fauske's model when $k$ is allowed to vary. The same is true for curve 3 and Moody's model. The results given by the models of Fauske and Moody are not very different because the curve $G$-$k$ is rather flat for the high values of the slip ratio.

**3.2.4 Three-equation models** The homogeneous model (Steltz, 1961; Tremblay and Andrews, 1971) give critical mass velocities lower than those given by the experimental data.

**3.2.5 Liquid–vapor flows at low qualitities** It is very difficult to predict critical mass velocities at low qualities since the results of the different models diverge when the quality decreases. This is illustrated in Fig. 18, where some critical-flow data, the homogeneous flow model, and the six-equation models are presented.

***Henry's model*** Henry (1968) developed a model to predict critical flow rates for equilibrium qualities less than 0.02. He used two mass balance equations, a

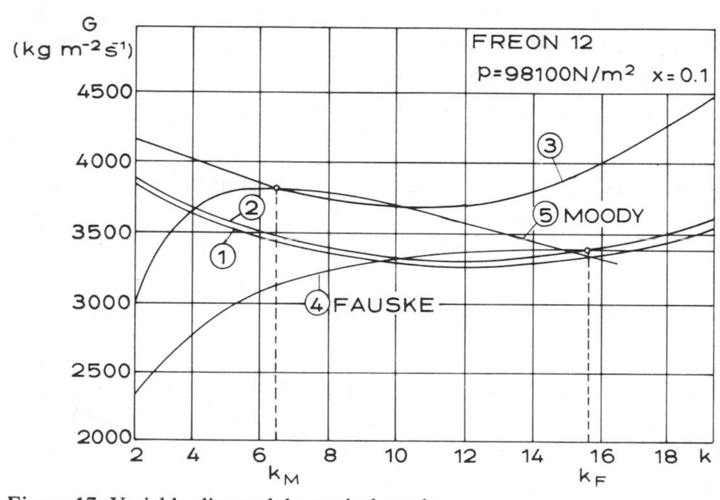


**Figure 17** Variable slip models: typical result.

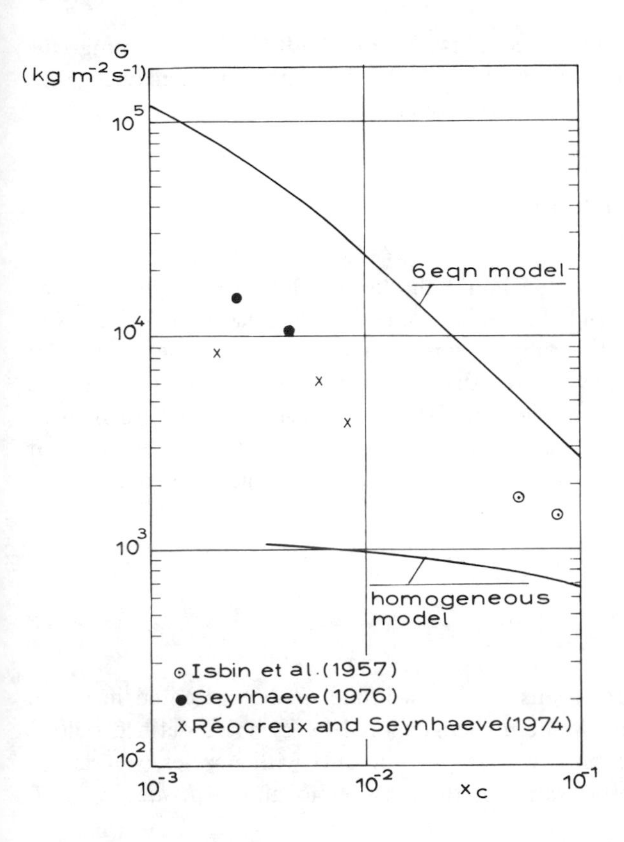


**Figure 18** Critical mass velocities at low qualities.

momentum mixture equation, and an energy mixture equation. The quality is related to the equilibrium quality by the expression

$$x = kNx_{\text{eq}}$$

where $N$ is a characteristic parameter of the system.

This parameter results from experimental data on critical flow. The phase change occurring in the flow is found to have a nonequilibrium nature, and the slip ratio is found to be close to unity.

***Other studies*** The studies of Flinta et al. (1973) and Flinta (1975) deal with the analysis of the nucleation process, which plays an important role in this problem and affects the critical flow rate significantly.

A nonequilibrium, axial, two-phase flow model has been developed by Bauer et al. (1978), who used the three mixture balance equations and an equation for the nonequilibrium quality:

$$\frac{\partial x}{\partial t} + w\,\frac{\partial x}{\partial z} = \frac{x_{\text{eq}} - x}{\theta}$$

where $\theta$ is a vaporization relaxation time. They found good agreement with Réocreux's data, except that they also found a so-called pseudocritical flow (see supra) instead of a critical flow.

Finally, let us mention the recent study of Städtke (1977) using the formulation of the theory of irreversible thermodynamics and analyzing the influence of the interfacial transfer terms on the critical flow.

### 3.4 The Flow of Subcooled Liquid through Orifices

Some aspects of the flow of subcooled liquid through orifices have been discussed in Sec. 2. Recent experimental and theoretical studies by Seynhaeve (1977) showed that the critical section is in the tube downstream of the orifice below the jet rupture. In the studies of Seynhaeve the homogeneous model predicts the experimental values of the critical flow rate with a precision better than 10%. Consequently, under these conditions the critical flow rate depends on the diameter of the tube downstream of the orifice. This particularity is also a conclusion of Ogasawara's (1969b) study.

## APPENDIX

In this chapter, we used the equations of the two-fluid model presented in Chap. 10. For simplicity some constitutive terms are denoted by symbols defined below. Because we are concerned not only with pipes but also with nozzles, we cannot drop the terms that include the derivative of the cross-sectional area $dA/dz$:

**A.1 Mass** 1. Phase equation:

$$\frac{\partial}{\partial t}(R_k \rho_k) + \frac{\partial}{\partial z}(R_k \rho_k w_k) = -\frac{R_k \rho_k w_k}{A}\frac{dA}{dz} + M_{kI} \tag{A.1}$$

where

$$M_{kI} \triangleq -\frac{1}{A}\overline{\int_C \dot{m}_k \frac{dC}{\mathbf{n}_k \cdot \mathbf{n}_{kI}}}$$

is the mass flux through the interface to phase $k$, per unit length of the channel and per unit area.

2. Interface equation:

$$M_{GI} = -M_{LI} \tag{A.2}$$

We put

$$M \triangleq M_{GI} = -M_{LI}$$

3. Mixture equation: In the mass mixture equation we set $\alpha \triangleq R_G$ and $1 - \alpha \triangleq R_L$.

$$\frac{\partial}{\partial t}[\alpha\rho_G + (1-\alpha)\rho_L] + \frac{\partial}{\partial z}[\alpha\rho_G w_G + (1-\alpha)\rho_L w_L]$$

$$= -\frac{1}{A}[\alpha\rho_G w_G + (1-\alpha)\rho_L w_L]\frac{dA}{dz} \tag{A.3}$$

**A.2 Momentum** 1. Phase equation:

$$\frac{\partial}{\partial t}(R_k\rho_k w_k) + \frac{\partial}{\partial z}(R_k\rho_k w_k^2) + R_k\frac{\partial p}{\partial z} = -\frac{R_k\rho_k w_k^2}{A}\frac{dA}{dz}$$

$$-R_k\rho_k g\cos\phi + (MV)_{kI} - \tau_{kI} - \tau_{kC} - \frac{1}{A}\phi_k \tag{A.4}$$

where $\phi$ is the angle between the direction of flow and the upward vertical;

$$(MV)_{kI} \triangleq -\frac{1}{A}\overline{\int_I \mathbf{n}_z\cdot(\dot{m}_k\mathbf{v}_k)\frac{dC}{\mathbf{n}_k\cdot\mathbf{n}_{kI}}}$$

is the momentum flux through the interface to phase $k$, per unit length of channel and per unit area, resulting from mass transfer;

$$\tau_{kI} \triangleq \frac{1}{A}\overline{\int_C \mathbf{n}_z\cdot(\mathbf{n}_k\cdot\tau_k)\frac{dC}{\mathbf{n}_k\cdot\mathbf{n}_{kI}}}$$

is the projection on the $0z$ axis of the resultant of the viscous stresses acting on phase $k$ at the interface, per unit length of channel and per unit area;

$$\tau_{kc} \triangleq \frac{1}{A}\int_{C_k} \mathbf{n}_z\cdot(\mathbf{n}_k\cdot\tau_k)\frac{dC}{\mathbf{n}_k\cdot\mathbf{n}_{kC}}$$

is the projection on the $0z$ axis of the resultant of the viscous stresses acting on phase $k$ at the wall, per unit length of channel and per unit area; $\phi_k$ takes into account the viscous stress derivatives and will be neglected.

2. Interface equation:

$$(MV)_{GI} - \tau_{GI} = -[(MV)_{LI} - \tau_{LI}] \tag{A.5}$$

we put

$$(MV) \triangleq (MV)_{GI} - \tau_{GI} = -[(MV)_{LI} - \tau_{LI}]$$

3. Mixture equation:

$$\frac{\partial}{\partial t}[\alpha\rho_G w_G + (1-\alpha)\rho_L w_L] + \frac{\partial}{\partial z}[\alpha\rho_G w_G^2 + (1-\alpha)\rho_L w_L^2] + \frac{\partial p}{\partial z}$$

$$= -\frac{1}{A}[\alpha\rho_G w_G^2 + (1-\alpha)\rho_L w_L^2]\frac{dA}{dz} - [\alpha\rho_G + (1-\alpha)\rho_L]g\cos\phi - (\tau_{GC} + \tau_{LC}) \tag{A.6}$$

**A.3 Total energy** 1. Phase equation:

$$\frac{\partial}{\partial t}\left[R_k\rho_k\left(h_k+\frac{1}{2}v_k^2\right)\right]+\frac{\partial}{\partial z}\left[R_k\rho_k\left(h_k+\frac{1}{2}v_k^2\right)w_k\right]-R_k\frac{\partial p}{\partial t}$$

$$=-\frac{R_k\rho_k(h_k+\frac{1}{2}v_k^2)w_k}{A}\frac{dA}{dz}-R_k\rho_k w_k g\cos\phi$$

$$+(MH)_{kI}+(EC)_{kI}+\chi_{kI}+q_{kI}+q_{kC}+\frac{1}{A}\psi_k \tag{A.7}$$

where

$$(MH)_{kI}\triangleq-\frac{1}{A}\overline{\int_I \dot{m}_k h_k\frac{dC}{\mathbf{n}_k\cdot\mathbf{n}_{kI}}}$$

is the enthalpy flux through the interface to phase $k$, per unit length of channel and per unit area, as a result of mass transfer;

$$(EC)_{kI}\triangleq-\frac{1}{A}\overline{\int_I \frac{1}{2}\dot{m}_k v_k^2\frac{dC}{\mathbf{n}_k\cdot\mathbf{n}_{kI}}}$$

is the kinetic energy flux through the interface to phase $k$, per unit length of channel and per unit area, as a result of mass transfer;

$$\chi_{kI}\triangleq\frac{1}{A}\overline{\int_I (\tau_k\cdot\mathbf{v}_k)\cdot\mathbf{n}_k\frac{dC}{\mathbf{n}_k\cdot\mathbf{n}_{kI}}}$$

is the power as a result of the viscous stress tensor of phase $k$ at the interface, per unit length of channel and per unit area;

$$q_{kI}\triangleq-\frac{1}{A}\overline{\int_I \mathbf{q}_k\cdot\mathbf{n}_k\frac{dC}{\mathbf{n}_k\cdot\mathbf{n}_{kI}}}$$

is the heat flux to phase $k$ through the interface, per unit length of channel and per unit area;

$$q_{kC}\triangleq-\frac{1}{A}\overline{\int_{C_k} \mathbf{q}_k\cdot\mathbf{n}_k\frac{dC}{\mathbf{n}_k\cdot\mathbf{n}_{kC}}}$$

is the heat flux to phase $k$ through the wall, per unit length of channel and per unit area.

The term $\psi_k$ takes into account longitudinal conduction terms and their derivatives and phase viscous-stress derivatives and the power of these viscous stresses; it will be neglected.

2. Interface equation:

$$(MH)_{GI} + (EC)_{GI} + \chi_{GI} + q_{GI} = -[(MH)_{LI} + (EC)_{LI} + \chi_{LI} + q_{LI}] \tag{A.8}$$

We put

$$(MH) \triangleq (MH)_{GI} + (EC)_{GI} + \chi_{GI} + q_{GI} = -[(MH)_{LI} + (EC)_{LI} + \chi_{LI} + q_{LI}]$$

3. Mixture equation:

$$\begin{aligned}
&\frac{\partial}{\partial t}\left[\alpha\rho_G\left(e_G + \frac{1}{2}v_G^2\right) + (1-\alpha)\rho_L\left(e_L + \frac{1}{2}v_L^2\right)\right] \\
&\quad + \frac{\partial}{\partial z}\left[\alpha_G\rho_G w_G\left(h_G + \frac{1}{2}v_G^2\right) + (1-\alpha)\rho_L w_L\left(h_L + \frac{1}{2}v_L^2\right)\right] \\
&\quad = -\frac{1}{A}\left[\alpha\rho_G w_G\left(h_G + \frac{1}{2}v_G^2\right) + (1-\alpha)\rho_L w_L\left(h_L + \frac{1}{2}v_L^2\right)\right]\frac{dA}{dz} \\
&\qquad - [\alpha\rho_G w_G + (1-\alpha)\rho_L w_L]g\cos\phi + (q_{GC} + q_{LC})
\end{aligned} \tag{A.9}$$

## NOMENCLATURE

| | |
|---|---|
| $A$ | cross-sectional area |
| $a_{ij}$ | elements of system matrix [Eq. (2)] |
| $\mathbf{a}_k$ | propagation velocity |
| $a_s$ | sound speed |
| $b_{ij}$ | elements of system matrix [Eq. (1)] |
| $C_1, C_2$ | contraction coefficients [Eq. (14)] |
| Ca′ | cavitation number |
| $c_i$ | elements of column vector [Eq. (1)] |
| $c_p, c_v$ | specific heats |
| $D$ | diameter |
| $d_{jk}$ | elements of matrix [Eq. (7)] |
| $(EC)$ | kinetic energy flux per unit length of channel and per unit area |
| $e$ | internal energy per unit mass |
| $F$ | force |
| $f_k$ | function defined in Eq. (7) |
| $G$ | mass velocity |
| $g$ | gravity constant |
| $h$ | enthalpy per unit mass |
| $K$ | function defined in Eq. (34) |
| $k$ | wave number |
| $k$ | slip ratio |
| $L$ | length of channel |
| $\mathfrak{L}$ | heat of vaporization |
| $M$ | mass flux per unit length of channel and per unit area |

| | |
|---|---|
| Ma | Mach number |
| $(MH)$ | enthalpy flux per unit length of channel and per unit area |
| $(MV)$ | momentum flux per unit length of channel and per unit area |
| $N_j$ | numerator defined in Sec. 1.1.2 |
| $n$ | exponent of the polytropic expansion |
| $\mathbf{n}$ | normal unit vector |
| $p$ | pressure |
| $q$ | heat flux per unit length of channel and per unit area |
| $R$ | ideal gas constant |
| $R_G$ | void fraction |
| $s$ | entropy per unit mass |
| $T$ | temperature |
| $t$ | time |
| $\mathbf{V}_k$ | characteristics |
| $w$ | velocity |
| $\mathbf{w}_k$ | entrainment velocity |
| $x$ | quality |
| $x_j$ | dependent variable |
| $z$ | axial coordinate |
| $\alpha$ | void fraction |
| $\gamma$ | ratio of the specific heats |
| $\Delta h$ | difference between the enthalpy and its saturation value |
| $\Delta T_s$ | superheat |
| $\theta$ | vaporization relaxation time |
| $\lambda$ | coefficient [Eq. (13)] |
| $\mu$ | coefficient [Eq. (13)] |
| $\nu$ | coefficient [Eq. (13)] |
| $\rho$ | density |
| $\tau$ | projection of the resultant of the viscous stresses, per unit length of channel and per unit area |
| $\boldsymbol{\tau}_k$ | viscous stress tensor of phase $k$ |
| $\Phi_l$ | Martinelli parameter |
| $\phi$ | angle of the flow with respect to the upward vertical |
| $\phi_k$ | term resulting from the viscous stress derivatives [Eq. (A.4)] |
| $\phi_l$ | function defined in Sec. 1.2.2 |
| $\chi$ | power resulting from the viscous stress tensor per unit length of channel and per unit area |
| $\psi_k$ | term related to the axial conduction [Eq. (A.7)] |
| $\psi_l$ | function defined in Sec. 1.2.2 |
| $\omega$ | wave pulsation |

**Subscripts**

| | |
|---|---|
| $C$ | contour at the wall |
| $c$ | critical |
| eq | equilibrium |
| exc | excess |

| | |
|---|---|
| $F$ | flashing |
| $f$ | friction |
| $G$ | gas phase |
| $I$ | interface |
| $i$ | inlet |
| $k$ | phase $G$ or $L$ |
| $kC$ | through the wall to phase $k$ |
| $kI$ | through the interface to phase $k$ |
| $L$ | liquid phase |
| $o$ | flowing alone |
| $R$ | reaction |
| $S$ | constant entropy |
| $SP$ | single-phase |
| sat | saturation |
| $TP$ | two-phase |
| up | upstream |

## REFERENCES

Bailey, J. F., Metastable Flow of Saturated Water, *Trans. ASME*, vol. 73, pp. 1109–1116, 1951.

Bauer, E. G., Houdayer, G. R., and Sureau, H. M., A Non-equilibrium Axial Flow Model and Application to Loss of Coolant Accident Analysis: The CLYSTERE System Code, in *Transient Two-Phase Flow, Proc. CSNI Specialists Meet., Toronto, Aug. 3 and 4, 1976*, eds. S. Banerjee and K. R. Weaver, vol. 1, pp. 429–457, AECL, 1978.

Bouré, J., The Critical Flow Phenomenon with Reference to Two-Phase Flow and Nuclear Reactor Systems, in *Thermal and Hydraulic Aspects of Nuclear Reactor Safety*, vol. 1, *Light Water Reactors*, eds. O. C. Jones and S. G. Bankoff, pp. 195–216, ASME, New York, 1977.

Bouré, J. A., Fritte, A. A., Giot, M. M., and Réocreux, M. L., A Contribution to the Theory of Critical Two-Phase Flow: On the Links between Maximum Flow-Rates, Choking, Sonic Velocities, Propagation and Transfer Phenomena in Single and Two-Phase Flow, *Energie Primaire*, vol. 11, pp. 1–27, 1975.

Bouré, J. A., Fritte, A. A., Giot, M. M., and Réocreux, M. L., Highlights of Two-Phase Critical Flow, *Int. J. Multiphase Flow*, vol. 3, pp. 1–22, 1976.

Burgess, M. H., Two-Phase Choked Flow in Vortex Diodes, European Two-Phase Flow Group Meet., Erlangen, May 31–June 4, 1976.

Chen, P. C. and Isbin, H. S., Two-Phase Flow through Apertures, *Energia Nucleare*, vol. 13, no. 7, pp. 347–358, 1966.

Chisholm, D., Calculation of Choking Flowrates during Steam-Water Flow through Pipes, NEL rept. 377, 1968.

Costa, J. and Charlety, P., Critical-Flow Experiments in a Forced-Convection Boiling Sodium Loop, in *Progress in Heat and Mass Transfer, Heat Transfer in Liquid Metals*, vol. 7, ed. O. E. Dwyer, pp. 429–439, Pergamon, New York, 1973.

Courant, R. and Friedrichs, K., *Supersonic Flow and Shock Waves*, Interscience, New York, 1948.

Cruver, J. E. and Moulton, R. W., Critical Flow of Liquid-Vapor Mixtures, *AIChE J.*, vol. 13, no. 1, pp. 52–60, 1967.

Duggins, R. K., Cavitating Flow in Pipes, in *Two-Phase Flow and Cavitation in Power Generation Systems*, pp. 217–229, Société Hydrotechnique de France, Paris, 1976.

Edmonds, D. K. and Smith, R. V., Comparison of Mass Limiting Two-Phase Flow in a Straight Tube and in a Nozzle, Symp. on Two-Phase Flow, Exeter, vol. 1, ed. P. M. C. Lacey, pp. G401–G414, Department of Chemical Engineering, University of Exeter, 1965.

Faletti, D. W. and Moulton, R. W., Two-Phase Critical Flow of Steam-Water Mixtures, *AIChE J.*, vol. 9, no. 2, pp. 247–253, 1963.

Fauske, H. K., Two-Phase Critical Flow with Application to Liquid-Metal Systems, ANL-6779, 1963a.

Fauske, H. K., A Theory for Predicting Pressure Gradients for Two-Phase Critical Flow, *Nucl. Sci. Eng.*, vol. 17, pp. 1–7, 1963b.

Fauske, H. K., Some Ideas about the Mechanism Causing Two-Phase Critical Flow, *Appl. Sci. Res., Sect. A*, vol. 13, pp. 149–160, 1964.

Fauske, H. K., Two-Phase Two- and One-Component Critical Flow, Symp. on Two-Phase Flow, Exeter, vol. 1, ed. P. M. C. Lacey, pp. 101–114, Department of Chemical Engineering, University of Exeter, 1965.

Fauske, H. K. and Min, T. C., A Study of the Flow of Saturated Freon-11 through Apertures and Short Tubes, ANL-6667, 1963.

Fauske, H. K., Quinn, D., and Jeans, W., Sodium Flashing Experiment, *Trans. ANS*, vol. 10, no. 2, p. 1973, 1967.

Flinta, J., Blowdown of Subcooled Hot Water, European Two-Phase Flow Group Meet., Haifa, June 2–5, 1975.

Flinta, J., Hernborg, G., Siden, L., and Skinstad, A., Blow-down through Nozzles of Different Types, European Two-Phase Flow Group Meet., Brussels, June 4–7, 1973.

Fritte, A., Vitesses de Propagation de Petites Perturbations dans les Ecoulements Diphasiques Liquide-Gaz, Thèse de doctorat, Departement Thermodynamique et Turbomachines, Université Catholique de Louvain, 1974.

Giot, M., Débits Critiques des Ecoulements Diphasiques, Thèse de doctorat, Departement Thermodynamique et Turbomachines, Université Catholique de Louvain, 1970.

Giot, M. and Fritte, A., Two-Phase Two- and One-Component Critical Flows with the Variable Slip Model, in *Progress in Heat and Mass Transfer*, vol. 6, *Proc. Int. Symp. on Two-Phase Systems*, eds. G. Hetsroni, S. Sideman, and J. P. Hartnett, pp. 651–670, Pergamon, New York, 1972.

Giot, M. and Meunier, D., Méthodes de Détermination du Débit Critique en Ecoulements Monophasiques et Diphasiques à un Constituant, *Energie Primaire*, vol. 4, no. 1–2, p. 23, 1968.

Hammitt, F. G., Robinson, M. J., and Lafferty, J. F., Choked-Flow Analogy for Very Low Quality Two-Phase Flows, *Nucl. Sci. Eng.*, vol. 29, pp. 131–142, 1967.

Henry, R. E., A Study of One- and Two-Component, Two-Phase Critical Flows at Low Qualities, ANL-7430, 1968.

Henry, R. E. and Fauske, H. K., The Two-Phase Critical Flow of One-Component Mixtures in Nozzles, Orifices, and Short Tubes, *J. Heat Transfer*, vol. 93, pp. 179–187, 1971.

Isbin, H., Critical Two-Phase Flow, Lecture Series on Boiling and Two-Phase Flow, University of California, May 27–28, pp. 173–184, 1965.

Isbin, H. S., Moy, J. E., and Da Cruz, A. J. R., Two-Phase, Steam-Water Critical Flow, *AIChE J.*, vol. 3, no. 3, pp. 361–365, 1957.

Ivandaev, A. I. and Nigmatulin, R. I., Elementary Theory of Critical (Maximal) Flow-Rates of Two-Phase Mixtures, *High Temp.*, vol. 10, pp. 946–953, 1972.

Katto, Y., Dynamics of Compressible Saturated Two-Phase Flow (Critical Flow), *Bull. JSME*, vol. 11, no. 48, pp. 1135–1144, 1968.

Katto, Y. and Sudo, Y., Study of Critical Flow (Completely Separated Gas-Liquid Two-Phase Flow), *Bull. JSME*, vol. 16, no. 101, pp. 1741–1749, 1973.

Klingebiel, W. J. and Moulton, R. W., Analysis of Flow Choking of Two-Phase, One-Component Mixtures, *AIChE J.*, vol. 17, no. 2, pp. 383–390, 1971.

Levy, S., Prediction of Two-Phase Critical Flow-Rate, *J. Heat Transfer*, pp. 53–58, 1965.

Malnes, D. and Van der Meer, F. F., Transient Blowdown Procedure, European Two-Phase Flow Group Meet., Rome, June 6–8, 1972.

Mauget, C. Contribution à l'Etude Expérimentale des Débits Critiques en Ecoulement Diphasique Eau-vapeur, Thèse de 3e Cycle, Université de Poitiers, Centre d'Etudes Aérodynamiques et Thermiques, 1977.

Meunier, D. and Fritte, A., Modèle à Glissement Variable pour l'Etude des Débits Critiques en Double Phase, Institut International du Froid, *Proc. Meet. Liege, Belgium, Sept. 9-11, 1969*, Com. II and VI, pp. 413-424, 1969.

Moody, F. J., Maximum Flow-Rate of a Single Component, Two-Phase Mixture, *J. Heat Transfer*, vol. 87, pp. 134-142, 1965.

Moody, F. J., Maximum Two-Phase Vessel Blowdown from Pipes, *J. Heat Transfer*, vol. 88, pp. 285-293, 1966.

Moody, F. J., A Pressure Pulse Model for Two-Phase Critical Flow and Sonic Velocity, *J. Heat Transfer*, vol. 91, pp. 371-384, 1969.

Muir, J. F. and Eichhorn, R., Compressible Flow of an Air-Water Mixture through a Vertical, Two-dimensional, Converging-diverging Nozzle, *Proc. 1963 Heat Transfer and Fluid Mech. Inst.*, eds. A. Roshko, B. Sturtevant, and D. R. Bartz, pp. 183-204, Stanford University Press, Stanford, Calif., 1963.

Muncaster, R. and Thomson, G. M., Equilibration and the Critical Flow of Flashing Mixtures in Short Nozzles, *Proc. 4th Int. Symp. on Fresh Water from the Sea*, vol. 1, eds. A. Delyannis and E. Delyannis, pp. 391-401, European Federation of Chemical Engineering, 1973.

Ogasawara, H., A Theoretical Prediction of Two-Phase Critical Flow, *Bull. JSME*, vol. 10, no. 38, pp. 279-290, 1967.

Ogasawara, H., A Theoretical Approach to Two-Phase Critical Flow, 3d Rept. The Critical Condition including Interphasic Slip, *Bull. JSME*, vol. 12, no. 52, pp. 827-836, 1969.

Ogasawara, H., A Theoretical Approach to Two-Phase Critical Flow, 4th Rept. Experiments on Saturated Water Discharging through Long Tubes, *Bull. JSME*, vol. 12, no. 52, pp. 837-846, 1969a.

Ogasawara, H., A Theoretical Approach to Two-Phase Critical Flow, 5th Rept. Several Problems on Discharging of Saturated Water through Orifices, *Bull. JSME*, vol. 12, no. 52, pp. 847-856, 1969b.

Réocreux, M., Contribution à l'Etude des Débits Critiques en Ecoulement Diphasique Eau-vapeur, Thèse de docteur ès sciences physiques, Université Scientifique et Médicale de Grenoble, 1974.

Réocreux, M. L., Experimental Study of Steam-Water Choked Flow, in *Transient Two-Phase Flow, Proc. CSNI Specialists Meet., Toronto, Aug. 3 and 4, 1976*, eds. S. Banerjee and K. R. Weaver, vol. 2, pp. 637-671, AECL, 1978.

Réocreux, M. and Seynhaeve, J. M., Ecoulements Diphasiques Eau-vapeur: Essais Comparatifs de Débits Critiques, *Energie Primaire*, vol. 10, no. 3-4, pp. 115-137, 1974.

Réocreux, M. L., Barrière, G., and Vernay, B., Etude Expérimentale des Débits Critiques en Ecoulement Diphasique Eau-vapeur à Faible Titre dans un Canal avec Divergent de 7 Degrés, CENG, rapport interne RTT 115, 1973.

Seynhaeve, J. M., Contribution à l'Etude Expérimentale des Débits Critiques en Milieu Eau-vapeur, à Faible Titre, in *Two-Phase Flow and Cavitation in Power Generation Systems*, pp. 241-253, Société Hydrotechnique de France, Paris, 1976.

Seynhaeve, J. M., Critical Flow through Orifices, European Two-Phase Flow Group Meet., Grenoble, June 6-8, 1977.

Seynhaeve, J. M., Giot, M. M., and Fritte, A. A., Non-equilibrium Effects on Critical Flow-Rates at Low Qualities, in *Transient Two-Phase Flow, Proc. CSNI Specialists Meet., Toronto, Aug. 3 and 4, 1976*, eds. S. Banerjee and K. R. Weaver, vol. 2, pp. 672-688, AECL, 1978.

Silver, R. S., Temperature and Pressure Phenomena in the Flow of Saturated Liquids, *Proc. R. S. London, Ser. A*, vol. 194, pp. 464-480, 1948.

Smith, R. V., Two-Phase Two-Component Critical Flow in a Venturi, *J. Basic Eng.*, vol. 94, pp. 147-155, 1972.

Smith, R. V., Critical Two-Phase Flow of Helium, *Cryogenics*, vol. 13, pp. 616-619, 1973.

Smith, R. V., Cousins, L. B., and Hewitt, G. F., Critical Flow in an Annular Venturi, *Proc. Symp. on Two-Phase Flow Dynamics, Eindhoven, Sept. 4-9*, EUR 4288e, vol. 1, pp. 487-505, 1967.

Smith, R. V., Cousins, L. B., and Hewitt, G. F., Two-Phase Two-Component Critical Flow in a Venturi, AERE-R 5736, 1968.

Städtke, H., Two-Phase Critical Flow of Initially Subcooled Water through Nozzles, European Two-Phase Flow Group Meet., Grenoble, June 6–8, 1977.

Steltz, W. G., The Critical and Two-Phase Flow of Steam, *J. Eng. Power*, vol. 83, pp. 145–154, 1961.

Sudo, Y. and Katto, Y., Mechanics of Critical Flow, in *Theoretical and Applied Mechanics, Proc. 22nd Jpn. Natl. Cong. for Appl. Mech., 1972*, vol. 22, pp. 67–78, University of Tokyo Press, 1974.

Tremblay, P. E. and Andrews, D. G., A Physical Basis for Two-Phase Pressure Gradient and Critical Flow Calculations, *Nucl. Sci. Eng.*, vol. 44, pp. 1–11, 1971.

Wallis, G. B. and Sullivan, D. A., Two-Phase Air-Water Nozzle Flow, *J. Basic Eng.*, vol. 94, pp. 788–794, 1972.

Zaloudek, F. R., The Low Pressure Critical Discharge of Steam-Water Mixtures from Pipes, HW-68934 REV, 1961.

Zaloudek, F. R., The Critical Flow of Hot Water through Short Tubes, HW-77594, 1963.

Zaloudek, F. R., Steam-Water Critical Flow from High Pressure Systems, HW-80535, 1964.

CHAPTER

# NINETEEN

# CODES AND TWO-PHASE FLOW HEAT TRANSFER CALCULATIONS

**Y. Y. Hsu**

## 1 PHYSICAL MODELING AND UNCERTAINTIES OF CODES

Since a nuclear reactor is very complicated, its behavior can be predicted only by large computer codes. Many codes have been developed. Some are components codes, or separate-effects codes, which are written with the specific purpose of predicting one aspect of reactor safety. There are codes for blowdown heat transfer, reflood, containment, etc. Then there are system codes that relate all of the component codes into an integrated body with each part connected by input and output of individual components. The relationship is illustrated in Fig. 1.

### 1.1 Physical Modeling of the Thermohydraulic Process

Because of the complicated nature of the physical phenomena of two-phase flow in reactor, a physical model has to be formulated to describe the real process. The simplest description is the so-called one-dimensional homogeneous, no-slip, equilibrium model. In this form the steam–water mixture is treated as one fluid at one temperature (saturated) flowing at one velocity in one dimension. In reality, the two phases do not mix very well, and each phase flows at its own velocity; the two

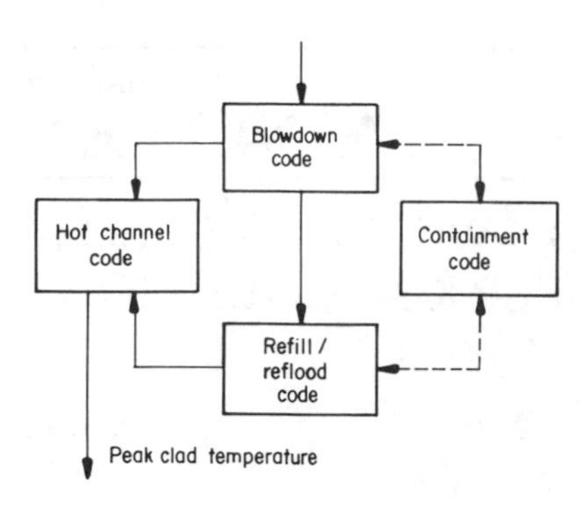


**Figure 1** Prediction of peak clad temperature (Fabic, 1976).

phases may not be at equilibrium; thus liquid temperature and vapor temperature are not the same. The distribution of each is different to form a different flow pattern; the flow may be multidimensional; and last, the process is transient.

Thus a complete description would require the writing of mass, momentum, and energy conservation equations for each phase tied together by constitutive equations to relate the interfacial transport between phases. Such a complete system is very difficult and costly to solve. Consequently, codes of various degrees of complication are in existence, ranging between the two extremes. These are classified according to $1V/2V$, $1T/2T$, $S$(slip)/$D$(drift). A tabulation of the different types of codes and the equations needed to describe each is shown in Table 1. A survey of existing codes was made by Fabic (1976), as shown in Table 2.

## 1.2 Approximations and Uncertainties of Codes

Physical modeling is already an approximate description of nature. Mathematical formulation of physical models by differential equations has to be further simplified by nodes and lumped parameters into difference equations, which are the third generation of the approximation. Computer calculation is yet another step of approximation. Furthermore, the constitutive equations used in the physical model are of the form of empirical correlations. Prominent examples are the heat transfer and pressure drop correlations. The net result of all these approximations is the anticipated uncertainties of the code calculations.

To ascertain the accuracy and limitations of codes, the codes have to go through a verification process. The ultimate verification is, naturally, comparison with experimental results. But even during the code development stage, there is sensitivity analysis and uncertainty analysis for verification. The sensitivity analysis is performed by assigning a variable multiplier ("dial") to a correlation to see how much variation is propagated to the ultimate parameter (which is clad temperature, for the case of heat transfer). This sensitivity test is essential in assessing the relative importance of each correlation so that efforts can be concentrated on the crucial correlations.

Another test is the uncertainty analysis. The methodology is essentially the same as sensitivity analysis, that is, through the assignment of dials. The dials

**Table 1 Practical two-phase flow models (assuming $p_G = p_L = p$)[a]**

| Model designation | Restrictions: No. | Restrictions: Imposed on | No. of field equations: Mass | Momentum | Energy | No. of interface transfer equations | External constitutive equations: No. | Type |
|---|---|---|---|---|---|---|---|---|
| $1V1T$ | | $V_G = V_L, h_L, h_G$ | | | | | 2 | $\bar{\tau}, \bar{q}$ |
| $1VS1T$ | 3 | $V_G/V_L, h_L, h_G$ | 1 | 1 | 1 | 0 | 3 | $\bar{\tau}, \bar{q}$, $V_G/V_L$ = slip |
| $1VD1T$ | | $V_G - V_L, h_L, h_G$ | | | | | 3 | $\bar{\tau}, \bar{q}$, $V_r \triangleq V_G - V_L$ or $V_{Gm} \triangleq V_G - V_m$ or $V_{Gj} \triangleq V_G - V_j$ |
| $1VT_KT_{sat}$ | | $V_G = V_L$, $h_L$ | 1 | 1 | 2 | | 3 | $\bar{\tau}, \bar{q}, q_G, E$ |
| $1VST_KT_{sat}$ | 2 | $V_G/V_L$, or | 1 | 1 | 2 | 1 | 4 | $\bar{\tau}, \bar{q}, q_G, V_G/V_L, E$ |
| $1VDT_KT_{sat}$ | | $V_G - V_L$, $h_G$ | 2 | 1 | 1 | | 4 | $\bar{\tau}, \bar{q}, \Gamma, V_r$ (or $V_{Gm}$ or $V_{Gj}$) |
| $2V1T$ | | $h_L, h_G$ | 1 | 2 | 1 | | 4 | $\tau_L, \tau_G, \bar{q}, M$ |
| $1V2T$ | | $V_G = V_L$ | 2 | 1 | 2 | | 5 | $\bar{\tau}, q_L, q_G, \Gamma, E$ |
| $1VD2T$ | 1 | $V_G - V_L$ | 2 | 1 | 2 | 2 | 6 | $\bar{\tau}, q_L, q_G, \Gamma, E, V_r$ (or $V_{Gm}$) |
| $2VT_KT_{sat}$ | | $h_L$ or $h_G$ | 2 | 2 | 1 | | 5 | $\tau_L, \tau_G, \bar{q}, \Gamma, M$ |
| $2VT_KT_{sat}$ | | $h_L$ or $h_G$ | 1 | 2 | 2 | | 6 | $\tau_L, \tau_G, \bar{q}, q_K, M, E$ |
| $2V2T$ | 0 | None | 2 | 2 | 2 | 3 | 7 | $\tau_L, \tau_G, q_L, q_G, \Gamma, M, E$ |

[a]From Fabic (1976).

**Table 2 Practical two-phase flow model designations**[a]

| Designation | Characteristics | Remaining computed dependent variables |
|---|---|---|
| $1V1T$ | Homogeneous, equilibrium (HEM) | $p, \bar{V}, \alpha$ |
| $1VS1T$ | Slip, equilibrium | $p, \bar{V}, \alpha$ |
| $1VD1T$ | Drift, equilibrium | $p, \bar{V}, \alpha$ |
| $1VT_KT_{sat}$ | Homogeneous, partial nonequilibrium | $p, \bar{V}, \alpha, h_L$ or $h_G$ |
| $1VST_KT_{sat}$ | Slip, partial nonequilibrium | $p, \bar{V}, \alpha, h_L$ or $h_G$ |
| $1VDT_KT_{sat}$ | Drift flux (DF), partial nonequilibrium | $p, \bar{V}, \alpha, h_L$ or $h_G$ |
| $2V1T$ | Two-fluid, equilibrium | $p, V_L, V_G, \alpha$ |
| $1V2T$ | Homogeneous, full nonequilibrium | $p, \bar{V}, h_L, h_G, \alpha$ |
| $1VD2T$ | Drift, full nonequilibrium | $p, \bar{V}, h_L, h_G, \alpha$ |
| $2VT_KT_{sat}$ | Two-fluid, partial nonequilibrium | $p, V_L, V_G, \alpha, h_L$ or $h_G$ |
| $2V2T$ | Two-fluid (TF), full nonequilibrium | $p, V_L, V_G, \alpha, h_L, h_G$ |

[a]From Fabic (1976).

assigned to uncertainty analysis are not arbitrary but are based upon the estimation of the error band for each correlation. The ultimate uncertainty is the cumulative error as a result of the contributions of errors from each correlation propagated through the whole code computation process. Ideally, the uncertainty bands of code calculations of ultimate parameters should coincide with the experimental error bands. If they do not, either the code or the measurements will have to be improved. It should be noted that uncertainty analysis is meaningful only for the best-estimate codes which are supposed to reflect nature. The evaluation model (EM) codes are meant to determine conservative bands and thus are expected to fall on one side of experimental data.

## 1.3 Code Requirements for Equation Selection

The code calculation involves a very large number of small steps. Any potential for mathematical or numerical instability should be avoided. Also, to economize computer time, decision points for switching logics should be kept to a minimum. The following conditions are the requirements from code developers when correlations are selected:

1. Avoid all possibility of singularity, or discontinuity. Singularities cause divergence and discontinuities, which can invite oscillations.
2. Keep needs for iteration at a minimum, not only to keep computer time low but avoid potentials for oscillation or even divergence. An example is the difficulty code developers encountered with the Tong-Young equation to compute $q_l$ (Nelson, 1976).
3. Keep the number of decision points at a minimum. Thus, whenever possible, a smooth transition between two modes should be used. An example of a good transition is shown by the transition boiling equation (Tong and Young, 1974),

$$q = q_l F_l + q_v(1 - F_l) \tag{1}$$

4. Ensure proper nodalization. Codes are often sensitive to nodalization. One example is the number of nodes assigned to the lower plenum during reflood. Although the total enthalpy of water in the lower plenum is the same no matter how many nodes are assigned, the subcooling of water entering the core is sensitive to the number of nodes because of the stratification of the fluids. Another example is the rewet analysis of Elias and Yadigaroglu (1977). The calculated rewetting velocity of a rod is sensitive to the number of heat transfer nodes that the clad is exposed to (Fig. 2).

## 2 EXAMPLES OF CODES FOR THERMOHYDRAULIC CALCULATIONS

Two codes will be discussed as examples for thermohydraulic calculations: RELAP-4 and TRAC. RELAP-4 is at present the workhorse for reactor safety thermohydraulic analysis. TRAC is an advanced code with much refinement over the present codes.

### 2.1 RELAP-4

RELAP-4 (Katsma et al., 1976; Fischer et al., 1978) is a $1V1T$ code. It includes RELAP-4/MOD-5, RELAP-4-EM, RELAP-4-Flood, and RELAP-4-Containment.

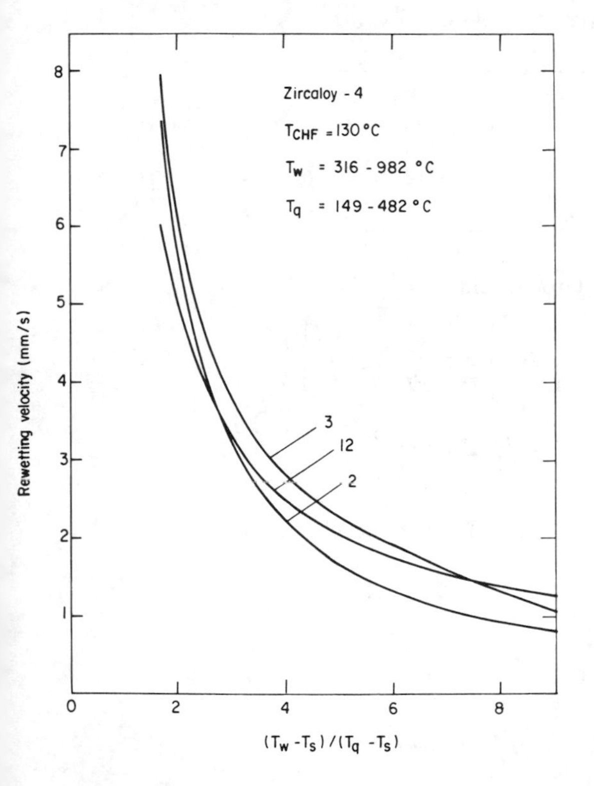


**Figure 2** Variation of the rewetting velocity with initial and quench temperatures and number of segments (Elias and Yadigaroglu, 1977).

RELAP-4/MOD-6 is to be released soon. MOD-6 is primarily a reflood code. Heat transfer correlations in RELAP-4/MOD-5 are given below (see Chap. 14). Figures 3 and 4 are the descriptions of heat transfer logic flow paths for MOD-5 and MOD-6, respectively. For RELAP-4/MOD-5 the heat transfer correlations are

Mode 1: Subcooled liquid forced convection (Dittus and Boelter)

$$h = 0.023 \frac{k}{D_e} \mathrm{Pr}_l^{0.4} \mathrm{Re}_l^{0.8}$$

Mode 2: Nucleate boiling (Thom)

$$q = 1.97 \times 10^{-3} (T_w - T_s)^2 e^{0.23p}$$

where $p$ is in megapascals, $T$ is in kelvins, and $q$ is in megawatts per square meter.

Mode 3: Forced convection vaporization (Schrock and Grossman)

$$h = (2.5)(0.023) \frac{k_f}{D_e} \mathrm{Pr}_f^{0.4} [\mathrm{Re}_f (1-x)]^{0.8}$$

$$\times \left[ \left( \frac{x}{1-x} \right)^{0.9} \left( \frac{\mu_g}{\mu_f} \right)^{0.1} \left( \frac{\rho_f}{\rho_g} \right)^{0.5} \right]^{0.75}$$

Mode 4: Transition boiling (McDonough, Milich, and King)

$$q = q_{\mathrm{CHF}} - C(p)(T_w - T_{w,\mathrm{CHF}})$$

| Pressure (bar) | $C$ (W/m$^2$ K) |
|---|---|
| 138 | 5560 |
| 83 | 6710 |
| 55 | 8530 |

Mode 5: Stable film boiling (Groeneveld)

$$h = a \frac{k_g}{D_e} \mathrm{Pr}_w^c \left\{ \mathrm{Re}_g \left[ x + \frac{\rho_g}{\rho_f}(1-x) \right] \right\}^b$$

$$\times \left[ 1.0 - 0.1(1-x)^{0.4} \left( \frac{\rho_f}{\rho_g} - 1 \right)^{0.4} \right]^d$$

| | Groeneveld Equation (5.9) | Groeneveld Equation (5.7) |
|---|---|---|
| $a =$ | 0.00327 | 0.052 |
| $b =$ | 0.901 | 0.688 |
| $c =$ | 1.32 | 1.26 |
| $d =$ | −1.50 | −1.06 |

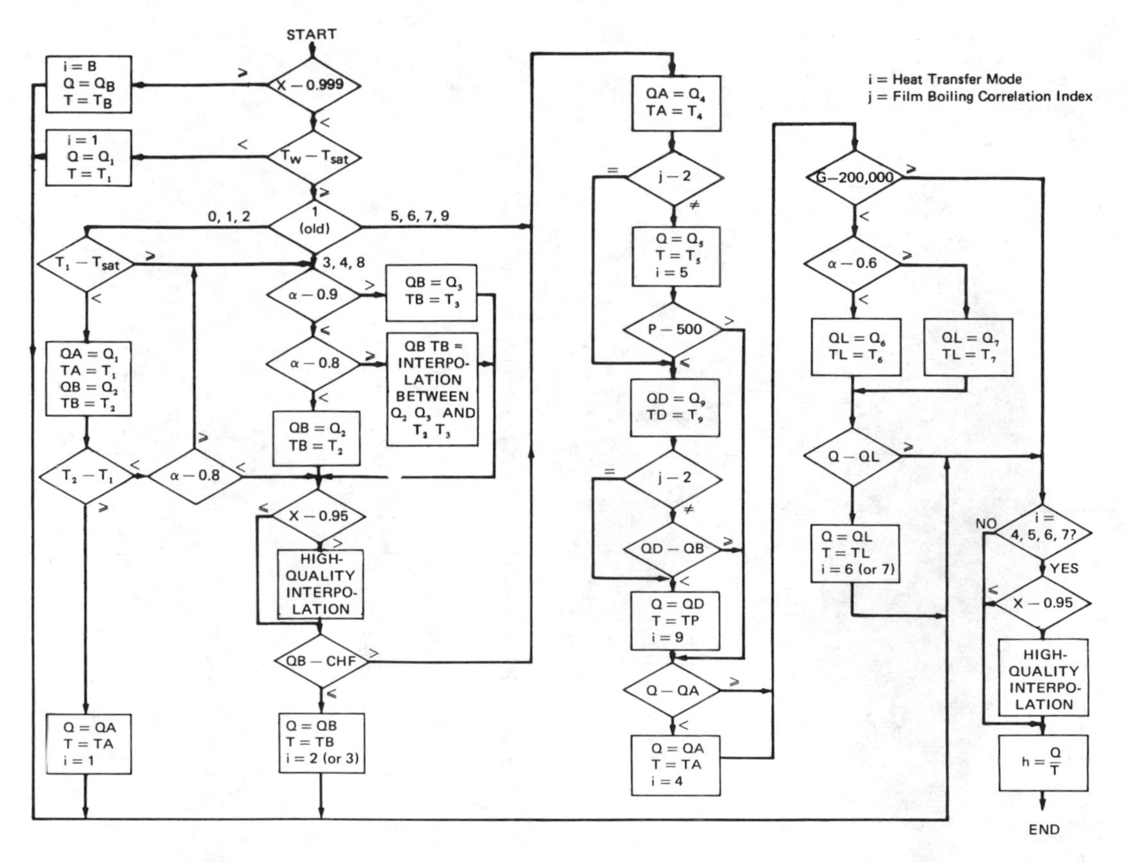


**Figure 3** (*a*) Selection of heat transfer correlations in RELAP-4/MOD-5. (English units used for pressure and mass velocity.)

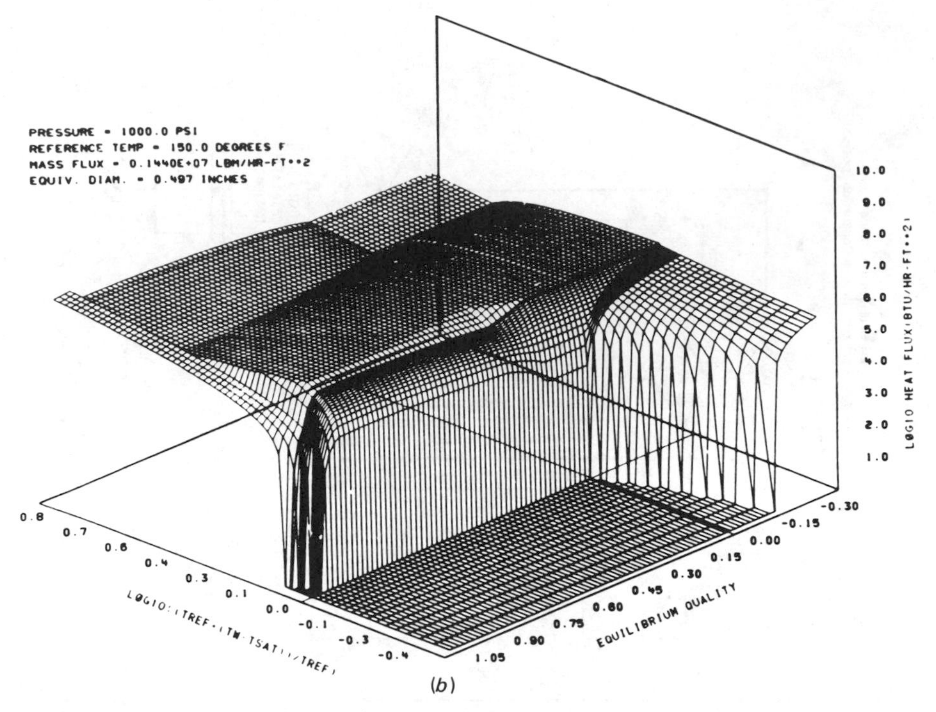


**Figure 3** (*Continued*) (*b*) RELAP-4/MOD-5 LOCA heat transfer surface.

Mode 6: Low-flow film boiling (modified Bromley)

$$h = 0.62 \left[ \frac{k_g^3 h_{fg} \rho_g g (\rho_f - \rho_g)}{\mu_g L \, \Delta T_{\text{sat}}} \right]^{0.25}$$

$$L = 2\pi \sqrt{\frac{\sigma}{g(\rho_f - \rho_g)}}$$

Mode 7: Free convection plus radiation

$$h = h_c + h_r$$

$$h_c = 0.4(\text{Gr}\,\text{Pr}_f)^{0.2}$$

$$\text{Gr} = \frac{L^3 \beta_g \rho_g^2 \, \Delta T_{\text{sat}}}{\mu_g^2}$$

$$L = \frac{D_e}{2}$$

$$h_r = 0.23 \, \frac{(1.714 \times 10^{-9})(T_w^4 - T_{\text{sat}}^4)}{\Delta T_{\text{sat}}}$$

Mode 8: Superheated vapor forced convection (Dittus and Boelter)

$$h = 0.023 \, \frac{k}{D_e} \, \text{Pr}^{0.4} \, \text{Re}^{0.8}$$

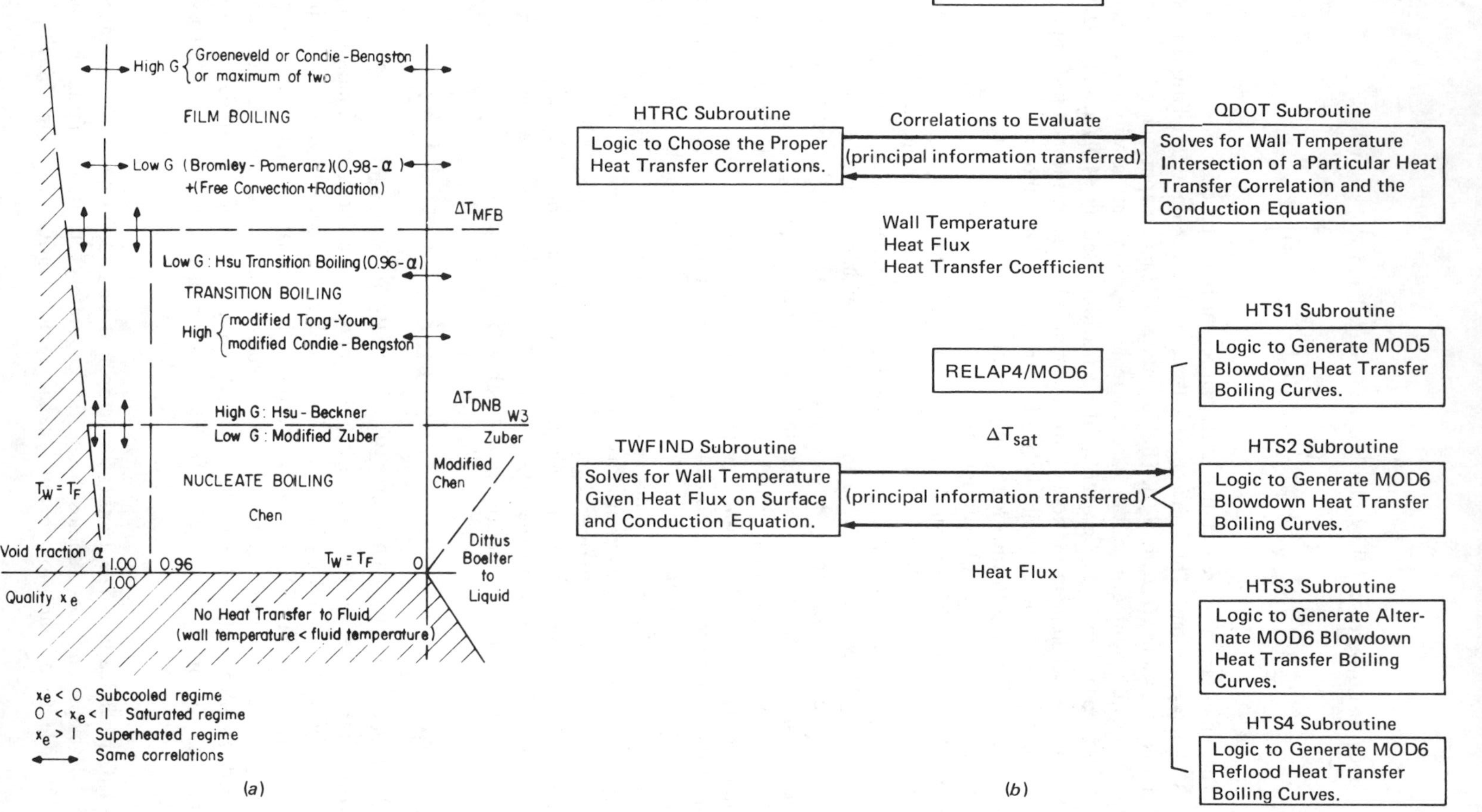


**Figure 4** (*a*) RELAP-4/MOD-6 HTS2 blowdown heat transfer correlations and their regions of application. (*b*) Comparison of RELAP-4/MOD-5 and MOD-6 heat transfer logic.

Mode 9: Low-pressure flow film boiling (Dougall and Rohsenow)

$$h = 0.023 \frac{k_g}{D_e} \mathrm{Pr}_g^{0.4} \left\{ \mathrm{Re}_g \left[ x + \frac{\rho_g}{\rho_f}(1-x) \right] \right\}^{0.8}$$

The important features of the heat transfer surface approach are the modular forms of the equations. Thus, the code will search for the right correlations to use to fit the conditions ($G$, $\alpha$, $p$, $T_w$). The advantage of such an approach as compared with the earlier approach is the ease of replacing an equation without a major change of code structure. Comparing the equations used in MOD-5 and MOD-6 shows the gradual implementation of BE recommendations mentioned in Chap. 14. Since the code revision can only be made during a certain window period, it will always lag the BE recommendation in implementation.

Clad-temperature profile during the blowdown period as calculated from RELAP/MOD-5 and MOD-6 can be compared with ORNL experimental data. With thermocouples located in both rod center and inside the clad and because of careful calibrations, ORNL temperature data may be more reliable than those of other major test facilities. However, due to the absence of in-core flow instrumentation, the local hydraulic parameters have to be interpolated from the inlet and outlet measurements through the code itself and thus are subject to the limitations of the code too. Such interpolation could be quite a bit in error for rapid transient conditions. The code calculation flow parameters are shown in Fig. 5 for inlet and outlet. But the error in these core flow parameters is uncertain. Figure 6 shows a comparison of experimental and MOD-5-calculated clad temperatures at various levels (Hedrick et al., 1977). It can be seen that the major uncertainty is during the rapid transient flow. This uncertainty causes a major deviation of predicted and experimental clad temperature. Even so, at the middle level where the flow conditions are less uncertain, the comparison is fairly favorable. Figures 7 and 8 show the comparison using the MOD-5 and MOD-6 programs. The predictions of MOD-6 seem an improvement over those of MOD-5.

## 2.2 TRAC

TRAC (Liles et al., 1977; TRAC-P1, 1978) is a $2V2T$ three-dimensional code with a drift-flux model used in some cases. It is a highly detailed and highly modular code and is very versatile. TRAC has been designated as the code to serve as the link between the German, U.S., and Japanese three-dimensional large-scale tests. It will be used to analyze and predict the test results of one test, and then the code verified by this test will be used to perform pretest analysis to define input conditions for another test (Fig. 9). The characteristics of TRAC are discussed below.

**2.2.1 General features** The transient reactor analysis code (TRAC) is an advanced code developed by Los Alamos Scientific Laboratory. It is a multidimensional, $2V2T$ code to account for thermodynamic nonequilibrium and slip velocity. In

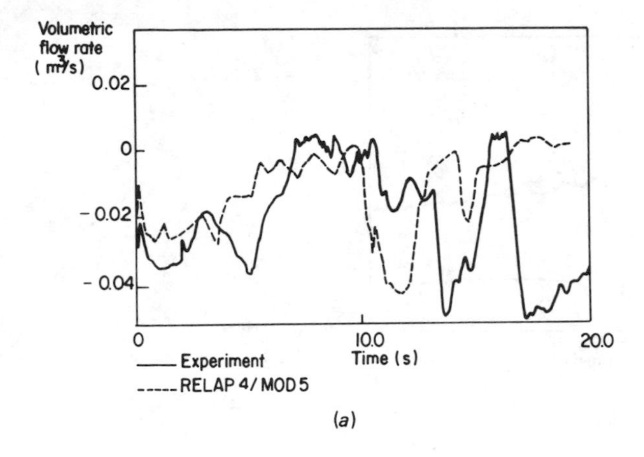


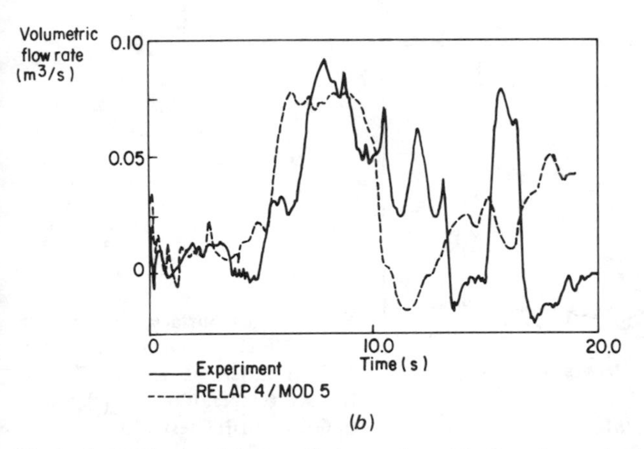


**Figure 5** (*a*) Vertical inlet spool piece volumetric flow (corrected) in ORNL PWR BDHT test 103 (Hedrick et al., 1977). (*b*) Vertical outlet spool piece volumetric flow in BDHT test 103.

addition, the constitutive equations are flow regime-dependent and the heat transfer equations of various modes are incorporated. It is modular both in components and in function, and thus the code is highly flexible.

The physical phenomena that can be treated by TRAC include:

1. ECC downcomer penetration and bypass, including the effects of countercurrent flow and hot walls
2. Lower-plenum refill with entrainment and phase separation effects
3. Bottom flood and falling-film reflood quench fronts
4. Multidimensional flow patterns in the core and plenum regions
5. Pool formation and fallback at the upper-core support-plate (UCSP) region
6. Deentrainment and pool formation in the upper plenum

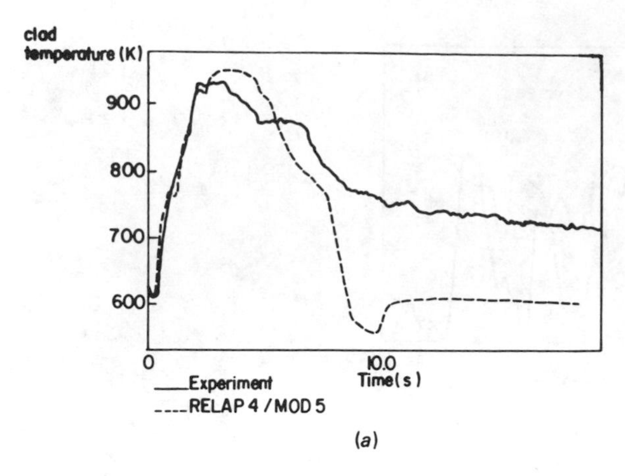


(a)

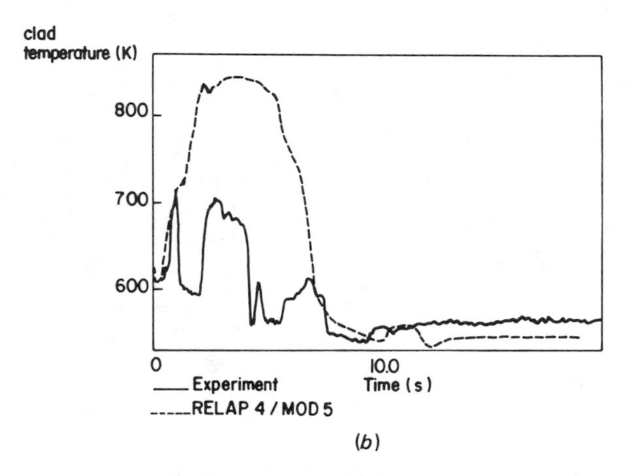


(b)

**Figure 6** (*a*) Surface temperature, rod 18, level G in ORNL BDHT test 103 (Hedrick et al., 1977). (*b*) Surface temperature, rod 18, level J in ORNL BDHT test 103.

7. Steam binding
8. Average-rod and hot-rod cladding temperature histories
9. Alternate ECC injection systems, including hot-leg and upper-head injection
10. Direct injection of subcooled ECC water, without the requirement for artificial mixing zones
11. Critical flow (choking)
12. Liquid carryover during reflood
13. Metal–water reaction
14. Water-hammer effects
15. Wall friction losses

**2.2.2 Hydrodynamics** TRAC is one-dimensional for components and three-dimensional for the vessel. The one-dimensional field equations for components used in TRAC are given below (TRAC-P1, 1978).

Mixture mass equation:

$$\frac{\partial}{\partial t}\rho_m + \frac{\partial}{\partial x}(\rho_m V_m) = 0 \tag{2}$$

Vapor mass equation:

$$\frac{\partial}{\partial t}(\alpha\rho_g) + \frac{\partial}{\partial x}(\alpha\rho_g V_m) + \frac{\partial}{\partial x}\left[\frac{\alpha\rho_g(1-\alpha)\rho_l V_r}{\rho_m}\right] = \Gamma \tag{3}$$

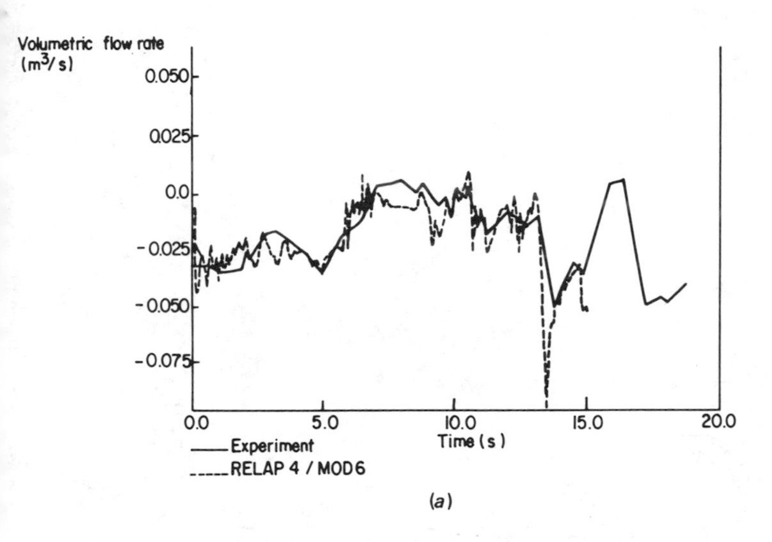


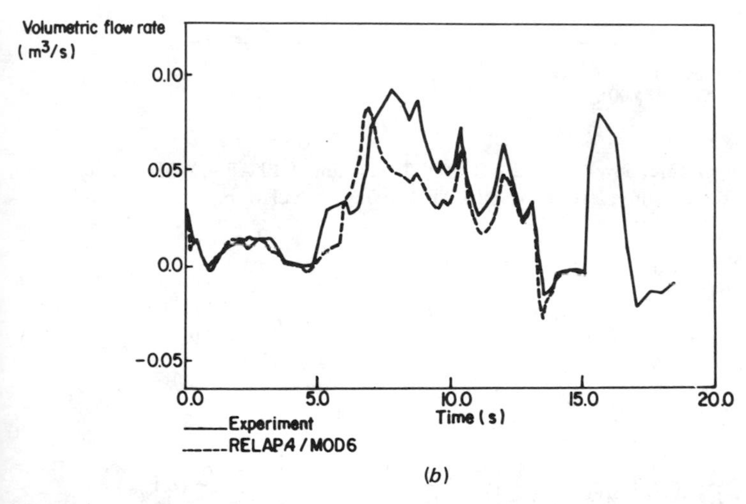


**Figure 7** (*a*) Vertical inlet spool piece volumetric flow with RELAP-4/MOD-6 calculation (Davis, 1977). (*b*) Volumetric flow, vertical outlet spool piece and RELAP-4/MOD-6.

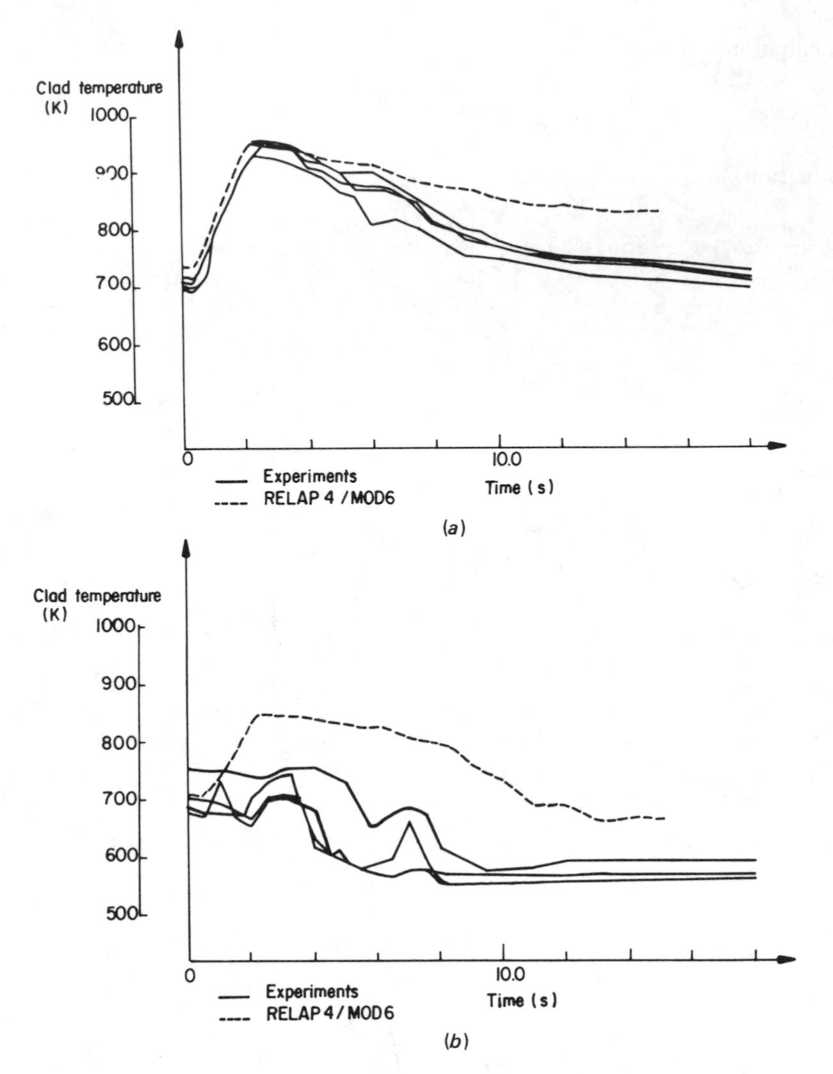


**Figure 8** (*a*) Clad temperature, level G in PWR BDHT test and RELAP-4/MOD-6 calculation (Davis, 1977). (*b*) Clad temperature, level J and RELAP-4/MOD-6 calculation.

Mixture equation of motion:

$$\frac{\partial}{\partial t} V_m + V_m \frac{\partial}{\partial x} V_m + \frac{1}{\rho_m} \frac{\partial}{\partial x} \left[ \frac{\alpha \rho_g (1-\alpha) \rho_l V_r^2}{\rho_m} \right] = - \frac{1}{\rho_m} \frac{\partial p}{\partial x} - K V_m |V_m| + g \quad (4)$$

Vapor thermal energy equation:

$$\frac{\partial}{\partial t}(\alpha \rho_g e_g) + \frac{\partial}{\partial x}(\alpha \rho_g V_m e_g) + \frac{\partial}{\partial x}\left[ \frac{\alpha \rho_g (1-\alpha) \rho_l V_r e_g}{\rho_m} \right] + p \frac{\partial}{\partial x}(\alpha V_m)$$

$$+ p \frac{\partial}{\partial x}\left[ \frac{\alpha(1-\alpha)\rho_l}{\rho_m} V_r \right] = q_{wg} + q_{ig} - p \frac{\partial \alpha}{\partial t} + \Gamma h_{lg} \quad (5)$$

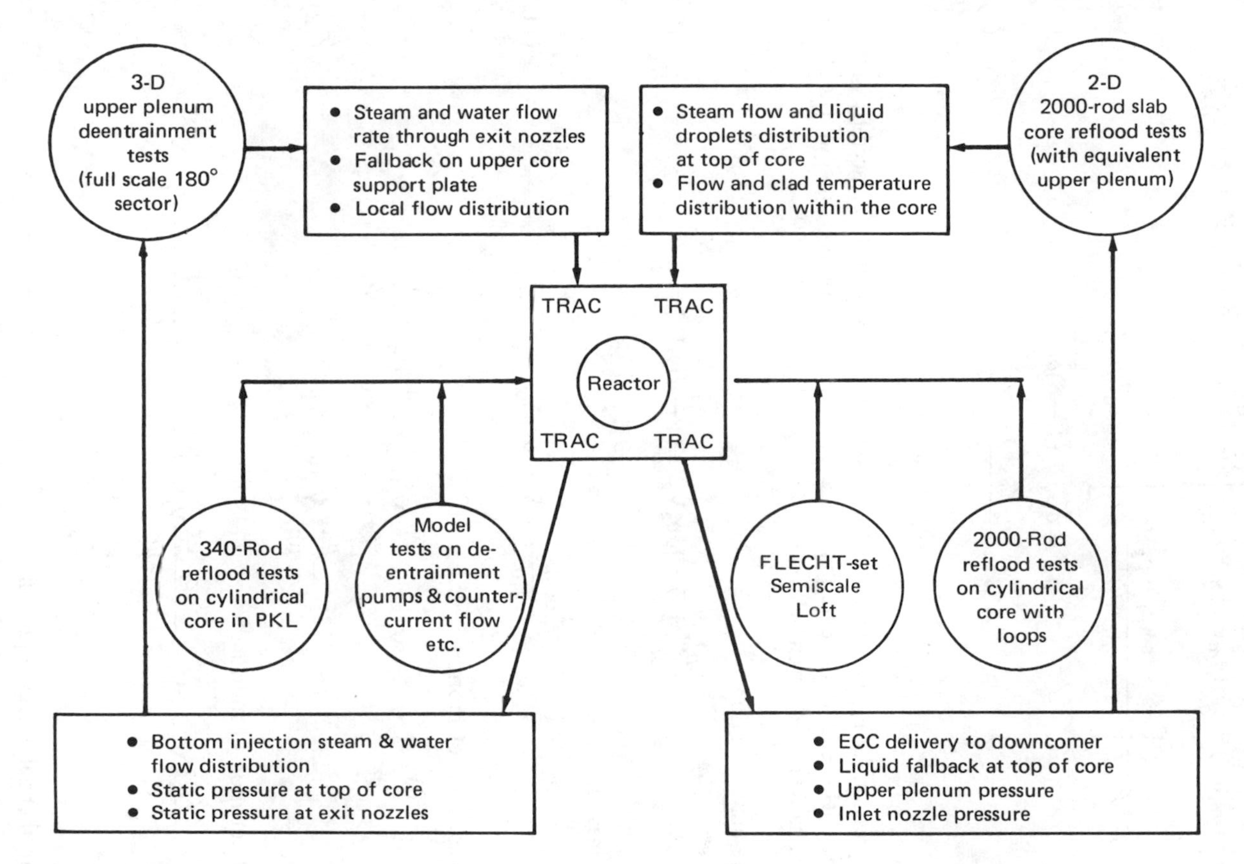


**Figure 9** Coordinated reflood and deentrainment tests.

Mixture thermal energy equation:

$$\frac{\partial}{\partial t}(\rho_m e_m) + \frac{\partial}{\partial x}(\rho_m e_m V_m) + \frac{\partial}{\partial x}\left[\frac{(1-\alpha)\rho_l \alpha \rho_g (e_g - e_l)}{\rho_m} V_r\right] + p\frac{\partial V_m}{\partial x}$$

$$+ p\frac{\partial}{\partial x}\left[\frac{\alpha(1-\alpha)(\rho_l - \rho_g)}{\rho_m} V_r\right] = q_{wg} + q_{wl} \tag{6}$$

where

$$\rho_m \triangleq \alpha\rho_g + (1-\alpha)\rho_l \tag{7}$$

$$V_m \triangleq \frac{\alpha\rho_g V_g + (1-\alpha)\rho_l V_l}{\rho_m} \tag{8}$$

$$V_r \triangleq V_g - V_l \tag{9}$$

The expression for $e_m$ is the same as that shown in Eq. (8) with $V$ replaced by $e$.

The field equations describing the two-phase, two-fluid flow in the three-dimensional vessel are given below (TRAC P-1, 1978).

Mixture mass equation:

$$\frac{\partial \rho_m}{\partial t} + \nabla \cdot [\alpha\rho_g \mathbf{V}_g + (1-\alpha)\rho_l \mathbf{V}_l] = 0 \tag{10}$$

Vapor mass equation:

$$\frac{\partial(\alpha\rho_g)}{\partial t} + \nabla \cdot (\alpha\rho_g \mathbf{V}_g) = \Gamma \tag{11}$$

Vapor equation of motion:

$$\frac{\partial \mathbf{V}_g}{\partial t} + \mathbf{V}_g \cdot \nabla \mathbf{V}_g = -\frac{c_i}{\alpha\rho_g}\mathbf{V}_r|\mathbf{V}_r| - \frac{1}{\rho_g}\nabla p - \frac{\Gamma}{\alpha\rho_g}(\mathbf{V}_g - \mathbf{V}_{ig})$$

$$-\frac{c_{wg}}{\alpha\rho_g}\mathbf{V}_g|\mathbf{V}_g| + \mathbf{g} \tag{12}$$

Liquid equation of motion:

$$\frac{\partial \mathbf{V}_l}{\partial t} + \mathbf{V}_l \cdot \nabla \mathbf{V}_l = \frac{c_i}{(1-\alpha)\rho_l}\mathbf{V}_r|\mathbf{V}_r| - \frac{1}{\rho_l}\nabla p + \frac{\Gamma}{(1-\alpha)\rho_l}(\mathbf{V}_l - \mathbf{V}_{il})$$

$$-\frac{c_{wl}\mathbf{V}_l|\mathbf{V}_l|}{(1-\alpha)\rho_l} + \mathbf{g} \tag{13}$$

Mixture thermal energy equation:

$$\frac{\partial[(1-\alpha)\rho_l e_l + \alpha\rho_g e_g]}{\partial t} + \nabla \cdot [(1-\alpha)\rho_l e_l \mathbf{V}_l + \alpha\rho_g e_g \mathbf{V}_g]$$

$$= -p\nabla \cdot [(1-\alpha)\mathbf{V}_l + \alpha\mathbf{V}_g] + q_{wg} + q_{wl} \tag{14}$$

Vapor thermal energy equation:

$$\frac{\partial(\alpha\rho_g e_g)}{\partial t} + \nabla \cdot (\alpha\rho_g \mathbf{V}_g e_g) = -p\,\frac{\partial\alpha}{\partial t} - p\nabla \cdot (\alpha \mathbf{V}_g) + q_{wg} + q_{ig} + \Gamma h_{lg} \qquad (15)$$

where $\rho_m$ and $V_r$ are identical to the definitions used in the drift-flux model [see Eqs. (7) and (9)].

The finite-difference equations are written in terms of a staggered difference scheme on the Eulerian mesh. State variables, such as pressure, internal energy, and void fraction, are obtained at the center of the mesh cell, and flow variables are obtained at the cell boundaries (Fig. 10). With this configuration, the donor-cell average is expressed as

$$\begin{aligned} \langle YV\rangle_{j+1/2} &= Y_j V_{j+1/2} \qquad \text{for } V_{j+1/2} \geqslant 0 \\ &= Y_{j+1} V_{j+1/2} \qquad \text{for } V_{j+1/2} < 0 \end{aligned} \qquad (16)$$

where $Y$ is any state variable or combination of state variables. An integer subscript indicates that a quantity is evaluated at a mesh-cell center and a half-integer denotes that it is obtained at a cell boundary. With this notation the finite-difference divergence operator is

$$\nabla_j(YV) = \frac{A_{j+1/2}\langle YV\rangle_{j+1/2} - A_{j-1/2}\langle YV\rangle_{j-1/2}}{\text{vol}_j} \qquad (17)$$

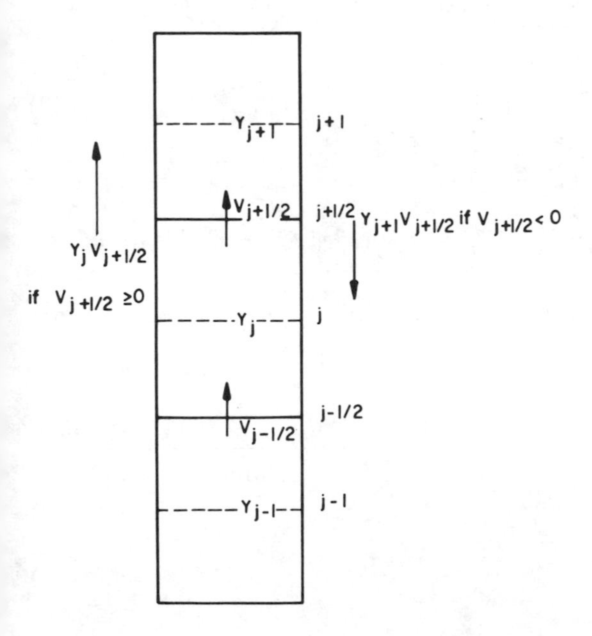


**Figure 10** Staggered meshes in TRAC.

where $A$ is the cross-sectional area, and $\mathrm{vol}_j$ is the volume of the $j$th cell. Slight variations of these donor-cell terms appear in the velocity equation of motion. Donor-cell averages are of the form

$$\begin{aligned} \langle YV_r^2\rangle_j &= Y_j V_{r,j-1/2}^2 \qquad \text{for } V_{r,j-1/2} \geqslant 0 \\ &= Y_j V_{r,j+1/2}^2 \qquad \text{for } V_{r,j+1/2} < 0 \end{aligned} \tag{18}$$

and the donor-cell form of the term $V_m \nabla V_m$ is

$$\begin{aligned} V_{m,j+1/2} \nabla_{j+1/2} V_m &= \frac{V_{m,j+1/2}(V_{m,j+1/2} - V_{m,j-1/2})}{\Delta x_j} \qquad \text{for } V_{m,j+1/2} \geqslant 0 \\ &= \frac{V_{m,j+1/2}(V_{m,j+3/2} - V_{m,j+1/2})}{\Delta x_{j+1}} \qquad \text{for } V_{m,j+1/2} < 0 \end{aligned} \tag{19}$$

The finite-difference equations for the *partially implicit* method used in TRAC are presented below.

Mixture mass equation:

$$\frac{(\rho_m^{n+1} - \rho_m^n)_j}{\Delta t} + \nabla_j(\rho_m^n V_m^{n+1}) = 0 \tag{20}$$

Vapor mass equation:

$$\frac{(\alpha^{n+1}\rho_g^{n+1} - \alpha^n \rho_g^n)_j}{\Delta t} + \nabla_j(\alpha^n \rho_g^n V_m^{n+1}) + \nabla_j(\rho_f^n V_r^n) = \Gamma^{n+1} \tag{21}$$

Mixture energy equation:

$$\begin{aligned} &\frac{(\rho_m^{n+1} e_m^{n+1} - \rho_m^n e_m^n)_j}{\Delta t} + \nabla_j(\rho_m^n e_m^n V_m^{n+1}) + \nabla_j[\rho_f^n(e_g^n - e_l^n)V_r^n] \\ &= -p_j^{n+1}\nabla_j\left[V_m^{n+1} + \rho_f^n\left(\frac{1}{\rho_g^n} - \frac{1}{\rho_l^n}\right)V_r^n\right] + q_{j,wg} + q_{j,wl} \end{aligned} \tag{22}$$

Vapor energy equation:

$$\begin{aligned} &\frac{[(\alpha\rho_g e_g)^{n+1} - (\alpha\rho_g e_g)^n]_j}{\Delta t} + \nabla_j(\alpha^n \rho_g^n e_g^n V_m^{n+1}) + \nabla_j(\rho_f^n e_g^n V_r^n) + p_j^{n+1}\nabla_j(\alpha^n V_m^{n+1}) \\ &+ p_j \nabla_j\left(\frac{\rho_f^n V_r^n}{\rho_g^n}\right) = \frac{-p_j^{n+1}(\alpha_j^{n+1} - \alpha_j^n)}{\Delta t} + (q_{wg} + q_{ig} + \Gamma h_{lg}^{n+1})_j \end{aligned} \tag{23}$$

Mixture equation of motion:

$$\begin{aligned} &\frac{(V_m^{n+1} - V_m^n)_{j+1/2}}{\Delta t} + V_{m,j+1/2}^n \nabla_{j+1/2} V_m^n = \frac{-(p_{j+1}^{n+1} - p_j^{n+1})/\overline{\Delta x}_{j+1/2} - \nabla_{j+1/2}(\rho_f V_r^2)^n}{\bar{\rho}_{m,j+1/2}^n} \\ &+ g^n - (K^n V_m^{n+1}|V_m^n|)_{j+1/2} \end{aligned} \tag{24}$$

where

$$\overline{\Delta x}_{j+1/2} = \frac{\Delta x_j + \Delta x_{j+1}}{2} \tag{25}$$

$$\rho_f^n = \frac{\alpha^n(1-\alpha^n)\rho_g^n\rho_l^n}{\rho_m^n} \tag{26}$$

$$\begin{aligned} \bar{\rho}_{m,j+1/2}^n &= \rho_{m,j}^n \quad \text{for } V_{j+1/2} \geqslant 0 \\ &= \rho_{m,j+1}^n \quad \text{for } V_{j+1/2} < 0 \end{aligned} \tag{27}$$

The superscript $n$ indicates that the quantity is evaluated at the "current" time and thus is known, whereas the superscript $n+1$ indicates that the variable is evaluated at the new time, and hence is an unknown for which the equations must be solved.

The *fully implicit* difference equations used in TRAC are presented below.

Mixture mass equation:

$$\frac{(\rho_m^{n+1}-\rho_m^n)_j}{\Delta t} + \nabla_j(\rho_m^{n+1}V_m^{n+1}) = 0 \tag{28}$$

Vapor mass equation:

$$\frac{(\alpha^{n+1}\rho_g^{n+1}-\alpha^n\rho_g^n)_j}{\Delta t} + \nabla_j(\alpha^{n+1}\rho_g^{n+1}V_m^{n+1}) + \nabla_j(\rho_f^{n+1}V_r^n) = \Gamma^{n+1} \tag{29}$$

Mixture energy equation:

$$\begin{aligned} &\frac{(\rho_m^{n+1}e_m^{n+1}-\rho_m^n e_m^n)_j}{\Delta t} + \nabla_j(\rho_m^{n+1}e_m^{n+1}V_m^{n+1}) + \nabla_j[\rho_f^{n+1}(e_g^{n+1}-e_l^{n+1})V_r^n] \\ &= -p_j^{n+1}\nabla_j\left[V_m^{n+1} + \rho_f^{n+1}\left(\frac{1}{\rho_g^{n+1}} - \frac{1}{\rho_l^{n+1}}\right)V_r^n\right] + q_{j,wg} + q_{j,wl} \end{aligned} \tag{30}$$

Vapor energy equation:

$$\begin{aligned} &\frac{[(\alpha\rho_g e_g)^{n+1}-(\alpha\rho_g e_g)^n]_j}{\Delta t} + \nabla_j(\alpha^{n+1}\rho_g^{n+1}e_g^{n+1}V_m^{n+1}) + \nabla_j(\rho_f^{n+1}e_g^{n+1}V_r^n) \\ &+ p_j^{n+1}\nabla_j(\alpha^{n+1}V_m^{n+1}) + p_j\nabla_j\left(\frac{\rho_f^{n+1}}{\rho_g^{n+1}}V_r^n\right) = \frac{-p_j^{n+1}(\alpha_j^{n+1}-\alpha_j^n)}{\Delta t} \\ &+ (q_{wg}+q_{ig}+\Gamma h_{lg}^{n+1})_j \end{aligned} \tag{31}$$

Mixture equation of motion:

$$\begin{aligned} &\frac{(V_m^{n+1}-V_m^n)_{j+1/2}}{\Delta t} + \frac{V_{m,j+1/2}^{n+1}(V_{m,j+3/2}^{n+1}-V_{m,j-1/2}^{n+1})}{\Delta x_j + \Delta x_{j+1}} \\ &= \frac{-(p_{j+1}^{n+1}-p_j^{n+1})/\overline{\Delta x}_{j+1/2} - \nabla_{j+1/2}[\rho_f^{n+1}(V_r^2)^n]_j}{\bar{\rho}_{m,j+1/2}^{n+1}} \\ &+ g^n - (K^n V_m^{n+1}|V_m^{n+1}|)_{j+1/2} \end{aligned} \tag{32}$$

The slip correlations are flow regime-dependent. For the flow regime map shown in Fig. 11 the following relative velocity correlations are used for nonhorizontal flow. These correlations have been shown to be in satisfactory agreement with

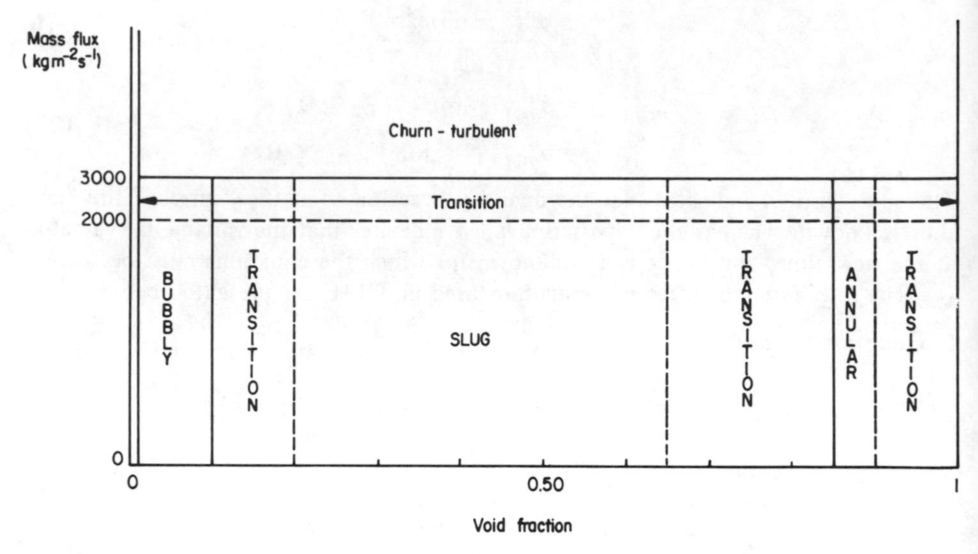


**Figure 11** TRAC flow regime map for slip correlations (TRAC P1, 1978).

experimentally measured relative velocities in steam–water flows for each particular flow regime.

Bubbly regime:

$$V_r = \frac{1.41}{1-\alpha}\left[\frac{\sigma g(\rho_l - \rho_g)}{\rho_l^2}\right]^{1/4} \tag{33}$$

Slug regime:

$$V_r = \frac{0.345}{1-\alpha}\left[\frac{g D_h(\rho_l - \rho_g)}{\rho_l}\right]^{1/2} \tag{34}$$

Churn-turbulent regime:

$$V_r = \frac{V_m}{(1 - C_0\alpha)/(C_0 - 1) + \alpha\rho_g/\rho_m} \tag{35}$$

where $C_0 = 1.1$ and $\alpha$ is restricted to a maximum value of 0.8, and

Annular regime:

$$V_r = \frac{V_m}{[\rho_g(76 - 75\alpha)/\rho_l\sqrt{\alpha}]^{1/2} + \alpha\rho_g/\rho_m} \tag{36}$$

The dashed lines in Fig. 11 are transition regions between flow regimes. In these transition regions, the relative velocity is linearly interpolated between values for the two adjacent flow regimes. The linear coefficient is determined by the vapor fraction if $G < 2000$ kg/m$^2$ s and also by the mass flux if $2000 \leqslant G \leqslant 3000$ kg/m$^2$ s.

For example, if $G < 2000$ and $\alpha = 0.18$, then $V_r = 0.2$ times Eq. (33) evaluated at $\alpha = 0.10$ plus 0.8 times Eq. (34) evaluated at $\alpha = 0.20$.

**2.2.3 Heat transfer** The heat transfer correlations include two classes: the conduction equations for solids and the convective equations for fluids, the latter including all the boiling two-phase flow modes. Of most interest in this group are the equations used for reflood, which are shown in the list below.

1. Single-phase liquid
    a. Laminar: constant Nusselt number
    b. Turbulent: Dittus-Boelter
2. Nucleate boiling and forced convection vaporization
    a. Chen
3. Critical heat flux
    a. Low flow: Zuber
    b. High flow: Biasi
4. Transition boiling
    a. Log-log interpolation
5. Minimum film boiling point
    a. Homogeneous nucleation
    b. Henry-Berenson
6. Film boiling
    a. Modified Bromley and Dougall-Rohsenow

**2.2.4 Examples** Application of TRAC is very versatile. Figure 12 shows the calculation of peak clad temperature for TRAC. Figure 13 shows the noding for a

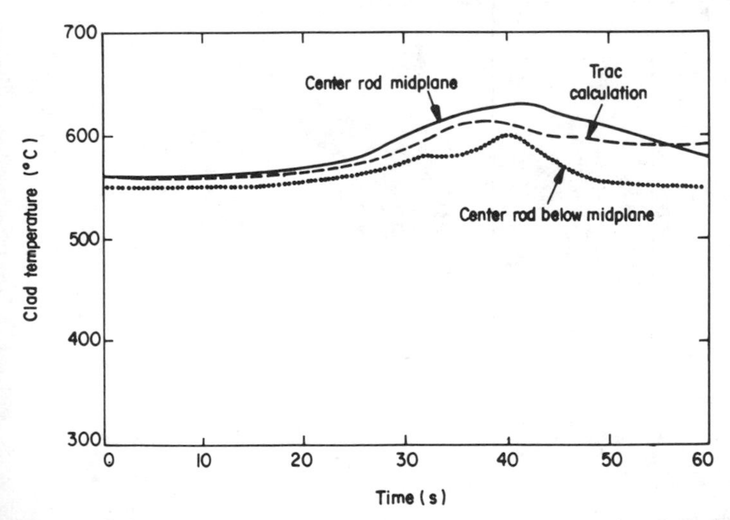


**Figure 12** Comparison of PKL test and TRAC calculation of peak clad temperature (TRAC-P1, 1978).

three-dimensional upper-plenum test facility in FR Germany. Note the block-off of half the cylindrical vessel to represent a 180° test vessel, which was the older version of design. The current design calls for a 360° vessel. The calculated void fractions in the first two levels above the upper core support plate are shown in Fig. 14.

## 3 CONCLUDING REMARKS

LOCA code developments have progressed rapidly in the past few years. Experimental tests have been used as a basis for code development and verification. Codes are now entering new phases of their mission in that they are not only used for test predictions, but also as a link between tests in different facilities and as a tool to

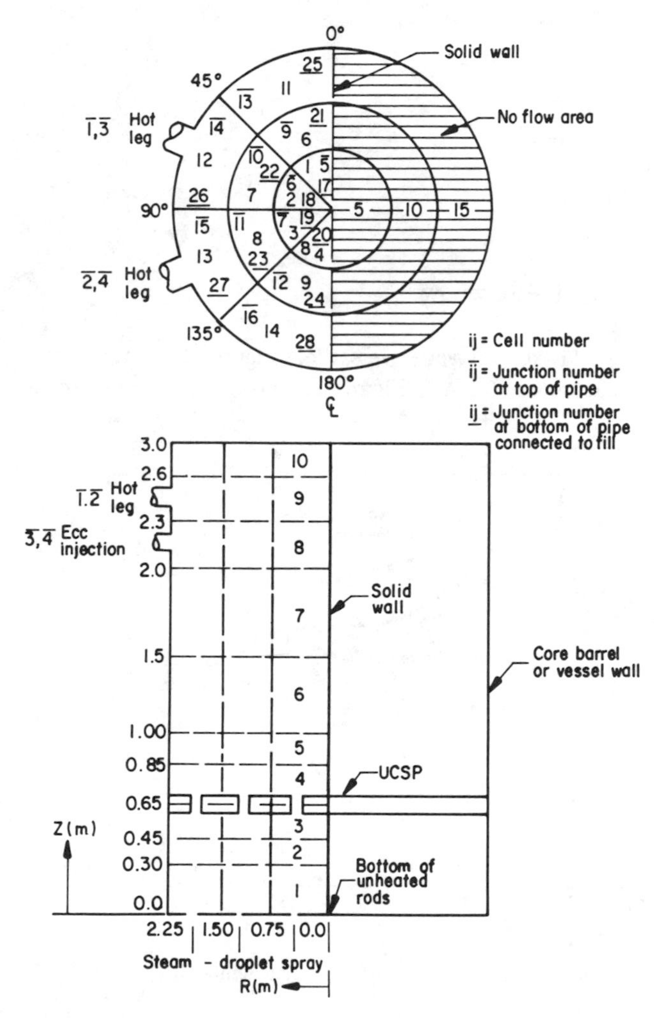


Figure 13 German upper-plenum vessel noding (TRAC-P1, 1978).

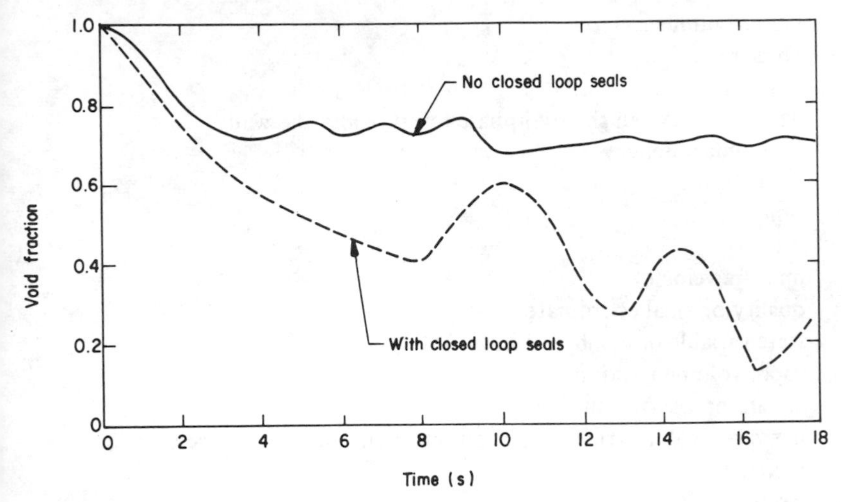


**Figure 14** Average void fraction in the first two levels above the upper core support plate as calculated by TRAC for the German three-dimensional facility.

determine local parameters in the core. If a new generation of in-core instruments can be successfully used with codes, the future LOCA experiments can be analyzed and evaluated to yield much more confidence and insight into the understanding of LOCA phenomena and to improve the confidence in reactor safety margins.

## NOMENCLATURE

| | |
|---|---|
| $A$ | area |
| $C$ | function used in the McDonough et al. correlation |
| $C_0$ | distribution parameter |
| $c$ | shear or friction coefficient in two-fluid equations |
| $D$ | diameter |
| $E$ | energy transfer between the phases |
| $e$ | specific internal energy |
| $F$ | weighing factor used in the Tong-Young equation |
| $G$ | mass flux |
| $g$ | acceleration resulting from gravity |
| Gr | Grashof number |
| $h$ | specific enthalpy or heat transfer coefficient |
| $h_{lg}$ or $h_{fg}$ | heat of vaporization |
| $j$ | superficial mixture velocity |
| $K$ | wall shear coefficient in drift-flux equations |
| $k$ | thermal conductivity |
| $L$ | characteristic length |
| $M$ | momentum transfer between the phases |

| | |
|---|---|
| Pr | Prandtl number |
| $p$ | pressure |
| $q$ | heat flux |
| $\bar{q}$ | heat flux between the two-phase mixture and the wall |
| Re | Reynolds number |
| $T$ | temperature |
| $t$ | time |
| $V$ | velocity |
| $\bar{V}$ | mixture velocity |
| $x$ | quality or axial coordinate |
| $Y$ | state variable or combination of state variable |
| $\alpha$ | vapor volume fraction |
| $\beta$ | expansion coefficient |
| $\Gamma$ | net vapor volumetric production rate resulting from phase change |
| $\Delta$ | increment |
| $\mu$ | viscosity |
| $\rho$ | density |
| $\sigma$ | surface tension |
| $\bar{\tau}$ | wall shear stress |

**Subscripts**

| | |
|---|---|
| CHF | critical heat flux |
| $e$ | equivalent |
| $f$ | liquid |
| $G$ | gas |
| $g$ | vapor |
| $h$ | hydraulic |
| $i$ | interface or one-dimensional cell index in heat transfer equations |
| $j$ | mixture quantity (weighed by volume) or one-dimensional cell index in hydrodynamic equations |
| $K$ | phase index |
| $L$ | liquid |
| $l$ | liquid |
| $m$ | mixture quantity (weighed by mass) |
| $q$ | quench |
| $r$ | relative quantity or radiation |
| $s$ | saturation |
| sat | saturation |
| $v$ | vapor |
| $w$ | wall |

**Abbreviations**

| | |
|---|---|
| BE | best estimate |
| CHF | critical heat flux |

| | |
|---|---|
| D | drift, dimension |
| DF | drift flux |
| ECC | emergency core cooling |
| EM | evaluation model |
| HEM | homogeneous equilibrium model |
| S | slip |
| T | temperature |
| TF | two fluid |
| V | velocity |
| vol | volume |

## REFERENCES

Davis, C. B., RELAP-4/MOD-6 Comparison with PWR-BDHT Test 103 Core Data, INEL PG-R-77-27, 1977.

Elias, E. and Yadigaroglu, G., A General One-dimensional Model for Conduction-controlled Rewetting of a Surface, *Nucl. Eng. Des.*, vol. 42, no. 2, pp. 185–194, 1977.

Fabic, S., Review of Existing Codes for Loss-of-Coolant Accident Analysis, *Adv. Nucl. Sci. Technol.*, vol. 20, pp. 365–404, 1976.

Fischer, S. R., et al., RELAP-4/MOD-6, A Computer Program for Transient Thermal-hydraulic Analysis of Nuclear Reactor and Related Systems, User's Manual INEL rept. CDAP TR 003, 1978.

Hedrick, R. A., Craddick, W. G., Turnage, K. G., and Hyman, C. R., PWR Blowdown Heat Transfer Separate Effects Program Data Evaluation Report–System Response for Thermal-hydraulic Test Facility Test Series 100, ORNL-NUREG-19, 1977.

Katsma, K. R., et al., RELAP-4/MOD-5, A Computer Program for the Transient Thermal-hydraulic Analysis of Nuclear Reactor and Related Systems, User's Manual, vol. 1, RELAP-4/MOD-5 Description, ANCR-NUREG-1335, 1976.

Liles, D., Kirchner, W., and Mahaffy, K., Fluid Mechanics and Heat Transfer Methods of TRAC, presented at 5th WRSR Information Meet., USNRC, 1977.

Nelson, R., INEL, private communication, 1976.

Tong, L. S. and Young, J. D., A Phenomenological Transition and Film Boiling Heat Transfer Correlation, *Heat Transfer 1974, Proc. 5th Int. Heat Transfer Conf.*, vol. 4, pp. 120–124, 1974.

TRAC-P1: An Advanced Best Estimate Computer Program for PWR LOCA Analysis: 1. Methods, Models, User Information and Programming Details, NUREG/CR-0-63, 1978.

CHAPTER

# TWENTY

# TWO-PHASE FLOW CALCULATIONS IN LIQUID-METAL FAST-BREEDER REACTORS

D. Grand

Two-phase flow problems appearing in safety studies of liquid-metal fast-breeder reactors (LMFBR) are presented in Chap. 2. It is beyond the scope of this chapter to survey all the models available to describe the phenomena during the LMFBR accident scenario. On the contrary, this chapter describes a few models, each corresponding to a given methodology. Thus the following subjects are covered:

1. One-dimensional boiling in a fixed geometry
2. Multidimensional boiling in a fixed geometry
3. One-dimensional clad melting and relocation
4. Multidimensional motion of molten materials (fuel and clad)

For each subject, different models are developed. Rather than cover all these models we give the main characteristics they have in common, sometimes going back to the initial simplest models.

The phenomena listed above may appear sequentially after an unprotected loss of flow (LOF). First the sodium coolant starts boiling in the subassembly. Then dryout may occur, and if it does would result in an increase of the temperature of the solid materials. The steel cladding melts around the hot spot and moves under gravity and vapor drag. The cladding may freeze in upper and lower parts of the subassembly where heat sinks are important. Extended melting and motion of materials like fuel and steel might develop because of continuing fission reaction and might concern the whole core leading to a hypothetical core-disruptive accident. Having recalled these sequences, we will now describe the results in modeling them.

In describing these models, we will focus on the physical aspects, and relate them to data given in the preceding chapters. Very little will be said about numerical methods that are not our present interest. However, these methods often are crucial to the use of these models.

## 1 SODIUM BOILING: SINGLE-CHANNEL APPROACH

In a first approach, one considers that the variations along the axis of the bundle are the most important and that variations of the variables in a cross section may be neglected. This viewpoint justifies the name of the equivalent single-channel approach. From the mathematical point of view, this leads to a one-dimensional problem.

The subassembly may be represented by the typical single channel with three axial sections given in Fig. 1.

For simplicity, we consider a coolant channel with a constant cross-sectional area. Two types of models will be discussed: geometrically structured flow models (SFM) and geometrically unstructured flow models (UFM). The two models differ in the assumptions concerning the two-phase flow pattern inside the tube or subassembly.

1. SFM assumes that the vapor and the liquid are well separated:
   a. A single vapor bubble growing in the heated section of the bundle
   b. Two liquid slugs upstream and downstream of the bubble and a thin liquid film squeezed between the vapor and the clad
2. Except in a few cases, two-phase flows have too complex and random a structure for the SFM model to be realistic. The absence of structure requires

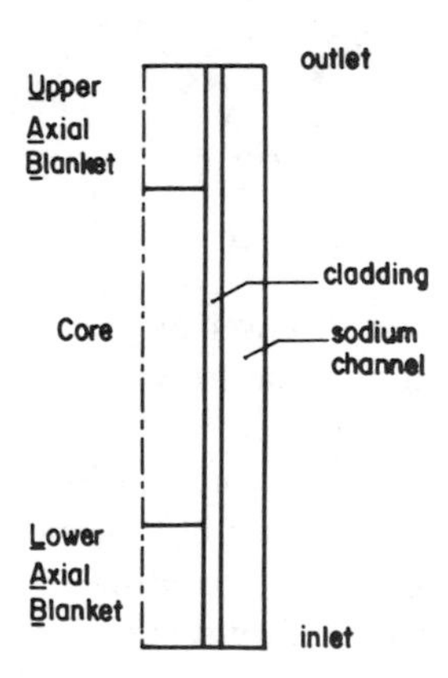


**Figure 1** Single-channel geometry.

the use of surface-averaged equations (UFM). The most rational method consists in using the two-fluid model. From the use of the two-fluid model different models may be deduced if some evolutions are specified.

## 1.1 Structured Flow Models

**1.1.1 Review** As just mentioned the two phases are assumed to be well separated. In the early versions of the SFM model the vapor was inside a single bubble. In the latest developments of these models, for example, last version of the SAS code, a multibubble approach is used, but no more than nine bubbles may be present at the same instant. The important fact is that the interfaces between liquid and vapor bubbles are assumed to be cylinders. Such models are deduced from experimental evidence using sodium pool boiling and forced convection transient boiling in tubes under high power input. In both studies a fast transient occurs, leading to the voiding of the tube in less than 1 s. Well-known codes using this approach include SAS, developed by Argonne National Laboratory (Dunn et al., 1974) and BLOW developed by Gesellschaft für Kernforschung (Wirtz, 1973).

In the references presenting these codes, the physical model and its mathematical resolution are often developed together, which does not make the reading easy and does not always show clearly the basic assumptions.

For purposes of simplicity we will use the paper of Cronenberg et al. (1971), which develops a single-bubble model; though a simple model, it possesses the main features of the SFM.

Before describing the model for the thermohydraulics of the coolant channel, we should note that the heat flux and temperature at the cladding surface constitute the link between the thermohydraulics of the coolant and the heat conduction inside the cylindrical fuel element. Either a lumped parameter model (assuming uniform temperature in a cross section of the fuel element) or the radial heat conduction equation is used as representative of the transient behavior of the fuel pin.

**1.1.2 Single-bubble model** For the thermohydraulics of the coolant, the following assumptions are made:

1. Constant density inside the liquid phase
2. Uniform pressure and temperature inside the vapor bubble, both in equilibrium with the surrounding liquid
3. The initial steady state is a single-phase liquid flow

Once boiling has been initiated, there are three axial regions in the coolant: the two upper and lower liquid slugs and the vapor bubble with its liquid film on the wall.

***The liquid slugs*** For a liquid with constant density and for a uniform channel, three balance equations are written:

Continuity:

$$\frac{\partial w}{\partial z} = 0 \tag{1}$$

Momentum:

$$\rho \frac{dw}{dt} = -\frac{dp}{dz} - \rho g - \frac{\mathcal{P}}{\mathcal{A}} \tau_w \tag{2}$$

where $\mathcal{P}$ and $\mathcal{A}$ are the wetted perimeter and cross-sectional area. $\tau_w$ is the wall shear stress.

Thermal equation:

$$\rho c \left( \frac{\partial T}{\partial t} + w \frac{\partial T}{\partial z} \right) = \phi_w \frac{\mathcal{P}}{\mathcal{A}} \tag{3}$$

with

$$\tau_w = C_f (\mathrm{Re}) \rho \frac{w^2}{2}$$

$$\phi_w = h(T_w - T)$$

where $C_f$ = friction factor
$T_w$ = temperature at the wall
$h$ = heat transfer coefficient

Since $w$ and consequently $dw/dt$ are independent of $z$, the integration of Eq. (2) between the top of the bubble ($z = z_t$) and the outlet ($z = H$) for the upper slug yields

$$\rho(H - z_t) \frac{dw}{dt} = -(p_{\text{out}} - p_v) - \rho g(H - z_t) - \frac{\mathcal{P}}{\mathcal{A}} C_f \rho \frac{w^2}{2} (H - z_t)$$

where $p_{\text{out}}$ and $p_v$ are the pressures at the outlet of the channel and in the vapor, respectively. Then an important hypothesis is made: The lower boundary of the upper slug is a material surface, and there is no mass transfer across it. This condition may be written

$$w = \frac{dz_t}{dt} \tag{4}$$

which states that the liquid velocity in the upper slug is equal to the speed of displacement of its lower boundary. Such a hypothesis is rather restrictive. It fails in many boiling situations.

One example where such a hypothesis fails is in a steady-state boiling situation, for which Costa (1977) gave numerous experimental examples. In this situation boiling occurs on a fixed fraction of the height of the channel. The boundaries of the two-phase region are stationary. However, there is a net flow from the inlet to the outlet of the channel. Obviously, the boundaries of the two-phase region are not material surfaces.

A second situation in which such a hypothesis is restrictive is the case of transients where thermal inertia is important. There is experimental evidence in many transient two-phase flows that thermal inertia of the solid structures (heated pin and can wall) play an important role in the progress of voiding. In such transients, the speed of displacement of the boiling boundary is related to the evolution of the axial temperature profile at the clad surface. In this case, the speed of displacement of the boiling boundary is different from the liquid velocity in the lower slug.

Equation (4) is valid only when the extension of the bubble is controlled by dynamic effects alone. When inserted in the momentum equation for the upper slug, this gives

$$\rho(H-z_t)\frac{d^2 z_t}{dt^2}=p_v-p_{\text{out}}-\rho g(H-z_t)-\frac{\mathcal{P}}{\mathfrak{a}}\frac{C_f}{2}\rho\left(\frac{dz_t}{dt}\right)^2(H-z_t) \tag{5}$$

A similar equation may be derived for the lower slug

$$\rho z_b\frac{d^2 z_b}{dt^2}=p_{\text{in}}-p_v-\rho g z_b-\frac{\mathcal{P}}{\mathfrak{a}}\frac{C_f}{2}\rho\left(\frac{dz_b}{dt}\right)^2 z_b \tag{6}$$

The motions of the slugs are then known if the pressure of the vapor bubble is known.

***The vapor region*** The vapor region includes the bubble itself and the liquid film at the clad. Cronenberg et al. (1971) gave an energy equation for the vapor region (see also Fig. 2):

$$\mathfrak{a}(-\mathbf{q}_t\cdot\mathbf{n}_t-\mathbf{q}_b\cdot\mathbf{n}_b)+\phi'_w\mathcal{P}=\mathfrak{a}\frac{d}{dt}\left[\rho_v h_{fg}(z_t-z_b)\right]+\sigma\mathcal{P}\rho_f c\frac{dT_f}{dt}(z_t-z_b) \tag{7}$$

where $-\mathbf{q}_t\cdot\mathbf{n}_t$ and $-\mathbf{q}_b\cdot\mathbf{n}_b$ = heat fluxes per unit area across the upper and lower boundaries of the bubble region, respectively ($\mathbf{n}_t$ and $\mathbf{n}_b$ being the unit normal vectors outwardly directed)

$\phi'_w$ = heat flux per unit height at the wall of the clad

$h_{fg}$ = heat of vaporization

$\sigma$ = liquid film thickness and $T_f$ its temperature equal to the temperature of the vapor (hypothesis of thermal equilibrium)

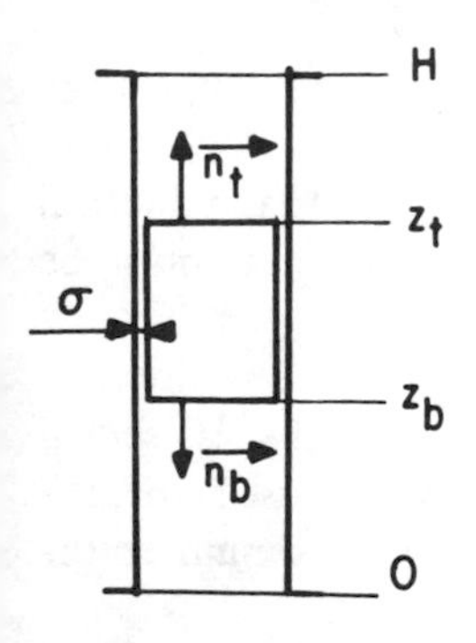


Figure 2 Vapor region.

They claimed that this derivation is made under the assumptions of negligible pressure work, negligible vapor heat capacity, and negligible dissipation.

The heat fluxes appearing in the left-hand side (lhs) of Eq. (7) are evaluated in either of the following ways:

1. By definition of the heat transfer coefficient across the film

$$\phi'_w = h_f \int_{z_b}^{z_t} [T_w(z, t) - T_g(t)] \, dz$$

which takes into account the axial profile of the clad outer surface temperature. The heat transfer coefficient $h_f$ is given by

$h_f = k/\sigma$ for vaporization
$h_f = 6.3 \times 10^4 \, W/m^2 \, K$ for condensation
$k$ = thermal conductivity of the liquid

2. $\mathbf{q}_t \cdot \mathbf{n}_t$ and $\mathbf{q}_b \cdot \mathbf{n}_b$ are calculated by solving the conduction equation in the liquid slugs in the vicinity of the interfaces.

The solution of Eq. (7) together with the equation of state for the vapor phase permits the calculation of the vapor pressure needed in Eqs. (5) and (6) if the film thickness $\sigma$ is specified.

The film thickness $\sigma$ is variable along $z$; at the boundaries with the liquid slugs $\sigma$ is set to an arbitrary value, with the initial film thickness $\sigma_0$ ($\sigma_0 = 0.15$ mm in Cronenberg et al., 1971). Its variation along the vapor bubble is given by a differential equation resulting from the following energy balance:

Heat transfer at the clad surface = heat of vaporization of the liquid film
+ energy storage inside the liquid film

This equation must result from the enthalpy equation averaged over the liquid film.

**1.1.3 Initiation of boiling** The initiation of boiling is one of the key points of the bubble-slug ejection model since the choices at this stage have important consequences on the subsequent development of the boiling pattern. There are two important parameters:

1. The initial superheat: Boiling is assumed to occur when the liquid temperature reaches some value usually prescribed by the user.
2. The initial film thickness: When the bubble originates, one assumes a cylindrical shape with a liquid film left on the wall. The liquid-film thickness must be prescribed.

Both parameters have a tremendous effect on the subsequent voiding, as shown in the following example. Figure 3 shows the channel geometry used for the calculation and the normalized power profile, characterized by a maximum at the center of the core and negligible power in the lower and upper blankets.

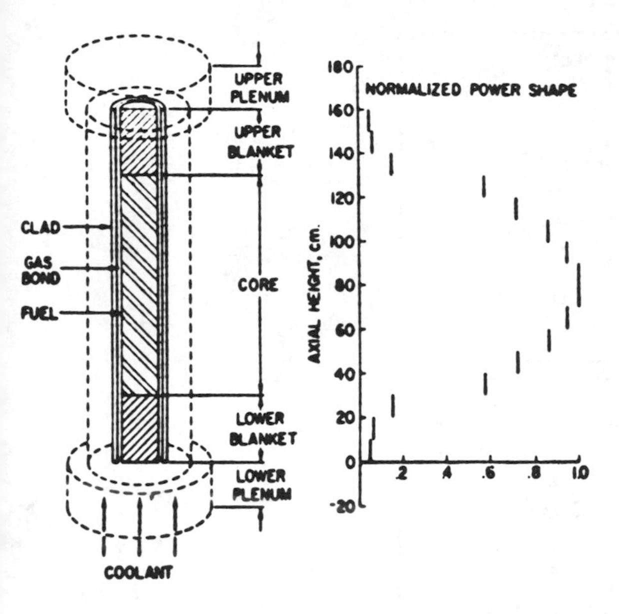


Figure 3 Geometry used for calculations (Cronenberg et al., 1971).

Figure 4 gives a comparison between two calculations made, one with the wall film (initial thickness $\sigma_0 = 0.15$ mm) the other without. When the wall film is present, Fig. 4 shows its axial variation.

The effect of initial superheat is shown in Fig. 5 for two different power inputs.

## 1.2 Unstructured Flow Models

Usually unstructured flow models have been developed for sodium boiling calculations, using the code already developed for water reactors, for example,

NALAP (Martin et al., 1975), developed at Brookhaven National Laboratory on the basis of the RELAP-3 code for pressurized-water reactor (PWR) system transients

FLINA (Rousseau, 1971) and FLINT (Grand and Latrobe, 1977), developed at the French AEC from FLICA initially made for water reactors

NASLIP (Moxon, 1977), a new code developed in the United Kingdom

However, sodium boiling near atmospheric pressure possesses features that make computation much more difficult than that for water boiling. The most obvious difficulty is the large density ratio between the liquid and its vapor. For this reason, early attempts for direct use of computational tools developed for water reactors failed. Since that time, efforts directed in two directions:

1. Elimination from the physical model of aspects important in water boiling but not in sodium boiling, for example, local boiling in the subcooled region: In

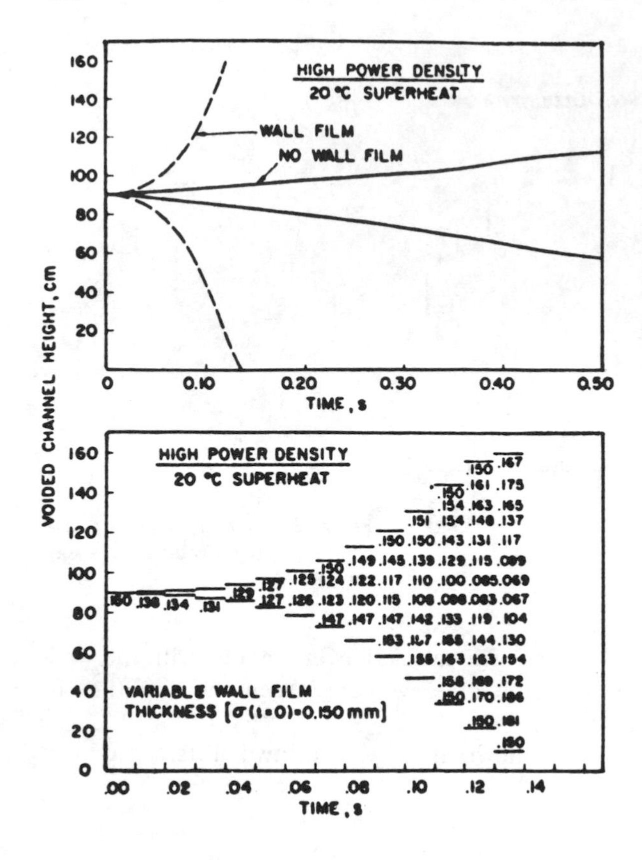


Figure 4 Effect of the liquid wall film (Cronenberg et al., 1971).

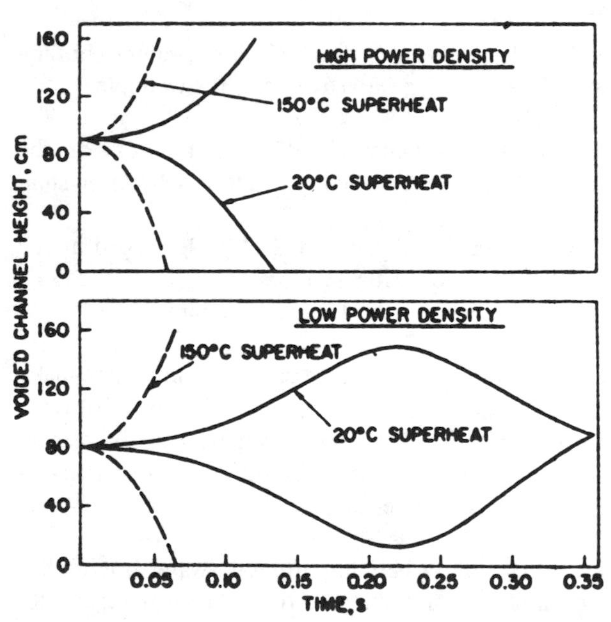


Figure 5 Effect of superheat, wall film present (Cronenberg et al., 1971).

sodium, this region occupies only a short length of the boiling zone, owing to the high thermal conductivity of sodium.

2. Development of reliable numerical schemes: One of the challenges for numerical methods is to deal with the sharp gradients of variables at the boiling boundaries.

To the author's knowledge all the codes of this type use an Eulerian formulation of the conservation equations, that is, on a fixed mesh. The codes use the balance equations of the two-fluid model with certain restrictions. Each of these restrictions helps replace one field equation by one algebraic relation.

1. Equal velocities–equal temperatures: The two phases are in thermodynamic and dynamic equilibrium. Thus, all thermodynamic quantities are evaluated along the saturation line ($T_v = T_l = T_{sat}(p)$). On the other hand,

$$w_g = w_f$$

the case of NALAP, which leads to the well-known homogeneous equilibrium model.

2. Unequal velocities–equal temperatures: The thermodynamic properties of the two phases are evaluated along the saturation line ($T_v = T_l = T_{sat}(P)$).

A correlation must be used for the evaluation of the slip ratio $\gamma = w_g/w_f$. A consensus seems obtained on the Lockhart-Martinelli correlation for the void fraction from which $\gamma$ may be deduced. The Lockhart-Martinelli two-phase friction multiplier is used for the evaluation of $\tau_w$. This is the procedure followed in NASLIP.

3. Unequal velocities–unequal temperatures: In FLINA and FLINT, the vapor is assumed to be in thermodynamic equilibrium ($T_v = T_{sat}$) but the liquid is not. Thus the enthalpy equation for the liquid must be solved. Of course, this brings up new terms for which models must be found: the energy exchange with the vapor and with the walls. For the void fraction, the evaluation is done as in the previous case.

Below we see results obtained with such a code for the thermohydraulic behavior of a channel during a loss of flow. The results are compared with experimental data gained in out-of-pile experiments made in support of the French LMFBR program. In the case studied, the channel is 0.8 m high and the heat power is uniform over its whole height.

In Fig. 6, the length of the boiling region is compared to time and the decrease of the mass flow rate. Notice the difference in time scales with the transients described in Sec. 1.1 (0.15 s compared with 22 s).

In Fig. 7 the axial profile of the heat flux at the outer surface of the cladding is plotted at different instants. In the liquid region, the continuous increase of the temperature at any given point reduces the heat flux delivered by the heated pin. In the two-phase region the characteristic temperature of the fluid is the saturation temperature. Thus the heated pin recovers steady-state conditions for which the power is entirely transmitted to the fluid.

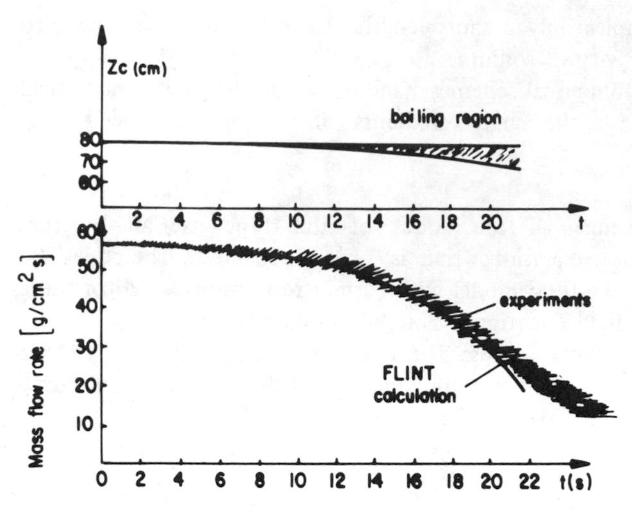


**Figure 6** Compared experimental and calculated results for the voiding rate and flow decreases (Grand and Latrobe, 1977).

# 2 SODIUM BOILING: MULTIDIMENSIONAL APPROACH

## 2.1 Code Classification

The single-channel approach described in Sec. 1 is an approximate description of a subassembly. The approach assumes uniform distribution of the flow variables in a cross section, whereas in a subassembly such uniformity is not encountered. Among the causes of heterogeneities in a cross section are

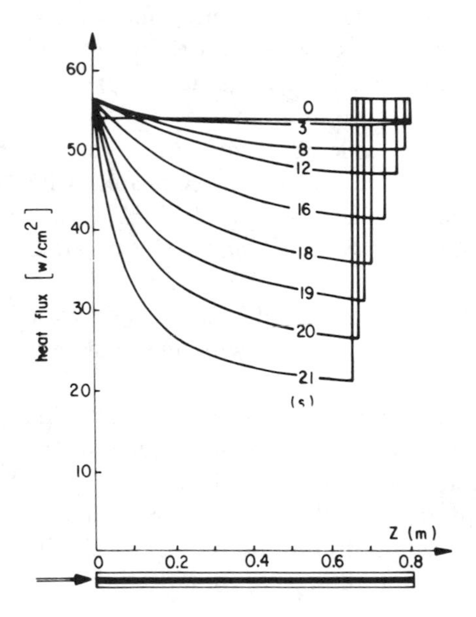


**Figure 7** Heat flux profiles (Grand and Latrobe, 1977).

1. The geometric differences between inner and peripheral subchannels. In a hexagonal duct, inner subchannels are demarcated by triangles and peripheral subchannels by rectangles. Such geometric differences will cause different flow resistances.
2. The nonuniformity of the power released by the pins.
3. The flow-sweeping effects induced by the helical wires used as spacers in most reactors (Phenix, Superphenix, FFTF, etc.).

Thus, the use of mathematical tools to handle these cross-sectional variations is necessary to achieve a correct description of the boiling in a subassembly.

As in the single-channel approach, workers have tried to use the efforts pursued in water reactors and extend their results to sodium boiling. For example, versions of FLICA, HAMBO, and COBRA (for a review of these codes, see Weisman and Bowring, 1975) have been derived that possess the physical properties of sodium. To date this approach has not been very successful for the same reason it was not successful in the single channel, that is, the high ratio of liquid and vapor densities encountered in sodium boiling at atmospheric pressure. Another generation of codes has been developed specific to sodium: SABRE (Moxon, 1977), HEV2D (Miao and Theofanous, 1977), and BACCHUS (Basque et al., 1978).

Another classification of the codes may be followed, depending on the control volume used for writing conservation laws: a subchannel or a geometric volume containing many pins and subchannels (a ring, for example). The two types of control volumes are shown in Fig. 8.

In the first approach, one finds codes like FLICA, HAMBO, COBRA, and SABRE. In the second approach, the arbitrary control volume containing both the coolant channels and the solid matrix belongs to the class of flows through porous media. If the control volume possesses axial symmetry (ring), the resulting formulation is two-dimensional. This is the case of sodium boiling calculations at this date (BACCHUS, HEV2D) in which only radial heterogeneities are retained. We will base our presentation on the second approach and give examples of results obtained.

## 2.2 The Physical Model

Let $\mathcal{V}$ be a fixed geometric volume, a cylinder parallel to the bundle axis and of arbitrary cross section. It contains fuel pins, spacer wires, and coolant channels (see Fig. 9). Note that

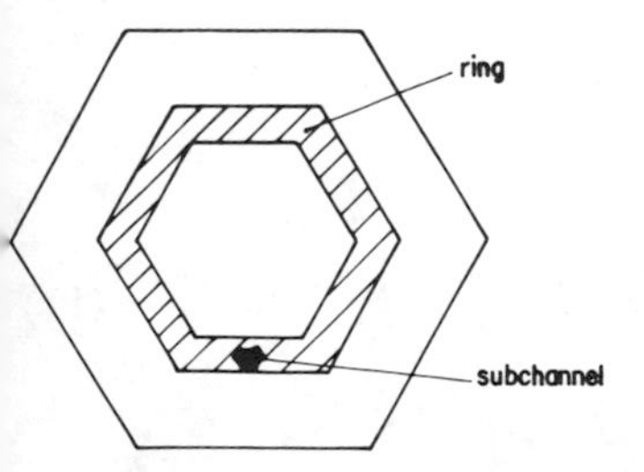


**Figure 8** Control volumes used in multidimensional modeling.

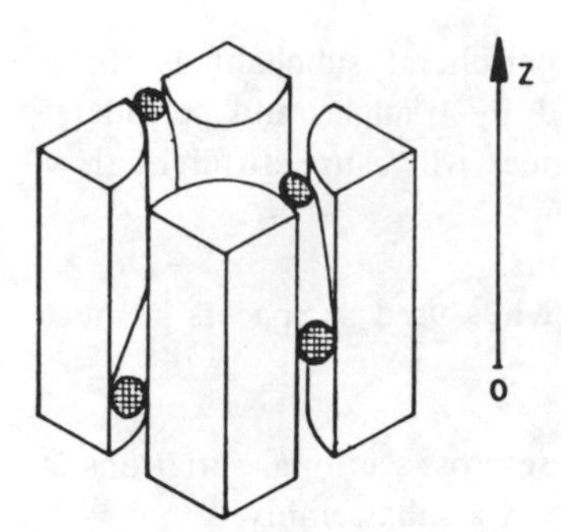


**Figure 9** Control volume.

$\mathcal{V}_f$ = volume of the coolant channels
$\mathcal{A}_f$ = area in the lateral surface occupied by the coolant channels
$\mathcal{A}_w$ = area of the fluid–solid interfaces contained in the control volume
**k** = unit vector in the axial direction $0z$

In the coolant the conservation equations may be written in their local form. They are simplified by the following assumptions:

The homogeneous equilibrium model is used for the two-phase flow.
The flow is steady.
The boundary-layer approximation is valid: uniform pressure in a cross section, negligible longitudinal diffusion.

With these approximations the local time-averaged equations may be written.

Mass:

$$\nabla \cdot \rho \mathbf{v} = 0 \tag{8}$$

Axial momentum:

$$\nabla \cdot (\rho \mathbf{v} w) = -\frac{\partial p}{\partial z} + \nabla \cdot (\boldsymbol{\tau} \cdot \mathbf{k}) - \rho g \tag{9}$$

Enthalpy:

$$\nabla \cdot (\rho \mathbf{v} h) = w \frac{\partial p}{\partial z} + \nabla \cdot \mathbf{q} \tag{10}$$

By integration of these local equations on the control volume we have:

Mass:

$$\frac{\partial}{\partial z} (\epsilon \langle \rho w \rangle) + \frac{1}{\mathcal{V}} \int_{\mathcal{A}_f} \dot{m}\, d\mathcal{A} = 0 \tag{11}$$

Axial momentum:

$$\frac{\partial}{\partial z}(\epsilon\langle\rho w^2\rangle) + \frac{1}{\mathcal{V}}\int_{\mathcal{a}_f} \dot{m}w \, d\mathcal{a} = -\frac{\partial}{\partial z}(\epsilon\langle p\rangle) + \frac{1}{\mathcal{V}}\int_{\mathcal{a}_w} (\tau \cdot \mathbf{k}) \cdot \mathbf{n} \, d\mathcal{a}$$

$$+ \frac{1}{\mathcal{V}}\int_{\mathcal{a}_f} (\tau \cdot \mathbf{k}) \cdot \mathbf{n} \, d\mathcal{a} - \epsilon\langle\rho\rangle g \tag{12}$$

Enthalpy:

$$\frac{\partial}{\partial z}(\epsilon\langle\rho wh\rangle) + \frac{1}{\mathcal{V}}\int_{\mathcal{a}_f} \dot{m}h \, d\mathcal{a} = \epsilon\langle w \frac{\partial p}{\partial z}\rangle + \frac{1}{\mathcal{V}}\int_{\mathcal{a}_w} \mathbf{q} \cdot \mathbf{n} \, d\mathcal{a}$$

$$+ \frac{1}{\mathcal{V}}\int_{\mathcal{a}_f} \mathbf{q} \cdot \mathbf{n} \, d\mathcal{a} \tag{13}$$

$$\langle f\rangle \triangleq \frac{1}{\mathcal{V}_f}\int_{\mathcal{V}_f} f \, d\mathcal{V} \tag{14}$$

where $\langle f\rangle$ denotes the average value over the fluid volume and the variables have the following meaning:

$\epsilon \triangleq \mathcal{V}_f/\mathcal{V}$ = porosity of the bundle or fraction of the volume occupied by the coolant channels
$w$ = axial component of the velocity of the coolant
$\dot{m}$ = mass flux density on the lateral surface
$\tau$ = sum of the viscous and turbulent stress tensors
$h$ = specific enthalpy

The lhs of the equations contain the convective terms related to axial and transverse flow.

In the right-hand side (rhs), we have the following terms:

| Momentum equation | Enthalpy equation |
|---|---|
| Pressure gradient | Pressure work |
| Friction at the fluid–solid interfaces | Heat flux at the solid–fluid interfaces |
| Momentum exchange with the surrounding fluid | Energy exchange with the surrounding fluid |

The momentum and energy exchanges are related to viscous and turbulent fluxes and flow sweeping by helical wires.

In the particular case of $\epsilon = 1$, the control volume is identical to the volume occupied by the fluid. Then Eqs. (11)–(13) are identical to the balance equations written for a subchannel.

For the closure of the preceding system [Eqs. (11)–(13)] we need

1. Restrictions on volume averages

$$\langle fg \rangle = \langle f \rangle \langle g \rangle$$

Subsequently, we will omit notation of the volume-averaging operator.

2. Exchanges with surrounding fluid. The exchanges are usually expressed by a gradient law:

$$\int_{\mathfrak{a}_f} \mathbf{q} \cdot \mathbf{n}\, d\mathfrak{a} = -\mathfrak{a}_f \alpha_e \nabla_t \rho h \tag{15}$$

$$\int_{\mathfrak{a}_f} (\tau \cdot \mathbf{k}) \cdot \mathbf{n}\, d\mathfrak{a} = -\mathfrak{a}_f \nu_e \nabla_t \rho w \tag{16}$$

where $\nabla_t$ is a difference approximation for the transverse component of the gradient, and $\alpha_e$ and $\nu_e$ stand for effective diffusivities, that is, they include transport owing to turbulent fluctuations and flow sweeping because of the wires.

3. Exchanges at the fluid-solid interfaces:

$$\int_{\mathfrak{a}_w} \mathbf{q} \cdot \mathbf{n}\, d\mathfrak{a} = \mathfrak{a}_w \phi_w$$

$$\int_{\mathfrak{a}_w} (\tau \cdot \mathbf{k}) \cdot \mathbf{n}\, d\mathfrak{a} = -\mathfrak{a}_w C_f \rho \frac{w^2}{2}$$

where $\phi_w$ is the heat flux density and $C_f$ is the friction factor.

4. Expression for the transverse transports:

$$\int_{\mathfrak{a}_f} \dot{m} h\, d\mathfrak{a} = \mathfrak{a}_f h^* \dot{m}$$

where $h^*$ stands for the value in the donor region, that is, the region where the cross flow originates.

Using these four assumptions, we may write the system equations [Eqs. (11)–(13)] as

$$\frac{\partial}{\partial z} (\epsilon \rho w) + \frac{\mathfrak{a}_f}{\mathfrak{V}} \dot{m} = 0 \tag{17}$$

$$\frac{\partial}{\partial z} (\epsilon \rho w^2) + \frac{\mathfrak{a}_f}{\mathfrak{V}} \dot{m} w^* = -\frac{\partial}{\partial z} (\epsilon p) - \frac{\mathfrak{a}_w}{\mathfrak{V}} C_f \rho \frac{w^2}{2}$$

$$- \frac{\mathfrak{a}_f}{\mathfrak{V}} \nu_e \nabla_t \rho w - \rho g \tag{18}$$

$$\frac{\partial}{\partial z}(\epsilon \rho w h) + \frac{\alpha_f}{\upsilon}\dot{m}h^* = \epsilon w \frac{\partial p}{\partial z} + \frac{\alpha_w}{\upsilon}\phi_w - \frac{\alpha_f}{\upsilon}\alpha_e \nabla_t \rho h \tag{19}$$

If the control volumes are concentric rings, we show that given the appropriate initial conditions at the inlet of the subassembly (pressure, velocity, and enthalpy), the preceding system [Eqs. (17)-(19)] may be solved in terms of the main dependent variables ($\dot{m}$, $h$, and $w$, averaged values over a ring, and $p$, averaged pressure over the cross section of the bundle). However, we must add to Eqs. (17)-(19) the continuity equation over the whole cross section.

We recall that here two-phase flow is represented by a homogeneous equilibrium model. Although simple, it can be used to assess the particularities of multidimensional flows resulting from cross flows and transverse exchanges. To achieve greater sophistication in the two-phase model, the difficulties of deriving the transverse exchange terms must be faced. For this reason HEV2D, like BACCHUS, is based on such a two-phase model.

## 2.3 Illustration of Boiling in a Rod Bundle

Figures 10 and 11 give results of BACCHUS in steady-state boiling.

In Fig. 10 a 19-rod bundle is considered. The quality contours are plotted. The

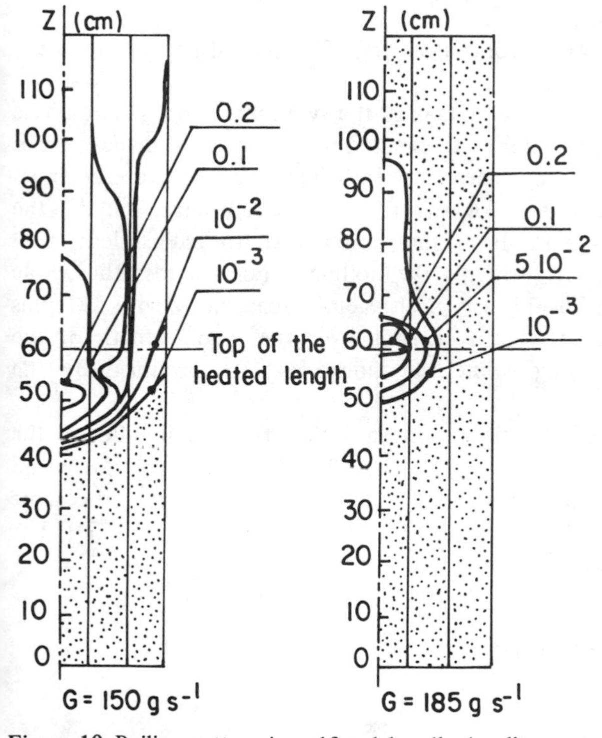


**Figure 10** Boiling pattern in a 19-rod bundle (quality contours), where the inlet temperature is 400°C and the power is 120 kW.

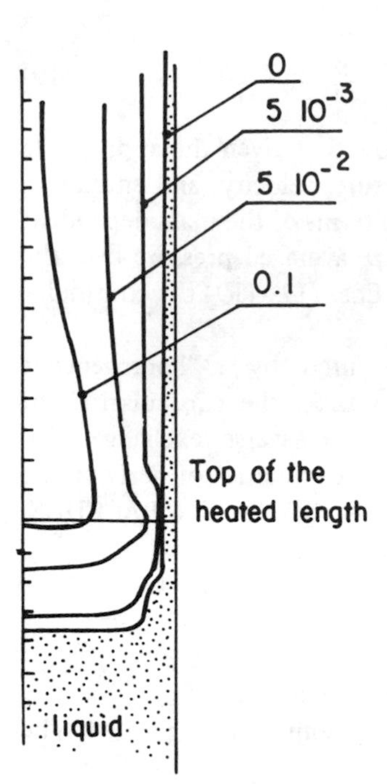


**Figure 11** Boiling pattern in a 271-rod bundle (quality contours).

19-rod bundle is represented by three concentric rings. The heated part occupies the first 60 cm.

For $G \triangleq \rho w = 185$ g/s, the boiling affects only the two inner rings. A two-phase region, centered on the axis at the end of the heated region, appears in the warmer inner rings. It is a localized boiling since the enthalpy averaged over the cross section does not reach saturation. Downstream, in the adiabatic length, the two-phase region condenses as a result of the mixing with the subcooled liquid flowing in the outer ring. For $G = 150$ g/s the boiling extends across the whole cross section of the bundle. In Fig. 11, a typical reactor subassembly with 271 pins and 10 rings is shown. A liquid bypass remains over the whole length of the subassembly in the peripheral ring, while the inner rings are occupied by the two-phase region.

To illustrate the impact of the flow pattern inside the rod bundle on the transient resulting from a loss of flow, we present in Fig. 12 a result from Miao and Theofanous (1977), who compared a transient calculation made with the same parameters for a single channel (HEV1D) and three concentric rings (HEV2D). The inlet flow transient is considerably slowed by the distribution effects.

## 3 ANALYSIS OF CLAD MOTION

### 3.1 Physical Description

During an unprotected LOF accident in an LMFBR, it is currently considered that, after sodium boiling and voiding, no sodium reentry occurs, leading to cladding

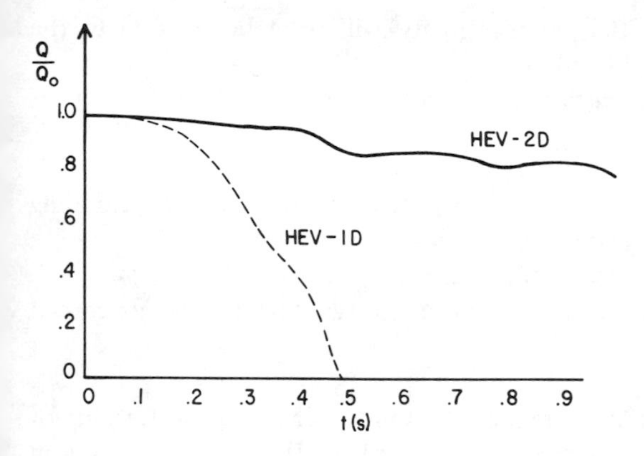


**Figure 12** Inlet flow transient, where $Q$ is the actual flow rate and $Q_0$ is the initial flow rate (Miao and Theofanous, 1977).

dryout, then clad melting, motion, and relocation. These phenomena and the possible plugging of coolant channels may have important effects on reactor reactivity and subsequent core damage. A representation of the voided channel is given in Fig. 13.

The vapor stream flows upward under the effect of the pressure difference. Since plenty of bypass flow area is available for liquid sodium in the core region, this pressure drop is roughly equal to the liquid sodium hydrostatic head:

$$\Delta p = \rho_l g l_0$$

As a result of this pressure difference a typical sodium vapor velocity at the onset of clad melting is 80 m/s. This quantity is below the critical vapor velocity (120 m/s), where droplets of the molten material would be entrained in the vapor stream (Ishii et al., 1976). It is thus expected that the molten clad moves as a film rather than as entrained droplets. The vapor velocity of 80 m/s is, on the other hand, in excess of the flooding velocity, which is 20 m/s (Ishii et al., 1976). Thus at the onset of melting, flooding occurs and has the effect of moving the film upward. However, as the interfacial resistance is increased by flooding and the molten clad

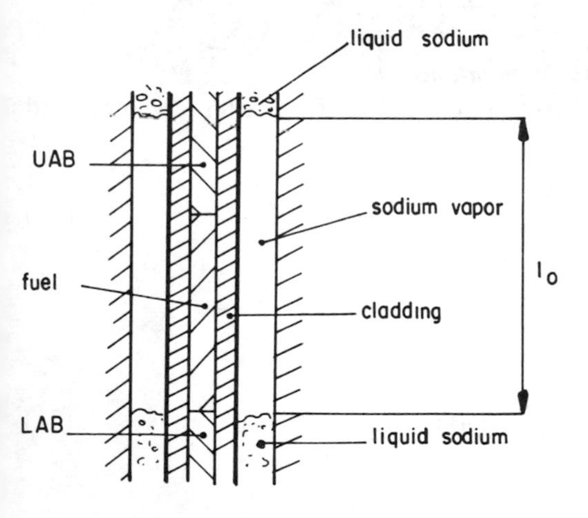


**Figure 13** Voided channel.

section extends with time, the initial vapor velocity will soon be reduced to the flooding velocity, and deflooding will follow.

Notice that the flooding problem is not conventional (compare the findings of Epstein, 1977):

1. Here a vapor stream exists first (with a velocity greater than the flooding velocity) and a molten film appears later.
2. The molten film length is increased with time.
3. Radial melting incoherencies may have an influence (we shall not be concerned with that problem here).

Another phenomenon that tends to reduce the vapor velocity is the freezing of the molten clad driven upward in the upper axial blanket. This freezing results in an obstruction of the vapor channel (plugging).

## 3.2 Example of Molten Clad Motion Analysis

This example of molten clad motion analysis was reported by Ishii et al. (1976) and Chen et al. (1977).

**3.2.1 Thermal and dynamic analyses** The thermal and dynamic aspects of the problem are decoupled: first the fuel and clad behavior is thermally analyzed with phase change. A model gives the upper and lower ends of the melted cladding section, $z_2(t)$ and $z_1(t)$, respectively. The difference, $\lambda(t) = z_2(t) - z_1(t)$, is the length of the melted cladding section (see Fig. 14). These values are inputs for the motion model of the molten clad itself.

The analysis of the motion of the molten clad is a purely dynamic analysis. Owing to the low velocities encountered in the channel, the sodium vapor is assumed to be incompressible. Of course, the same hypothesis holds for the liquid film. Conservation equations for each phase (vapor and liquid film) are given in Sec. 3.2.2.

**3.2.2 Area-averaged equations** *Continuity equations*

1. For molten film: Under the assumptions of uniform film thickness and negligible curvature effects of the pin, we obtain

$$\delta = \delta_c \frac{\lambda}{\lambda_m} \tag{20}$$

where $\delta_c$ = initial clad thickness
$\delta$ = molten clad film thickness

2. For sodium vapor:

$$w_g = \frac{\alpha_{gi}}{\alpha_g} w_{gi}(t) \tag{21}$$

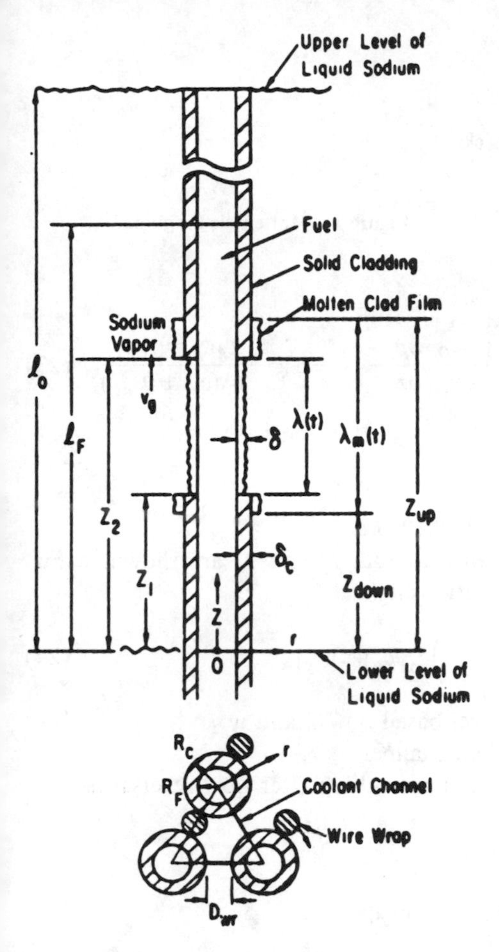


**Figure 14** Cladding-motion model (Chen et al., 1977).

where $w_g$ = vapor velocity
$w_{gi}$ = inlet vapor velocity
$\mathfrak{a}_g$ = vapor flow area
$\mathfrak{a}_{gi}$ = vapor flow area at the inlet

***Momentum equations*** Momentum equations are written for each phase, for vapor and liquid film. Vapor and liquid being incompressible, these equations simplify to

1. For sodium vapor:

$$\frac{\partial}{\partial t}(\alpha w_g) + \frac{\partial}{\partial z}(\alpha w_g^2) = -\frac{\alpha}{\rho_g}\frac{\partial p}{\partial z} - \alpha g - \frac{\tau_g \mathcal{P}_g}{\rho_g(\mathfrak{a}_g + \mathfrak{a}_{mc})} \tag{22}$$

where $\alpha$ = void fraction of the vapor phase, defined by $\mathfrak{a}_g/(\mathfrak{a}_g + \mathfrak{a}_{mc})$
$\mathfrak{a}_{mc}$ = molten clad film area, as shown in Fig. 15
$\mathcal{P}_g$ = wetted perimeter for the vapor phase
$\tau_g$ = shear stress at the clad outer surface

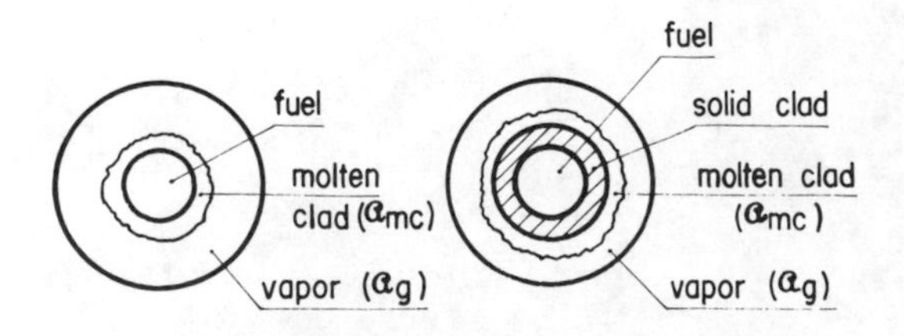


**Figure 15** Molten clad region.

2. For molten clad:

$$\frac{\partial}{\partial t}(1-\alpha)w_c + \frac{\partial}{\partial z}(1-\alpha)w_c^2 = -\frac{1-\alpha}{\rho_c}\frac{\partial p}{\partial z} - (1-\alpha)g + \frac{\tau_g \mathcal{P}_g + \tau_c \mathcal{P}_c}{\rho_c(\alpha_g + \alpha_{mc})} \qquad (23)$$

where $w_c$ = cladding film velocity
$\rho_c$ = cladding density
$\mathcal{P}_c$ = wetted perimeter for the flow
$\tau_c$ = shear stress at the boundary with the solid

The shear stresses acting on the molten clad film $\tau_c$ and $\tau_g$ are shown in Fig. 16. They are expressed by the following relationships:

$$\tau_g = \frac{f_s}{4}\left(1 + 300\epsilon\,\frac{\delta}{D}\right)\frac{1}{2}\,\rho_g w_g |w_g| \qquad (24)$$

where $f_s$ = friction factor for the vapor phase, based on $D_0$ and $w_{gi}$
$D_0$ = hydraulic diameter for the original channel
$\epsilon$ = deflooding parameter including film smoothing after flow reversal (see Sec. 3.2.4)
$D$ = hydraulic diameter for the flow

and

$$\tau_c = -\frac{f}{4}\,\frac{1}{2}\,\rho_c w_c |w_c|$$

where $f$ is the wall friction factor for the cladding.

**3.2.3 Integral equations** The area-averaged equations are then integrated over the channel height.

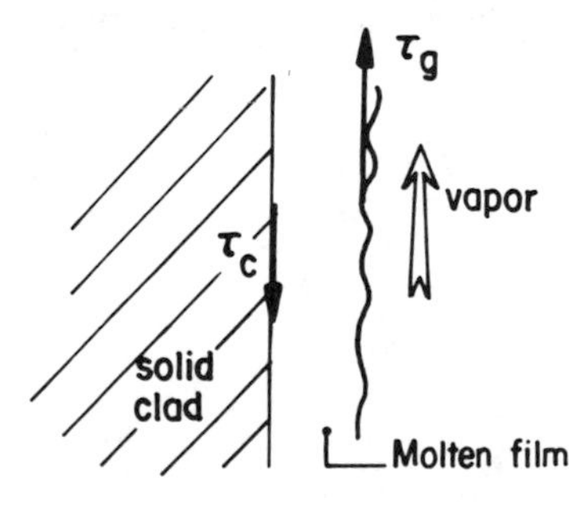


**Figure 16** Shear stresses acting on the molten clad film.

***Vapor*** The total pressure gradient in the vapor is obtained by summing pressure gradients over the four sections encountered by the vapor stream (see Fig. 17). In the process of summation, the convective terms cancel out. Moreover two assumptions are made:

1. The transient term is neglected since its order of magnitude is of the order of 1% of the hydrostatic head of liquid sodium.
2. The film thickness is negligible in comparison with the hydraulic diameter of the channel.

Then, the integration of the momentum equation for the vapor over the length of the channel gives

$$\Delta p = \rho_g g l_0 + \frac{f_s}{2D_0} \rho_g w_{gi}^2 l_0 \left(1 + 75\epsilon \frac{|1-\alpha_0|}{\alpha_0} \frac{\lambda}{l_0}\right) + \Delta p_{fr} \tag{25}$$

where $\Delta p_{fr}$ is the pressure drop at the blockage of the freezing clad. $\alpha_0$ is a geometric ratio defined as

$$\alpha_0 = \frac{a_{gi}}{a_\infty} = \frac{\text{inlet flow area}}{\text{flow area with stripped fuel pins}}$$

The derivation of this equation from Eq. (22) is detailed by Ishii et al. (1976).

The blockage caused by the freezing of the clad in the cooler upper blanket region is treated as a singular pressure drop through a nozzle:

$$\Delta p_{fr} = K_{fr} \rho_g \frac{w_{gi}^2}{2} \tag{26}$$

The contraction coefficient increases during the solidification process, that is, when the frozen layer gets thicker. The width of frozen layer is calculated with a one-dimensional radial conduction model. When the blockage is completed $K_{fr}$ becomes infinite.

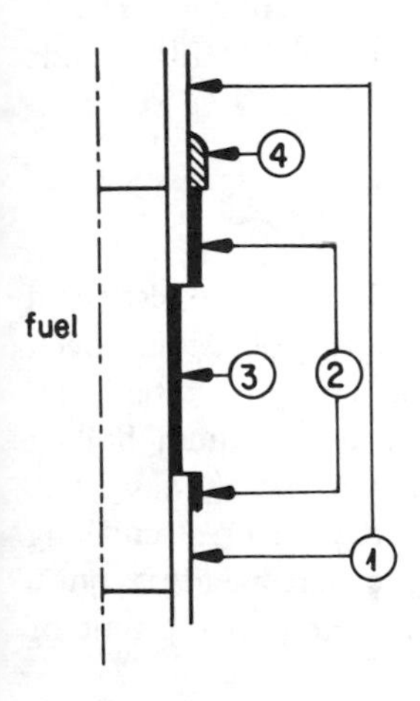


**Figure 17** The four regions. 1, section of intact cladding; 2, section where the molten film is overlapping the solid cladding; 3, section of clad melting; and 4, section of clad freezing.

***Liquid film*** For the liquid film, the same integration process pursued under the assumption of negligible film thickness gives

$$\frac{\rho_c}{\lambda}\frac{dw_c\lambda}{dt} = -(\rho_c - \rho_g)g - \frac{2\mu_c w_c}{\delta_c^2}\left(\frac{\lambda_m}{\lambda}\right)^2 + \frac{\Delta p - \rho g l_0}{(1-\alpha_0)l_0}\,\frac{1 + 75\epsilon(1-\alpha_0) + \alpha_0(\lambda_m/\lambda - 1)}{1 + 75\epsilon(1-\alpha_0)(\lambda/\alpha_0 l_0) + (D_0/l_0)(K_{fr}/f_s)} \tag{27}$$

Again the reader is referred to Ishii et al. (1976) for the derivation of Eq. (27).

To solve Eq. (27) it is necessary to specify the relation between the molten clad length ($\lambda_m = z_{\text{up}} - z_{\text{down}}$) and the film velocity. $z_{\text{up}}$ and $z_{\text{down}}$ are calculated from the time integral of the mean clad-film velocity. They depend, of course, on the penetration of the melting fronts $z_1(t)$ and $z_2(t)$ which are, as we mentioned earlier, input parameters of the cladding motion model.

**3.2.4 Further considerations** ***Deflooding*** Because of the increase of interfacial resistance for the vapor and possible flow blockage caused by freezing of the molten clad, the initial gas velocity (greater than flooding velocity) is gradually decreased. Because of uncertainties of the validity of existing flooding correlations when applied to liquid metals in transient situations, the authors of the model assume that the interfacial friction factor for rough surface may be applied until the film motion reverses flow. In this case, $\epsilon = 1$ in Eq. (24).

After flow reversal has occurred ($t = t_r$) the roughness of the interface decreases gradually according to the law

$$\epsilon = \left(\frac{\Delta p - \rho g l_0}{f_s \rho_g w_{gi}^2(t_r)} - 1 - \frac{K_{fr} D_0}{f_s f_0}\right)\frac{\alpha_0 l_0}{75(1-\alpha_0)\lambda} \tag{28}$$

until $\epsilon$ reaches zero, where the film is completely smooth.

***Clad slumping and bottom plugging*** Owing to the rapidity of the bottom-plugging process, a detailed analysis of plug formation is not called for. Thus a simple solidification model has been used (Ishii et al., 1976) that involves conduction heat transfer to the cold wall alone. When the blockage is total, the liquid film drains down by gravity only and accumulates at the bottom.

## 3.3 Example of Results

The same calculation dealt with here is the simulation of the R5 TREAT test (Ishii et al., 1976). Results are shown in Fig. 18. The effect of chugging on $\Delta p$ has been averaged out and $\Delta p$ has been approximated by the liquid sodium hydrostatic head.

The time of clad failure in the experiment was 2.2 s after sodium boiling initiation, and, in the calculation, the time of cladding-motion initiation was 2.8 s.

The model indicates an upper penetration above the active fuel 5.3 cm long with a smaller experimental value ($\leqslant 1$ cm). However, the agreement is quite satisfactory, considering uncertainties in the test-bundle pressure drop at the time of

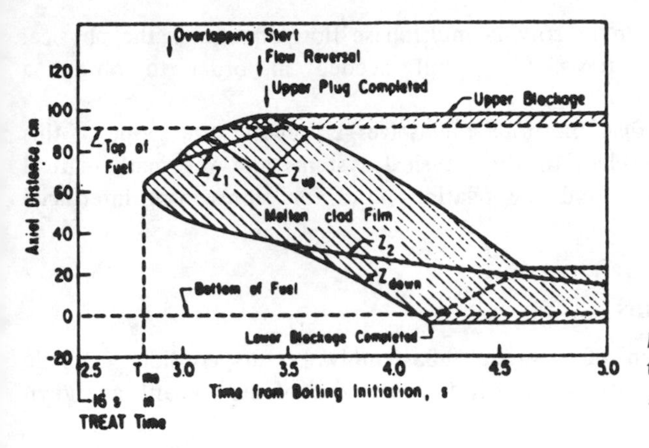


Figure 18 Cladding motion with the standard pressure drop ($\Delta p = \rho_l g l_o$) (Chen et al., 1977).

clad melting and possible partial remelting of the blockage before freezing and observation.

An initial lower blockage of 4.4 cm was calculated and, on shutdown, the film was calculated to have drained completely and accumulated 24 cm above the top of the lower blockage. These results compare well with posttest results.

## 4 TWO-DIMENSIONAL CALCULATIONS OF INTERPENETRATING MATERIALS

We now consider a methodology recently developed to calculate the transition phase of the hypothetical core disruptive accident, a method that can describe the transport of different materials under different states (solid, liquid, and gaseous). This capability has been made possible with the development of numerical procedures having the potentiality to solve a large set of closely linked partial differential equations. In this development, a leading role has been played by Harlow and Amsden, with the development of the ICE method (Harlow and Amsden, 1971) and the KACHINA method (Harlow and Amsden, 1975). The interesting aspect of these methods is that they provide efficient algorithms and iterative methods for the treatment of coupling between equations.

ICE is designed to treat a homogeneous, highly compressible fluid. The heart of the method is the implicit coupling between the equation of state and the pressure gradient in the momentum equations, and the continuity equation.

KACHINA is devoted to the treatment of a two-phase, two-component system and implicitly treats the coupling between the equation of state for the compressible phase, pressure gradients, and drag terms in the momentum equations for each phase.

With this choice of implicit coupling to which iterative methods are applied, one avoids the severe limitations on time steps resulting from the high sound speeds. However, questions such as the sound-speed definitions in multiphase flows and their representation in mathematical models are still open. More generally, even if this methodology is a big step in solving multiphase flow equations, the physical understanding of interfacial transfers is still needed in order to obtain a representative set of equations.

The study of the numerical methods themselves is beyond the scope of this section. We will restrict ourselves to the physical systems and the mathematical model they are able to handle and the relations used for representing interfacial transfers.

## 4.1 Two Physical Problems

**4.1.1 Problem 1: Destruction of a single subassembly** Let us consider a single subassembly, closed at both ends and that has been voided of the coolant, as shown in Fig. 19.

If the nuclear reaction is still present, the loss of cooling will result in a temperature increase of the solid materials: nuclear fuel and clad (cylindrical shell containing the fuel). These materials begin to melt and subsequently to vaporize in the hottest region of the fissile zone, and they are transported toward both ends of the subassembly, in the fertile regions. There the volumetric heating from the nuclear reaction is lower and they may freeze. If the power burst is controlled and limited as in laboratory experiments, a final situation may be reached within the limits of the subassembly. This condition is shown in Fig. 19, where the middle of the pins is destroyed and the molten materials accumulated in lower and upper blankets of the subassembly.

Between initial and final stages, the two components present (fuel and steel) may be in one of the three phases: solid, liquid, or vapor. The vapor phase also contains the fission products, which are noncondensable (Fig. 20).

The concentration of the two components in each phase is directly controlled by the mass exchanges, which may occur as

1. Melting and freezing between the solid and liquid phases
2. Vaporization and condensation between the liquid and vapor phases

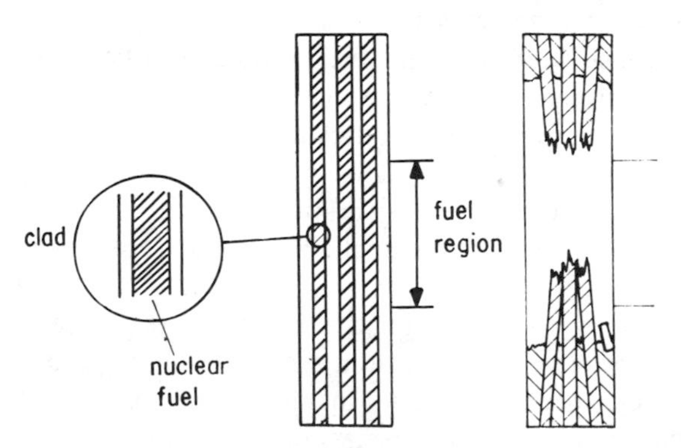


**Figure 19** Initial and final stages of a subassembly destruction.

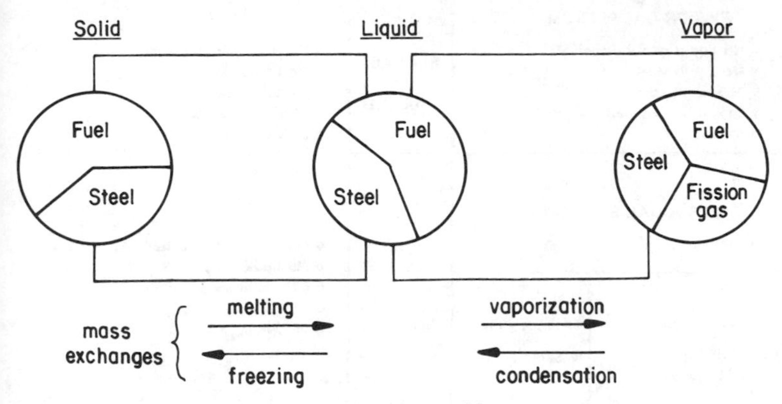


**Figure 20** Evolution of materials of a subassembly.

These mass exchanges are themselves determined by local conditions of pressure, temperature, etc., and thus are coupled with the transport processes in the subassembly.

**4.1.2 Problem 2: Hypothetical core-disruptive accident** In a hypothetical core-disruptive accident the power burst is such that a fault occurring in the hottest subassemblies does not remain in the limits of these subassemblies but propagates to the others as a result of the rupture of the housings; then the calculation must cover the whole reactor vessel. The SIMMER-1 code (Bell et al., 1976) was developed to predict the dynamics of extreme hypothetical accident sequences during which extended core motion is expected.

Figure 21 gives the initial configuration, which has been studied with SIMMER-1 in connection with the U.S. LMFBR program.

In Fig. 22 each component in each phase and the interfacial mass exchanges, which are present during extended core motion, are shown. Here, sodium is present in both liquid and vapor phases, while steel and fuel may be present in the three phases: solid, liquid, and vapor.

## 4.2 Mathematical Formulation of the SIMMER-1 Code

In this section we present the balance equations used in SIMMER-1 for the mathematical formulation of the two preceding problems. However, even at the stage of the balance equations, uncertainties remain concerning their derivation. Thus it should be kept in mind that the mathematical formulation used in the SIMMER-1 code may be subject to changes. It is simply one step toward the solution of difficult problems; it is not the ultimate answer. Smith et al. (1976) themselves draw attention to this by calling it the "first version."

The basis of the derivation is the choice of an Eulerian formulation. The high degree of interpenetration and mixing between phases and components excludes the use of a Lagrangian mesh following the particles.

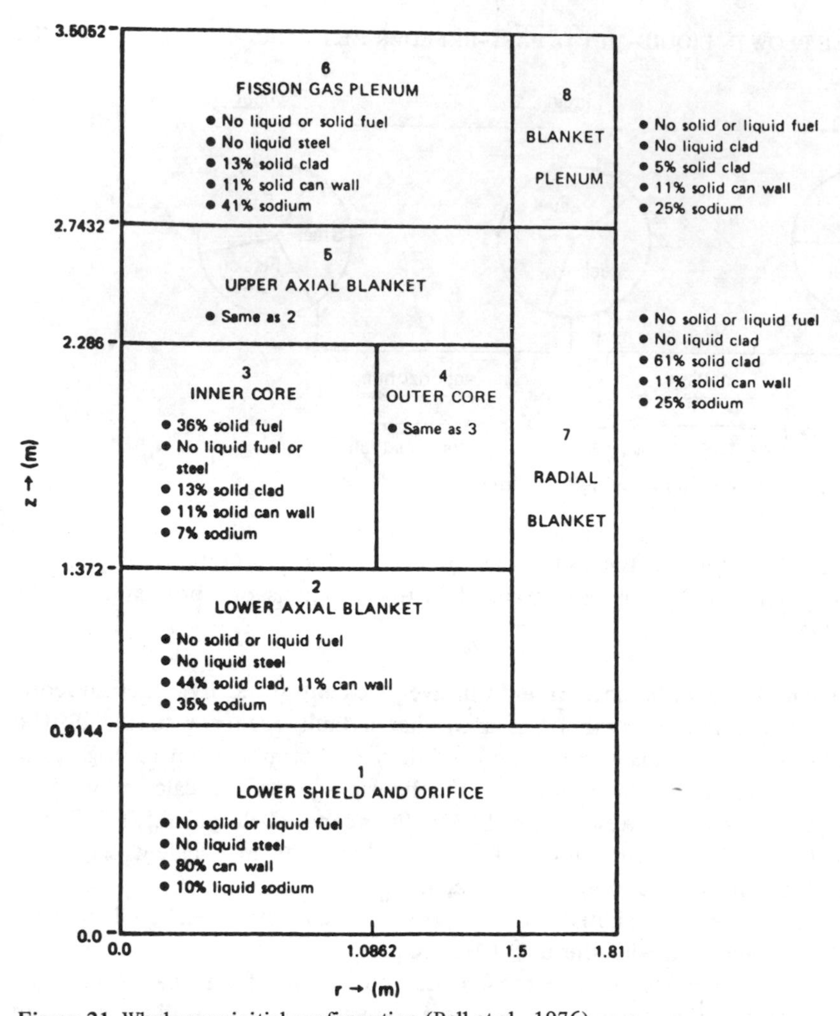


**Figure 21** Whole core initial configuration (Bell et al., 1976).

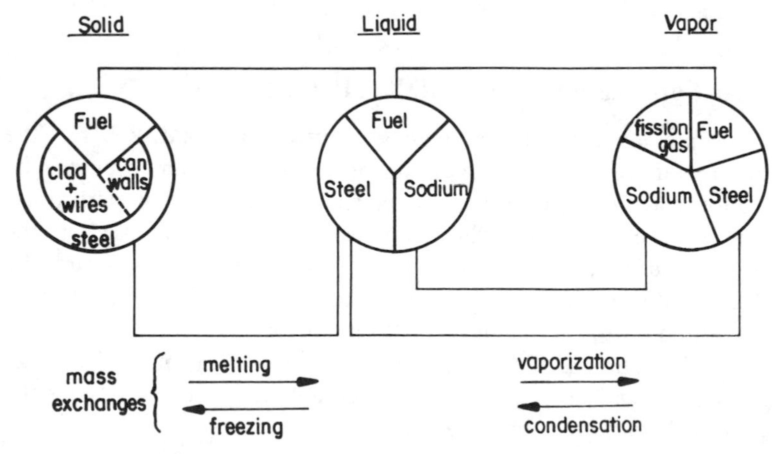


**Figure 22** Evolution of materials of the core.

Cylindrical coordinates are chosen. Although nothing is said by the authors about the averaging operators used in the derivation of the equations, we think that double space and time averaging are used:

1. Volume averaging on the volume element of the cylindrical coordinates, with constants $\Delta r$, $\Delta\theta$, and $\Delta z$.
2. Time averaging on a time interval having a magnitude large enough compared with turbulent fluctuations and small enough compared with the overall flow fluctuations.

The following equations describe the phase $p$ of component $c$:

### 4.2.1 Mass conservation

$$\frac{\partial}{\partial t}(R_p^c \rho_p^c) + \nabla \cdot (R_p^c \rho_p^c \mathbf{v}_p^c) = \Gamma_p^c \tag{29}$$

where $R_p^c$ = fraction of the control volume occupied by phase $p$ of component $c$
$\rho_p^c$ = density
$\mathbf{v}_p^c$ = velocity vector
$\Gamma_p^c$ = rate of production of $p$th phase of component $c$ from the phase changes at the interfaces (This is the only source term in the absence of chemical reaction between the components.)

The interfacial mass transfer condition gives

$$\sum_p \Gamma_p^c = 0 \qquad \forall_c \tag{30}$$

If $\gamma_{p,p'}$ denotes phase transition from phase $p$ to phase $p'$, then

$$\Gamma_p^c = \sum_{p' \neq p} \gamma_{p',p}^c - \sum_{p' \neq p} \gamma_{p,p'}^c \tag{31}$$

which simply states that $\Gamma_p^c$ is the sum of all phase transitions leading to a gain for phase $p$ minus all phase transitions leading to a loss for phase $p$. This definition evidently satisfies Eq. (30).

The following example is offered: two-phase liquid-vapor, one component:

$$\begin{aligned} \Gamma_l^c &= \gamma_{v,l}^c - \gamma_{l,v}^c \\ \Gamma_v^c &= \gamma_{l,v}^c - \gamma_{v,l}^c \end{aligned} \quad \Rightarrow \quad \Gamma_l^c + \Gamma_v^c = 0$$

### 4.2.2 Momentum equation

$$\begin{aligned} \frac{\partial}{\partial t}(R_p^c \rho_p^c \mathbf{v}_p^c) + \nabla \cdot (R_p^c \rho_p^c \mathbf{v}_p^c \mathbf{v}_p^c) &= -R_p^c \nabla p + R_p^c \rho_p^c \mathbf{g} \\ &+ \nabla \cdot R_p^c \tau_p^c + \mathbf{M}_p^c + \sum_{(c',p') \neq (c,p)} \sum D_{pp'}^{cc'}(\mathbf{v}_{p'}^{c'} - \mathbf{v}_p^c) \end{aligned} \tag{32}$$

The terms on the rhs are the pressure gradient, body force, stresses inside phase $p$ of component $c$, and two interfacial exchange terms: $\mathbf{M}_p^c$, momentum exchange caused by phase transitions, and $D_{pp'}^{cc'}(\mathbf{v}_{p'}^{c'} - \mathbf{v}_p^c)$, momentum exchange caused by interfacial stresses.

**4.2.3 Internal energy equation** The internal energy equation is preferred to the total energy equation for numerical reasons. Contrary to the total energy equation, there is not one unique internal energy equation but rather different forms depending on how it is derived and the assumptions needed in this process. The one used in SIMMER is

$$\frac{\partial}{\partial t}(R_p^c \rho_p^c u_p^c) + \nabla \cdot (R_p^c \rho_p^c \mu_p^c \mathbf{v}_p^c) = \frac{R_p^c p}{\rho_p^c}\left(\frac{\partial \rho_p^c}{\partial t} + \mathbf{v}_p^c \cdot \nabla \rho_p^c\right)$$

$$+ \nabla \cdot R_p^c \mathbf{q}_p^c + R_p^c \rho_p^c Q_p^c + \Lambda_p^c + \sum_{(c',p') \neq (c,p)} \sum H_{pp'}^{cc'}(T_{p'}^{c'} - T_p^c) \tag{33}$$

The terms on the rhs are the pressure work, heat diffusion inside the field, heat source caused by the nuclear reaction, and two interfacial exchange terms: $\Lambda_p^c$, the heat source from the phase changes, and $H_{pp'}^{cc'}(T_{p'}^{c'} - T_p^c)$, the heat flux through the interfaces.

For closure of the system at this level, one needs algebraic relations for the unknown terms:

1. Exchange terms corresponding to phase transitions:
   a. Mass: $\Gamma_p^c = \Sigma_{p' \neq p} \gamma_{p',p}^c - \Sigma_{p' \neq p} \gamma_{p,p'}^c$
   b. Momentum: $\mathbf{M}_p^c$
   c. Internal energy: $\Lambda_p^c$
2. Drag coefficients at the interfaces $D_{pp'}^{cc'}$
3. Heat transfer coefficient at the interfaces $H_{pp'}^{cc'}$
4. Stress tensor $\tau_p^c$
5. Heat flux $\mathbf{q}_p^c$

These relations should be given in terms of the main variables, $p$, $u_p^c$, and $\mathbf{v}_p^c$.

## 4.3 Further Assumptions and Constitutive Laws

**4.3.1 Restrictions** At this date, further restrictions are made to reduce the number of equations. Let us consider the ones used in SIMMER-1 (Smith et al., 1976).

1. Velocities of different components in the same phase are assumed to be equal:

$$\mathbf{v}_p^{c_1} = \mathbf{v}_p^{c_2} \qquad \forall\, c_1 \text{ and } c_2$$

Such an assumption is easily put in default: a large volume of melted fuel may have a velocity very different from that of the surrounding liquid sodium. If it is adopted, it is simply because it is essential to arrive at a solution. Consequently, each phase is modeled by a single momentum equation.

2. The components in the vapor phase are assumed to be a completely mixed continuum in thermal equilibrium. The vapor mixture is thus modeled by a single energy equation.

3. The solid phase is fixed in space. There is no transport of solid constituents. The only way to transport solids from one place to another is thus through the following process: melting of the solid → transport in liquid phase → freezing. Consequently, no momentum equation is needed for the solid phase, and other transport equations are simplified.

4. Calculations are conducted in two-dimensional axisymmetrical cases.

Let us consider the consequences of these assumptions on the number of equations needed to solve problem 2, for example. Initially there would have been a total of 42 equations:

| | Solid | Liquid | Vapor | Total |
|---|---|---|---|---|
| Continuity | 3 | 3 | 3 | 9 |
| Momentum | 2 (× 3) | 3 (× 3) | 3 (× 3) | 8 (× 3) |
| Energy | 3 | 3 | 3 | 9 |
| Total | 12 | 15 | 15 | 42 |

However, after the preceding restrictions the number of equations is reduced to 20. This remains a large system.

| | Solid | Liquid | Vapor | Total |
|---|---|---|---|---|
| Continuity | 3 | 3 | 3 | 9 |
| Momentum | 0 | 1 (× 2) | 1 (× 2) | 2 (× 2) |
| Energy | 3 | 3 | 1 | 7 |
| Total | 6 | 8 | 6 | 20 |

**4.3.2 Constitutive laws** Even more than in the case of balance equations, the constitutive laws used for the interfacial phenomena need future improvements. According to the authors themselves, models used at the present time are very simplistic, and an important effort will be required in the future to develop better models for these phenomena.

***Phase transitions*** Before we give the mass transfer term, let us see how the momentum and energy transfer resulting from phase transitions are related to it.

1. Momentum transfer: The current procedure consists in relating the momentum transfer resulting from phase transitions $\mathbf{M}_p^c$ to the mass transfer $\Gamma_p^c$, by using the velocity of the donor phase, that is,

$$\mathbf{M}_p^c = \sum_{p' \neq p} \gamma_{p',p}^c \mathbf{v}_{p'} - \sum_{p' \neq p} \gamma_{p,p'}^c \mathbf{v}_p \quad (34)$$

Such a formulation ensures the conservation of momentum since if one term is present in the equation for phase $p$, it is present with the other sign in the equation for phase $p'$.

2. Energy transfer: In KACHINA, pressure-work and $\Lambda_l^c$ are rewritten for the liquid–vapor interface (Travis and Rivard, 1976):

For the liquid phase:

$$\frac{R_l^c p}{\rho_l^c}\left(\frac{\partial \rho_l^c}{\partial t} + \mathbf{v}_l \cdot \nabla \rho_l^c\right) + \Lambda_l^c(v \leftrightharpoons l) = -p\left(\frac{\partial R_l^c}{\partial t} + \nabla \cdot \mathbf{v}_l R_l^c\right) + (\gamma_{v,l}^c - \gamma_{l,v}^c)h_g^c \quad (35)$$

For the vapor phase:

$$\frac{R_v^c p}{\rho_v^c}\left(\frac{\partial \rho_v^c}{\partial t} + \mathbf{v}_v \cdot \nabla \rho_v^c\right) + \Lambda_v^c(l \leftrightharpoons v) = -p\left(\frac{\partial R_v^c}{\partial t} + \nabla \cdot \mathbf{v}_v R_v^c\right) + (\gamma_{l,v}^c - \gamma_{v,l}^c)h_g^c \quad (36)$$

However, Lal and Carlson (1976) used for the same two terms (superscript $c$ being omitted throughout):

For the liquid phase:

$$-p\nabla R_l \mathbf{v}_l + (\gamma_{v,l} - \gamma_{l,v})u_f + \frac{\rho_f c_f}{\rho_f c_f + \rho_g c_{pg}}(\gamma_{v,l} - \gamma_{l,v})h_{fg} \quad (37)$$

For the vapor phase:

$$-p\nabla R_v \mathbf{v}_v + (\gamma_{l,v} - \gamma_{v,l})(u_g + pv_{fg}) + \frac{\rho_g c_{pg}}{\rho_f c_f + \rho_g c_{pg}}(\gamma_{v,l} - \gamma_{l,v})h_{fg} \quad (38)$$

where subscripts $f$ and $g$ = saturated liquid and vapor phase
$h$ = specific enthalpy
$h_{fg}$ = heat of vaporization
$v_{fg}$ = change in specific volume owing to condensation

For the liquid–solid interface Lal and Carlson (1976) used:

For the liquid phase:

$$\Lambda_l(s \leftrightharpoons l) = (\gamma_{s,l} - \gamma_{l,s})u_f - \frac{\rho_f c_f}{\rho_f c_f + \rho_s c_s}(\gamma_{s,l} - \gamma_{l,s})h_{sf} \quad (39)$$

For the solid phase:

$$\Lambda_s(l \leftrightharpoons s) = (\gamma_{l,s} - \gamma_{s,l})u_s - \frac{\rho_s c_s}{\rho_f c_f + \rho_s c_s}(\gamma_{s,l} - \gamma_{l,s})h_{sf} \quad (40)$$

3. Mass transfer: In the LASL codes, the phase transition model needed for the evaluation of $\gamma_{l,v}$, $\gamma_{l,s}$, is as follows. For the evaporation–condensation phase:

$$\gamma_{l,v} = \frac{a_i k_l (T_l - T_{sat})}{h_{fg} L} \qquad \text{if } T_l > T_{sat}$$

$$= 0 \qquad \text{otherwise} \tag{41}$$

$$\gamma_{v,l} = \frac{a_i k_l (T_{sat} - T_v)}{h_{fg} L} \qquad \text{if } T_v < T_{sat}$$

$$= 0 \qquad \text{otherwise} \tag{42}$$

where $a_i$ is the interfacial area per unit volume, and $L$ is a characteristic thickness of the thermal boundary layer developing in the liquid phase near the interfaces. This model states that evaporation and condensation processes are controlled by the heat flux in the liquid phase in the vicinity of the interfaces. It needs correlations for the interfacial area per unit volume and the characteristic width of the thermal boundary layer.

For the melting-freezing phase these rates are determined in the following way:

Freezing rate: the liquid energy does not fall below the liquidus energy.
Melting rate: the solid energy does not rise above the solidus energy.

***Drag coefficients*** For momentum conservation across the interfaces, the drag coefficient must obey the following constraint:

$$D_{pp'} = D_{p'p}$$

For the liquid-vapor interfaces, an expression is used by Harlow and Amsden (1975) that follows this constraint:

$$D_{lv} = D_{vl} = \frac{3}{8}(r_l + r_v)^2 \frac{C_{lv}^D \rho_l \rho_v}{r_l r_v} \frac{|\mathbf{v}_l - \mathbf{v}_v|}{r_v \rho_v + r_l \rho_l} \tag{43}$$

where $r_v$ and $r_l$ are the characteristic radii of vapor bubbles and liquid globes, respectively.

For liquid-solid and vapor-solid interfaces the usual friction factors using the equivalent diameter are defined.

***Heat transfer coefficients*** No detail is given for the calculation of the heat exchange coefficient $H_{pp'}^{cc'}$.

***Stress tensor and heat flux*** The stress tensor is often neglected in comparison with friction at the interfaces or on the fixed structures. The heat flux is calculated using the effective diffusivity concept.

## 4.4 Example of Results

Bell et al. (1976) presented some results concerning the two problems mentioned at the beginning of the chapter and obtained with SIMMER-1.

**4.4.1 Problem 1: The single subassembly** The initial configuration of the single subassembly is shown in Fig. 23. The subassembly is closed at both ends and filled with sodium vapor. The transient is initiated by a power burst of 1 ms, and then the power is maintained to three times nominal power for 4 s, the time of the calculation. The subassembly is thermally insulated.

Distribution of the solid fuel is given in Fig. 24 for different instants. The fuel melts first in the midplane of the core, and melting extends symmetrically. Fuel freezes first in the lower axial blanket (LAB), which constitutes a heat sink; then after 2 s frozen fuel appears in the upper axial blanket (UAB).

The motion of the liquid fuel is shown in Fig. 25. The molten fuel appears in the core midplane and extends toward the bottom (slumping) at 0.4 s. At 0.8 s it penetrates in the LAB, where it partially freezes. A maximum in the liquid profile thus appears at the interface between the core and the LAB. This forms a liquid slug that is driven upward by the vapor pressure of steel and fuel between 0.8 and 1.2 s. The liquid fuel moves inside the UAB where it starts freezing at 2.0 s. The

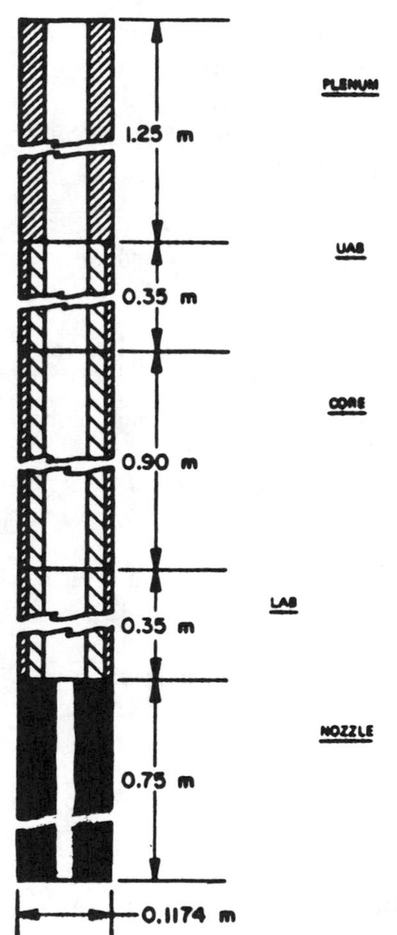


Figure 23 Subassembly initial configuration (Bell et al., 1976).

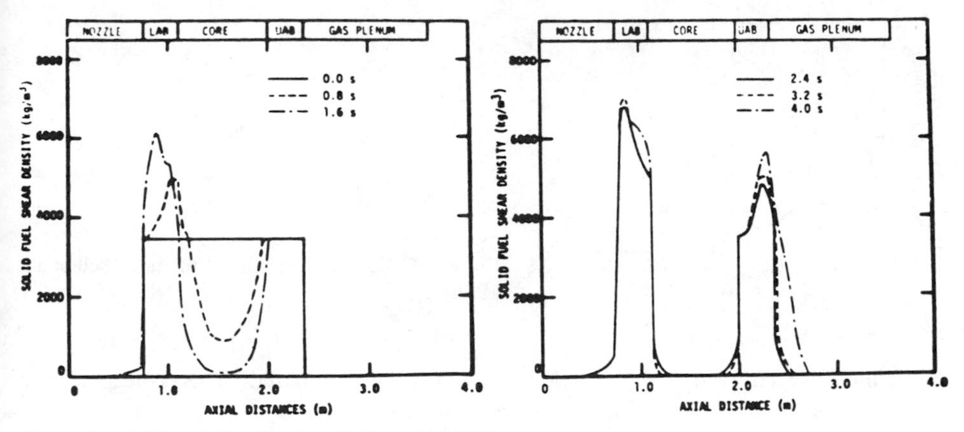


**Figure 24** Solid fuel distribution (Bell et al., 1976).

distribution of the liquid fuel becomes more and more difficult to interpret in the following seconds but continues to move in the UAB and LAB, where it increases the amount of frozen fuel.

**4.4.2 Problem 2: Hypothetical core-disruptive accident** The initial configuration has been presented in Fig. 21. It must be kept in mind that the initial configuration and the subsequent transient evolution concern one type of reactor specific of the U.S. program. Parameter ranges and thus physical evolutions might be different in other situations. Moreover, this calculation, although very interesting, is limited by the physical models, which are crude and might be far from real phenomena.

The transient evolution starts at delayed critical at 10 times nominal power and is driven along an assumed continuous reactivity ramp of 20 dollars/s. Once sodium boiling and significant voiding of the core have started, neutronics and thermo-

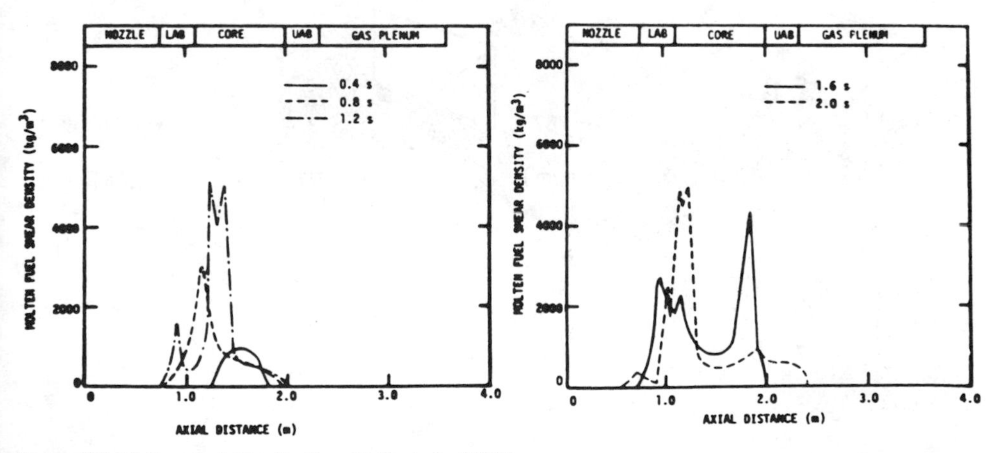


**Figure 25** Molten fuel distribution (Bell et al., 1976).

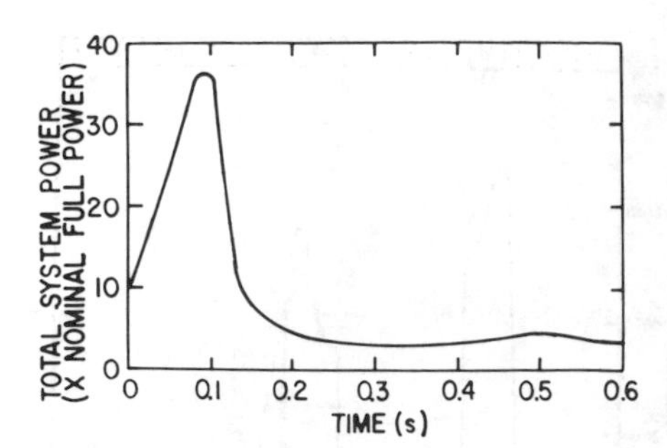


**Figure 26** Total power related to time (Bell et al., 1976).

hydraulics are coupled and the reactivity must be calculated. This aspect of the neutronic calculation is not our present concern.

The sequence of computed events following the initial time is (Bell et al., 1976)

1. 0–0.07 s: core voiding of the sodium.
2. 0.07–0.13 s: melting of fuel and clad that enter the flow channels and mix with the liquid sodium left in the core. A mild interaction occurs that develops an overpressure at the core midplane.
3. 0.13–0.25 s: as a consequence of this overpressure, the molten fuel-steel-sodium mixture is pushed away from the core midplane and is driven into the upper and lower axial blankets. In these regions, the molten fuel is preferentially frozen out of the mixture. At the same time, the total power has significantly decreased (Fig. 26).
4. 0.2–0.25 s: the pressure within the core drops continuously. The effect of gravity begins to be noticeable, causing a slumping of the molten fuel toward the bottom of the core.
5. 0.25–0.64 s (end of the calculation): the fission power is sufficiently reduced so

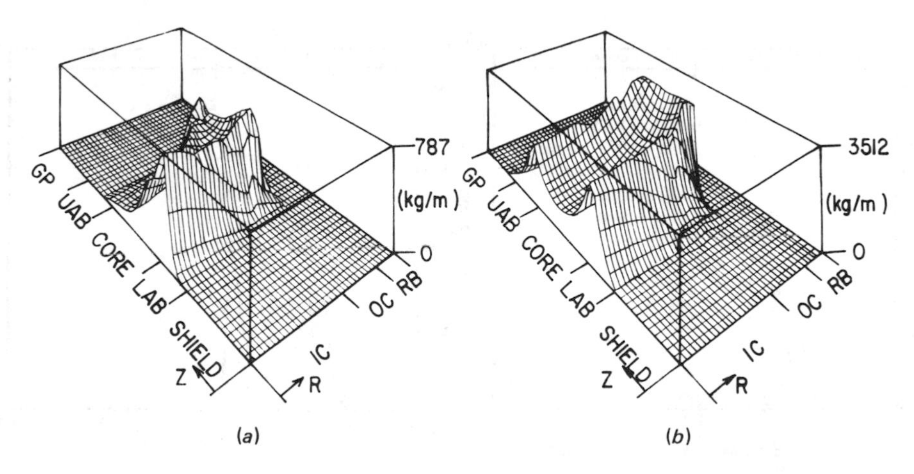


**Figure 27** Fuel distributions. (*a*) Molten fuel. (*b*) Solid fuel. (Bell et al., 1976.

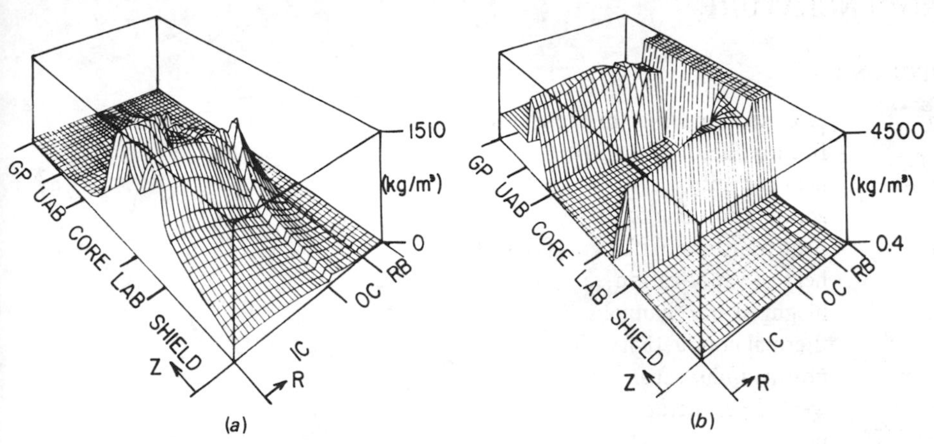


**Figure 28** Steel distributions. (*a*) Molten steel. (*b*) Solid steel. (Bell et al., 1976.)

that heat transfer from the molten fuel to the other materials induces a freezing rate in the molten fuel that offsets the melting rate from fission heating.

Figure 27 shows the calculated distributions of molten and solid fuel at the termination of the analysis (0.64 s). The amount of molten fuel that remains is relatively small (Fig. 27*a*). It has slumped to the LAB core interface. Most of the fuel is solid. Figure 27*b* shows that it has massively frozen in the axial blankets.

The distribution of molten steel is shown in Fig. 28*a*. Molten steel exists all the way down to the bottom of the core. Solid clad distribution is shown in Fig. 28*b*. It has entirely disappeared from the core region.

The loss of material from the core midplane is evident from the distributions shown in Figs. 27 and 28. It is also indicated by the high vapor fraction in that region shown in Fig. 28.

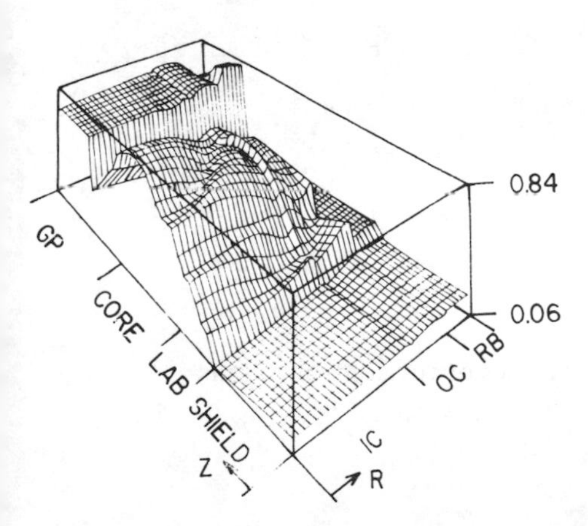


**Figure 29** Vapor volume fraction (Bell et al., 1976).

# NOMENCLATURE

## SECTION 1

| | |
|---|---|
| $\mathcal{A}$ | cross-sectional area of the channel |
| $C_f$ | friction factor [Eq. (3)] |
| $c$ | heat capacity |
| $g$ | acceleration due to gravity |
| $h$ | heat transfer coefficient [Eq. (3)] |
| $h_{fg}$ | heat of vaporization |
| $H$ | height of the channel |
| $k$ | thermal conductivity |
| $\mathbf{n}$ | normal unit vector |
| $\mathcal{P}$ | wetted perimeter |
| $p$ | pressure |
| $\mathbf{q}$ | heat flux |
| Re | Reynolds number |
| $t$ | time |
| $T$ | temperature |
| $w$ | vertical component of the velocity |
| $z$ | vertical coordinate |
| $\gamma$ | slip ratio |
| $\rho$ | density |
| $\sigma$ | film thickness |
| $\tau_w$ | wall shear stress |
| $\phi_w$ | wall heat flux |

### Subscripts

| | |
|---|---|
| $b$ | bottom of the vapor bubble |
| $f$ | saturated liquid |
| $g$ | saturated vapor |
| in | inlet |
| $l$ | liquid |
| out | outlet |
| sat | along the saturation line |
| $t$ | top of the vapor bubble |
| $v$ | vapor |
| $w$ | wall |

## SECTION 2

| | |
|---|---|
| $\mathcal{A}_f$ | coolant channels' area in the lateral sides of the control volume |
| $\mathcal{A}_w$ | area of fluid-solid interfaces |
| $C_f$ | friction factor |
| $g$ | acceleration due to gravity |

| | |
|---|---|
| $G$ | mass velocity |
| $h$ | specific enthalpy |
| $\mathbf{k}$ | unit vector in axial direction |
| $\dot{m}$ | mass flux across the lateral sides of the control volume (Fig. 9) |
| $\mathbf{n}$ | unit normal vector |
| $p$ | pressure |
| $\mathbf{q}$ | heat flux |
| $\mathcal{V}$ | control volume |
| $\mathcal{V}_f$ | volume of coolant channels |
| $\mathbf{v}$ | velocity |
| $w$ | vertical component of the velocity |
| $z$ | vertical coordinate |
| $\alpha_e$ | effective heat diffusivity |
| $\epsilon$ | porosity |
| $\nu_e$ | effective kinematic viscosity |
| $\rho$ | density |
| $\tau$ | viscous stress tensor |
| $\phi_w$ | wall heat flux |

**Subscripts**

| | |
|---|---|
| $f$ | coolant channels |
| $w$ | wall |

**Superscript**

| | |
|---|---|
| * | enthalpy or velocity entering the control volume |

**Operators**

| | |
|---|---|
| $\langle\ \rangle$ | volume-averaging operator over $\mathcal{V}_f$ [Eq. (14)] |
| $\nabla_t$ | difference approximation [Eq. (15)] |

**SECTION 3**

| | |
|---|---|
| $\mathcal{A}$ | cross-sectional area |
| $\mathcal{A}_\infty$ | flow area without cladding |
| $D$ | hydraulic diameter |
| $D_0$ | hydraulic diameter for the original channel |
| $f_s$ | resistance coefficient for vapor flow [Eq. (24)] |
| $f_c$ | wall resistance coefficient for the cladding |
| g | acceleration due to gravity |
| $K_{tr}$ | blockage-resistance coefficient [Eq. (26)] |
| $l_0$ | height of the voided channel |
| $\mathcal{P}$ | wetted perimeter |
| $p$ | pressure |
| $t$ | time |
| $t_r$ | time of flow reversal |

$w_g$ axial component of the vapor velocity
$w_c$ axial component of the cladding-film velocity
$z$ axial coordinate
$\alpha$ void fraction [Eq. (22)]
$\delta$ molten clad film thickness
$\delta_c$ initial clad thickness
$\epsilon$ deflooding parameter [Eq. (24)]
$\lambda$ length of the melted clad section
$\lambda_m$ length of the molten clad film
$\mu_c$ dynamic viscosity of the molten clad
$\rho$ density
$\tau_g$ interfacial shear stress (Fig. 16)
$\tau_c$ liquid-cladding shear stress (Fig. 16)

**Subscripts**

$fr$ blockage
$g$ sodium vapor
$i$ inlet
$l$ liquid
$mc$ molten clad

**SECTION 4**

$\mathbf{a}_i$ interfacial area per unit volume
$c$ specific heat capacity
$C_{lv}^D$ constant in the expression of the drag coefficient (Sec. 4.3.2)
$D$ drag coefficient
$H$ heat exchange coefficient
$h$ specific enthalpy
$h_{fg}$ heat of vaporization
$h_{sf}$ heat of fusion
$k$ thermal conductivity
$L$ characteristic length
$\mathbf{M}$ momentum exchange as a result of phase transitions [Eq. (32)]
$p$ pressure
$\mathbf{q}_c$ heat flux
$Q_p^c$ specific heat source
$r$ radial coordinate
$R_p^c$ volume fraction of phase $p$ of component $c$
$t$ time
$T$ temperature
$u$ internal energy
$\mathbf{v}$ velocity
$v_{fg}$ change in specific volume during condensation
$z$ axial coordinate
$\Gamma_p^c$ mass rate gained by phase $p$ of component $c$ [Eq. (29)]

| | |
|---|---|
| $\gamma_{p',p}$ | phase transition rate from phase $p'$ to phase $p$ |
| $\theta$ | circumferential coordinate |
| $\Lambda$ | energy exchange due to phase transitions |
| $\rho$ | density |
| $\tau$ | viscous stress tensor |

**Subscripts**

| | |
|---|---|
| $f$ | saturated liquid |
| $g$ | saturated vapor |
| $l$ | liquid |
| $p$ | phase index |
| $v$ | vapor |

**Superscript**

| | |
|---|---|
| $c$ | component index |

## REFERENCES

Basque, G., Grand, D., and Menant, B., Theoretical Analysis and Experimental Evidence of Three Types of Thermohydraulic Incoherency in Undisturbed Cluster Geometry, 8th Meet. of the Liquid Metal Boiling Working Group, Mol, Belgium, 1978.

Bell, C. R., Bleiweis, P. B., and Boudreau, J. E., Analysis of LMFBR Disruption and Accident Phenomena Using the SIMMER-1 Code, *Proc. ANS/ENS Int. Meet. on Fast Reactor Safety Related Physics, Chicago, Oct. 5-8*, CONF-761001, pp. 1203–1215, 1976.

Chen, W. L., Ishii, M., and Grolmes, M. A., Parametric Study of the Molten Clad Motion Based on One Dimensional Model, *Nucl. Eng. Des.*, vol. 41, pp. 1–12, 1977.

Costa, J., Contribution to the Study of Sodium Boiling during Slow Pump Coast Down in LMFBR Subassemblies, in *Thermal and Hydraulic Aspects of Nuclear Reactor Safety*, vol. 2. *Liquid Metal Fast Breeder Reactors*, eds. O. C. Jones and S. G. Bankoff, pp. 155–170, ASME, New York, 1977.

Cronenberg, A. W., Fauske, H. K., Bankoff, S. G., and Eggen, D. T., A Single-Bubble Model for Sodium Expulsion from a Heated Channel, *Nucl. Eng. Des.*, vol. 16, pp. 285–293, 1971.

Dunn, F. E., Fischer, G. J., Heames, T. J., Pizzica, P. A., McNeal, N. A., Bohl, W. R., and Prastein, S. M., The SAS2A LMFBR Accident-Analysis Computer Code, ANL-8138, 1974.

Epstein, M., Melting, Boiling and Freezing, the "Transition Phase" in Fast Reactor Safety Analyses, in *Thermal and Hydraulic Aspects of Nuclear Reactor Safety*, vol. 2. *Liquid Metal Fast Breeder Reactors*, eds. O. C. Jones and S. G. Bankoff, pp. 171–193, ASME, New York, 1977.

Grand, D. and Latrobe, A., FLINT–A Code for Slow Flow Transients in a Single Channel, 7th Meet. of the Liquid Metal Boiling Working Group, Petten, Holland, 1977.

Harlow, F. H. and Amsden, A. A., A Numerical Fluid Dynamics Calculation Method for All Flow Speeds, *J. Comput. Phys.*, vol. 8, pp. 197–213, 1971.

Harlow, F. H. and Amsden, A. A., Flow of Interpenetrating Material Phases, *J. Comput. Phys.*, vol. 18, pp. 440–464, 1975.

Ishii, M., Chen, W. L., and Grolmes, M. A., Molten Clad Motion Model for Fast Reactor Loss-of-Flow Accident, *Nucl. Sci. Eng.*, vol. 60, pp. 435–451, 1976.

Lal, D. and Carlson, R. W., Two-dimensional Fuel Motion Analysis during Loss-of-Flow in a Liquid Metal Cooled Fast Breeder Reactor, ORO-4958-4, 1976.

Martin, B. A., Agrawal, A. K., Albright, D. C., Epel, L. G., and Maise, G., NALAP–An LMFBR System Transient Code, BNL-50457, 1975.

Miao, C. and Theofanous, T., Numerical Studies on Transient Two-dimensional Boiling in LMFBR Subassemblies, 7th Meet. of the Liquid Metal Boiling Working Group, Petten, Holland, 1977.
Moxon, D., Progress with U.K. Codes for the Analysis of Boiling Transients in Single Channels, 7th Meet. of the Liquid Metal Boiling Working Group, Petten, Holland, 1977.
Rousseau, J. C., The FLINA Code, European Two-Phase Flow Group Meet., Riso, Denmark, 1971.
Smith, L. L., Boudreau, J. E., Bell, C. R., Bleiweis, P. B., Barnes, J. F., and Travis, J. R., SIMMER-1: An LMFBR Disrupted Core Analysis Code, *Proc. ANS/ENS Int. Meet Fast Reactor Safety Related Physics, Chicago, Oct. 5-8*, CONF-761001, pp. 1195-1202, 1976.
Travis, J. R. and Rivard, W. C., Multiphase Fluid Dynamics with Applications in LMFBR Safety Analysis, *Proc. ANS/ENS Int. Meet. Fast Reactor Safety Related Physics, Chicago, Oct. 5-8*, CONF-761001, pp. 1511-1518, 1976.
Weisman, J. and Bowring, R. W., Methods for Detailed Thermal and Hydraulic Analysis of Water-cooled Reactors, *Nucl. Sci. Eng.*, vol. 57, pp. 255-276, 1975.
Wirtz, P., Ein Beitrag zur theoretischen Beschreibung des Siedens unter Störfallbedingungen in Natriumgekühlten schnellen Reaktoren, KFK 1858, 1973.

# INDEX